Laser Ionization Mass Analysis

Edited by

AKOS VERTES

George Washington University
Washington, D.C.

RENAAT GIJBELS
FRED ADAMS

University of Antwerp (UIA)
Antwerp, Belgium

A WILEY-INTERSCIENCE PUBLICATION

JOHN WILEY & SONS, INC.

New York / Chichester / Brisbane / Toronto / Singapore

This text is printed on acid-free paper.

Library of Congress Cataloging in Publication Data:

Laser ionization mass analysis / editors, Akos Vertes, Renaat Gijbels,
Fred Adams.
p. cm.—(Chemical analysis ; v. 124)
Includes bibliographical references and index.
ISBN 0-471-53673-3
1. Mass spectroscopy. 2. Laser spectroscopy. I. Vertes, Akos,
1952– . II. Gijbels, R. (Renaat) III. Adams, Fred. IV. Series.
QD96.M3L37 1993 92-34983
543'.0873—dc20 CIP

Printed in the United States of America

10 9 8 7 6 5 4 3 2 1

CONTENTS

CONTRIBUTORS

Fred Adams, Department of Chemistry, University of Antwerp (UIA), Antwerp, Belgium

Johanna Sabine Becker, Central Department of Chemical Analysis, Research Centre Jülich GmbH; Jülich, Germany

Hans-Joachim Dietze, Central Department of Chemical Analysis, Research Centre Jülich GmbH, Jülich, Germany

Russell W. Dreyfus, IBM Research Division, T. J. Watson Research Center, Yorktown Heights, New York

Renaat Gijbels, Department of Chemistry, University of Antwerp (UIA), Antwerp, Belgium

David M. Lubman, Department of Chemistry, The University of Michigan, Ann Arbor, Michigan

George G. Managadze, Space Research Institute, Russian Academy of Sciences, Moscow, Russia

Lieselotte Moenke-Blankenburg, Department of Chemistry, Martin-Luther-University Halle-Wittenberg, Halle, Germany

Claude R. Phipps, Chemical and Laser Sciences Division, Los Alamos National Laboratory, Los Alamos, New Mexico

Igor Yu. Shutyaev, Space Research Institute, Russian Academy of Sciences, Moscow, Russia

Wim Van Roy, Department of Chemistry, University of Antwerp (UIA), Antwerp, Belgium

Luc Van Vaeck, Department of Chemistry, University of Antwerp (UIA), Antwerp, Belgium

Akos Vertes, Department of Chemistry, George Washington University, Washington, D.C.

FOREWORD

The ability to produce short bursts of coherent photon radiation with control of photon density, wavelength, and time is truly one of the remarkable achievements of the twentieth century. What we have done with the laser pulse to probe new features of nature is also remarkable in breadth, ranging from studying phenomena simulating near stellar conditions to unraveling details of the structure of a life-supporting protein at a chemical resolution of a single hydrogen atom. Since the introduction of the laser, most applications have evolved smoothly based on the established principles of linear and nonlinear optics. Some of the important analytical applications are discussed in this volume: the laser microprobe, multiphoton ionization, ablation of solids, and the mass spectrometry of large biomolecules, a subject that is a new feature of laser technology and which is appropriately highlighted.

Twenty years ago, important advances were made in the solution of a long-standing problem in mass spectrometry on how to produce gas phase molecular ions of involatile thermally unstable biopolymers from a solid matrix. Results from particle bombardment studies using high and low energy primary ions demonstrated that the process could be carried out if the sample is subjected to a short intense energy pulse, the kind that is created by energy deposition when a fast ion penetrates a solid.

It was obvious to those of us working on this problem at that time that a pulsed laser should not only induce the same effect but it should also be far superior to particle bombardment because of the greater element of control. Someone would surely soon try it. The demonstration of feasibility came in 1978, and the feeling of my group at that time was that laser desorption mass spectrometry would become the method of choice for biomolecule analysis and that our efforts in continuing to develop particle bombardment mass spectrometry would be phased out.

However, after these initial encouraging results, laser desorption mass spectrometry applied to biomolecule analysis did not progress as quickly as had been anticipated, and during the next decade, in the "race for highest mass," laser desorption was out of the running.

In 1988, a seemingly straightforward incremental change in the way the experiment is done suddenly transformed overnight an "also-ran" technique into a leader. The story of the evolution of MALDI is an inspiration to all of us who have devoted our scientific activity to instrumentation and methods development. This event and what quickly followed most certainly have inspired the Editors of this volume to bring together the latest developments in *Laser Ionization Mass Analysis* at this time.

RONALD D. MACFARLANE

College Station, Texas
January 1993

CHEMICAL ANALYSIS

A SERIES OF MONOGRAPHS ON ANALYTICAL CHEMISTRY AND ITS APPLICATIONS

J. D. Winefordner, *Series Editor*
I. M. Kolthoff, *Editor Emeritus*

Vol. 1. **The Analytical Chemistry of Industrial Poisons, Hazards, and Solvents.** *Second Edition.* By the late Morris B. Jacobs

Vol. 2. **Chromatographic Adsorption Analysis.** By Harold H. Strain (*out of print*)

Vol. 3. **Photometric Determination of Traces of Metals.** *Fourth Edition*
Part I: General Aspects. By E. B. Sandell and Hiroshi Onishi
Part IIA: Individual Metals, Aluminum to Lithium. By Hiroshi Onishi
Part IIB: Individual Metals, Magnesium to Zirconium. By Hiroshi Onishi

Vol. 4. **Organic Reagents Used in Gravimetric and Volumetric Analysis.** By John F. Flagg (*out of print*)

Vol. 5. **Aquametry: A Treatise on Methods for the Determination of Water.** *Second Edition (in three parts).* By John Mitchell, Jr. and Donald Milton Smith

Vol. 6. **Analysis of Insecticides and Acaricides.** By Francis A. Gunther and Roger C. Blinn (*out of print*)

Vol. 7. **Chemical Analysis of Industrial Solvents.** By the late Morris B. Jacobs and Leopold Schetlan

Vol. 8. **Colorimetric Determination of Nonmetals.** *Second Edition.* Edited by the late David F. Boltz and James A. Howell

Vol. 9. **Analytical Chemistry of Titanium Metals and Compounds.** By Maurice Codell

Vol. 10. **The Chemical Analysis of Air Pollutants.** By the late Morris B. Jacobs

Vol. 11. **X-Ray Spectrochemical Analysis.** *Second Edition.* By L. S. Birks

Vol. 12. **Systematic Analysis of Surface-Active Agents.** *Second Edition.* By Milton J. Rosen and Henry A. Goldsmith

Vol. 13. **Alternating Current Polarography and Tensammetry.** By B. Breyer and H. H. Bauer

Vol. 14. **Flame Photometry.** By R. Herrmann and J Alkemade

Vol. 15. **The Titration of Organic Compounds** (*in two parts*). By M. R. F. Ashworth

Vol. 16. **Complexation in Analytical Chemistry: A Guide for the Critical Selection of Analytical Methods Based on Complexation Reactions.** By the late Anders Ringbom

Vol. 17. **Electron Probe Microanalysis.** *Second Edition.* By L. S. Birks

Vol. 18. **Organic Complexing Reagents: Structure, Behavior, and Application to Inorganic Analysis.** By D. D. Perrin

Vol. 19. **Thermal Analysis.** *Third Edition.* By Wesley Wm. Wendlandt

Vol. 20. **Amperometric Titrations.** By John T. Stock

Vol. 21. **Reflectance Spectroscopy.** By Wesley Wm. Wendlandt and Harry G. Hecht

Vol. 22. **The Analytical Toxicology of Industrial Inorganic Poisons.** By the late Morris B. Jacobs

Vol. 23. **The Formation and Properties of Precipitates.** By Alan G. Walton

Vol. 24. **Kinetics in Analytical Chemistry.** By Harry B. Mark, Jr. and Garry A. Rechnitz

Vol. 25. **Atomic Absorption Spectroscopy.** *Second Edition.* By Morris Slavin

Vol. 26. **Characterization of Organometallic Compounds** (*in two parts*). Edited by Minoru Tsutsui

Vol. 27. **Rock and Mineral Analysis.** *Second Edition.* By Wesley M. Johnson and John A. Maxwell

Vol. 28. **The Analytical Chemistry of Nitrogen and Its Compounds** (*in two parts*). Edited by C. A. Streuli and Philip R. Averell

Vol. 29. **The Analytical Chemistry of Sulfur and Its Compounds** (*in three parts*). By J. H. Karchmer

Vol. 30. **Ultramicro Elemental Analysis.** By Günther Tölg

Vol. 31. **Photometric Organic Analysis** (*in two parts*). By Eugene Sawicki

Vol. 32. **Determination uf Organic Compounds: Methods and Procedures.** By Frederick T. Weiss

Vol. 33. **Masking and Demasking of Chemical Reactions.** By D. D. Perrin

Vol. 34. **Neutron Activation Analysis.** By D. De Soete, R. Gijbels, and J. Hoste

Vol. 35. **Laser Raman Spectroscopy.** By Marvin C. Tobin

Vol. 36. **Emission Spectrochemical Analysis.** By Morris Slavin

Vol. 37. **Analytical Chemistry of Phosphorus Compounds.** Edited by M. Halmann

Vol. 38. **Luminescence Spectrometry in Analytical Chemistry.** By J. D. Winefordner, S. G. Schulman and T. C. O'Haver

Vol. 39. **Activation Analysis with Neutron Generators.** By Sam S. Nargolwalla and Edwin P. Przybylowicz

Vol. 40. **Determination of Gaseous Elements in Metals.** Edited by Lynn L. Lewis, Laben M. Melnick, and Ben D. Holt

Vol. 41. **Analysis of Silicones.** Edited by A. Lee Smith

Vol. 42. **Foundations of Ultracentrifugal Analysis.** By H. Fujita

Vol. 43. **Chemical Infrared Fourier Transform Spectroscopy.** By Peter R. Griffiths

Vol. 44. **Microscale Manipulations in Chemistry.** By T. S. Ma and V. Horak

Vol. 45. **Thermometric Titrations.** By J. Barthel

Vol. 46. **Trace Analysis: Spectroscopic Methods for Elements.** Edited by J. D. Winefordner

Vol. 47. **Contamination Control in Trace Element Analysis.** By Morris Zief and James W. Mitchell

Vol. 48. **Analytical Applications of NMR.** By D. E. Leyden and R. H. Cox

Vol. 49. **Measurement of Dissolved Oxygen.** By Michael L. Hitchman

Vol. 50. **Analytical Laser Spectroscopy.** Edited by Nicolo Omenetto

Vol. 51. **Trace Element Analysis of Geological Materials.** By Roger D. Reeves and Robert R. Brooks

Vol. 52. **Chemical Analysis by Microwave Rotational Spectrscopy.** By Ravi Varma and Lawrence W. Hrubesh

Vol. 53. **Information Theory As Applied to Chemical Analysis.** By Karel Eckschlager and Vladimir Štěpánek

Vol. 54. **Applied Infrared Spectroscopy: Fundamentals, Techniques, and Analytical Problem-solving.** By A. Lee Smith

Vol. 55. **Archaeological Chemistry.** By Zvi Goffer

Vol. 56. **Immobilized Enzymes in Analytical and Clinical Chemistry.** By P. W. Carr and L. D. Bowers

Vol. 57. **Photoacoustics and Photoacoustic Spectroscopy.** By Allan Rosencwaig

Vol. 58. **Analysis of Pesticide Residues.** Edited by H. Anson Moye

Vol. 59. **Affinity Chromatography.** By William H. Scouten

Vol. 60. **Quality Control in Analytical Chemistry.** By G. Kateman and F. W. Pijpers

Vol. 61. **Direct Characterization of Fineparticles.** By Brian H. Kaye

Vol. 62. **Flow Injection Analysis.** By J. Ruzicka and E. H. Hansen

Vol. 63. **Applied Electron Spectroscopy for Chemical Analysis.** Edited by Hassan Windawi and Floyd Ho

Vol. 64. **Analytical Aspects of Environmental Chemistry.** Edited by David F. S. Natusch and Philip K. Hopke

Vol. 65. **The Interpretation of Analytical Chemical Data by the Use of Cluster Analysis.** By D. Luc Massart and Leonard Kaufman

Vol. 66. **Solid Phase Biochemistry: Analytical and Synthetic Aspects.** Edited by William H. Scouten

Vol. 67. **An Introduction to Photoelectron Spectroscopy.** By Pradip K. Ghosh

Vol. 68. **Room Temperature Phosphorimetry for Chemical Analysis.** By Tuan Vo-Dinh

Vol. 69. **Potentiometry and Potentiometric Titrations.** By E. P. Serjeant

Vol. 70. **Design and Application of Process Analyzer Systems.** By Paul E. Mix

Vol. 71. **Analysis of Organic and Biological Surfaces.** Edited by Patrick Echlin

Vol. 72. **Small Bore Liquid Chromatography Columns: Their Properties and Uses.** Edited by Raymond P. W. Scott

Vol. 73. **Modern Methods of Particle Size Analysis.** Edited by Howard G. Barth

Vol. 74. **Auger Electron Spectroscopy.** By Michael Thompson, M. D. Baker, Alec Christie, and J. F. Tyson

Vol. 75. **Spot Test Analysis: Clinical, Environmental, Forensic and Geochemical Applications.** By Ervin Jungreis

Vol. 76. **Receptor Modeling in Environmental Chemistry.** By Philip K. Hopke

Vol. 77. **Molecular Luminescence Spectroscopy: Methods and Applications** (*in three parts*). Edited by Stephen G. Schulman

Vol. 78. **Inorganic Chromatographic Analysis.** Edited by John C. MacDonald

Vol. 79. **Analytical Solution Calorimetry.** Edited by J. K. Grime

Vol. 80. **Selected Methods of Trace Metal Analysis: Biological and Environmental Samples.** By Jon C. VanLoon

Vol. 81. **The Analysis of Extraterrestrial Materials.** By Isidore Adler

Vol. 82. **Chemometrics.** By Muhammad A. Sharaf, Deborah L. Illman, and Bruce R. Kowalski

Vol. 83. **Fourier Transform Infrared Spectrometry.** By Peter R. Griffiths and James A. de Haseth

Vol. 84. **Trace Analysis: Spectroscopic Methods for Molecules.** Edited by Gary Christian and James B. Callis

Vol. 85. **Ultratrace Analysis of Pharmaceuticals and Other Compounds of Interest.** Edited by S. Ahuja

Vol. 86. **Secondary Ion Mass Spectrometry: Basic Concepts, Instrumental Aspects, Applications and Trends.** By A. Benninghoven, F. G. Rüdenauer, and H. W. Werner

Vol. 87. **Analytical Applications of Lasers.** Edited by Edward H. Piepmeier

Vol. 88. **Applied Geochemical Analysis.** By C. O. Ingamells and F. F. Pitard

Vol. 89. **Detectors for Liquid Chromatography.** Edited by Edward S. Yeung

Vol. 90. **Inductively Coupled Plasma Emission Spectroscopy: Part I: Methodology, Instrumentation, and Performance; Part II: Applications and Fundamentals.** Edited by J. M. Boumans

Vol. 91. **Applications of New Mass Spectrometry Techniques in Pesticide Chemistry.** Edited by Joseph Rosen

Vol. 92. **X-Ray Absorption: Principles, Applications, Techniques of EXAFS, SEXAFS, and XANES.** Edited by D. C. Konnigsberger

Vol. 93. **Quantitative Structure–Chromatographic Retention Relationships.** By Roman Kaliszan

Vol. 94. **Laser Remote Chemical Analysis.** Edited by Raymond M. Measures

Vol. 95. **Inorganic Mass Spectrometry.** Edited by F. Adams, R. Gijbels, and R. Van Grieken

Vol. 96. **Kinetic Aspects of Analytical Chemistry.** By Horacio A. Mottola

Vol. 97. **Two-Dimensional NMR Spectroscopy.** By Jan Schraml and Jon M. Bellama

Vol. 98. **High Performance Liquid Chromatography.** Edited by Phyllis R. Brown and Richard A. Hartwick

Vol. 99. **X-Ray Fluorescence Spectrometry.** By Ron Jenkins

Vol. 100. **Analytical Aspects of Drug Testing.** Edited by Dale G. Deutsch

Vol. 101. **Chemical Analysis of Polycyclic Aromatic Compounds.** Edited by Tuan Vo-Dinh

Vol. 102. **Quadrupole Storage Mass Spectrometry.** By Raymond E. March and Richard J. Hughes

Vol. 103. **Determination of Molecular Weight.** Edited by Anthony R. Cooper

Vol. 104. **Selectivity and Detectability Optimizations in HPLC.** By Satinder Ahuja

Vol. 105. **Laser Microanalysis.** By Lieselotte Moenke-Blankenburg

Vol. 106. **Clinical Chemistry.** Edited by E. Howard Taylor

Vol. 107. **Multielement Detection Systems for Spectrochemical Analysis.** By Kenneth W. Busch and Marianna A. Busch

Vol. 108. **Planar Chromatography in the Life Sciences.** Edited by Joseph C. Touchstone

Vol. 109. **Fluorometric Analysis in Biomedical Chemistry: Trends and Techniques Including HPLC Applications.** By Norio Ichinose, George Schwedt, Frank Michael Schnepel, and Kyoko Adochi

Vol. 110. **An Introduction to Laboratory Automation.** By Victor Cerdá and Guillermo Ramis

Vol. 111. **Gas Chromatography: Biochemical, Biomedical, and Clinical Applications.** Edited by Ray E. Clement

CHAPTER

1

INTRODUCTION

AKOS VERTES

Department of Chemistry
George Washington University
Washington, D.C.

RENAAT GIJBELS and FRED ADAMS

Department of Chemistry
University of Antwerp (UIA)
Antwerp, Belgium

After two decades of incubation, laser ionization mass analysis is rapidly gaining importance in several areas of analytical chemistry. Although in the early 1970s lasers were already claimed to be capable of directly probing and ionizing solid samples, the high hopes of a molecular microprobe have been disproved both by methodological difficulties and by poor sales statistics of several commercial instruments. It is the more mature understanding of laser–solid interaction acquired during the early applications of laser ionization that led to the redefinition of its place in analytical chemistry and to the discovery of several new methods. The pulsed laser in the new context is viewed as a tool of depositing a tunable amount of energy into a chosen degree of freedom of the material, rather than as a fast local surface heating device. Our book is centered on this concept, with special emphasis on proven or promising applications.

Laser-based methods also are at the forefront of advanced material synthesis and processing. Owing to increased efforts in the past few years we have witnessed several breakthroughs in employing the unique features of laser-driven energy deposition in solids. Laser–solid interaction is utilized in diamond synthesis, high-temperature superconductor preparation, desorption of large biomolecules, and sampling of solids for chemical analysis. Let us browse through these discoveries and their promising applications.

Laser Ionization Mass Analysis, Edited by Akos Vertes, Renaat Gijbels, and Fred Adams. Chemical Analysis Series, Vol. 124.
ISBN 0-471-53673-3

Because of the unique mechanical, thermal, and electric properties of diamond, synthetic diamond production is an area of large technological importance. Earlier efforts included high-pressure and low-pressure plasma, ion beam, and combustion flame methods of preparation. Pulsed laser deposition led to the first successful synthesis of high-quality diamond films on metal substrate (Narayan et al., 1991). This result presented a landmark in the continuing effort to introduce diamond into wide areas of microelectronic applications.

Laser ablation also is part of the pulsed laser deposition technique, a promising contender for producing films of multicomponent materials such as high-transition-temperature superconductors. In the preparation of superconducting $YBa_2Cu_3O_{7-x}$ layers, laser processing also offers unique features in the oxygen atmosphere annealing step. The extreme fast heating and cooling rates achieved with lasers make it possible to reach the necessary surface temperature without substantially changing the bulk temperature. This selectivity has its obvious advantages in dealing with complex microstructures.

Another breakthrough related to laser–solid interaction came from the field of analytical chemistry. For decades, for gas phase investigation of large (bio)molecules, there has been an obstacle: above a certain molecular weight, it was impossible to volatilize these substances, therefore precluding the application of powerful analytical techniques such as mass spectrometry in their investigation. The introduction of matrix-assisted laser desorption made it possible to bring about the necessary phase transition. As a result, mass spectrometric investigation now provides unprecedented accuracy in molecular mass determination up to several hundred thousand atomic mass units (Karas et al., 1989; Beavis and Chait, 1990; Spengler and Cotter, 1990).

The general application of laser sampling of solids for analytical purposes is also in the introductory phase. Preliminary results are available for laser sampling in combination with inductively coupled plasma mass spectrometry (Denoyer et al., 1991) and with glow discharge mass spectrometry (Harrison and Bentz, 1988) for the analysis of metals, glasses, ceramics, and other hard-to-analyze insulators. The technique is far from maturity, but indications of its superior analytical sensitivity and versatility are already apparent.

The relation between these applications of laser–solid interaction is clear if we display the laser irradiances, I_0, and light absorption coefficients, α, involved as illustrated in Figure 1.1.

Although preparative and analytical applications of laser–solid interaction are based on the same physical phenomenon, little or no interaction exists between these fast-growing areas. Even the two sets of journals (preparative applications: *Applied Physics Letters*, *Journal of Applied Physics*, etc.; analytical applications: *Analytical Chemistry*, *International Journal for Mass*

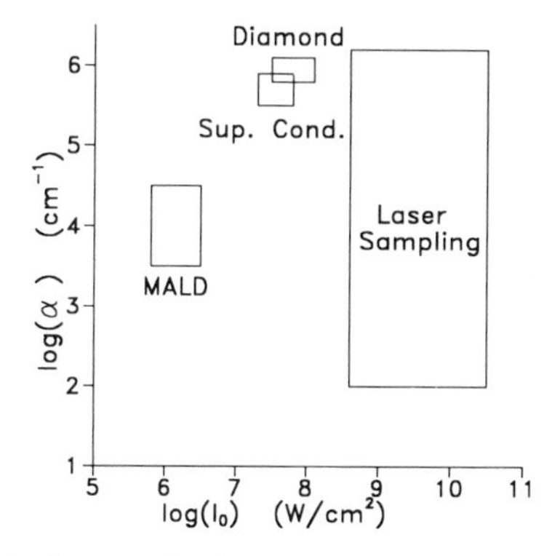

Figure 1.1. Map of deposited power density in some important applications of laser-induced processes: MALD, matrix-assisted laser desorption; Sup. Cond., pulsed laser deposition of high-temperature superconductor layers; Diamond, synthetic diamond formation; and laser sampling.

Spectrometry, etc.) and conferences are almost completely disjunct, and cross-references in the publications are scarce. Obviously the result of this parallelism is considerable waste of time and energy. On the other hand, recognition and investigation of the common elements of laser–solid interaction in the aforementioned areas can lead to cross-fertilization. One of our objectives in this volume is to blend the different fields together. That is why we have invited leading scientists from both groups to contribute.

There are continuing efforts from both the "preparative" and the "analytical" community to understand and describe the basic processes involved. Laser ablation of copper—the successful substrate for diamond film synthesis—has recently been investigated from the point of view of technological (Dreyfus, 1991) and analytical (Kimbrell and Yeung, 1988; Balazs et al., 1991) processes. Among other techniques mass spectrometry, fast photography, spectroscopy, scanning electron microscopy, and laser-based diagnostic techniques have been used to uncover the ionic, cluster, and particulate constituents of laser-induced plumes above different targets. Theoretical studies ranged from gas dynamic treatment of Knudsen layer formation (Kelly and Dreyfus, 1988) to the numerical solution of the equations of heat conduction and of plume hydrodynamics (Vertes et al., 1989; Nakamiya et al., 1990) to a two-step (isothermal + adiabatic) plasma expansion model (Singh and Narayan, 1990).

Obviously, the significance of basic studies in preparative and analytical applications of laser–solid interaction should not be underestimated. New materials and novel processing and analytical methods are expected to emerge in this area. Also the mechanisms of a number of underlying processes will no doubt eventually be unveiled.

In this volume we outline the large and fast-growing area of analytical applications: ion generation for mass analysis. As editors of this multiauthor volume, we have tried to maintain a coherent approach based on the following structure.

In Chapter 2 an elaborate overview of instrumentation is given. Starting with the history of lasers in mass spectrometry we see how the design concepts evolved in the organic and inorganic applications. Microprobing with lasers captured the imagination of instrument builders for about two decades. Separation of volatilization and ionization has been suggested as a way to increase sensitivity and selectivity of the chemical analysis. To further enrich the analytical capabilities, various combination techniques have been introduced. In these methods ion or neutral beams are used complementing the laser beam at different phases of the measurement. Chapter 2 also deals with various types of mass analyzers. Considering the special features of the laser-generated ion swarms, devoted constructions had to be developed. Magnetic sector mass spectrometers, time-of-flight mass spectrometers, Fourier transform mass spectrometers, and other less common constructions are discussed.

Chapter 3 deals with methods utilizing low and medium laser irradiance. Because molecular structure is usually not destroyed in these experiments, this irradiance range is the domain of organic analysis. The up-and-coming laser ionization method is the matrix-assisted technique, as is obvious from the programs of more than a dozen related conferences. The virtue of matrix assistance is to volatilize and ionize extremely large molecules (up to 300,000 Da as of now) by an easy and extremely fast sample preparation technique. The method is currently in the phase of vigorous development. It now seems possible to analyze mixtures of large molecules without separation and to monitor the time development of enzymatic degradation. Proteins, nucleotides, nucleosides, lipids, carbohydrates, and possibly industrial polymers are within the scope of this new technique.

With the careful selection of the ionization conditions it is possible to obtain structural information on small organic molecules. Large numbers of organic compound groups are studied and show characteristic fragmentation schemes. Comparison with electron impact fragmentation pathways is the topic of numerous current studies. Accumulating knowledge on the structure–fragmentation relationship makes this technique an important and fast tool for organic chemists.

Another promising method is based on the recognition that optimal conditions of volatilization and ionization/fragmentation may require substantially different laser parameters. While volatilization often can be achieved by infrared radiation, selective fragmentation and/or ionization can only be carried out by ultraviolet laser pulses. Optionally, a gas jet can be applied to entrain and cool the desorbed molecules. By this or a related method, ultimately one can hope for the controlled fragmentation of complex organic molecules.

In Chapter 4 the high laser irradiance regime is addressed. Owing to the elevated energy deposition in the laser–target interaction, few intramolecular bonds can survive; therefore, this regime is mainly bound to elemental analysis. Laser ablation and plasma formation—used also for material processing—exhibit special features in laser light coupling. These mechanisms have been studied both experimentally and theoretically. Laser-induced fluorescence, Langmuir-probe, and time-of-flight analyzers are used to determine density and velocity distributions of the different species in photoablated plumes. Theoretical considerations reveal spatial and temporal details of the laser-generated microplasma.

Solid sampling for elemental analysis by inductively coupled plasma or by glow discharge mass spectrometry is a promising technique, especially for the analysis of insulating materials. The plume generated by the laser shot is introduced directly to the glow discharge cell or entrained into the inductively coupled plasma torch by a gas flow. Crater characterization and analyte transport studies are underway to improve the analytical features of this new method. Although commercial laser samplers are already available from different vendors, the field of analytical applications is far from being fully exploited.

The idea of inorganic trace analysis by laser plasma ionization has been present since the conception of laser microprobe analysis. Uniform sensitivity for most of the elements and detection limits in the ppm range combined with better than 5 μm lateral resolution have established the worthiness of this technique in the family of elemental microprobes. Instrumental considerations, analytical performance optimization, evaluation procedures, and interesting applications are discussed in Part C of Chapter 4.

Finally Chapter 5, highlights some exotic applications of laser ionization mass spectrometry. Owing to their exceptional analytical characteristics, laser ionization mass spectrometers were placed on two interplanetary space stations by scientists of the former Soviet Union. Both stations were aimed at one of the satellites of Mars named Phobos. Although, unfortunately, both space probes were lost during their trip to Mars, the care and thought that went into the design of some of the scientific payload may serve more "earthly" applications in the future. In Chapter 5 we gain a unique insight into the development of these on-board laser ionization mass spectrometers and into the operation of the prestigious Space Research Institute in Moscow.

REFERENCES

Balazs, L., Gijbels, R., and Vertes, A. (1991). *Anal. Chem.* **63**, 314–320.

Beavis, R. C., and Chait, B. T. (1990). *Anal. Chem.* **62**, 1836–1840.

Denoyer, E. R., Fredeen, K. J., and Hager, J. W. (1991). *Anal. Chem.* **63**, 445A–457A.

Dreyfus, R. W. (1991). *J. Appl. Phys.* **69**, 1721–1729.

Harrison, W. W., and Bentz, B. L. (1988). *Prog. Anal. Spectrosc.* **11**, 53–110.

Karas, M., Bahr, U., and Hillenkamp, F. (1989). *Int. J. Mass Spectrom. Ion Processes* **92**, 231–242.

Kelly, R., and Dreyfus, R. W. (1988). *Nucl. Instrum. Methods* **B32**, 341–348.

Kimbrell S. M., and Yeung, E. S. (1988). *Spectrochim. Acta* **43B**, 529–534.

Nakamiya, J., Ebihara, K., John, P. K., and Tong, B. Y. (1990). *Mater. Res. Soc. Proc.* **191**, 109–114.

Narayan, J., Godbole, O. P., and White, C. W. (1991). *Science* **252**, 416–418.

Singh, R. K., and Narayan, J. (1990). *Phys. Rev.* **B41**, 8843–8859.

Spengler, B., and Cotter, R. J. (1990). *Anal. Chem.* **62**, 793–796.

Vertes, A., Juhasz, P., DeWolf, M., and Gijbels, R. (1989). *Int. J. Mass Spectrom. Ion Processes* **94**, 63–85.

CHAPTER

2

LASERS IN MASS SPECTROMETRY: ORGANIC AND INORGANIC INSTRUMENTATION

L. VAN VAECK, W. VAN ROY, R. GIJBELS, and F. ADAMS

Department of Chemistry
University of Antwerp (UIA)
Antwerp, Belgium

2.1. INTRODUCTION

The recent progress of laser technology has added innovating and fascinating perspectives to the concept of ion generation and decomposition. Prime assets of lasers in mass spectrometry (MS) for analytical applications are selectivity as a result of the mass monitoring principle, sensitivity, ultimately down to a few ions, and the extended dynamic range. The linear dependence of the signal on the temporary analyte concentration in the source covers several orders of magnitude.

Modern lasers are extremely versatile and convenient means to introduce light and energy into the ion source. The emitted radiation is monochromatic, coherent, intense, and directional. The combination of all these virtues offers a lot of experimental flexibility. Laser radiation can be readily transported over relatively long distances, without noticeable loss of intensity. Hence, the beam can easily be introduced into the confined space of an MS source. Photons do not disturb the electrical ion extraction fields. Focusing of the beam down to the diffraction limit by relatively simple optical components enables us to achieve extremely high photon fluxes. Unlike charged particle beams, attenuation remains quite straightforward. The narrow wavelength bandwidth permits us to exploit the ultimate selectivity of the physical processes under conditions of high irradiance, unavailable with classical monochromators.

Photons of a given wavelength can address selectively transitions of an analyte molecule and not the ones of the other constituents in complex

Laser Ionization Mass Analysis, Edited by Akos Vertes, Renaat Gijbels, and Fred Adams. Chemical Analysis Series, Vol. 124.
ISBN 0-471-53673-3

mixtures. Nevertheless, photon ionization in the gas phase is conventionally hindered by the experimental compromise between wavelength selection and intensity, preventing the use of saturation conditions in the MS source. Lasers have triggered a revival of the photon ionization approach in the gas phase, both for elemental and organic applications. Examples are the resonance ionization spectroscopy (RIS) technique, providing ultraselective detection of single atoms in the presence of matrix components. Organic chemists have employed photon ionization in the gas phase as a means of transferring a well-defined energy packet to the molecules, so that access is obtained to the actually imparted excitation energy of the molecular ion.

Numerous experiments now exploit laser multiphoton ionization coupled with a method to evaporate the analyte in the gas phase, for instance, laser desorption (LD) or gas chromatography (GC). The breakthrough of lasers in MS, however, is primarily related to the irradiation of solids. Analytical chemists appreciate the potential to achieve high-energy excitation of solids without charging effects of nonconducting samples, as opposed to, for instance, secondary ion mass spectrometry (SIMS). A complete phenomenological description of the important physical processes governing the laser–solid interaction is not yet available. Lasers are used to evaporate the analyte for subsequent laser postionization in the gas phase, or alternatively they are used to achieve a one-step desorption–ionization process. Laser interaction of solids enables us to achieve an ultrafast heating rate of the material up to extremely high temperatures. The steep gradient allows intact release of thermolabile compounds from the condensed phase. As a result, laser irradiation of organic solids allows the characterization of polar, large, and involatile molecules, which is impossible for the conductive heating of conventional introduction systems. Additionally, focusing the laser beam to a diffraction limited micrometer spot size has yielded the first successful all-round microprobe for inorganic and organic MS applications. Isotope selective ionization can be performed by the use of laser-based postionization approaches. Imparting precise amounts of internal energy to gas phase neutrals allows selective and/or efficiently controlled dissociation into specific fragments.

The rapid proliferation of laser-based experiments has resulted in a still-growing spectrum of analytical applications. A lot of interest has been raised by the advent of the matrix-assisted desorption technique, capable of generating giant ions up to 300,000 Da in mass from peptides, enzymes, and other biomolecules. Laser ablation is highly regarded for the sampling of solids in conjunction with postionization such as in the inductively coupled plasma (ICP), photon ionization in the gas phase, as well as the more conventional electron or chemical ionization techniques. Lasers are profitably used at different levels in the MS experiment. Good examples of the numerous approaches are found in the literature, reported in conjunction with ion storage Fourier transform (FT) MS: laser ablation, one-step desorption and ionization, postionization of neutrals, photofragmentation in the analyzer cell as an alternative for collision-induced dissociation (CID), photochemical transformation of ions as a means of structural analysis, highly selective chemical ionization by laser generated Fe^+ ions as reactant gas to distinguish alkenes, etc.

It is clear that the implementation of lasers in MS experiments represents an extremely broad topic. Nevertheless, a detailed discussion of specific experiments requires that detailed features of the mass spectrometric and ion detection systems have to be considered. For instance, laser microprobes based on time-of-flight (TOF) MS vs. FTMS yield different information, as a result of the incomparability of the time domains for sampling and analysis of the laser-generated ions in both applications. Hence, this chapter will emphasize the instrumental design and operational principles. Applications will only be mentioned to indicate the aim of a given development.

The astonishing number of fundamentally different analytical applications makes a complete coverage of the use of lasers in MS hard to achieve. We shall therefore describe diverse examples from early and recent papers to highlight the main directions in which we foresee substantial progress in the near future. Inevitably, the choices made show a personal bias.

2.2. EARLY APPLICATIONS IN ELEMENTAL ANALYSIS OF SOLIDS

The utilization of lasers in the field of inorganic mass spectrometric analysis started during the early 1960s. At that time, ionization of solids for trace element determinations was conventionally performed by the radio-frequency (RF) spark source. Dielectric specimens require the preparation of electrically conducting electrodes by mixing the sample with a conducting powder. This obviously implied dilution of the original sample and potentially led to trace contamination, with detrimental effects on the analytical detection limits. Quantization at a 10 ppb level could be achieved, but sample handling problems practically required an amount of analyte on the order of at least 10^{-8} g. Hence, it is not surprising that the initial reports on laser ionization for elemental determinations particularly emphasized the capability to deal with solids of any kind, i.e., from metals to dielectric materials, and of any morphology, for instance, thin films, bulk specimens or powders. Particularly appealing was the potential to examine extremely small amounts of analyte, combined with the prospects of layer-by-layer and local analysis, ultimately down to the micrometer scale.

The availability of given lasers strongly determines the applied energy regime on the solid. The power delivered by normal pulsed lasers ranges from 10^3 to 10^6 W, with an irradiation time between 10^{-3} and 10^{-4} s. The introduction of the Q-switching technology allowing the use of giant pulses yields 10^6–10^8 W and a duration between 10^{-7} and 10^{-8} s. Typical crater sizes range from 10^{-2} to 10^{-5} cm^2.

One of the very first attempts to combine lasers with a double-focusing (spark source) MS was reported by Honig and Woolston (1963). The instrumental setup is schematically represented in Figure 2.1. Note that in this design attention is paid from the beginning to the thermally emitted ions as well as to the vaporized neutrals. A ruby laser producing 50 μs pulses was mounted on a double-focusing Mattauch-Herzog AEI MS 7 instrument with photoplate detection. With the normal extracting field between the source slit and the target, the majority of the laser ionized species was single charged. A low-voltage discharge creates an electron flux toward the neutrals in the laser plume above the target, so that singly and multiply charged ions are observed. The energy spread is substantially larger than for the directly emitted ions. The performance of the laser analysis has been promising: around 2.10^{14} ions are generated from a metal sample, and a crater is produced of about 150 μm diameter and 125 μm deep, containing ca. 2.10^{17} atoms. The high ion current, however, causes severe space broadening effects. The idea of laser vaporization in conjunction with electron ionization prior

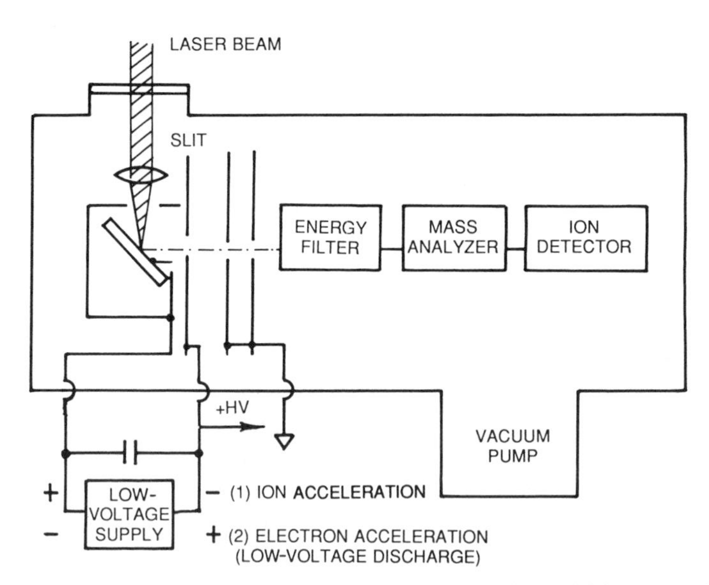

Figure 2.1. Schematic diagram of the modified ion source on an AEI MS 7 instrument for laser ionization of inorganic solids. Reprinted from Honig and Woolston (1963) with permission of the American Institute of Physics.

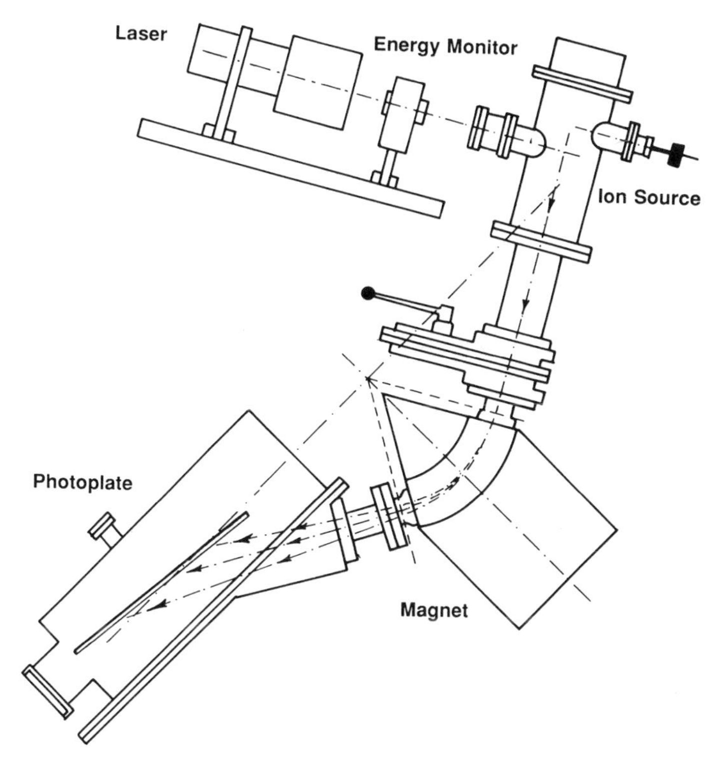

Figure 2.2. Magnetic sector instrument developed by Eloy and colleagues for the mass analysis of laser-generated ions from solid materials. The spectra are registered on a photoplate. Reprinted from Eloy (1969) with permission of the Société française de Chimie.

to use of a magnetic MS analyzer was soon exploited for the examination of different graphite type solids (Berkowitz and Chupka, 1964).

In the same period, Eloy (1969) started research that is still continuing on a magnetic sector instrument without electrostatic sector. Figure 2.2 shows a schematic diagram of that instrument. In this early version, photoplate registration was used. A specific ion source design was elaborated to optimize the sampling of the low-energetic ions (Eloy and Dumas, 1966). A pulsed ruby laser delivering 0.2 J in 4 μs allows detection limits in the percentage range and an accuracy of better than 10% on isotope ratios. The crater diameters are typically under 250 μm, and the depth ranges from 0.6 to 0.8 μm (Eloy, 1969). Figure 2.3 compares the results for chalcopyrite inclusions in a mineral, obtained by laser probe MS (LPMS) and by spark source MS. In the latter case, no local analysis is feasible. The laser analysis allows detection of minor isotopes down to the 0.5% level, whereas the absence of

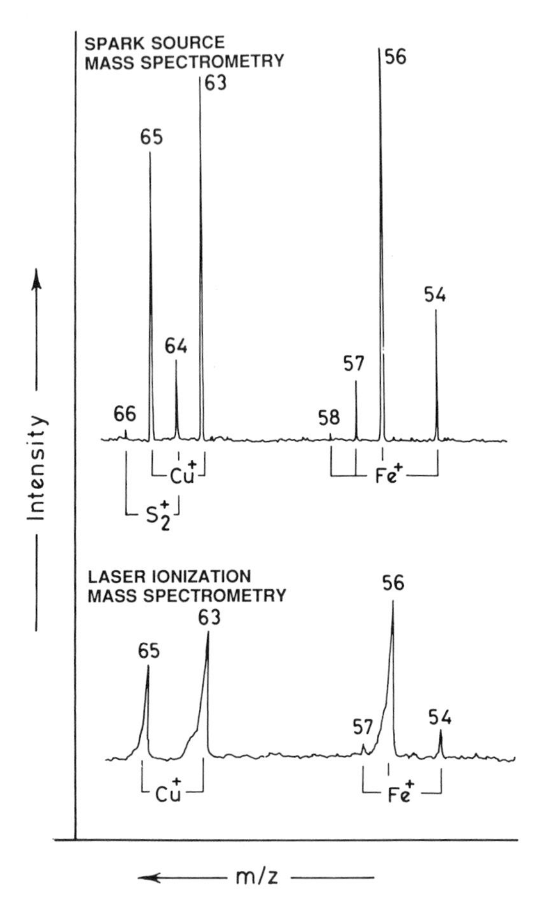

Figure 2.3. Comparative data from chalcopyrite recorded by spark source and laser ionization with a single magnetic sector analyzer. Reprinted from Eloy (1971) with permission of Elsevier Science Publishers.

some signals, such as the sulfur lines, clearly demonstrates that these signals arise from the surrounding bulk material (Eloy, 1971).

The application of a TOF-MS as an alternative to the low-transmission magnetic sector analyzer very soon entered the scene. Fenner and Daly (1966) designed an instrument that is schematically represented in Figure 2.4. The main operational principles and functional devices, incorporated later in commercial laser microprobes, were already included here, albeit in a rather crude form. A "giant pulse" ruby laser, delivering 30 mJ in 30 ns, is focused through a long working distance microscope objective on an area of $2 \times 10^{-5}\,\text{cm}^2$. The ions are extracted in transmission, i.e., from the opposite

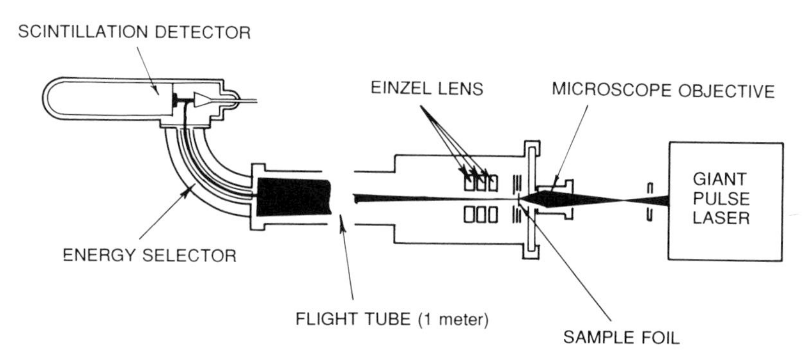

Figure 2.4. Early setup for laser ionization using so-called giant pulses in conjunction with TOF-MS. Reprinted from Fenner and Daly (1966) with permission of the American Institute of Physics.

side in comparison to the impinging laser beam. An einzel lens is provided at the entrance of the 1 m flight tube. A 90° deflection electrostatic filter improves the mass resolution from 6 to 30 at m/z 60, but the transmission decreases dramatically by 4 orders of magnitude. The ion yield is about 10^{-2} per atom evaporated from a variety of foils such as copper, carbon, boron, aluminum, silver, gold, or lead. The main problem resides in the initial time spread, up to 20 μs, and the velocity distribution of ions. In a later design, the high field extraction is no longer applied because of the huge ion currents generated. Standard deviations are substantial: up to 10% on major components but more than 50% for minor constituents in the 10% range. However, the prime assets have been the panoramic detection capability and the full elemental coverage achieved for a sample in the size range of 10^{-8}–10^{-10} g (Fenner and Daly, 1968).

As with the development of magnetic sectors, here also several investigators attempted to make effective use of the neutral components within the laser-generated plume. Pioneering work was performed at Pennsylvania State University on a modified Bendix 12-107 TOF-MS (Vastola and Pirone, 1968). The sample is mounted on the normal direct introduction probe, and a microscope, introducing a moderately focused ruby laser beam, is added along the same axis but at the opposite side. The analyte is situated just below the path of the standard electron beam in the ionization chamber. In this way, the importance of the directly emitted ions can be measured in relation to the abundance of the neutrals. Extensive studies on the Group Va and VIa elements and their compounds show the merits of this approach in respect to increased sensitivity and selectivity (Ban and Knox, 1969; Knox, 1968). Detailed information about the fast and transient phenomena of the ion production after laser pulses can be obtained by fully exploiting time-lag

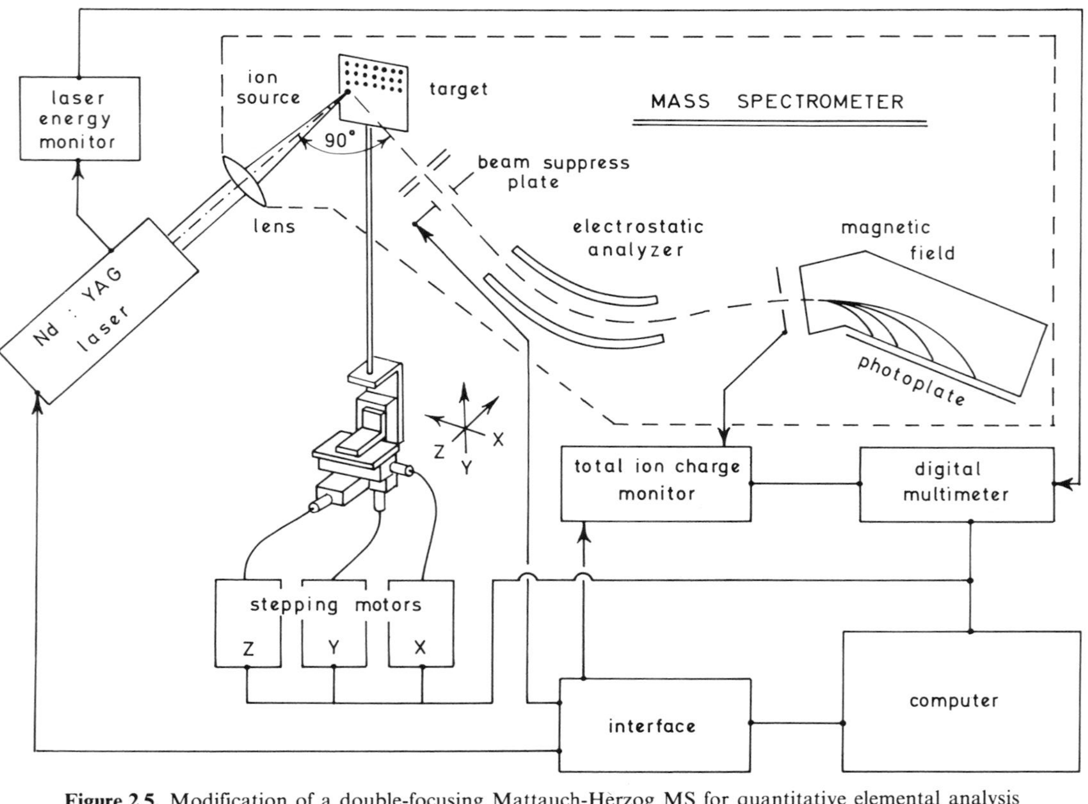

Figure 2.5. Modification of a double-focusing Mattauch-Herzog MS for quantitative elemental analysis by laser ionization of solids. Reprinted from Jochum et al. (1988) with permission of Springer-Verlag.

focusing techniques in combination with the possibilities from additional grids in the source and electron impact (EI) ionization (Ban and Knox, 1969). In the early experiments, photographs of the traces on a wideband oscilloscope were used for recording the data. Such procedures remain cumbersome and long hindered the practical use of TOF systems, despite their advantage of high transmission and simplicity of construction. In fact, it took almost another 10 years before convenient electronic equipment for digital signal registration became available and the renaissance of TOF-MS was initiated.

Meanwhile, research on laser ionization for inorganic MS of solids remained focused almost exclusively on the possibility of complementing or substituting for the currently used spark source. The energy spread of the laser plasma ions, ranging up to a few hundred electron volts, necessitates the use of double-focusing instruments in order to obtain the required mass resolution. From 1978 onwards, several publications have described the modification of existing spark source MSs on a laboratory scale. To our knowledge, only one instrument has ever been commercialized (EMAL Technical Information Bulletin, 1982; Briukhanov et al., 1983). Figure 2.5 illustrates a typical modified spark source instrument (Jochum et al., 1988). Table 2.1 summarizes the characteristics of representative setups. With its focalization on a flat plane, the Mattauch-Herzog-type analyzer permits photoplate registration and hence panoramic multielement determinations (Van Vaeck and Gijbels, 1989). Electrical detection offers better sensitivity in sequential measurements but lower mass resolution. Repetitive laser irradiation in combination with sample scanning is applied to overcome the low transmission of double-focusing sector analysers.

Most experiments are conducted with the infrared (IR) wavelength of 1.06 μm. Use of a CO_2 laser source causes severe artifacts by selective evaporation of sample constituents as a result of the characteristically tailed energy deposition (Bingham and Salter, 1976a, b). Application of Nd:YAG irradiation at high power density on the sample in the order of 10^9 $W \cdot cm^{-2}$ yields nearly constant elemental sensitivities throughout the periodic table. Figure 2.6 compares relative sensitivity factors (the elemental sensitivity in relation to a reference element, here iron) obtained from the analysis of a steel standard and an ocean basalt sample by laser and spark source MS (Matus et al., 1988).

Moreover, it has been observed that the relative ion yield remains unaffected by the precise laser energy or by the surrounding target material. It has been found that the laser focus in relation to the actual sample surface critically affects the production of the double-charged ions. The use of optics ensuring a long depth of focus obviously improves the reproducibility and analytical usefulness of the method for long exposures since the power density along the surface remains comparable as erosion proceeds. The relative

Table 2.1. Comparative Survey of Instruments for Laser Ionization in Elemental Mass Spectrometry[a]

	Bingham and Salter (1976a)			Conzemius et al. (1981)	Jansen and Witmer (1982)	Dietze et al. (1983)[b]	Jochum et al. (1988)	EMAL (1982)	Dietze and Becker (1985)
Laser	CO_2	Ruby	Nd:YAG	Nd:YAG	Nd:YAG	Nd:YAG	Nd:YAG	n.r.	Nd:YAG
Wavelength (μm)	10.6	0.694	1.06	1.06	1.06	1.06	1.06	1.06	1.06
Pulse τ (ns)	>100	20	15	200	15	100	15	15	40
Repetition rate (Hz)	<1	0.1	50	10 k	1–50	100	1–50	1–50	5–50
Energy per pulse (mJ)	300	10	10	1	10–100	1–10	15	3–5	60
Power density (W/cm^2)	$<10^9$	10^8–10^{11}	10^9–10^{11}	2×10^9	n.r.	5×10^9	2×10^{10}	n.r.	5×10^8
Diameter spot (μm)	20–500								
Crater diameter (μm)	300	20	20–300	25	20	50–500	50	50	10–500
Crater depth (μm)	10	5	0.5	3	1–10	50–500	3	1	

Mode	45°/45°	90°/60°	45°/45°	45°/45°	45°/45°	60°/90°	45°/45°
Mass spectrometer	AEI MS7	Graf II-2	AEI MS702	AEI MS702	AEI MS57		MX 3301
Transmission	n.r.	10^{-6}	n.r.	10^{-6}	n.r.	n.r.	10^{-6}
Resolution by m/z	3,000(60)	2,000	3,000	3,000	10,000	2,000	17,000
Photoplate detection	+	+	+	+	+	+	+
SEM detection	−	+	−	−	−	−	+
+/− polarity	+	+	+	+	+	+	+
s/m charged ions	s/m	s/m	s/m	s/m	s	s/m	s/m
Absolute detection limit (g)	10^{-11}	2×10^{-10} (photo) 2×10^{-13} (SEM)	10^{-12}	10^{-12}	n.r.	n.r.	10^{-12}
Mass range	6–240	7–250	7–250	7–250	7–250	n.r.	10–360

Source: Reprinted from Van Vaeck and Gijbels (1989) with permission of San Francisco Press, Inc.

[a] n.r., not reported; SEM, secondary electron multiplier; s/m, singly or multiply charged ions.

[b] Measurements were carried out at KFKI, Budapest.

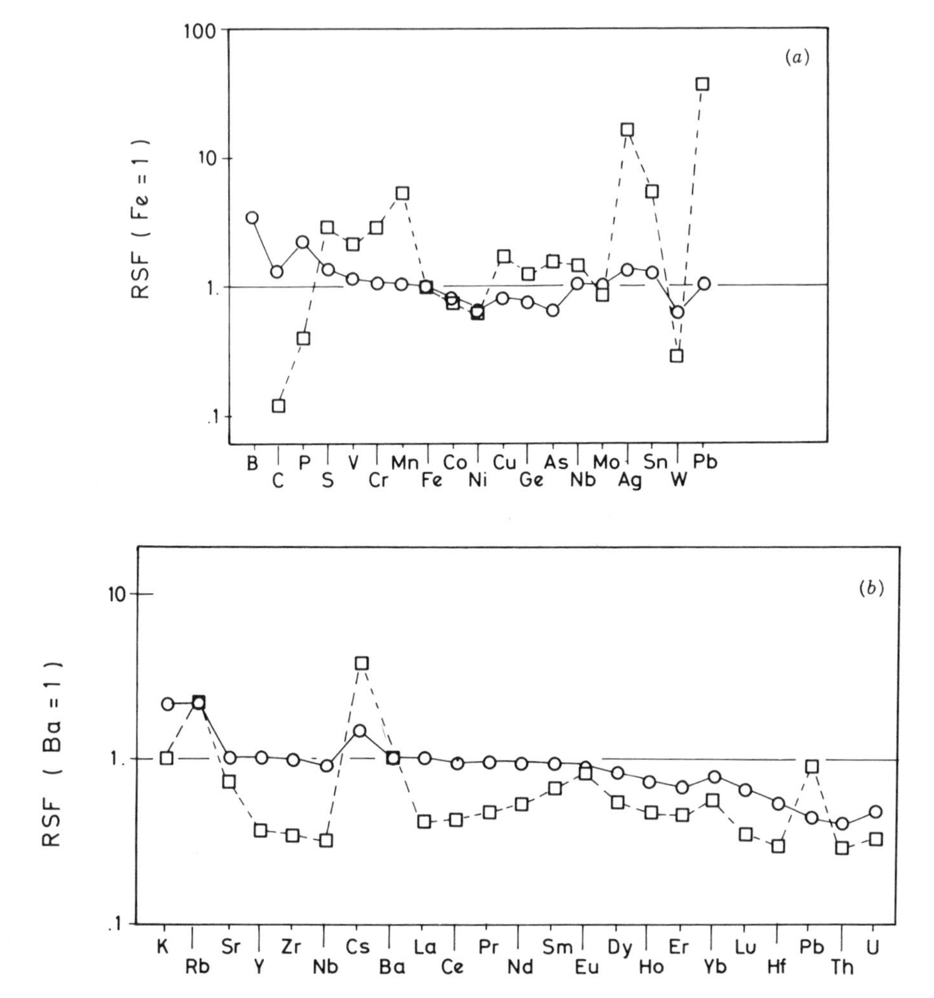

Figure 2.6. Comparison of the relative sensitivity factors (RSF) obtained from the analysis of a steel (a) and basalt (b) sample by laser ionization (○) and spark source (□) MS. Reprinted from Matus et al. (1988) with permission of Elsevier Science Publishers.

importance of the doubly charged ions can be used to critically tune the focusing plane during analysis and/or to calibrate the local energy regime for a particular sample matrix (Conzemius et al., 1984). Laser beam spot sizes of 20–50 μm enable us to scale down the size of analyte by 2 orders of magnitude, i.e., 10^{-10} g for the laser vs. 10^{-8} g for the spark. Rapid scanning of the sample allows reasonable surface selectivity with a crater depth in the range of 0.1 μm (Jansen and Witmer, 1982).

It should be noted that the idea of TOF analysis has remained appealing for more dynamic studies, for instance, those aiming at surface characterization and layer-by-layer analysis. The capabilities for quantitative determinations have clearly been inferior to those of double-focusing sector instruments. However, since laser interaction with solids may produce completely unexpected ionic and neutral species, the panoramic registration combined with the high sensitivity have made TOF analysis highly appreciated during the initial stage of laser ionization development. A lot of research has been devoted on the one hand to the optimization of data collection in function of the time variable ion production, characteristic of laser ionization of solids, and on the other hand to the problems arising from the broad kinetic distribution of the generated ions.

As to the first aspect, pioneering work was performed by Lincoln (1965). Time-resolved spectra for selected ions were obtained simultaneously with a resolution in the microsecond range by a six-gated electron multiplier, coupled to a number of oscilloscopes or a fast multichannel strip chart display (Lincoln, 1969). Applications reported later concern, for instance the pyrolytic treatment of several graphites (Lincoln, 1974). The broad kinetic energy distributions of the laser-generated ions necessitated a lot of research to make the TOF applications practically feasible. Several types of energy filters have been described. Bernal et al. (1966) designed a simple grid system, mounted directly in front of the electron multiplier, allowing independent measurements of the m/z of the ions and of their initial kinetic energy. Both parameters are characterized for the ionization by a single laser pulse. Several excellent reviews cover applications extensively (Harding-Barlow et al., 1973; Knox, 1971; Kovalev et al., 1978; Conzemius and Capellen, 1980; Conzemius et al., 1983).

To conclude this section, the original approach followed by Eloy and co-workers for the development of the LPMS should be mentioned. In this design, mass analysis has been performed in a single-focusing magnetic sector, ingeniously equipped with an electro-optical converter to substitute the photoplate registration. In this way, the sequential arrival of ions as a function of their m/z is recorded. Otherwise stated, TOF-type mass spectra are obtained and the advantages of better sensitivity by electrical detection can be combined with panoramic registration capabilities. One of the more recent versions of this instrument uses a laser focused down to 3 μm. Hence, that setup will be described in detail in Section 2.4 on laser microprobes.

2.3. EARLY APPLICATIONS IN ORGANIC MASS SPECTROMETRY

Organic chemists have been intensive users of MS techniques from the 1950s onward, but primarily for structural characterization and not for quantitative analysis. Conventional instruments are inevitably based on ionization of the analyte in the gas phase. Volatilization, required for fluid and solid samples, is of course detrimental to large and polar molecules. Continuing research into so-called soft ionization methods, seeking to produce molecular ions from intact structures, initially yielded the field desorption (FD) technique. In this approach, the solid residues from solutions are placed directly inside the ion source. Subsequent evaporation and ionization eliminates the losses of the analyte, which normally occur with the external introduction devices. Impressive results have been obtained but extensive and delicate sample pretreatment procedures have hindered the popularity of the method. Hence, the easiness of directing a laser beam on a solid inside the confined space of an ion source has given rise to a convenient method of heating microquantities of samples. Additionally, extremely steep temperature gradients are realized in the solid state, which allow the release of

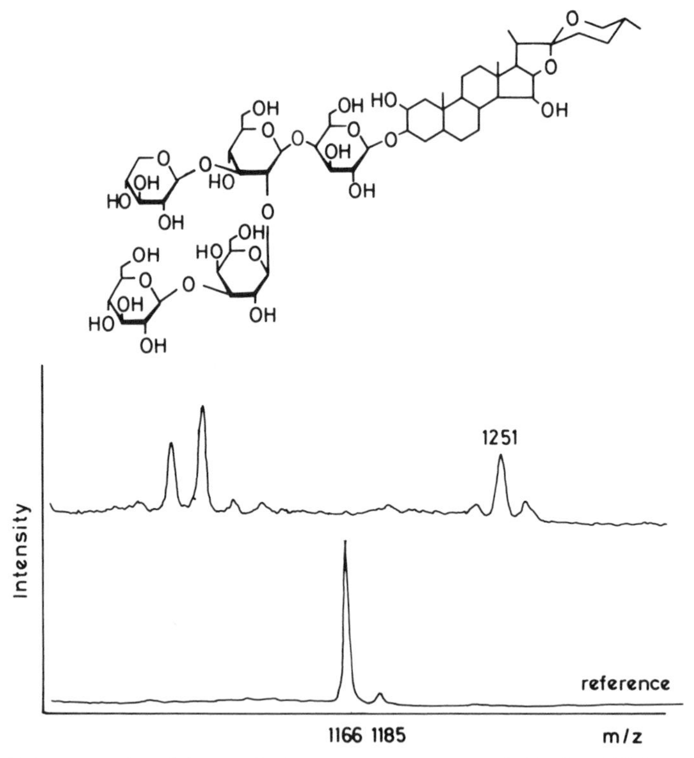

Figure 2.7. Parent region of the positive ion mass spectrum recorded by LD-MS from digitonin. The molecular weight is 1228. Reprinted from Posthumus et al. (1978) with permission of the American Chemical Society.

intact thermolabiles. Indeed, the rate constant for desorption usually exceeds that for decomposition. Hence, a substantial fraction of involatile molecules can be released without degradation whenever the high temperature makes the influence of the activation energy term in the Arrhenius equation negligible.

The field of organic laser MS was pioneered at Pennsylvania State University on the modified Bendix TOF equipment mentioned in Section 2.2. Initial results have concerned the analysis of polycyclic aromatic hydrocarbons (PAHs), alkyl aromatic compounds, and amino acids. A solid sample is irradiated with a normal pulsed ruby laser, focused to an 800 μm spot. Ion emission is observed for a 200–400 μs period during the peak of the laser burst, whereas neutral species, subsequently ionized by an electron beam, are detected up to several hundred microseconds after the laser irradiation. The PAHs yielded radical molecular ions, while amino acids such as leucine produced only fragments (Vastola and Pirone, 1968). Nevertheless, it is Vastola and associates' subsequent publication (1970) on the analysis of hexylsulfonates that is generally considered to be the first report on laser desorption (LD) MS. The salts were detected in the cationized form in the virtual absence of thermal degradation or fragmentation. Surprisingly, the idea was then practically abandoned for almost one decade until Posthumus et al. (1978) reported striking results on the LD-MS of thermolabile biomolecules such as oligosaccharides, small peptides, and glycosides. Figure 2.7 illustrates the parent region of the original mass spectrum for digitonin (Posthumus et al., 1978). The intact molecule is detected as a cationized structure, and accompanying signals in the mass spectrum refer to the cleavage of the terminal xylose and ribose residues. These experiments were responsible for the subsequent interest in LD-MS applications to the analysis of nonvolatile and relatively high-molecular-weight organic compounds.

Since that time, several groups in different laboratories have constructed numerous instruments, most of which have remained experimental in nature. So far, only the FT-based version has been commercialized. Table 2.2 shows a comparative survey of the major characteristics of various examples of reported LD-MS instruments (Van Vaeck and Gijbels, 1989). In contrast to the inorganic applications, a substantially larger array of lasers and MSs has been used. The following subsections will discuss selected setups in detail to highlight the possible influence of the actually selected MS on the obtainable type of information.

2.3.1. Magnetic Sector LD-MS Instruments

Both low and high-mass-resolution instruments have been used. The availability of these mass analyzers has represented perhaps the main

Table 2.2. Instruments for Laser Desorption–Ionization in Organic Mass Spectrometry

	Posthumus et al. (1978)		Stoll and Röllgen (1979)	Cotter (1980)	Van Breemen et al. (1983)
Laser type	TEA-CO_2	Nd:glass	CO_2	TEA-CO_2	
Mode	Pulsed	Pulsed	Continuous	Pulsed	
Wavelength (μm)	10.6	1.06	10.6	10.6	
Pulse duration (ns)	150	10^5	n.a.	40	40
Energy (W)	n.a.	n.a.	3	n.a.	n.a.
Energy per pulse (J)	0.1	1	n.a.	0.7	0.1–0.5
Spot diameter (μm)	$\geqslant 10^2$	10^2	$> 10^2$	10^3	10^3
Typical power density (W/cm^2)	10^6	10^7–10^8	10–10^4	5.10^6	10^6
Geometry[b]	90/0	90/0	0/90	70/0	90/0
Sample and support	Thin layer on metal	Thin layer on quartz	Thick on metal	Thin layer on polymer	
Mass spectrometer description	Magnetic sector	Double-foc. MS	Quadrupole scan: 0.1 s for full sp.	Double-foc. MS Cl source	Linear TOF
Mass range	1,500	500	800	500	4,000
Mass resolution at m/z	300 at 600	2,000 at 400	20 at 365	500	800
Detector type	OMA	Photo-plate	SEM + MCA	SEM + MCA	SEM + MCA
Ions	+/−	+	+/−	+	+

Source: Reprinted from Van Vaeck and Gijbels (1989) with permission of San Francisco Press, Inc.

[a] *Abbreviations:* OMA, optical multichannel analyzer (channelplate + vidicon + multichannel analyzer); MCA, multichannel analyzer; SEM, secondary electron multiplier; TOF, time-of-flight mass spectrometer; MW, molecular weight; p/m, pulse per mass; foc., focusing; abl., ablation; inz., ionization; n.a., not applicable; n.r., not reported.

Heresh et al. (1980)	Zakett et al. (1981)	Hardin and Vestal (1981)	Hardin et al. (1984)	Wilkins et al. (1985)	Grotemeyer et al. (1986a)	LMMS (1973–1988)
Nd:YAG	Nd:YAG	Dye	Nd:YAG	TEA-CO_2	abl. Nd:YAG inz.:dye	Nd:YAG
Pulsed	*Q*-switch	Pulsed	*Q*-switch	Pulsed	Pulsed	*Q*-switch
1.06	1.06	0.483	1.06	1.06	abl. 1.06 inz. 0.266–0.3	0.266
8.10^4	10	5–7	5–7	40	5	15
n.a.	n.a.	n.a.	n.a.	n.a.	n.a.	n.a.
0.1	0.1	10^4	10^4	0.3	inz. 2.10^3	2.10^3
5.10^2	10^3	10^3	10^2	10^3	10^3	0.5–5
10^4–10^6	10^8	10^7	10^8	10^8	n.r.	10^{10}–10^{11}
45/45	0/90	30/90	45/90	80/90	n.a.	(–)90/90 45/90
Thin on glass	Thin on metal	Thin layer on moving belt LCMS interface		Thin on metal	Thin on metal	Thin layer on bulk
Double-foc. MS	Double-foc. reversed geometry (MIKES)	Quadrupole scan rate 5 p/m + TOF	1 p/m	FTMS 3 T magnet	TOF + reflector	TOF + reflector
n.r.	500	500	1,200	7,000	(?)[c]	(?)[c]
1,000	n.r.	500	1,200	10^4 at 1,000	6,500 at 100	500–800; 500 at 208
SEM	Integrating electro-meter	SEM + boxcar integ. SEM + MCA	SEM + MCA	n.a.	SEM MCA	SEM MCA
+	+	+	+	+/–	+/–	+/–

[b]Geometry: angles (degrees) between laser and ion beam vs. sample surface according to Conzemius and Capellen (1980).

[c]Mass range unlimited unless by mass resolution and/or detection.

motivation for their choice as a mass separation device. Transmission is low, particularly in the high resolution mode, while registration of a complete mass spectrum requires slow scanning cycles up to several minutes. Hence, repetitive laser irradiation is mandatory and total sample consumption is relatively high in comparison with the modified spark source instruments of Section 2.2. No layer-by-layer analysis has been attempted. It is generally claimed that a single laser interaction with the solid does not cause visible damage under microscopical observation. However, these experiments have primarily aimed at the qualitative investigation of organic compounds, not amenable to other MS techniques available at that time. As a result, precise data on laser focus, material consumption, and sensitivity have not always been reported. Besides, samples have usually been prepared by evaporation of a solution on a substrate or by electrospray deposition, and no extensive investigation of surface coverage or layer thickness has been attempted. The influence of the substrate, namely, different metals, quartz, and polymers such as Vespel or Teflon, has not been clearly elucidated.

The experiments at the FOM Institute of Amsterdam (Stichting voor

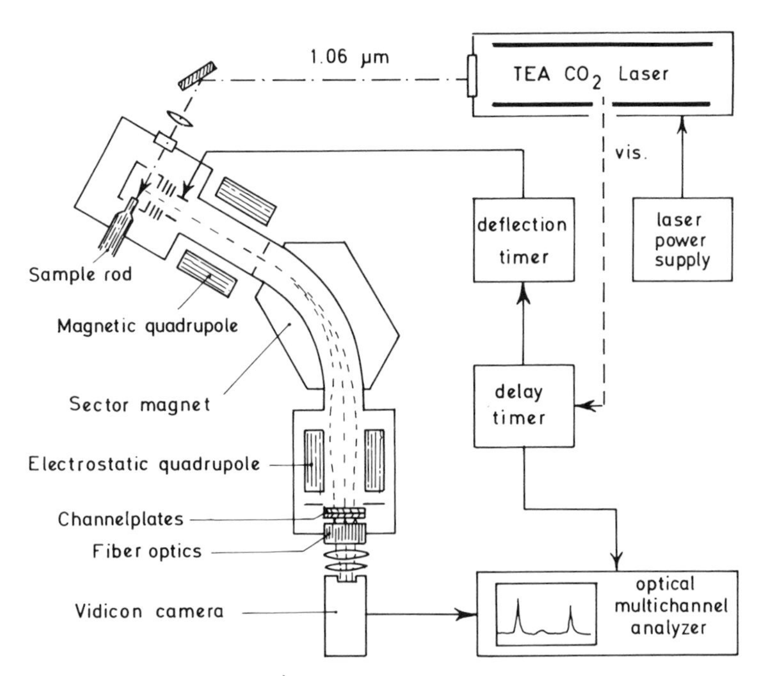

Figure 2.8. Block diagram of a single-focusing LD-MS instrument used in the original experiments at the FOM Institute of Amsterdam. Reprinted from Posthumus et al. (1978) with permission of the American Chemical Society.

Fundamenteel Onderzoek der Materie) have been performed primarily on a home-built single-focusing instrument, illustrated in Figure 2.8 (Posthumus et al., 1978). Sample irradiation is performed by a TEA CO_2 laser (GenTec, Dalton, Quebec, Canada), producing 150 ns pulses at 10.6 μm. Material consumption in these conditions can be as high as 10^{-6}–10^{-7} g. The 9 kG magnetic sector provides a m/z range up to 1500. Detection is performed with a channeltron electron multiplier array (CEMA) and vidicon camera. The image is resolved by the scanning electron beam in 500 lines. The video signals are integrated and digitized in a so-called optical multichannel analyzer (Tuithoff et al., 1975). The sector is complemented with a magnetic quadrupole lens at the source side and an electrostatic one before the CEMA. The latter lens changes the mass dispersion. The magnetic quadrupole acts as an image correcting lens to produce sharp lines on the detector. In this way, ions with an m/z ratio of 1.6 can be registered simultaneously. The mass resolution is about 300 at the original m/z ratio of 1.2 but decreases to 200 at an m/z ratio of 1.6 (Tuithoff and Boerboom, 1976). The detection system is integrating in nature, and hence a gating electrode in front of the magnet is required. The deflection timer allows windows of 100 ns.

Then, an electrostatic sector with a decelerating lens in front and a reacceleration lens behind is inserted between the quadrupole and the detector to perform kinetic energy measurements. Figure 2.9 shows the energy distribution plots for cationized sucrose and the most prominent fragment at m/z 185 accompanied by the results for the sodium cations as a function of the applied laser energy (Van der Peyl et al., 1984). High-resolution data have been recorded at the same institute on a Varian MAT SM1 double-focusing MS, equipped with a Nd:YAG glass laser. Of course, the lower transmission of this instrument requires a substantially increased number of laser shots to obtain a detectable blackening of the photoplate.

An AEI MS 702 (Kratos, Manchester, UK) has been converted to permit repetitive LD on the same spot of the sample as a means to improve the intensity statistics (Heresh et al., 1980). A fairly constant signal up to 100 pulses is obtained, at least when the edge of a relatively thick sample layer is analyzed. Scratched edges are clearly superior to smooth layers. Isolated particles provide more intense ion currents but are rapidly consumed. The problems of slow scanning are partly circumvented by resorting to peak switching the electrostatic sector.

Finally, Zakett et al. (1981) have reported the application of LD-MS/MS. An AEI MS 902 (Kratos, Manchester, UK), reconfigured and transformed into a reversed-geometry mass-analyzed ion-kinetic energy spectrometer (MIKES), has been used (Beynon et al., 1973). The modified chemical ionization (CI) source permits laser ionization. Pulsed irradiation at 10 Hz is applied while the magnet is scanned at 1 amu/s. The long time constant

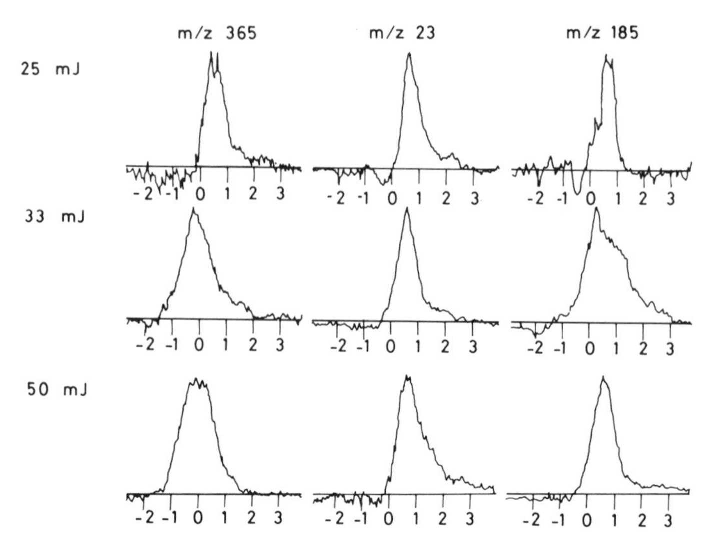

Figure 2.9. Kinetic energy distributions of laser-desorbed ions as a function of the laser energy. The indicated *m/z* numbers refer to cationized sucrose (*m/z* 365), sodium (*m/z* 23), and cationized fragments (*m/z* 185). Reprinted from Van der Peyl et al. (1984) with permission of Elsevier Science Publishers.

of the electrometer (0.5 s) smooths the pulsed nature of the ion current. Acquisition of a daughter spectrum typically requires signal accumulation over 30 min for a mass range from 50 to 450 *m/z*. The instrument is used for structural analysis of laser-produced ions in the gas phase from sucrose, alcohols, and naturally occurring quaternary compounds (Davis et al., 1983; McLuckey et al., 1982; Pierce et al., 1982).

It is obvious from these examples that the magnetic sector instruments and, in particular, the double-focusing analyzers are largely handicapped by their low transmission. Additionally, the scan-type mode of operation is inherently slow, making the time domains of ion formation and signal registration inherently disproportionate.

2.3.2. Quadrupole LD-MS Instruments

The use of purely electrostatic fields for mass separation makes quadrupole analyzers much more adequate for dynamic measurements. Scan rates are practically limited by the detection time per mass. A range of 100 *m/z* can be covered in only 0.1 s. Obviously, mass resolution is sacrificed when scan rates increase. Figure 2.10 illustrates the setup developed at the University of Bonn in Germany (Stoll and Röllgen, 1979). With the low-intensity

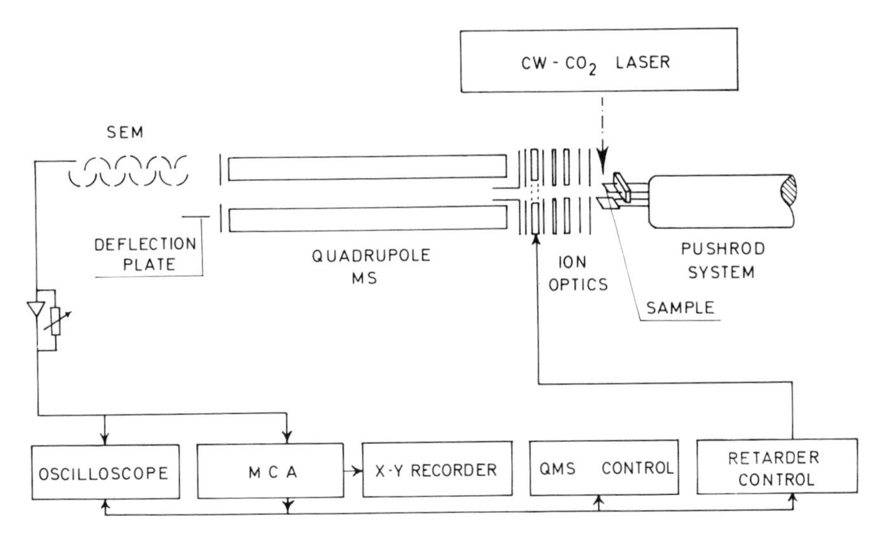

Figure 2.10. Functional scheme of the quadrupole-based LD-MS constructed at the University of Bonn. Reprinted from Stoll and Röllgen (1979) with permission of John Wiley & Sons, Ltd.

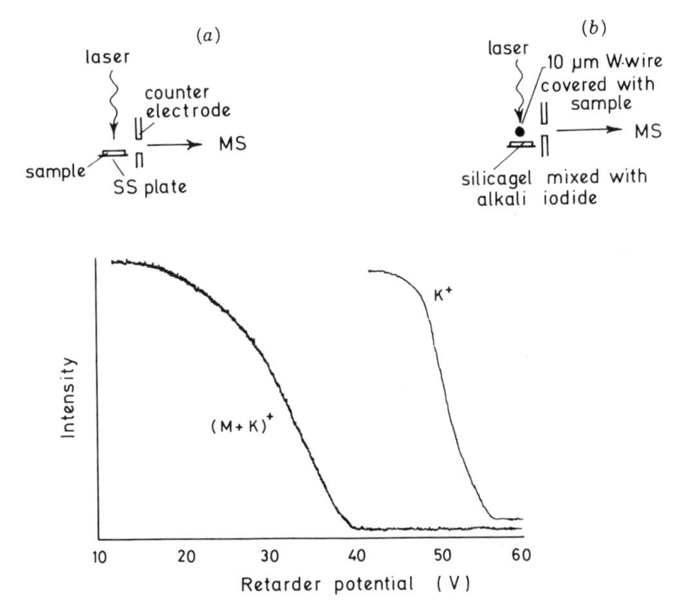

Figure 2.11. Experimental setup with physically distinct sources of (a) sucrose and (b) alkali salts to study the gas phase cationization of organic molecules and the corresponding cumulative kinetic energy distributions, measured in the quadrupole LD-MS at the University of Bonn. Reprinted from Stoll and Röllgen (1982) with permission of the *Verlag der Zeitschrift für Naturforschung*.

continuous wave (CW) laser, the ion emission of solids lasts for only a few seconds and the use of a dynamic MS is mandatory. Signal accumulation is fairly simple under conditions of CW irradiation.

The ion source is further equipped with additional lenses to allow study of the kinetic energy distributions. The sample is coated on a wire emitter or metal holder at 20 V. The first counter electrode is earthed. A retarding potential can be applied on the gold meshes in front of the quadrupole. The mass filter is then operated in the single ion monitoring mode. The transmitted ion current as a function of the voltage difference characterizes the energy distribution. A typical so-called retardation curve is shown in Figure 2.11 (Stoll and Röllgen, 1982). Sodium palmitate is spotted on an electrically heated wire and exposed to a K^+ ion emitter. The alkali emitter is biased at 60 V, whereas the wire supporting the sample is at 30 V. The retardation curves for the cationized molecules and K^+ ions reflect the difference in energy distribution width as well as a displacement of the origin according to the offset voltage between the sample wire and the alkali emitter. Hence, alkali attachment to molecules impinging on the alkali emitter can be excluded and the gas phase reaction is responsible for the formation of $(M + K)^+$ (Stoll and Röllgen, 1979).

Hardin, Vestal, and colleagues have nicely exploited the different possibilities of the quadrupole technology, while the benefits of LD are evidenced in a very practical application of liquid chromatography (LC) MS (see Hardin et al., 1984; also Hardin and Vestal, 1981). Figure 2.12 shows the existing interface with the moving belt (Hardin et al., 1984). The latter provides the required sample renewal for repetitive laser firing at 100–200 Hz, depending on the belt speed, i.e. about $0.25\,cm{\cdot}s^{-1}$ (Hardin et al., 1984). The LC effluent is injected through a capillary tube at 200 °C and thermosprayed on the belt in vacuum. A LC flow rate of $1\,mL{\cdot}min^{-1}$ is acceptable, even with water. Further solvent evaporation takes place during the transfer to the high vacuum region, where laser irradiation is performed at 200 Hz. Two quadrupoles are mounted, the first of which acts as a lens by the application of RF voltages only. The second is operated as a mass filter in the normal mode at a scan rate of $100\,Da{\cdot}s^{-1}$. As to sensitivity, injected amounts on the column in the order of 1 μg are certainly enough to detect parent ions for compounds such as erythromycin or gramicidin, whereas fast LC elution permits to measure cationized cytidine ions from a 10 ng injection.

In an early application only one quadrupole is used (Hardin and Vestal, 1981). By using the mass filter in the single ion monitoring mode, TOF measurements of mass selected ions are performed. A fast pulse amplifier/discriminator replaces the current detector circuitry, and a time-to-pulse height converter, coupled to a multichannel analyzer (MCA), allows determination of the flight time. The resolution of the quadrupole is increased so that the

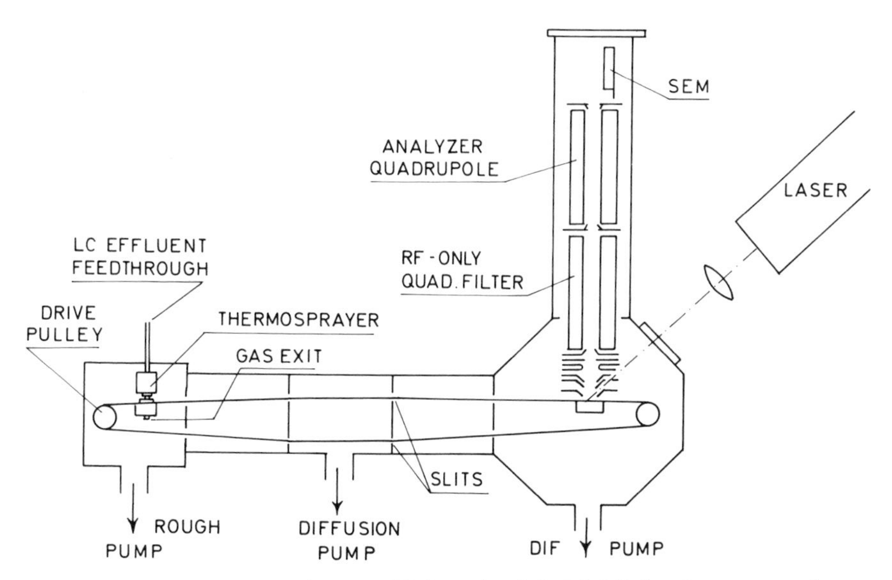

Figure 2.12. Laser desorption in a modified moving belt LC-MS interface connected to a quadrupole MS. The laser enters through the window at the right and intersects the belt at a 45° angle. Reprinted from Hardin et al. (1984) with permission of the American Chemical Society.

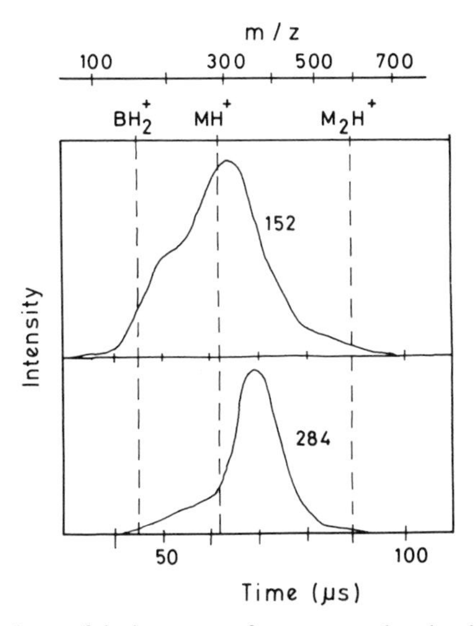

Figure 2.13. Time dependence of the ion current for protonated molecules and purine fragments from guanosine, measured with a quadrupole MS connected to the moving belt interface. Reprinted from Hardin and Vestal (1981) with permission of the American Chemical Society.

ion current reduces to the 1 count per pulse level. Of course, the overall length of a quadrupole, acting as field-free region, is rather limited but still sufficient to permit examination of the kinetic energy distributions as to mean and spread. The data yield important indications about the ion formation process, in particular the role of metastable transitions. During decomposition, the product ions maintain the energy distribution of the precursors and hence their flight time increases. Figure 2.13 shows the time profiles for selected m/z, which correspond to the TOF distributions (Hardin and Vestal, 1981). The data concern the $(M + H)^+$ and the major fragment, namely, the protonated purine moiety. Comparison of the respective maxima with the calculated values for zero kinetic energy shows that both species largely originate from metastable decompositions. The protonated dimers probably act as precursors for the $(M + H)^+$ ions, although 10% of the total intensity still arises from $(M + H)^+$ formation at or near to the belt surface. About 70% of the fragment intensity is related to the decomposition of the same dimeric form, and 30% from the $(M + H)^+$ monomeric precursors. This leaves a negligible role for the direct formation of the fragments at the sample surface. Of course, only the fragments generated before the quadrupole field can be detected. It is nevertheless likely that an important fraction of the ionized molecules is lost in the additional RF-only filter, which increases the time delay between ion formation and mass analysis.

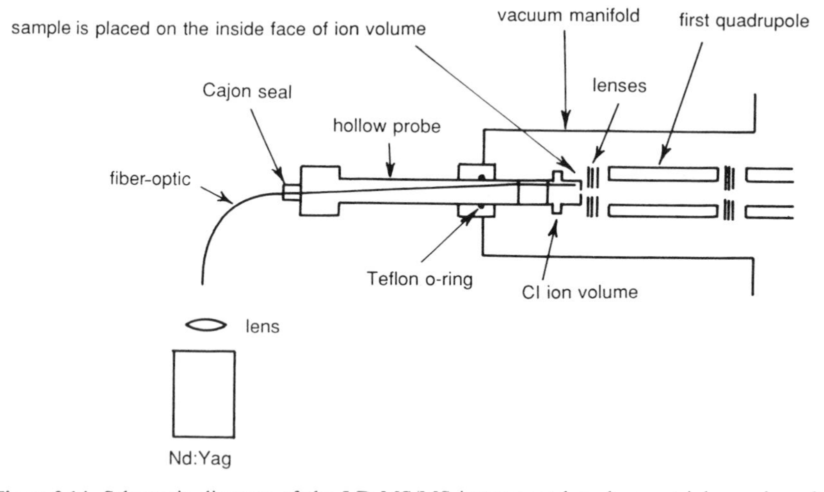

Figure 2.14. Schematic diagram of the LD-MS/MS instrument based on a triple-quadrupole instrument with chemical ionization source, modified for laser beam introduction by means of an optical fiber. Reprinted from Emary et al. (1987) with permission of the American Chemical Society.

Recently, the quadrupole approach has been extended to the field of tandem MS (Emary et al., 1987). The sample is placed inside a conventional chemical ionization (CI) source, as shown in Figure 2.14. A Nd:YAG laser, operated at 1.06 μm, delivers 15 mJ per pulse through a fiber optic to an area of 0.01 cm^2 at a rate of 10 Hz. No additional modifications of the Finnigan TSQ system are required. The first quadrupole selects the parent ions, whereas the second one serves as the collision chamber, filled with an inert target gas. The third mass filter allows registration of the daughter spectrum. Sample quantities of 1–10 μg permit recording of a full mass spectrum at a scan rate of 1 m/z per laser pulse. However, collision-induced dissociation (CID) experiments necessitate the accumulation of 5–10 scans. No auxiliary ionization is performed for salts and easily cationized molecules, otherwise CI with isobutane is applied. Fragmentation patterns of organic cations, cationized sugars, and peptides are elucidated (Busch et al., 1984; Wright et al., 1985; Emary et al., 1987).

2.3.3. Time-of Flight LD-MS Instruments

It is obvious that the TOF-MS technology is much better suited than quadrupoles or magnetic sectors to the study of laser-produced ions. The time domain for data registration in a current setup with a 1 m flight tube corresponds to the time scale of modern pulsed lasers. Major limitations of the quadrupole and magnetic sector instruments in respect to panoramic detection, fast scanning, and low transmission are avoided by the TOF principle itself. Nevertheless, signal processing requires a detector circuitry with high-frequency bandwidth. Originally, unavailability of fast transient recorders necessitated the use of pulsed ion extraction and the measurement of the signal at each m/z during consecutive TOF cycles. The time-lag focusing technique has enabled operators to optimize the resolution of the analyzer in regard to the time and spatial distribution of ion formation as well as the energy spread (Wiley and McLaren, 1955; Opsal et al., 1985). However, complete mass spectrum registration within a single TOF cycle is now easily accomplished with recent progress in high-resolution waveform recorders and time-to-digital converters.

Most of the LD experiments with TOF have been accomplished on commercially available laser microprobes, which are described separately. The current discussion will be confined to the instrument developed at Johns Hopkins University in Baltimore, Maryland (Van Breemen et al., 1983). A block diagram is shown in Figure 2.15. A standard CVC 2000 TOF MS (Rochester, New York) is fitted with an extended 2 m flight tube and a non focused CO_2 laser, irradiating the sample on a Vespel rod perpendicularly. The electron beam is maintained and directed 3–4 mm above the sample

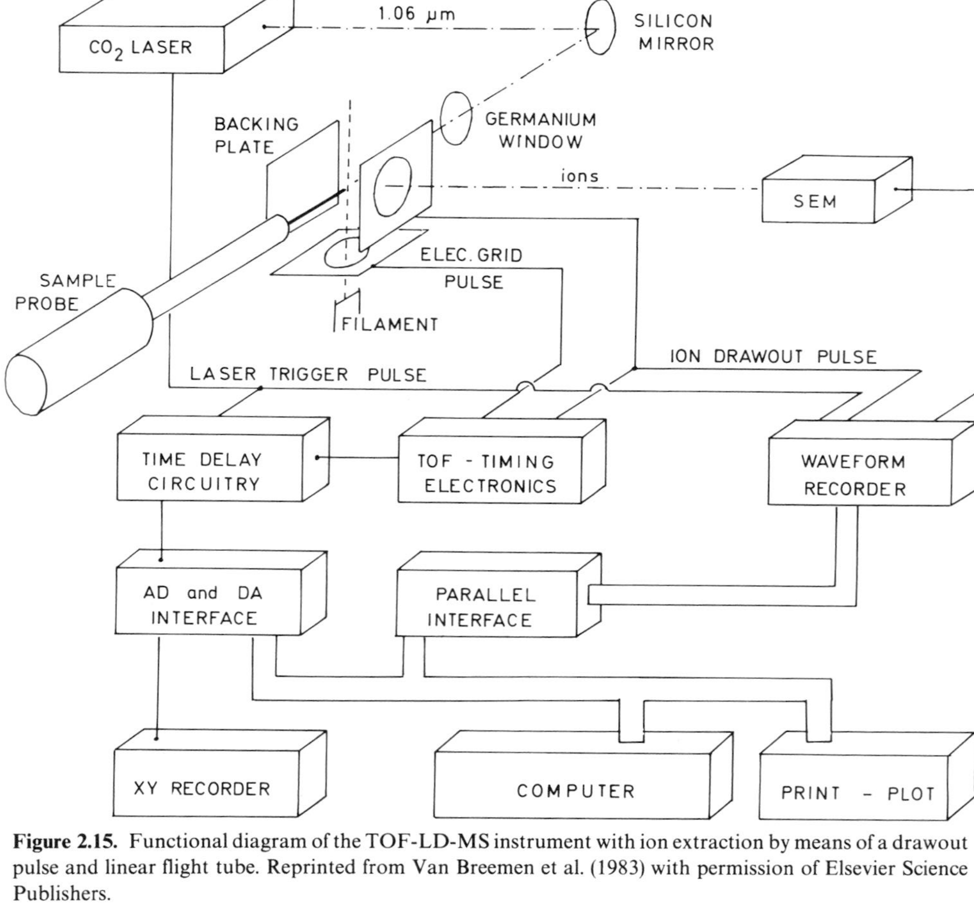

Figure 2.15. Functional diagram of the TOF-LD-MS instrument with ion extraction by means of a drawout pulse and linear flight tube. Reprinted from Van Breemen et al. (1983) with permission of Elsevier Science Publishers.

surface to avoid electron-stimulated desorption. The mass analysis in TOF-MS requires compact ion bunches at the entrance of the flight tube. It is shown that the ion emission continues over several seconds after a single laser irradiation. The required time definition of the ion bunches is then ensured by the application of drawout pulses on the gating electrode in front of the field-free region. Data registration includes a fast waveform recorder after the electron multiplier. Variation of the delays between the laser and the drawout pulses allows examination of the ion production as a function of time, while the relative importance of consecutive phenomena can be monitored as well.

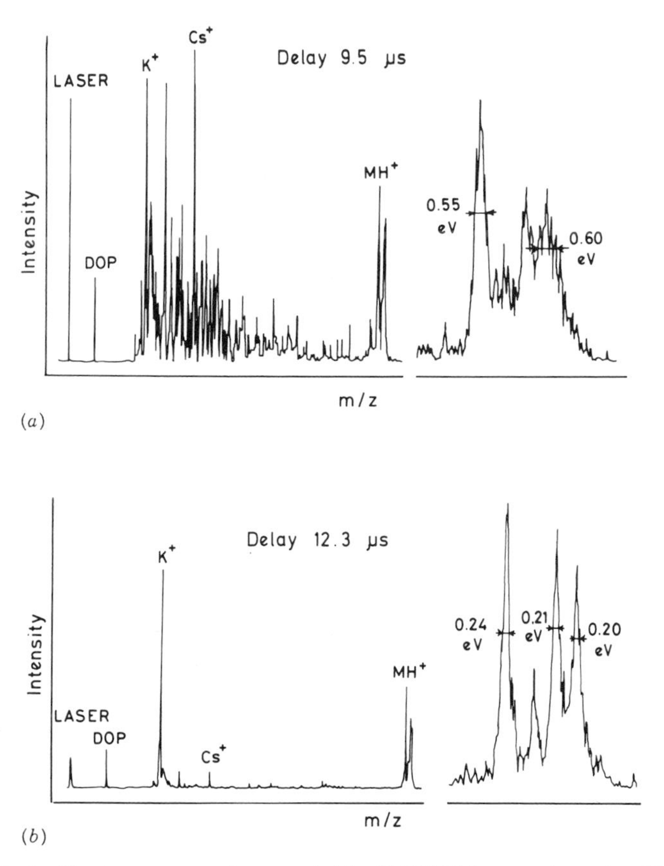

Figure 2.16. Positive ion mass spectra recorded by TOF-MS from cyclosporin with different delays between laser irradiation and ion extraction (a) 9.5 µs delay; (b) 12.3 µs delay. DOP, drawout pulse. Reprinted from Cotter and Tabet (1983) with permission of Elsevier Science Publishers.

Striking results are obtained as regards the prompt emission of intact cations from quaternary ammonium salts, but the release of thermal degradation products only starts later (Van Breemen et al., 1983). Moreover, the kinetic energy distribution in a TOF system is directly related to the peak width on condition that the uncertainty on the ion formation time and region remains low or at least constant. The energy spread has been monitored as a function of the delay between the laser and ion extraction pulse. With exception of the promptly formed ions, energy spreads are extremely small. This observation suggests that the analytical use of ions, generated later, would allow us to take advantage of improved mass resolution (Tabet and Cotter, 1983, 1984). Figure 2.16 shows the positive ion mass spectra of cyclosporin recorded with different delays between the laser irradiation and the drawout (Cotter and Tabet, 1983). The longer residence time of the ions in the source causes disappearance of the lower fragments while the peak shape of the parent signal improves. The decreased peak width reflects the lower energy spread, and the reduced tailing is assigned to the decreased contribution of metastable transitions. As a result, these data have renewed interest in time-lag focusing techniques, even for ions produced by desorption directly from thin solid surfaces. This approach becomes particularly profitable when ionization takes place over a broad time period, when the surface is not well defined, and/or when fragmentation information can be improved by allowing metastable decomposition to occur prior to ion extraction and mass analysis by TOF (Cotter, 1989).

2.3.4. LD in FTMS Instruments

The MS characterization of involatile and particularly high-mass molecules obviously comprises two subproblems: first, volatilization of the sample prior to or during the ionization must occur without thermal decomposition; secondly, the MS analyzer must provide sufficient mass resolution and an extended m/z range. As an ion storage system, Fourier transform (FT) ion cyclotron resonance (ICR) MS offers impressive capabilities for mass separation while the covered m/z range exceeds 10 kDa. The combination with pulsed laser ionization is particularly interesting because the production of discrete ion bunches matches the discontinuous mode of operation in FTMS whereas the extremely low material consumption does not compromise the high-vacuum conditions in the analyzer cell.

The initial experiments conducted by McCrery et al. (1982) soon yielded a commercial prototype of LD-FTMS, available from Nicolet (now *Extrel*) in Madison, Wisconsin. Initially, a single cell is coupled to a relatively obvious optical interface (Wilkins et al., 1985). Later, the advent of the dual cell instrument provided a simple and flexible solution to the requirement of low

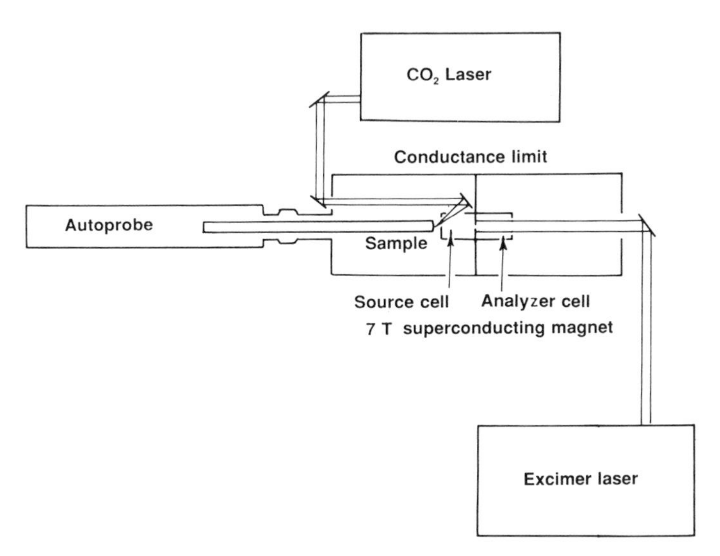

Figure 2.17. Schematic diagram of an FTMS instrument with a double cell, modified for laser desorption-ionization of the sample and photodecomposition of the ions in the cell. Reprinted from Nuwaysir and Wilkins (1990) with pemission of Oxford University Press, Inc.

pressure in the analyzer region, essential to maintain the high-resolution capabilities of FTMS (Cody et al., 1990). Figure 2.17 shows a functional diagram of the instrument, optionally equipped with an additional excimer laser to induce photodecomposition of the ions in the cell (Nuwaysir and Wilkins, 1990). The sample is excentrically mounted on a rotatable probe tip to provide fresh material for consecutive laser shots.

Figure 2.18 illustrates a typical sequence of events for laser desorption experiments (Cody et al., 1990). The quench pulse clears the cell by a high-power excitation sweep. The source cell trapping plate is then grounded during the subsequent laser interaction. If the conductance limit between the analyzer and the source cell is also grounded, ions are transferred directly to the low-pressure compartment. The performances of FTMS strongly depend on the number of ions in the measuring cell. Hence, one or more ejection sweeps can be applied to remove unwanted ions. Additional delays permit pumping off of neutrals and gases before the ions are excited to a coherently moving packet. Detection is based on the image currents induced on the receiver plates. The transient signal is then digitized and mathematically converted from the time to the frequency domain, which corresponds to the actual mass spectrum. Full information about the operation of FTMS is available in several excellent books and reviews (Marshall and Verdun,

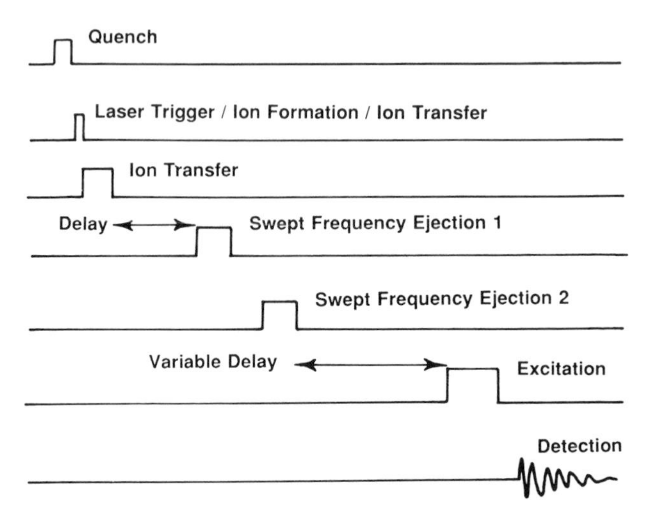

Figure 2.18. Experimental sequence of pulses used for a typical laser desorption experiment in FTMS. Reprinted from Cody et al. (1990) with permission of Oxford University Press, Inc.

1990; Buchanan, 1987; Wilkins et al., 1989; Marshall, 1989; Wanczek, 1984; Russell, 1986).

FTMS offers particularly interesting assets for laser-based ion formation. The instrument allows panoramic registration of the entire m/z range from only one laser-generated ion bunch (Cody and Kissinger, 1987; Brown and Wilkins, 1987). Moreover, ions do not collide on a multiplier as in most MS systems but remain available for subsequent experiments such as CID and MS^n. However, there are some drawbacks as well. Application of low trapping potentials limits the recovery of laser-generated ions with high kinetic energy distributions. Additionally, ultrahigh mass resolution requires very long data acquisition times, in the order of seconds. Molecular ions from very large biomolecules tend to be short lived and hence more appropriate to the time domain of an ion transport MS such as TOF or magnetic sector instruments (Marshall and Verdun, 1990). The latter systems suffer from the high-mass limitation of their m/z scale. The high-mass and high-resolution capabilities, i.e., about 60,000 at m/z 6000, are evidenced by molecular ion distributions up to m/z 8000 measured from, e.g., polyethylene and polypropylene glycols in Figure 2.19 (Ijames and Wilkins, 1988). In this respect, LD-MS becomes a valuable complement to the well-known static SIMS (secondary ion MS) methodology in the field of polymer analysis (Ijames and Wilkins, 1988; Cody et al., 1990; McCrery et al., 1982). Finally, the versatility of the FTMS combined ion source–analyzer not only offers various possibilities for post-

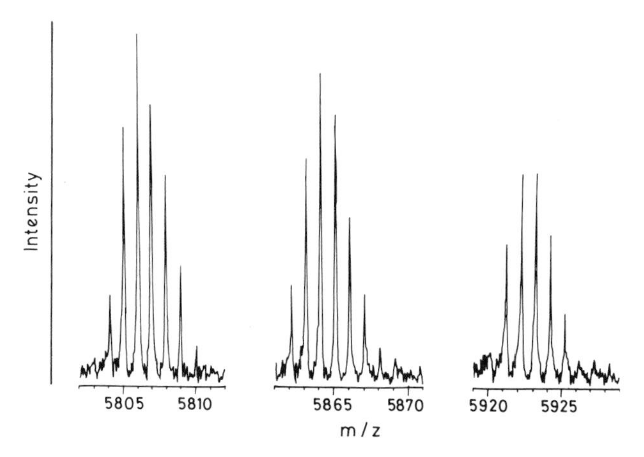

Figure 2.19. LD-FTMS spectrum with a mass resolution of 60,000 for the signals of the oligomers from poly(propylene glycol)-4000. Reprinted from Ijames and Wilkins (1988) with permission of the American Chemical Society.

ionization of laser-desorbed neutrals but also further experiments such as photoionization in the gas phase.

2.3.5. Ion Storage LD-MS Instruments

Another type of ion trapping analyzer is the three-dimensional quadrupole first described by Paul and Steinwedel (1953). Nevertheless, the current interest issues from the introduction for the mass selective instability mode of operation (Stafford et al., 1989). Theory and application are described (Todd, 1981; Griffiths and Heesterman, 1990). The now-commercialized quadrupole ion trap is particularly appealing in its possibility of doing multiple-stage MS, the lack of pressure problems, and the low cost of manufacturing and operation. As FTMS, the ion trap also requires the generation of discrete ion bunches and hence excellently matches the conditions of pulsed laser desorption–ionization.

Particularly interesting results are reported by Glisch et al. (1989). The standard ring electrode of a Finnigan MAT ion trap mass spectrometer (ITMS™) (San Jose, California) is perforated to permit introduction of a sample probe on the one side and a 1.06 μm beam of a Nd:YAG on the other. Figure 2.20 gives the schematic diagram and the scan functions for an LD-MS/MS experiment (Glisch et al., 1989). Repetitive firing at 10 Hz is used. The mass spectrum is recorded by ramping the RF amplitude at a rate corresponding to 5555 Da·s^{-1}. Here also, parent isolation can be performed by selective ejection of the lower ions. An appropriate amplitude dc voltage

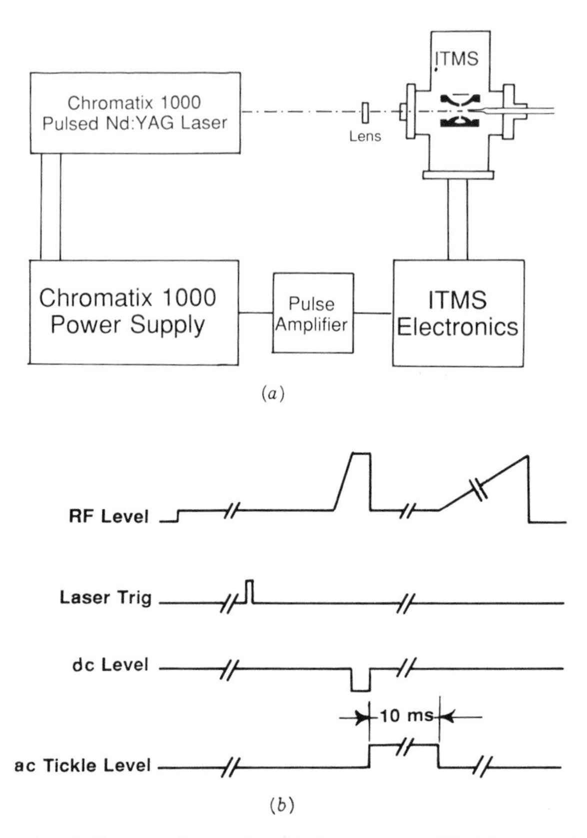

Figure 2.20. Functional diagram of a quadrupole ion trap, modified for LD of solids (a), and the typical scan function for a multistage experiment (b). Reprinted from Glish et al. (1989) with permission of Elsevier Science Publishers.

on the ring electrode combined with an ac field on the electrode at the secular frequency of the parents allows increase of the kinetic energy to facilitate the dissociation by collisions with the bath gas. The efficiency during MS^n experiments is high. All the parent attenuation is accounted for by the formation of the fragments with minimal loss of ions from the trap. Figure 2.21 shows a three-generation experiment performed on trimethylphenylammonium iodide (Glisch et al., 1989).

In contrast to inorganic applications, organic LD-MS experiments involve the use of a substantially larger range of laser and MS analyzers. In fact, almost any conceivable combination has been tried out. Surprisingly the recorded mass spectra are all largely comparable in spite of the widely different experimental conditions. As to the laser parameters in Table 2.2, wavelength ranges from 10.6 μm to 483 nm, pulse duration from 100 μs to

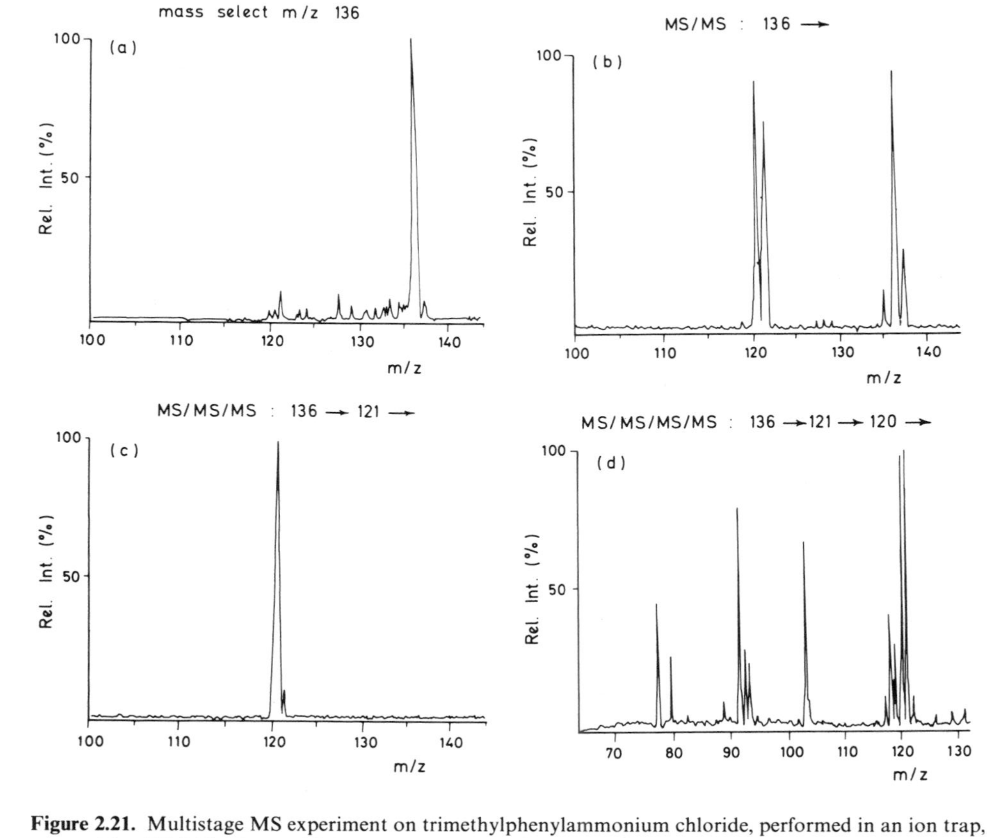

Figure 2.21. Multistage MS experiment on trimethylphenylammonium chloride, performed in an ion trap, adapted for LD-MS. The positive mass spectrum (a) is recorded after selection of *m/z* 136 and delaying 10 ms prior to scanning the quadrupole trap. The successive daughter ion mass spectra (b–d) are recorded after selection of *m/z* 136, 121, and 120, respectively. Reprinted from Glish et al. (1989) with permission of Elsevier Science Publishers.

5 ns, power density on the sample between 10 and 10^8 W·cm^{-2}, the time domain between microseconds and seconds (Van Vaeck and Gijbels, 1989). It is not yet clear which parameter is now really responsible for the rather subtle differences observed between some data. Additionally, it should be realized that most of these experiments are carried out with more or less extensive modifications of existing MSs. As a result, the comparison of data is not obvious in many cases and purely instrumental parameters make interpretation of observed phenomena in terms of fundamental desorption-ionization (DI) processes not trivial at all. The data, obtained by LD-FTMS for polymers, cannot be recorded on a magnetic sector because the mass range is inevitably too limited. It should be recognized as well that the interest in LD has suffered under the advent of the fast atom bombardment (FAB) technique, which is indeed much more convenient to implement on most existing MSs.

2.4. LASER MICROPROBE MASS SPECTROMETRY (LMMS)

The growing interest in local analysis with a spatial resolution on the order of 1 μm stimulated the development of a real laser microprobe. Essentially three types of instruments have been constructed. The approach followed by Eloy and colleagues (see Eloy, 1971) is somewhat unique in that a magnetic sector in used to perform TOF-type measurements. We will describe it (next, in Section 2.4.1) as a laser probe mass spectrograph. Most laser microprobe mass spectrometry results are, however, obtained on commercially available instruments, which are based on the straightforward application of TOF technology. Recently, the merits of FTMS in respect to mass resolution and versatility in performing extensive ion studies have been exploited as well in the field of microanalysis. All but the first instrument have been commercialized. The laser microprobe technique is referred to by a variety of acronyms, such as LAMMA or LIMA (both being registered trademarks) and LAMMS, LIMS, or LMMS. We will use the last notation, complemented with TOF or FT to indicate the type of MS analyzer being discussed.

2.4.1. Laser Probe Mass Spectrograph (LPMS)

The LPMS is schematically represented in Figure 2.22 (Stefani, 1988). Several laser sources have been used. A Nd:YAG operated at 353 nm can be focused down to 3 μm, but the spot can be increased up to 100 μm. The LPMS represents, in respect to spatial resolution on the sample, some major progress in comparison with the spark source MS and closely approximates the performance of commercial microprobes. The power density on the

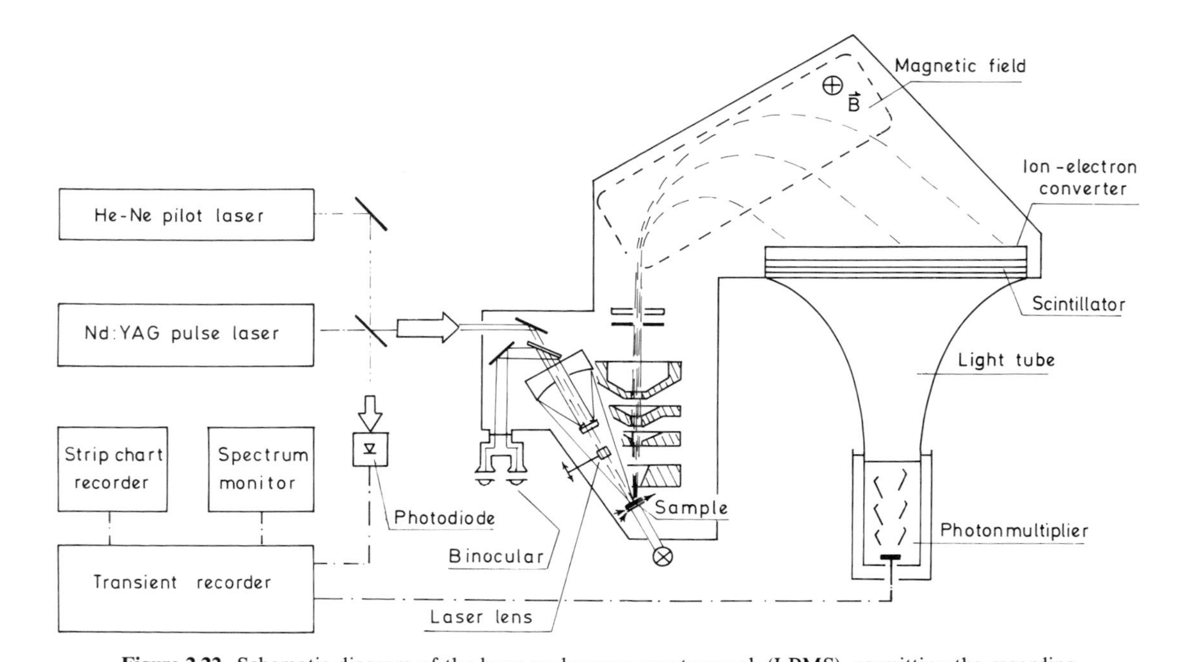

Figure 2.22. Schematic diagram of the laser probe mass spectrograph (LPMS), permitting the recording of TOF-type mass spectra in a magnetic sector instrument with a spatial resolution on the order of 3 μm. Reprinted from Stefani (1988) with permission of Editions Scientifiques Elsevier.

sample ranges from 10^8 to $10^{10}\,W\cdot cm^{-2}$. The laser solid interaction takes place in an equipotential region to ensure the free expansion of the plasma without auxiliary ionization. The presence of a high extraction field for positive ions causes, for instance, extensive bombardment of the laser vapor by reversely accelerated electrons. The LPMS provides a set of additional lenses for the acceleration of ions prior to mass analysis in the magnet. The operating conditions in the ion source permit application of the local thermal equilibrium (LTE) model as a very good approximation to quantify the results (Eloy and Dumas, 1966; Eloy 1985, 1986). The m/z separation is performed by a magnetic sector instrument without electrostatic sector. The original photoplate registration is substituted for by a patented optoelectrical device.

In a static magnetic field, ions follow a more or less curved trajectory as a function of m/z. At the same time, the covered distance before detection increases from low to high mass. Hence, the precise arrival time reflects the m/z value similarly to the position of the line on the photoplate. The electro-optical converter produces a real TOF-type mass spectrum, measured in a magnetic sector. This ingenious approach offers the better sensitivity of electrical vs. photoplate registration while the advantage of panoramic detection is maintained. Detection limits around 10^{-6} g compare favorably with those mentioned for other methods in Table 2.1. The final mass resolution, however, is only 100 for Ag, and the dynamic range is limited (about 500) (Chamel and Eloy, 1983). Moreover, systematic deviations of isotope ratios are observed. Part of the problem is assigned to the limitations of the available transient recorder. Quantitative data show a reproducibility from 5% onward, reaching 50% under unfavorable conditions. Nevertheless, the extremely low volume of evaporated analyte should be considered when comparison is attempted with modified spark source instruments. In fact, LPMS has evidenced the practical feasibility and potential of TOF measurements in a magnetic sector MS for at least semiquantitative analysis of conducting and insulating solids on a microscopic scale.

2.4.2. Laser Microprobe with TOF-MS

The most important breakthrough has been achieved by the commercialized instruments from Leybold Heraeus (Cologne, Germany) and Cambridge Mass Spectrometry (Cambridge, UK). Pulsed ultraviolet (UV) lasers, ultimately focused down to the diffraction-limited spot of 0.5 μm at 266 nm, have been combined with TOF analyzers, ensuring a high transmission and hence good sensitivity. A complete mass spectrum is registered from each laser shot, typically corresponding to 10^{-12} g of solid sample. In terms of analytical utility, the technique has become a milestone in both organic and inorganic MS, enabling element localization and quantification, inorganic speciation, and structural characterization of organic molecules in a given micro-object with often minimal sample preparation efforts (Van Vaeck and Gijbels, 1990a,b; Verbueken et al., 1988; Michiels et al., 1984).

Three instruments are commercialized: the LAMMA® 500 and 1000 from Leybold Heraeus and the LIMA® 2A, initially from Cambridge Mass Spectrometry, later from Kratos (Manchester, UK). The first is developed for thin specimens (less than 1 μm,) and operates in the transmission mode (Kaufmann, 1982). The LAMMA 1000 uses a reflection geometry and hence suits surface analysis of bulk materials (Heinen et al., 1983). The LIMA 2A is designed to be compatible with the two modes of operation, i.e., transmission for thin sections, reflection for surface characterization of bulk samples (Dingle, 1981; Dingle et al., 1982; Ruckmann et al., 1984). All these instruments share essentially the same functional principles, apart from the laser focusing and viewing optics as well as the specimen-mounting carriages.

Figure 2.23 shows the schematic diagram of the LIMA 2A instrument in its original configuration (LIMA Technical Documentation, 1985). The sample is mounted in the high vacuum of the MS. High-precision manipulators allow us to position the local region of interest exactly under the spot of the visible He–Ne pilot laser, aligned colinearly with the ionizing UV pulsed beam. The frequency-quadrupled output from a *Q*-switched Nd:YAG is most widely used for ion generation. The wavelength is 266 nm, and the pulse duration is 15 ns. During the LIMA development a lot of work has been done with the frequency-doubled Nd:YAG at 532 nm, but the results are rather disappointing, both for elements and for organic compounds (Clarke et al., 1984). The power density in the focus on the specimen ranges during ionization from about 10^7 to $10^{11}\ \mathrm{W \cdot cm^{-2}}$, depending on the mode of operation, which has to be adapted to the sample morphology. Organic compounds are measured under threshold conditions, whereas the application of high-energy pulses favors detection of elemental constituents.

The focusing optics comprise a Cassegrain-type lens, allowing the continuous selection of an appropriate spot size by a zoom-type operation. Hence, material consumption does not suffer from chromatic aberration and suits operation at different wavelengths without further adjustments. The mirrors are perforated to let the ion beam pass. A high-precision carriage is provided to interchange the specimen holder and optical parts, enabling us to switch from the transmission to the reflection mode without interruption of the vacuum. The spot size is about 1–3 μm, i.e., substantially larger than the diffraction limit of 0.5 μm at 266 nm.

Figure 2.24 illustrates the sample chamber of the LAMMA 500 transmission-type instrument (Vogt et al., 1981). More or less standard microscope

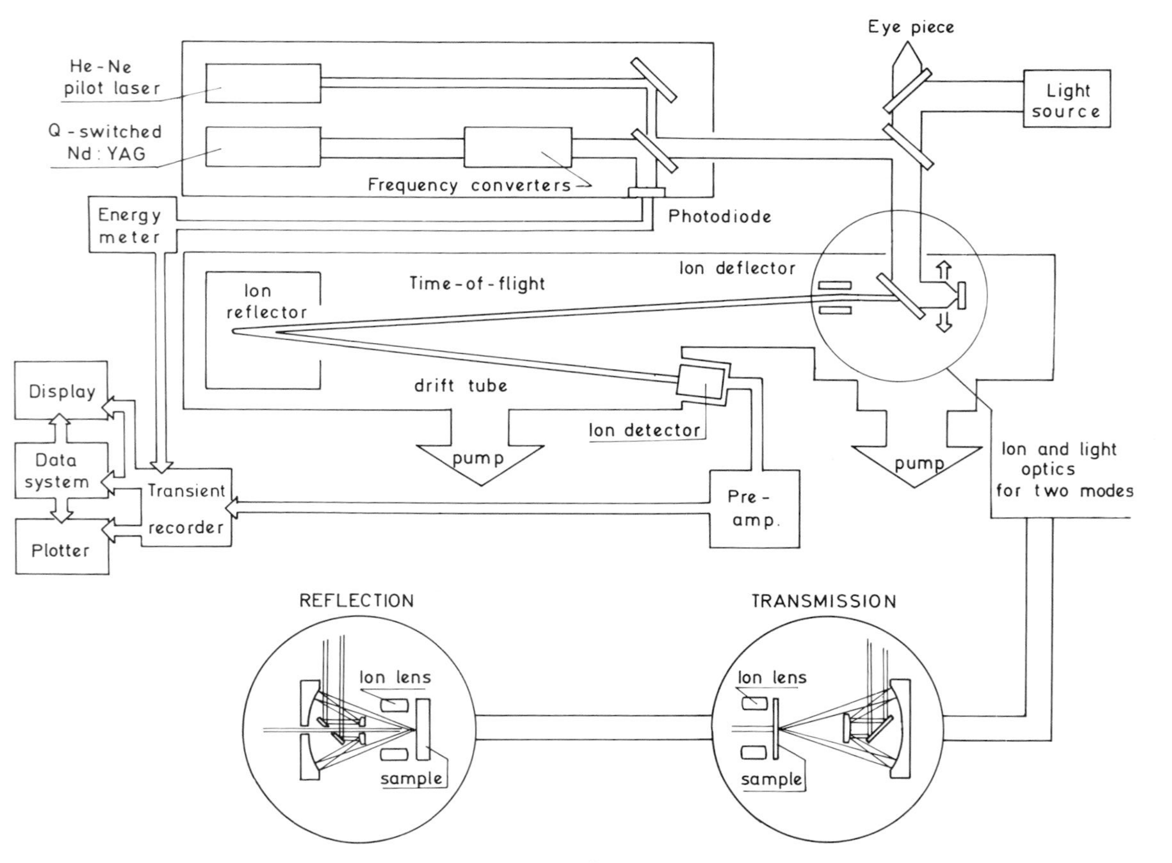

Figure 2.23. Block diagram of the commercial LIMA® 2A with TOF-MS and ion reflector. The inserts show the arrangement of the ion and light optics for analysis in the transmission and reflection geometry. Reprinted from LIMA Technical Documentation (1985) with permission of Kratos Analytical.

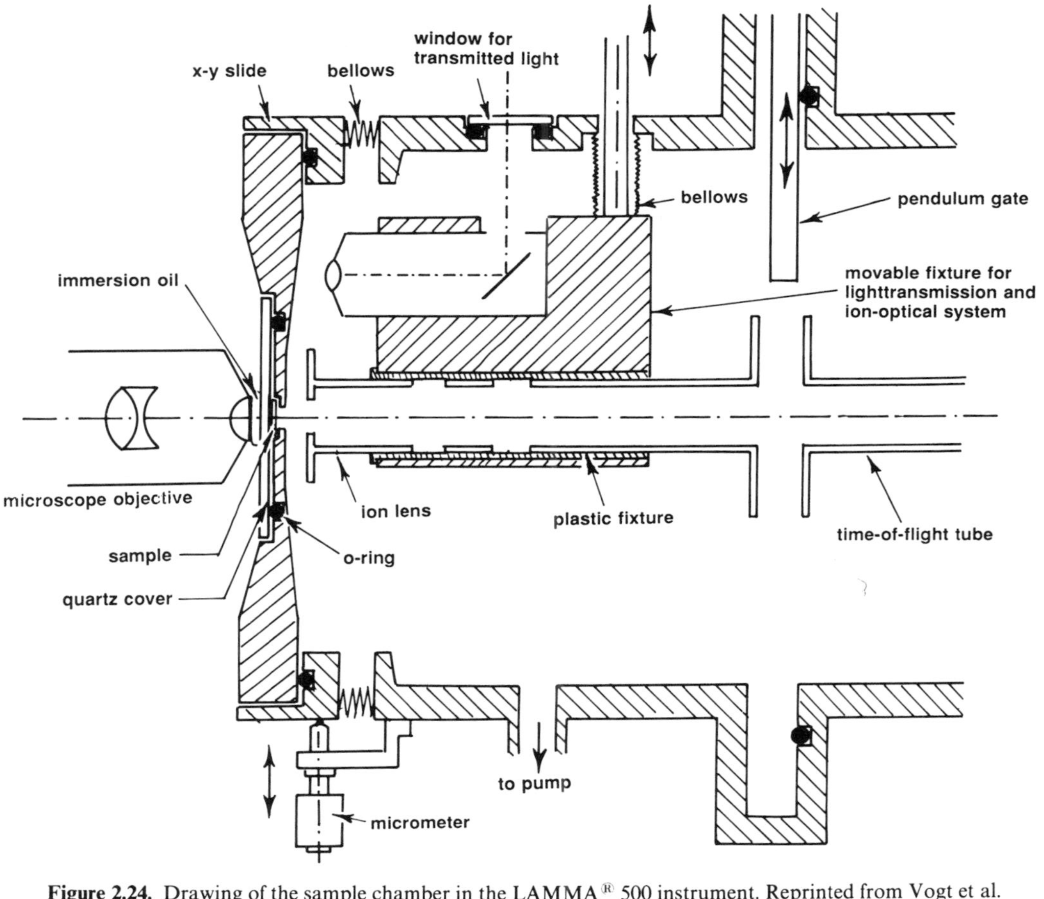

Figure 2.24. Drawing of the sample chamber in the LAMMA® 500 instrument. Reprinted from Vogt et al. (1981) with permission of Springer-Verlag.

objectives are used. The same microscope optics are used for final laser focusing and sample observation. A quartz window serves as vacuum seal. The specimen can be observed under incident or transmitted illumination. In the latter case the ion lens is electropneumatically substituted for by a mirror reflecting the beam from a current light source. The possibilities from a normal Kohler illumination system are not fully available. The spot size on the sample depends on the microscope objective: with the 100 × lens, 0.5 μm is feasible. In the LAMMA 1000 the sample is irradiated under an angle of 30° (Heinen et al., 1983). A second objective along the same axis as the MS, i.e., perpendicularly oriented to the sample, is mounted for critical observation. The lens is again electropneumatically retracted to clear the ion optic path during analysis. The lateral resolution is 1–3 μm, i.e., well above the diffraction limit. This arises from practical design parameters such as working distance, focal length, and lens diameter.

The ion packet is then accelerated by a high field, typically 600 V cm^{-1}, and transferred into the TOF drift tube. Therein the ions disperse depending on their initial velocity, which is inversely proportional to $(m/z)^{0.5}$. A reflector-type ion mirror allows decrease of the line width by compensation of the differences in the kinetic energy (Mamyrin et al., 1973; Mamyrin and Schmikk, 1979). The reflector is so designed as to improve the mass resolution by compensation for the kinetic energy distribution width and hence the spread in the arrival time of the ions with given mass at the detector. Basically, the way this device works is that the higher energy (faster) ions penetrate deeper into the electrostatic field and take longer to turn around than do the lower energy (slower) ions. The electrostatic potentials of the device are adjusted so that the fast ions lag behind the slow ones after reflection but catch up again to the slow ions at the point of detection after a second drift section. Hence, bunching of the ions at each m/z ratio takes place at the detector. The reflector can be used to determine the energy distributions (Mauney and Adams, 1984).

The mass resolution in a TOF system can be approximated by the drift time over the full width at half maximum ($t_d/2\Delta t$). It depends on the time dispersion during ion formation, the exact place in the accelerating field where the ions are formed, and the kinetic energy distribution (Cotter, 1989). Ideally, a resolution up to several thousands should be feasible in the described configuration. However, practical performances are more modest: the separation power is confined to about 850 for elemental ions and 500 for organic compounds (Kaufmann et al., 1979; Van Vaeck et al., 1986a).

The detector circuitry represents a key element in the TOF-LMMS instrumentation. First of all, signals arrive on the detector with a width between 25 and 100 ns over a period up to 40 μs. Transient recorders (TR) with high-frequency bandwidth are required. In the initial design, devices only allowed a reduced dynamic range at 50 MHz (Simons, 1983; Fletcher and Simons, 1985). Recent equipment permits correct digitization within 1% at 100 or even at 200 MHz. Secondly, it should be realized that the arrival of 10,000 ions in a peak width of 25 ns is equivalent to an output current of 60 mA. At such current levels, the amplification gain of electron multipliers drastically lowers by, e.g., space charge effects between the dynodes. Hence, large isotope ratios cannot be measured accurately unless special precautions are made (Van Vaeck and Gijbels, 1990a). Some authors have elaborated solutions by more or less complicated signal correction procedures (Simons, 1983; Musselman et al., 1988). Alternatively, two TR can be used in parallel (Fletcher and Simons, 1985). Nevertheless, quantization is not obvious (Mauney, 1985; Adams and Mauney, 1986).

The commercial TOF-LMMS offers a remarkable versatility enabling us to analyze almost any kind of solid samples. The reflection geometry is particularly profitable in material science applications because rather large samples up to a few centimeters long can be handled without any problem. In contrast, the transmission-type measurement requires the use of thin samples of typically 1 μm. The preparation of micrometer sections is not a trivial task. The currently used methods involve embedding of the sample in, e.g., epoxy resin prior to sectioning. This approach entails a real risk that the original sample composition may be seriously altered by leaching and/or redistribution of given constituents (Van Vaeck et al., 1986c). Hence, cryo-sectioning techniques are certainly adequate but experimentally not at all easy to perform, particularly when thin sections of hard materials are required.

However, the problem of section preparation for analysis in transmission geometry can be circumvented in many cases with some ingenuity. The analysis of thick geological materials has been performed by a two-step procedure (Kaufmann et al., 1980). A first laser shot serves to ablate some material, which is then deposited on the quartz cover glass. The actual mass spectrum is recorded using a second laser interaction on the deposited film. The procedure implies the risk of selective volatilization and/or redistribution of constituents in the analyzed deposit. Very volatile samples, incompatible with the 10^{-6} torr conditions inside the MS, are mounted on a polymer film that replaces the quartz window and serves as the vacuum seal (Holm et al., 1984). A single laser irradiation is then used to produce a small molecular leak through which the ionized species penetrate inside the MS. This method prevents the use of the high-magnification microscope objectives, which require the application of immersion oil. Moreover, the fine tuning of the laser power density on the sample, depending on the characteristics of the analyte, is not obvious at all. It has been reported that the use of extremely thick layers can be profitable to exploit the shock-wave-driven ion formation of organic compounds (Lindner and Seydel, 1985).

Laser microprobes primarily aim at spot analysis. No real imaging capabilities are offered. Moreover, the method remains essentially manual in nature, as the operator needs to select the spot for analysis each time and to readjust the laser focus. The latter operation is very critical for the locally applied power density and the energy regime in which the ionization takes place (Van Vaeck et al., 1990). This parameter strongly affects the general characteristics of the information obtained in the mass spectra (Hercules et al., 1987). Nevertheless, Hercules and co-workers automated their LAMMA 1000 instrument to perform mapping of extended sample areas such as thin-layer chromatography (TLC) plates or coal maceral samples (Wilk and Hercules, 1987; Morelli, 1990; Kubis et al., 1989). The micromanipulators were motorized, and full computer control over sample positioning and data acquisition was implemented. A raster of up to 100×100 points was covered. The spot size of the laser ranged between 2 and 5 μm under focused conditions. However, no correction to compensate for irregularities of the sample surface was attempted.

TOF-LMMS particularly well suits the characterization of surface constituents. This applies not only to the analysis in reflection geometry but also to the transmission mode (Wouters et al., 1988; Holm and Holtkamp, 1989). It has been shown that 97% of the elemental cluster ion current from carbon foils originates from the upper 20 nm of a 45 nm specimen, despite the fact that the sample is perforated upon analysis (Bruynseels and Van Grieken, 1986). Similar observations apply to organic compounds: structural ions are obtained only from the upper 20 nm when a 1 μm thick sample is perforated (Van Vaeck et al., 1988a).

Several hardware modifications of the commercial microprobe have been reported. The Nd:YAG system is frequently substituted for by other lasers. Tunable dye lasers permit the use of selected wavelengths to match the resonant absorption of the analyte (Verdun et al., 1987; Krier et al., 1985). As a result, the detection limits for analytes in well-characterized matrices can be greatly improved. General applicability on a routine basis is less obvious. Hillenkamp and co-workers have experimented with excimer lasers at shorter wavelengths to exploit the decrease in the threshold irradiance (Bahr and Hillenkamp, 1985; Karas et al., 1985, 1987; Spengler et al., 1987).

Some disadvantages of the actual TOF-MS must be considered. First, there is the poor visibility during sample observation. Subtle details and contrast, which may be clear under a modern routine microscope, are indistinguishable. This is particularly disadvantageous for biological experiments. UV illumination of the sample improves the resolution of the microscope optics by a factor of 2.5 (Kaufmann, 1982). Nevertheless, a substantial part of the gain is lost again in the closed-circuit TV. Hence, the benefits of the conversion are not entirely clear unless fluorescence of the sample can be exploited by the illumination system. Furthermore it is extremely difficult to assess the local energy regime, imposed to the spot area of the analyzed sample. Laser focus in relation to the morphology of the micro-object, i.e., thin film, particles, etc. is critical. Also, optimization of the methodology is severely hindered by lack of knowledge about the laser-induced ion formation mechanisms, which determine the promptly generated ions. Finally, a better mass resolution is highly desirable since the most appropriate working area of LMMS for organic compounds typically lies in the higher m/z range. The absence of a gating electrode in front of the flight tube necessitates the use of promptly formed ions, i.e., generated essentially during the laser pulse duration of 15 ns. Prolonged ion emission degrades the mass resolution and causes problems with the calibration of the m/z scale (Van Vaeck et al., 1986b). The time definition of ion generation at a given irradiance depends of course on the characteristics of the analyte itself (Van Vaeck et al., 1989a). The operator has no other means than the laser power density to optimize the time domain for ionization (Van Vaeck et al., 1990). Higher irradiances may improve the situation but also cause more intense photofragmentation of organic molecules.

Nevertheless, it is sometimes stated that LMMS applications are nearly unlimited. Surprisingly, even at the present state of the art, this statement tends to come very close to the truth. The proceedings from the triennial users' meetings permit us to readily survey the remarkable versatility of the method (Hillenkamp and Kaufmann, 1981; Seydel and Lindner, 1983; Adams and Van Vaeck, 1986; Russell, 1989). The following nonexhaustive list illustrates the applicability of TOF-LMMS to a wide variety of practical problems:

- Qualitative localization and/or semiquantitative determination of elements on a subcellular level in biological tissues (Verbueken et al., 1987; Schmidt, 1989; Jacob et al., 1984; Vandeputte et al., 1990a,b)
- Single-cell analysis yields fingerprints of bacterial strains, which are used for taxonomy and chemotherapy evaluation (Lindner and Seydel, 1989)
- Metallurgical applications with, e.g., corrosion studies, analysis of light elements in alloys, study of segregates formed during brazing (Kohler et al., 1986; Heinen et al., 1984; Mattews et al., 1989)
- Semiconductor analysis—direct characterization of ionic contaminants on integrated circuits or assessment of the lateral diffusion of dopants (Daniel et al., 1988; Heimbrook et al., 1989)

- Study of bulk and surface constituents of environmental aerosol particles and coal mine dust samples (De Waele and Adams, 1988; Tourmann and Kaufmann, 1989; Wouters et al., 1990)
- Determination of the surface composition of natural and chemically modified asbestos fibers (Bruynseels et al., 1985; De Waele et al., 1985, 1987)
- Characterization of organic ligand complexes, encapsulated in layered silicates (Casal et al., 1988)
- Direct in situ analysis of organic molecules, e.g., natural products in soybean samples, pigments in microlichen, phytoalexins in carnation sections (Moesta et al., 1982; Mathey et al., 1987, 1989)
- Identification of cloth tissues by means of the fingerprints from dyes and fabric softeners for forensic purposes (Schmidt and Brinkmann, 1989)
- Study of the volatilization and transformation of analytes during heating in graphite furnace atomic absorption spectrometry by speciation of inorganic constituents in the emanating aerosols (Güçer et al., 1989)
- Structural characterization of thermolabile drugs and metabolites, permitting generation of detailed information not available from conventional and soft ionization MS, under conditions of microscopical material consumption (Van Vaeck et al., 1988b, 1989b,c)
- Detection of giant ions up to 300 kDa from biomolecules such as enzymes proteins by the use of matrix-assisted desorption (Karas and Hillenkamp, 1989; Karas et al., 1989; Tanaka et al., 1988).

TOF-LMMS has marked a milestone in the field of microanalysis. Partly as a result of their commercial availability, these instruments have evidenced the feasibility and merits of the characterization of nearly all kinds of solid surfaces on a micrometer scale in respect to organic and inorganic constituents. Organic applications are particularly important, as there are few if any alternatives for microanalysis. They are described later in this volume (Chapter 3, Part B).

At this moment, the method particularly suits identification of organic and inorganic compounds. Detailed speciation capabilities for inorganic molecules still remain unsurpassed by other MS methods, whereas extensive structural information is obtained for organic compounds. Finally, the instrument has allowed researchers to demonstrate the generation and detection of giant ions from biological molecules. These achievements are indebted, of course, to the unique potential of the laser–solid interaction but also to the high transparency of TOF analyzers, allowing a high yield transmission of ions irrespective of their m/z. In spite of the numerous and promising results, several applications also evidence the need for a substantially higher mass resolution as well as more extensive possibilities for ion studies by, e.g., CID and MS^n experiments. Consequently, it is not surprising that recent research efforts have been directed toward the development of FTMS-based laser microprobe MS systems (discussed next).

2.4.3. Laser Microprobe with FTMS

It is an appealing idea to increase the analytical utility of the laser microbeam irradiation of solids by combination with one of the most powerful MS analyzers known at this moment, the FTMS. Two research groups at IBM (Endicott, New York) have modified the current Nicolet instruments (Madison, Wisconsin). Figure 2.25 shows a schematic diagram of the single-cell version (Brenna et al., 1988). Irradiation of a small spot requires the use of an Nd:YAG at the 266 nm wavelength of the fourth harmonic. The parabolic mirror is replaced by a flat one and a focusing lens, providing a laser spot on the sample of 5–8 μm. The same optics are used for sample observation. The sample stage is modified to allow fine positioning in the lateral and z-axis direction (Brenna et al., 1988). A mass resolution of up to 400,000 is reached for the $^{208}Pb^+$ isotope (Brenna, 1989). It should be noted that these instruments remain fully compatible with the other modes of operation: electron and/or chemical ionization can be applied in conjunction with laser desorption to recover a significant part of the laser generated neutrals.

The completely different approach of Muller's group at the University of Metz (France) has yielded a really dedicated FT-LMMS, which is based on a Nicolet 2000 (Madison, Wisconsin). The block diagram is given in Figure 2.26 (Muller et al., 1989). A particularly attractive solution to the problem of optical interfacing is elaborated by the use of a central focusing objective with long working distance, which is then surrounded by a circular Cassegrain system. The inner lens is used for tight focusing of the ionizing laser beam; the outer one for the postionization or measurements, involving the irradiation of large spots. Sample observation can be performed through the central optics or through the peripheral Cassegrain. The latter device offers a larger viewing field. The optics are inserted through the outer plate of the analyzer. The beam is directed through the enlarged orifice in the central conductance limit wall and impinges perpendicularly on the sample in the source cell. As a result, the analyzed spot is circular instead of ellipsoidal as in the other designs. Spot diameter is about 3–4 μm. Sample positioning is performed by a patented stage inside the vacuum with Pelletier memory alloys to allow precise micrometer displacements. It should be noted that the electron

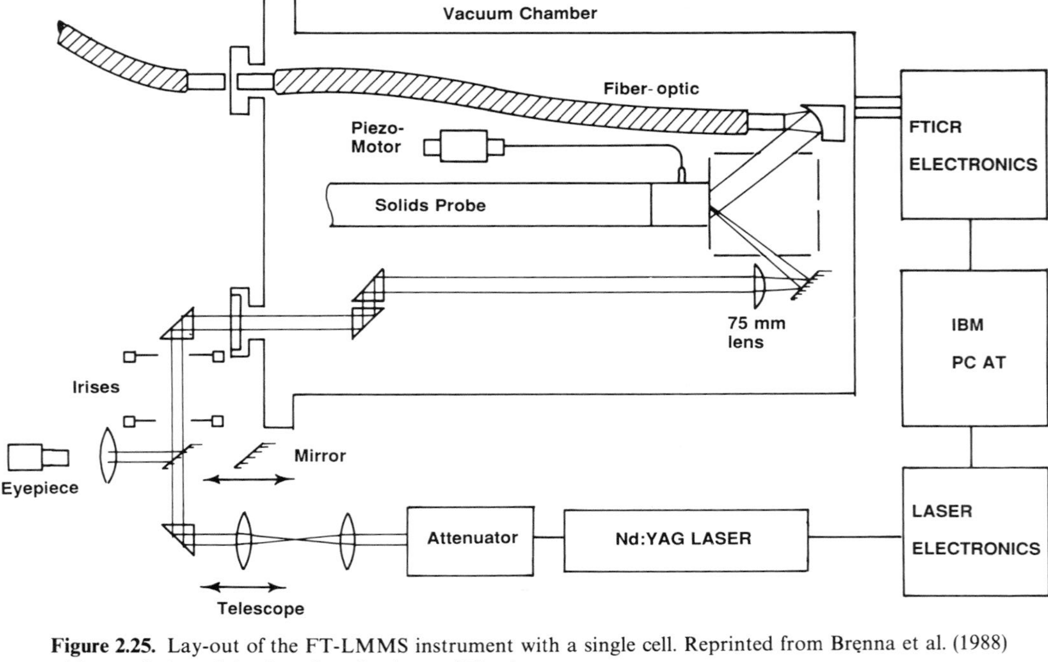

Figure 2.25. Lay-out of the FT-LMMS instrument with a single cell. Reprinted from Bręnna et al. (1988) with permission of the American Institute of Physics.

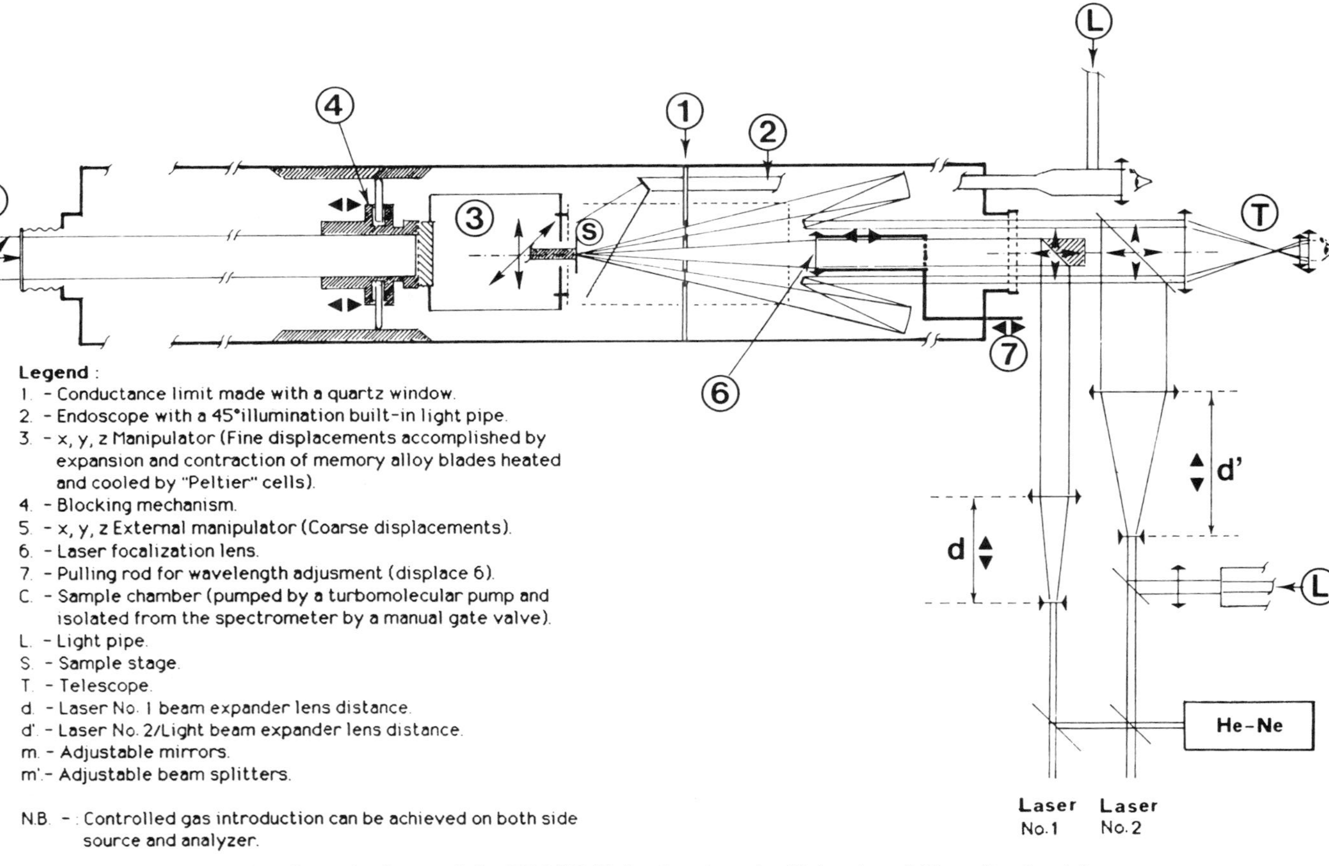

Figure 2.26. General scheme of the FT-LMMS developed at the University of Metz. Reprinted from Muller et al. (1989) with permission of San Francisco Press, Inc.

ionization beam is no longer available but the second laser beam through the peripheral optics allows photoionization in the source cell in conjunction with photofragmentation in the analyzer. Several examples in the field of materials sciences have demonstrated the merits of FT-LMMS in separating numerous interferences of elemental and cluster ions (Pelletier et al., 1988, 1989). The high-resolution capabilities are evidenced by the demanding separation of $^{88}Sr^{2+}$ from $^{44}Ca^{+}$ during a single excimer laser shot on a ceramic sample. Resolving power up to 160,000 is achieved under these conditions. The mass assignment of ions routinely attains an accuracy within 1–3 ppm. However, it is important to elaborate an adequate compromise between mass resolution and sensitivity. Recently, Extrel (Madison, Wisconsin) has started commercialization of FT-LMMS, primarily based on the development at IBM, while Bruker-Spectrospin (Fällanden, Switzerland) has developed an instrument using the external source technology.

The data from FT-LMMS and TOF-LMMS are not directly comparable, even when the laser irradiation is performed under similar conditions. Examples are explicitly discussed in Part B of Chapter 3 of this book. There are several possible reasons for this, all of which arise from the largely different conditions during ion formation, sampling and mass analysis (Brenna, 1989). The laser-generated species are subjected in the TOF microprobes to a high-extraction field. In contrast, the trapping potential in an FTMS is low, up to 10 V. As a result, ions with high kinetic energies can be recovered in TOF MS setups but are simply lost to the walls in FT-LMMS. Moreover, the ions trapped in small orbits remain available for second-order interactions and ion-molecule complex formation for a relatively long time. The total duration of a TOF cycle is on the order of 100 μs, whereas an FTMS, depending on the required resolution, takes milliseconds up to seconds. Short-lived species, e.g., parent ions from large organic molecules, cannot be measured as in TOF-MS. Literature data suggest that organic ions are not always compatible with the long time domain of FTMS. It should be noted that current magnets also impose a limit of about 15 kDa on the m/z range. Nevertheless, although a lot of research will be required to allow routine measurements, it is obvious that the FTMS technique also has the potential to perform the task of fundamental and practical microanalytical research.

2.5. LASER ABLATION IN TWO-STEP DESORPTION AND IONIZATION

Various kinds of experiments perform desorption and ionization (DI) in physically distinct regions of the ion source. For the sake of clarity, the numerous approaches are described in separate sections. The subsequent ones will discuss laser desorption with different types of postionization, except laser irradiation, and the combination of two lasers, one for desorption and one for ionization. The current section will focus on laser ionization of gases and analytes brought into the gas phase by non-laser-based techniques. Although the discussions do not aim at complete coverage of the respective areas, the diverse examples cited will highlight organic and inorganic applications, thereby illustrating the way in which lasers now serve to link up these traditionally distinct fields of analytical chemistry.

2.5.1. Inorganic Applications

The ultimate goal of analytical chemists remains to forge an instrument capable of (1) detection down to the level of a single atom or molecule, (2) with high discrimination between various isotopes, and (3) a broad dynamic range. The development of resonance ionization spectrometry (RIS) in conjunction with MS (RIMS) has produced important progress with respect to the efficient use of analytes in the gas phase. In regard to the analysis of solids, laser ionization can be applied to the secondary neutral atoms and molecules, removed from a surface by stimulating radiation such as an ion or electron beam. The advantage of this approach as opposed to the one-step DI lies in the fact that most of the emitted particles are neutrals, not ions. Hence, the relative variations on the local gas phase populations are less affected by the chemical composition of the local matrix.

The advantages of postionization of secondary neutrals were exploited by Honig as long ago as 1958. However, the concept has only recently taken a practical form for surface analysis. The actual postionization can be performed under resonant conditions or alternatively by the less sensitive multiphoton ionization (MUPI) process. The former technique is called *sputter-initiated resonance ionization spectrometry* (SIRIS), whereas the second is generally known as *surface analysis by laser ionization* (SALI). A third approach involves *surface analysis by resonance ionization of sputtered atoms* (SARISA).

Resonance Ionization MS (RIMS). The basic idea of RIMS concerns the formation of ions from atoms in the gas phase by the absorption of photons that energetically match selected quantum states of these atoms (Hurst et al., 1979). Apart from wavelength selectivity, lasers offer the advantage of sufficient power to promote a bound-state electron to a free state under saturated conditions (Donohue et al., 1983). Current literature provides extensive documentation on the various RIS schemes (Payne et al., 1982; Kramer et al., 1979). All elements of the periodic chart except He and Ne can be ionized by commercially available lasers, but just a few elements require only a single photon absorption for the different steps. Apart from the preceding discussion additional information about the RIS technique is readily available from several reviews (e.g., see Young et al., 1979; Hurst, 1981).

The RIMS technique obviously offers elemental or molecular selectivity of optical spectroscopy in combination with extremely sensitive detection capabilities of an MS system (Fassett and Kingston, 1985; Moore et al., 1984). Various instrumental approaches have been described. Extensive reviews are available (Smith et al., 1989; Young et al., 1989a). Virtually all types of analyzers have been used for RIMS studies. Gas phase atoms are continuously supplied by a heated filament, a graphite furnace, or an ICP torch, whereas atom sources with a pulsed laser or particle beam are used as well. Several lasers have been used, ranging from CW to pulsed sources such as tunable dye lasers, pumped by Nd:YAG or excimer systems, argon ion stabilized ring dye lasers, and semiconductor diode lasers. The major types of MS analyzers have been tried as well. TOF instruments are ideally combined with pulsed lasers in terms of the duty cycle matching to the MS characteristics. However, magnetic sector analyzers provide superior resolution capabilities but need modification of the detector circuitry to allow combination with pulsed laser excitation. The improved MS performance of recent quadrupoles enhances the possibility of performing RIMS with more compact and less costly analyzers.

Figure 2.27 illustrates the block diagram of a typical setup for sequential multielement analysis by RIMS and thermal atomization (Fassett et al., 1983). The initial gas phase production of analyte atoms is obtained by heating the dried salt residue from a solution on a rhenium filament or directly by sublimation of an elemental constituent of the filament itself. The laser wavelength is tuned to minimize the interference from accompanying elements in the gas atom reservoir. The beam, about 2 mm wide, is positioned about 5 mm above the filament and delivers 1–3 mJ in 4 ns pulses at a repetition rate of 10 Hz. The MS is a 60° 6 inch magnetic sector analyzer, originally designed for thermal ionization MS and high-precision isotope measurements. The ion source is modified to accommodate the laser entrance and exit windows. Detection is accomplished with a conventional electron multiplier at low voltage to preserve the dynamic range. No pulse counting is attempted because of the laser repetition rate. The preamplified signal is fed into a boxcar averager with an acceptance window of 200 ns. The trigger is provided by the photodiode in the laser system. The thermally produced ions as well as the ions generated by laser interaction with the solid surface of the filament are both discriminated by selection of time delays for signal collection. Indeed, the flight time of these unwanted species is longer than that of the laser-ionized atoms in the gas phase. As a result, the instrument

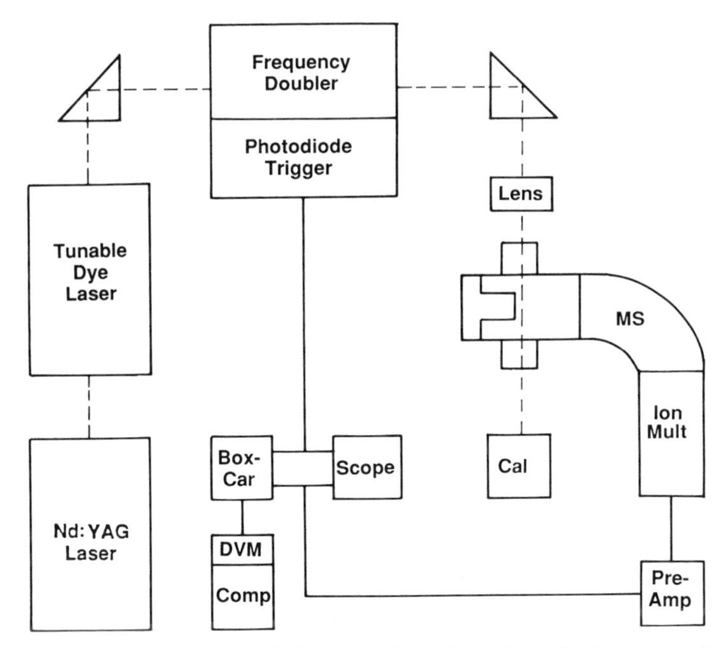

Figure 2.27. Block diagram of an RIMS system with a thermal atomization source. Reprinted from Fassett et al. (1983) with permission of the American Chemical Society.

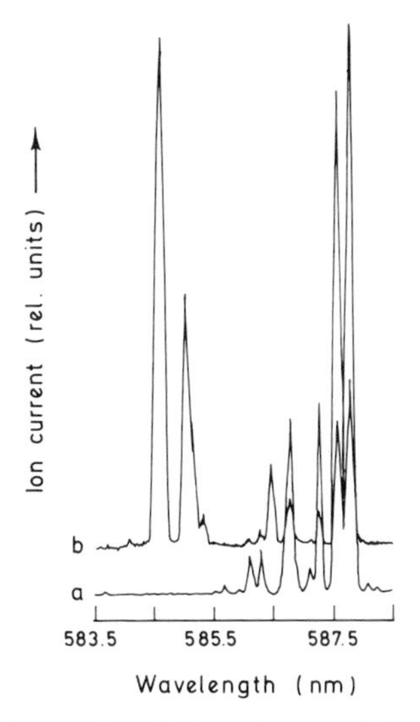

Figure 2.28. RIMS analysis of samarium (Sm) and promethium (Pm). Trace (a) refers to m/z 152, characterizing Sm; trace (b) to m/z 147, the sum of Sm and Pm. Reprinted from Shaw et al. (1988) with permission of the American Chemical Society.

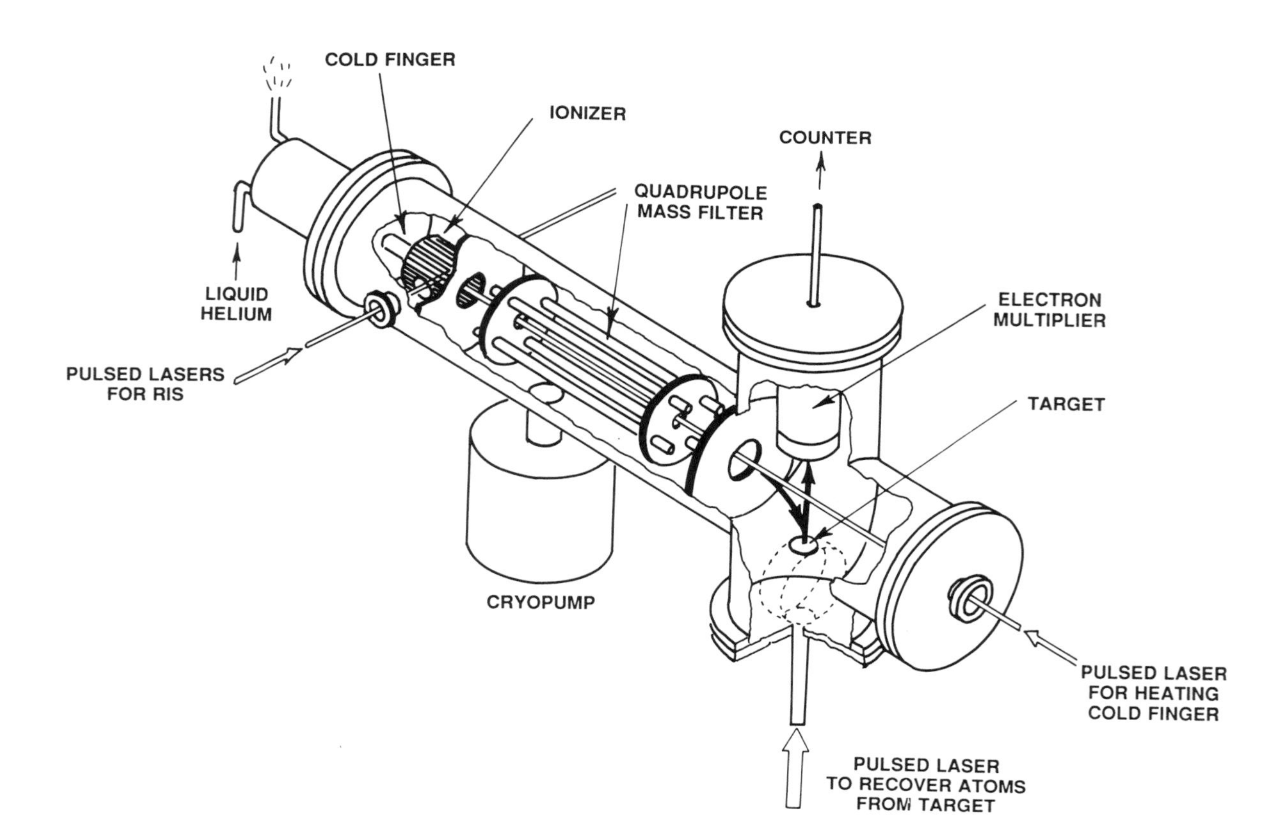

Figure 2.29. Schematic diagram of an isotopically selected rare gas atom counter with RIMS and quadrupole analyzer. Reprinted from Chen et al. (1990) with permission of Oxford University Press, Inc.

provides dual mass selectivity: the magnetic sector filters out the thermally produced ions except a single *m/z*, and the TOF filtering circuitry removes the thermally produced ions at the selected mass.

One of the strengths of the RIMS technique lies in the isotope-selective ionization, which eliminates the task of separating isobaric interferences from the MS. Hence, the measurement of promethium (Pm), which represents a real challenge to MS analytical chemists, is chosen as an example (Shaw et al., 1988). The man-made element Pm is unique among the lanthanides in that it has no naturally occurring isotopes. Reference samples are usually made from samarium (Sm), and frequently these standards still contain the starting material. Also β-decay of Pm produces Sm daughters. The situation can be handled successfully in RIMS. Tuning of the MS analyzer, on the one hand, and of the wavelength for the ionizing laser, on the other, gives a really two-dimensional selectivity. The lower trace in Figure 2.28 shows the response detected at *m/z* 152 i.e., the mass of the most abundant Sm isotope (Shaw et al., 1988).The laser is then scanned over a range between 583 and 588 nm, which is adequate for the Sm transitions. The upper trace is obtained by monitoring *m/z* 147 and shows a strong doublet at 584.6 and 585 nm, ascribed to Pm. These lines lie in a blank window, left by the lines from Sm. As a result, the elemental discrimination factor is at least 1,000. Additional wavelength ranges were reported later (Young et al., 1989b).

An elegant quadrupole-based application is described by Chen et al. (1990) from Oak Ridge National Laboratory . The experimental schematic for the isotopically selective counting of rare gas atoms is shown in Figure 2.29. The laser-generated ions are mass filtered and implanted on a target. The ions that are not transmitted are neutralized by collision with the wall and return to the gaseous sample. Hence, the laser irradiation is continued until complete implantation is achieved. Nevertheless, even high-resolution quadrupoles transmit $1{:}10^6$ ions of $M \pm 1$ when M is set. Better isotope separation is then obtained by pumping the nonselected isotopes off and subsequent recovery of the implanted ions on the cold finger by e.g. laser irradiation of the target. The cycle is then repeated, which allows an extremely high degree of isotope separation. The target is replaced in the final cycle by a Cu–Be disk, and the number of ions is counted by a Johnston electron multiplier with high gain.

Sputter-Initiated Resonance Ionization Spectroscopy (SIRIS). Figure 2.30 shows a schematic diagram of the SIRIS instrument, developed at Atom Sciences (Oak Ridge, Tennessee) (Parks, 1990). A pulsed 10 kV argon ion beam is produced by a commercial duoplasmatron microbeam ion gun system. The sputtering beam impinges on the sample surface under 60°. The irradiated spot can be varied between 5 and 100 μm. The pulse duration is

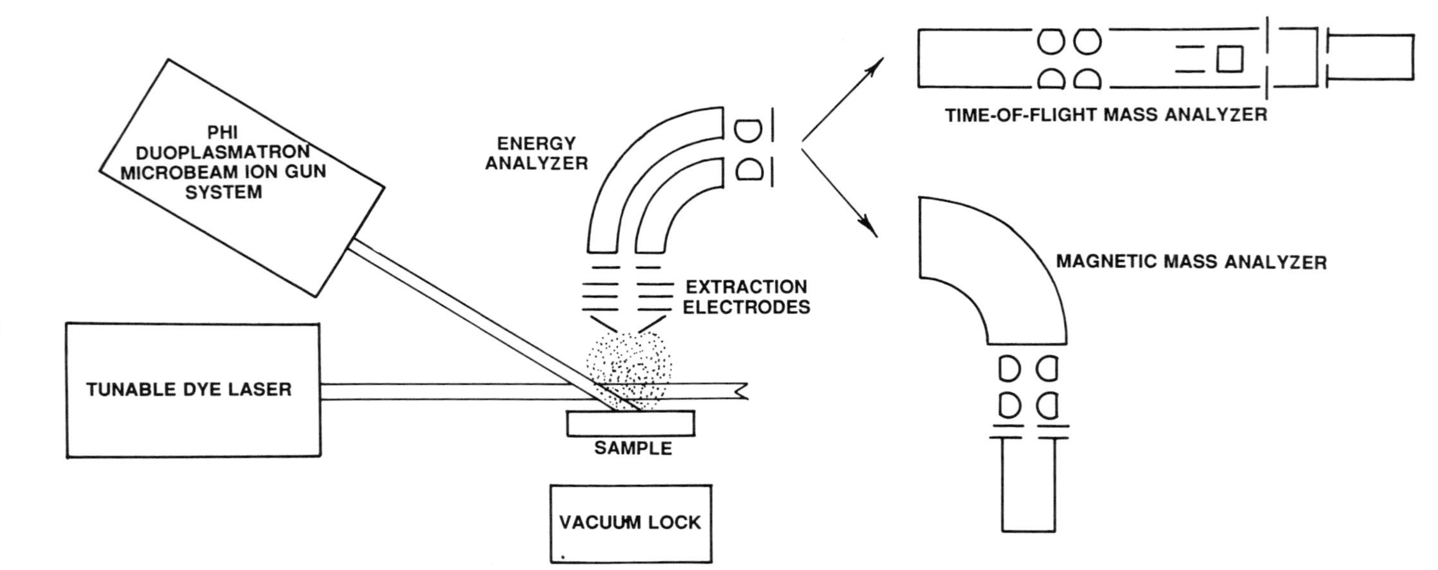

Figure 2.30. Block diagram of the apparatus for sputter initiated resonance ionization spectroscopy (SIRIS). Reprinted from Parks (1990) with permission of Oxford University Press, Inc.

1 μs. Ionization is performed by an Nd:YAG pumped tunable dye laser system. Frequency doubling permits coverage of the range between 270 and 540 nm. Mixing with the YAG fundamental even extends the wavelength range down to 220 nm. The bandwidth of the laser system exceeds the natural linewidth of the atoms, and no isotope selectivity is feasible at the ion formation stage. The ions are then separated in a double-focusing magnetic sector MS. The electrostatic sector permits rejection of the directly generated secondary ions as opposed to RIS ions. Both types of ions are indeed formed in distinct regions of the accelerating field. Moreover, time selection, synchronized to the laser pulse, can be used to make or refine that discrimination. Alternatively, the magnetic sector analyzer can be substituted for a TOF-MS to eliminate the dependence of isotope ratio measurements on the shot-to-shot fluctuations. In contrast to the magnetic sector, individual peak intensities for different masses are then monitored within the same laser-pulse-produced ion packet. It should be remembered that pulsed ion bombardment of the specimen is applied in the first place. Hence, SIRIS experiments typically do not remove more than one equivalent monolayer from the material during the entire measuring period. As a result, SIRIS remains essentially a surface technique. If depth profiling is needed, sputter erosion is performed with a continuously bombarding beam.

The SIRIS instrument nicely complements the SALI technique (discussed later). SIRIS has superior selectivity and sensitivity but remains essentially a single-element technique. Indeed, RIS discriminates inherently against all other elements except the one that is selected. Isotope information is available, and the sensitivity reaches the ppb range. In contrast, SALI exploits the nonselective ionization by a UV laser, of which the wavelength is chosen depending on the application to achieve a uniform ionization. The MS, instead of the ionizing radiation, now has the task of making the chemical distinction. As a result, survey analysis of—at least in principle—all constituents from a given evaporated layer is feasible.

Surface Analysis by Resonance Ionization of Sputtered Atoms (SARISA). The SARISA technique was developed at Argonne National Laboratories (Argonne, Illinois). Figure 2.31 schematizes the major subunits of the instrument (Pellin et al., 1987). The primary ion gun produces a 3.5 keV Ar^+ beam, focused down to about 250 μm, while rastering can be performed over a 2×2 mm area. The target is kept at high potential during the primary ion bombardment. Note that the same ion optics are used to direct the primary ions perpendicularly on the surface of the sample and afterward to transport the ionized neutrals toward the detector. A Nd:YAG pumped dye laser performs the RIS. Mass analysis is achieved by an energy and angle refocusing (EAR) TOF system. The unique lens design focuses the ions through the

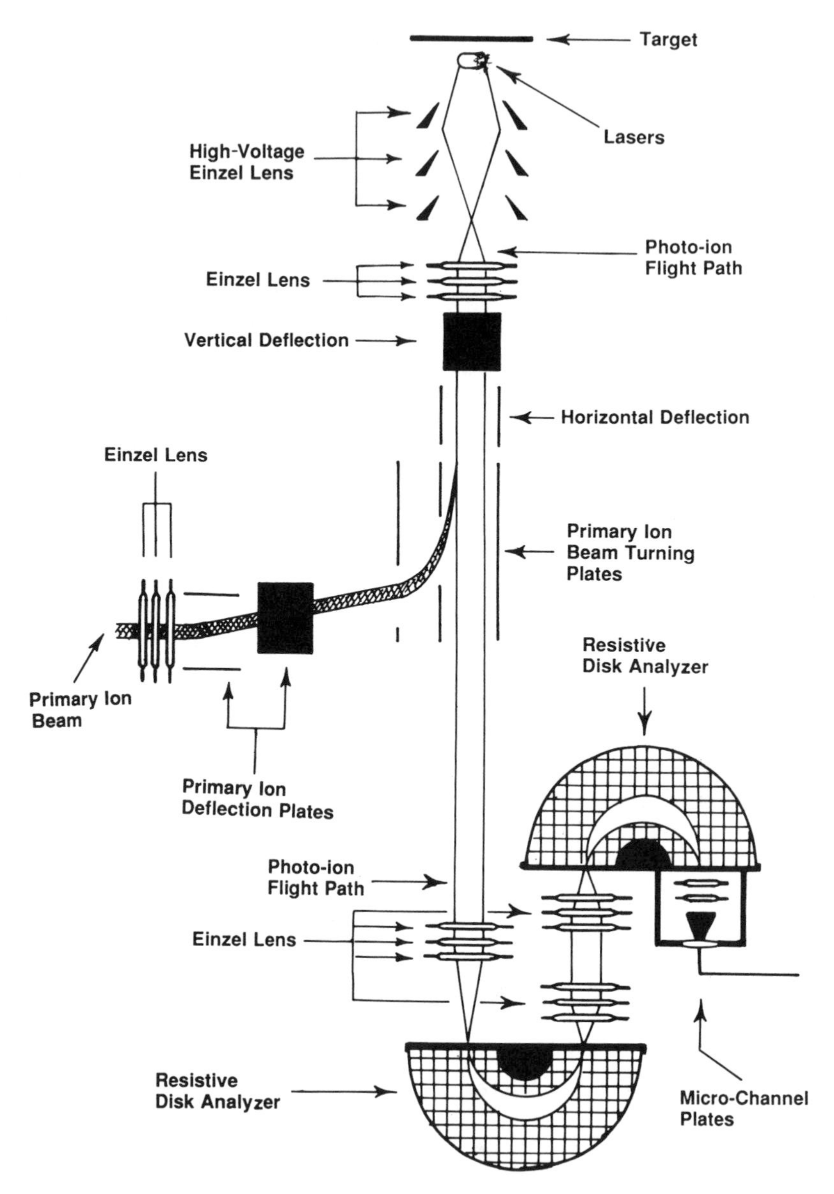

Figure 2.31. The SARISA III–EAR-TOF system for detection of sputtered neutral atoms. Reprinted from Pellin et al. (1987) with permission of Elsevier Science Publishers.

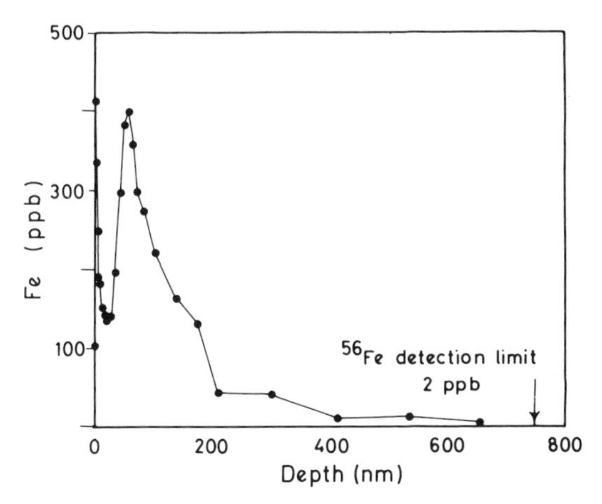

Figure 2.32. SARISA III determination of ^{56}Fe in silicon as a function of depth. Reprinted from Pellin et al. (1987) with permission of Elsevier Science Publishers.

primary ion steering plates at 0 V on the resistive disk analyzers with a large angular and energy acceptance window (Siegel and Vasile, 1981; Liebl, 1983). The ions then strike a chevron pair of microchannel plates. The subsequent electron pulse is then detected either by a gated pulse counting system or by a MCA charge digitizer. The combination of the two spherical analyzers provides two key functions: improvement of the TOF mass resolution, and suppression of the high-energy ions produced during the sputter process and ejected when the target is kept at high potential.

Figure 2.32 illustrates a depth profile of an ^{56}Fe implant in a silicon sample (Pellin et al., 1987). This example corresponds to an analytically demanding problem with substantial importance in semiconductor research. Iron is indeed a deep-level impurity that significantly affects the electrical characteristics of Si at the ppb level. SARISA now permits us to elegantly handle the isobaric interference between $^{56}Fe^+$ and $^{28}Si_2^+$, which may cause serious complications in conventional SIMS instruments. The indicated detection limit of 2 ppb suffers from the relative insensitivity of Fe under the actually used laser ionization conditions. The statistical error associated with each measurement is smaller than the symbol in the plot. The material removed in each individual measurement is less than 2% of a monolayer. Hence, the method is typically considered as a damage-free surface technique. The fact that sputtered neutrals are analyzed makes the method less susceptible to differences in matrices and chemical effects, arising in the SIMS technique, such as the enhanced response by oxygen buildup at an interface (Parks, 1990).

Surface Analysis by Laser Ionization (SALI). The SALI method is based on nonresonant multiphoton ionization (MUPI) of sputtered neutrals, released by primary ion bombardment. Mass analysis is performed by a TOF instrument with an ion reflector. The separation of the material removal and ionization steps leads to great flexibility in the choice of the probe beam for desorption. Primary ions, electrons or photons can be used. The high sensitivity enables us to keep the surface damage negligible, since only a minuscule fraction of one atomic layer is consumed to record the mass spectral information. The use of MUPI instead of RIS implies a lower degree of selectivity and sensitivity, but the general applicability to a large variety of analytical tasks is significantly increased. SALI typically suits multielement monitoring with a rather uniform ionization yield as opposed to RIS, which essentially remains a single-element method.

A schematic diagram of the SALI instrument is shown in Figure 2.33 (Becker and Gillen, 1984). Typically, an ion beam of 5 kV Ar^+ is used for sputtering. Pulsed ion bombardment is applied in the so-called static SIMS mode for submonolayer sampling. Material removal is then strictly confined to the actual measurement periods. Dynamic analysis using a continuous primary ion beam serves to erode many atomic layers and allows depth profiling. The lateral resolution of the probe beam is about 0.1 μm but can be improved with a liquid metal ion gun. Sampling depth in the static mode is one to two atomic layers. Depth profiling of inorganic materials can be achieved with a resolution of 2.5 nm after sputtering to a 100 nm distance from the initial surface (Pallix et al., 1988a). A laser can be used for material removal as well, but then the method belongs to our discussion in Section 2.7, below. The ionizing beam from an excimer UV laser passes about 1 mm above the surface of the sample. The wavelength is usually 193 or 248 nm; the pulse duration, 10 ns; the focused power density, $10^{10}\ W\cdot cm^{-2}$. These conditions permit complete ionization within the focal volume for species that are nonresonantly ionized by two-photon absorption. Atoms requiring the absorption of more photons, such as N, O, F, He, Ne, and Ar at 193 nm, are less efficiently ionized and hence measured by means of molecular ions. This means that a practically uniform detection efficiency is achieved and that a high degree of quantization can be reached throughout the entire periodic table.

Mass analysis is performed by a TOF instrument with an ion reflector. This device improves the resolution to about 1000. Moreover, careful adjustment of the back-reflecting potential allows separation of the secondary ions, which are collected, from the photoions, which are directed toward the detector. This possibility arises again from the fact that secondary ions and photoions are formed in distinct regions of the accelerating field. Detection is achieved by microchannel plates, which can accept a high flux of particles

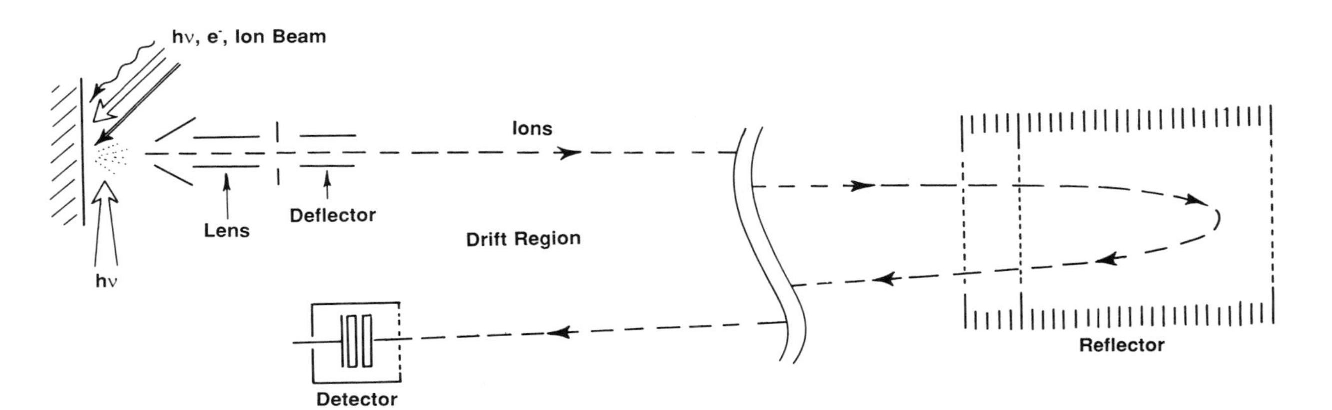

Figure 2.33. Functional diagram of the instrument for surface analysis by laser ionization (SALI). Reprinted from Becker and Gillen (1984) with permission of the American Chemical Society.

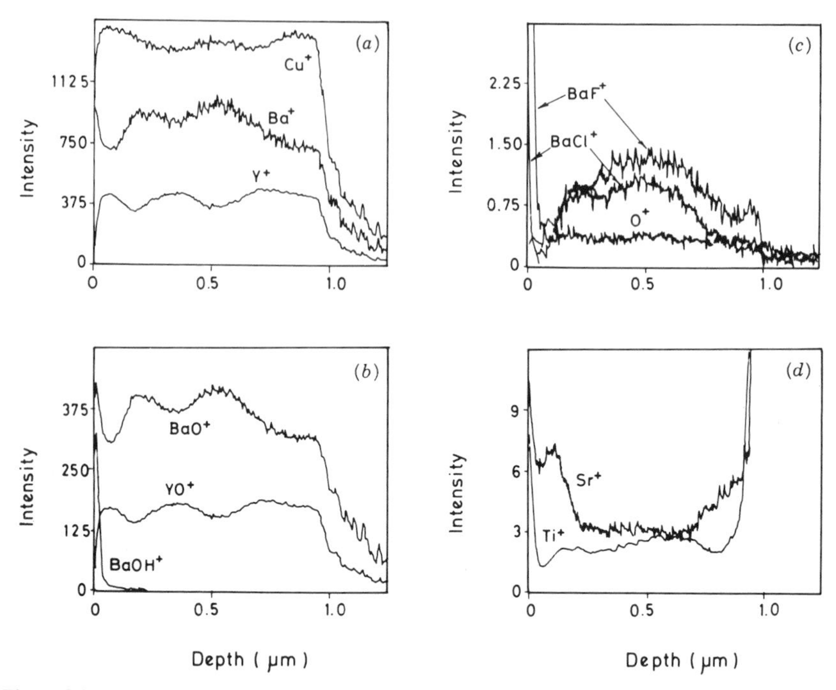

Figure 2.34. SALI depth profiles of a 1 μm thick superconductor with nominal composition $YBa_2Cu_3O_{7-\delta}$ deposited on $SrTiO_3$. (a)–(d), depth profiles of different species. Reprinted from Pallix et al. (1988b) with permission of John Wiley & Sons, Inc.

before saturation is reached. The signal is then fed into either a transient recorder for large signals or a time-to-digital converter for pulse counting of low-intensity ion fluxes. Modern devices of the latter type offer better time resolution, i.e., 1 ns vs. 5–10 ns for current transient recorders.

The analytical capabilities of SALI can be illustrated by some applications in the field of high T_c superconductors. Figure 2.34 gives the depth profiles of a superconducting 1 μm thin film with average composition of $YBa_2Cu_3O_{7-\delta}$ grown by electron beam codeposition on $SrTiO_3$ (Pallix et al., 1988b). As is typically found with SALI, the measured relative intensities for nearly all elemental components are within a factor of 2 of the true value. However, highly accurate determination of concentrations still need comparison with standard samples. Figure 2.35 illustrates the informative nature

Figure 2.35. Positive ion mass spectrum recorded by SALI-TOF-MS under static conditions from the central depth of a 180 nm thick superconducting thin film with nominal composition of $YBa_2Cu_3O_{7-\delta}$. (a)–(c), different segments of the mass spectrum. Reprinted from Pallix et al. (1988c) with permission of the American Institute of Physics.

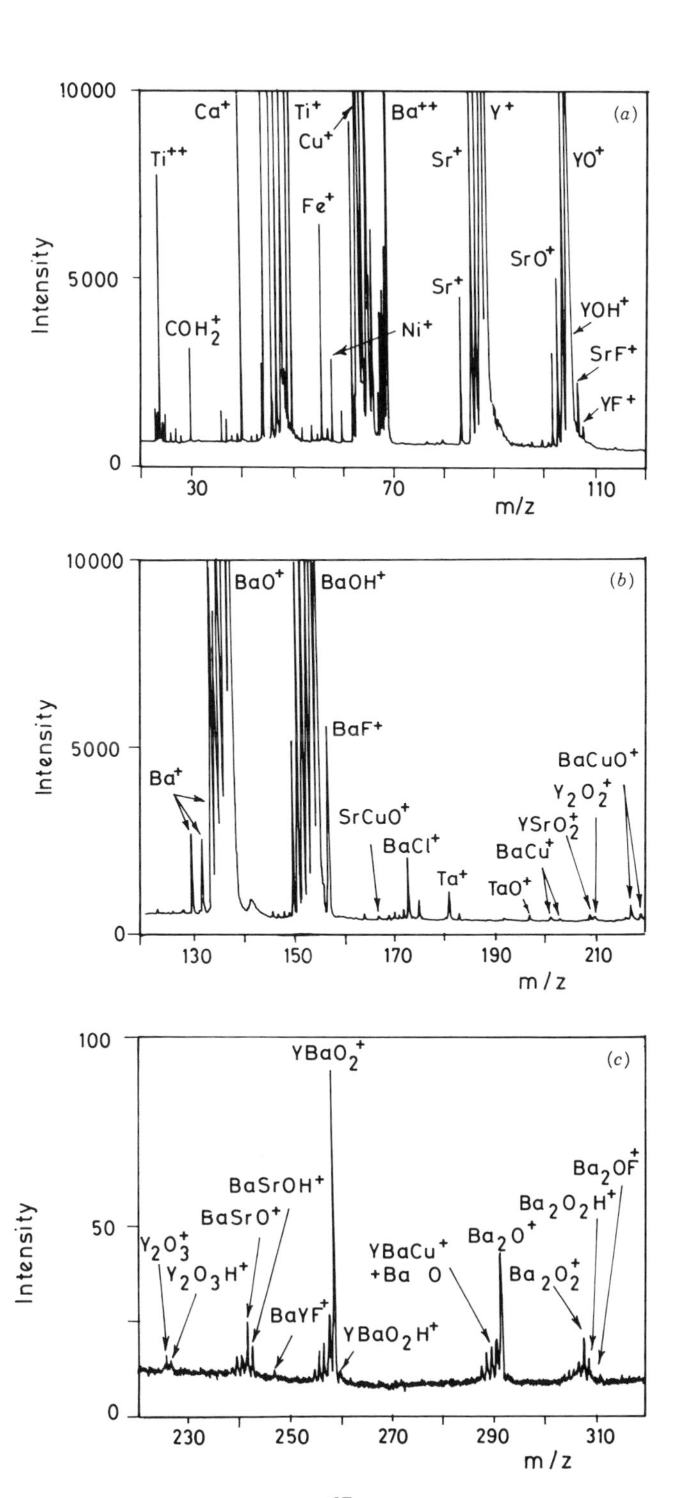

Intensity
m/z
(a)
Ti++
COH2+
Ca+
Ti+
Cu+
Fe+
Ni+
Ba++
Sr+
Y+
Sr+
SrO+
YO+
YOH+
SrF+
YF+
(b)
Ba+
BaO+
BaOH+
BaF+
SrCuO+
BaCl+
Ta+
TaO+
BaCu+
YSrO2+
Y2O2+
BaCuO+
(c)
Y2O3+
Y2O3H+
BaSrO+
BaSrOH+
YBaO2+
BaYF+
YBaO2H+
YBaCu+
+Ba O
Ba2O+
Ba2O2+
Ba2O2H+
Ba2OF+

of the SALI data with respect to identification and quantitative assessment of the local composition (Pallix et al., 1988c). The results are obtained in the static mode by accumulation of 5000 pulses. The broad dynamic range and the isotopic information allow detailed characterization of the different phases within the analyzed material, if required as a function of depth. Excellent reviews are available for interested readers (Becker, 1986, 1990).

2.5.2. Organic Applications

It is clear that the same approaches of MUPI and RIS are applicable to the field of organic analysis as well. However, the number of examples is much more limited. Indeed, laser ionization requires a suitable method to generate the neutrals in the gas phase. Many sample-probing techniques, such as primary ion bombardment, electron beam excitation of solids, or conductive heating, are destructive for most organic molecules. In fact laser irradiation of solids is one of the most appropriate means to make molecules available in the gas phase, starting from solid samples; these experiments will be considered in Section 2.7.

Gas phase generation of organic neutrals by nonlaser techniques and subsequent resonance-enhanced multiphoton ionization (REMPI) is primarily done by coupling the ion source either directly to the GC or supercritical fluid chromatography (SFC) instrument or by means of a molecular-beam-producing interface. The latter device permits rapid cooling by supersonic adiabatic expansion of the analyte in a light carrier gas. Hence, relaxation of the internal degrees of freedom through two-body collisions ensures that only low rotational and vibrational energy levels remain populated in the ground state molecule. As a result, transitions between much better defined states can be promoted and the selectivity of the UV ionization is considerably increased in comparison to that at room-temperature conditions (Lubman, 1987). As to solid state analysis, it should be noted that some of the aforementioned techniques are applicable to organic solids in selected cases. Specifically the SALI approach has yielded fine results and deserves the following brief discussion.

Surface Analysis by Laser Ionization (SALI). Molecular ions suffer from excessive photofragmentation under MUPI conditions used for elemental analysis. Hence, single-photon ionization is performed. The output of a Nd:YAG is frequency tripled twice to the ninth harmonic at 118 nm by using a gas cell of Xe phase matched with Ar after the currently used standard optic crystals for frequency conversion (Becker, 1990).

Figure 2.36 shows the positive ion mass spectrum of methylguanine, taken by SALI TOF MS (Schühle et al., 1988). A signal-to-noise ratio of 10:1

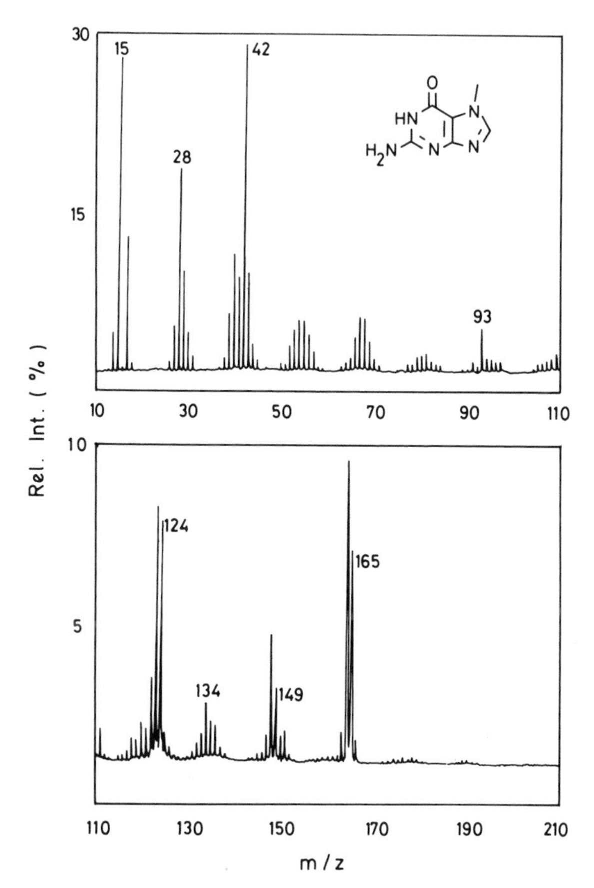

Figure 2.36. Positive ion mass spectrum of 7-methylguanine, recorded by SALI-TOF-MS after accumulation of 1000 pulses. Reprinted from Schühle et al. (1988) with permission of the American Chemical Society.

can be reached for a single-pulse measurement. Apart from the $(M + H)^+$ and $M^{+\cdot}$ at *m/z* 165 and 164, several structurally relevant ions are present. The signals at *m/z* 149 and 134 are easily associated with the cleavage of NH_2 and CH_3 groups. The peak at *m/z* 124 is ascribed to the decomposition of the imidazole ring (Schühle et al., 1988).

Low-mass fragments are detected from polymers. The mass spectra are reminiscent of the lower part of static SIMS data. The high-mass range signals from the cationized oligomers are not present in SALI (Pallix et al., 1989). In fact these spectra contain some interesting indications about the sputtering/cleavage process of these materials. Nevertheless, a lot of research is certainly still required to reach a comparable degree of practical applicability with, say, static TOF–SIMS for polymer applications.

Resonance-Enhanced Multiphoton Ionization (REMPI) in Chromatographic Methods. The multidimensional analysis of polycyclic aromatic hydrocarbons (PAHs) in environmental samples by GC-REMPI-MS provides a good example highlighting how chemical species discrimination can be pushed far above the usual limits of chromatographic separation, simply by virtue of the ionization process (Dobson et al., 1986). A capillary GC is interfaced with a so-called dynamic gas sample cell, where the laser–analyte interaction takes place. The multidimensional nature of the approach comes from the ability to perform on-line monitoring following analytical processes and products from any of these events: (a) photoelectrons by means of the total electron capture detector (TECD); (b) photoions via the TOF-MS; and (c) laser-induced fluorescence (LIF) by the optical emission probe.

Figure 2.37 illustrates the simplified schematic diagram of the instrument along with a functional block diagram of the dynamic gas cell (Dobson et al., 1986). The laser ionization is performed by a 10 Hz, 532 nm Nd:YAG coupled to a tunable dye laser, of which the frequency-doubled output covers a range of 282 to 350 nm. Delivered energies lie between 0.1 and 2.6 mJ for 10–15 ns pulses. The GC separation is performed on a current setup with a 15 m capillary column, split injection, and temperature programming. A

▶

Figure 2.37. Schematic diagram of an instrument for multidimensional analysis of a gas chromatographic effluent by REMPI-TOF-MS, laser-induced fluorescence (LIF), and flame ionization detector (FID). Reprinted from Dobson et al. (1986) with permission of the American Chemical Society.

Symbol	Component
A. Laser system	
FD	Frequency doubler
PM	Power meter
B. Detection system	
TOF-MS	TOF mass spectrometer
MEM	Magnetic electron multiplier
PA_L	Preamplifier
Tr. Dig. SA	Transient digitizer/ signal averager
PA_1	Source preamplifier
PA_2	Active filter preamplifier
PD	Fast photodiode circuit
Delay	Pulse delay generator
GI_1 & GI_2	Gated integrators
OSC	Oscilloscope
PMT	Photomultiplier
C. Sampling system	
GC	Gas chromatograph
FID	Flame ionization detector
CC	Capillary column
OS	Outlet splitter
HTL	Heated transfer line
HI	Heated interface
ASL	Alternate sampling line
D. Miscellaneous components	
F	Optical filters
P	Pellin Broca prism
L	Fused silica lenses
LB	Light baffles
BT	Optical beam trap

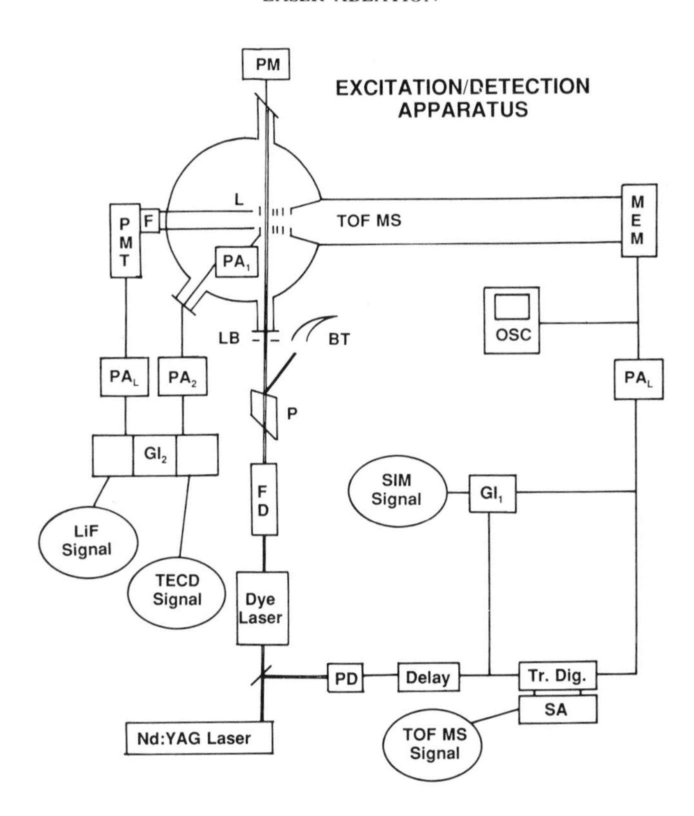
PM
EXCITATION/DETECTION APPARATUS
L
P M T
F
TOF MS
M E M
PA1
LB
BT
OSC
PAL
PA2
PAL
P
GI2
F D
SIM Signal
GI1
LiF Signal
TECD Signal
Dye Laser
PD
Delay
Tr. Dig.
SA
Nd:YAG Laser
TOF MS Signal

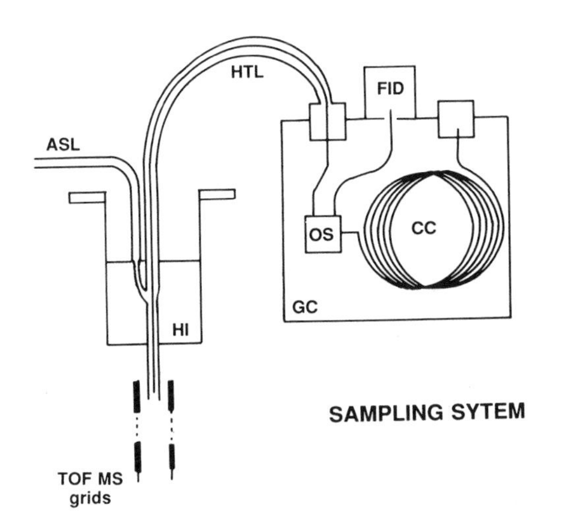
HTL
FID
ASL
OS
CC
GC
HI
SAMPLING SYTEM
TOF MS grids

vitreous silica outlet splitter leads 70% to the flame ionization detector (FID) and 30% to the low-pressure dynamic gas sample cell. All transfer lines are heated. The TECD provides independently quantitative information on the REMPI process and determines the suitable moments to start data collection with the TOF-MS in the same way as in previous GC-MS systems. The TOF-MS is an in-house-modified CVC MA 2000 (Rochester, New York) instrument. Ion bunching of the initially continuous sample supply is performed by drawout pulsing grids. A mass resolution of more than 300 is obtained with the 2 m flight tube. Detection is achieved by a magnetic electron multiplier. The signal is processed either by a gated integrator for selective

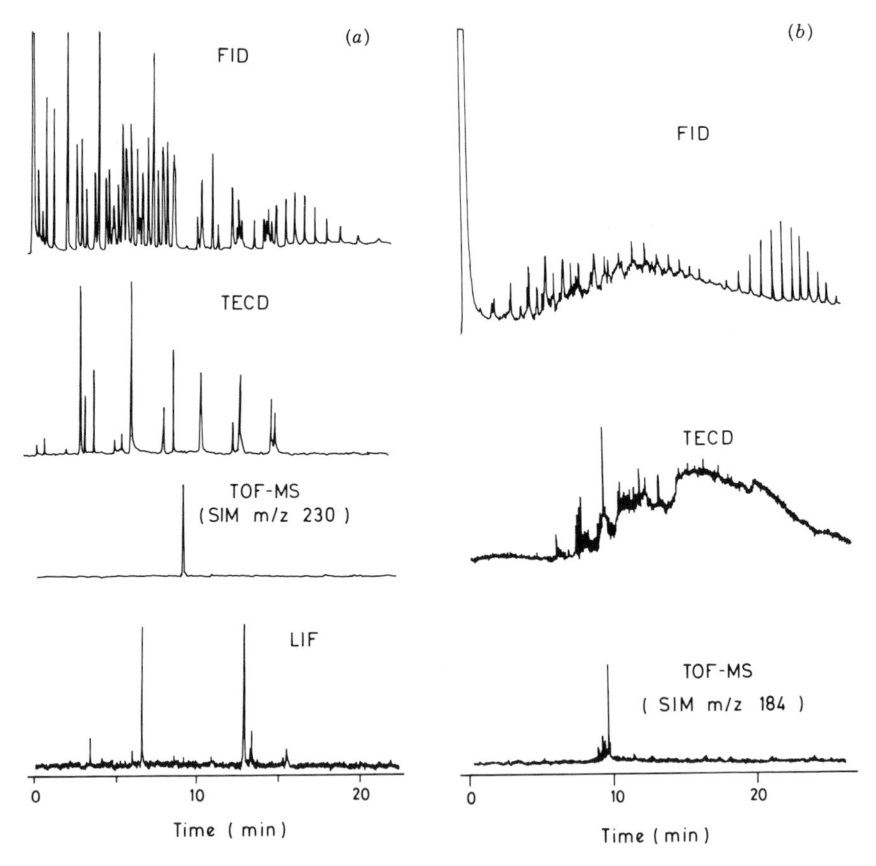

Figure 2.38. Chromatogram simplification for a 64-component mixture by application of multidimensional analysis using FID, TECD, LIF, and TOF MS (a) and application of the method to a Paraho shale oil (b). Reprinted from Dobson et al. (1986) with permission of the American Chemical Society.

ion monitoring or by a 200 MHz transient recorder for measurement of any mass spectral window of interest during the elution of selected chromatographic peaks. Complete mass spectra are recorded within 0.5 s.

Figure 2.38 illustrates the consecutive increase in molecular and isomeric specificity when a mixture of 64 components, including PAHs, aromatic compounds, aliphatic hydrocarbons, alcohols, sulfones and esters, is analyzed by FID, TECD with a laser wavelength of 285 nm, LIF, and TOF-MS in the selected ion monitoring (SIM) mode (Dobson et al., 1986). Typical detection limits for PAHs are in the low-picogram range for TOF-MS, whereas the sensitivity of TECD is typically 10 times less. Since it is not possible to tune the laser wavelength to each eluting compound, the results are recorded under the conditions of an optimized compromise. The applicability of the approach to a real environmental sample is evidenced by the data on the neutral fraction of a Paraho shale oil extract in the same figure.

The combination of efficient laser action, the selectivity of capillary column GC, and the high transparency of a TOF-MS is quite appealing. Recent papers indicate the possibility of exploiting the extremely soft ionization capabilities for identification purposes of, for instance, alkylbenzenes (Opsal and Reilly, 1988) or of reaching extremely favorable detection limits, such as 1.5 fg for tetraethyltin (Colby et al., 1990). Laser irradiation with pico- and nanosecond pulses allows ionization of compounds that undergo rapid relaxation by nonradiative processes. This approach may improve the detection sensitivity of, e.g., nitro-containing molecules, phenols, or chlorinated aromatics in environmental samples (Wilkerson et al., 1989; Wilkerson and Reilly, 1990). Although TOF analyzers are certainly the most popular type in these kinds of studies, GC-FTMS with laser ionization has also been tried out. High-quality mass spectra with a resolution of more than 20,000 and accurate mass assignment in the ppm range can be obtained with detection sensitivities on the order of 10 pg for PAHs (Sack et al., 1985). Such numbers can be considered adequate for purposes of trace analysis, but this type of analyzer is not ideal for large-scale monitoring of environmental pollutants.

REMPI with Supersonic Molecular Beam Interfaces. When a large polyatomic molecule in a bath of light carrier gas is injected under high pressure into a vacuum of 10^{-5} torr or less through an orifice, substantial cooling of the internal energy occurs. The physical background has been extensively described in the literature (Smalley et al., 1977; Anderson et al., 1966; Hayes and Small, 1983). The initial thermal population of a large number of vibrational and rotational states is collapsed to selected rotational and vibrational states. Hence, the partition function for excitation is much improved in comparison to the situation at room temperature. Moreover, these supersonic beams are also beneficial for the TOF mass resolution.

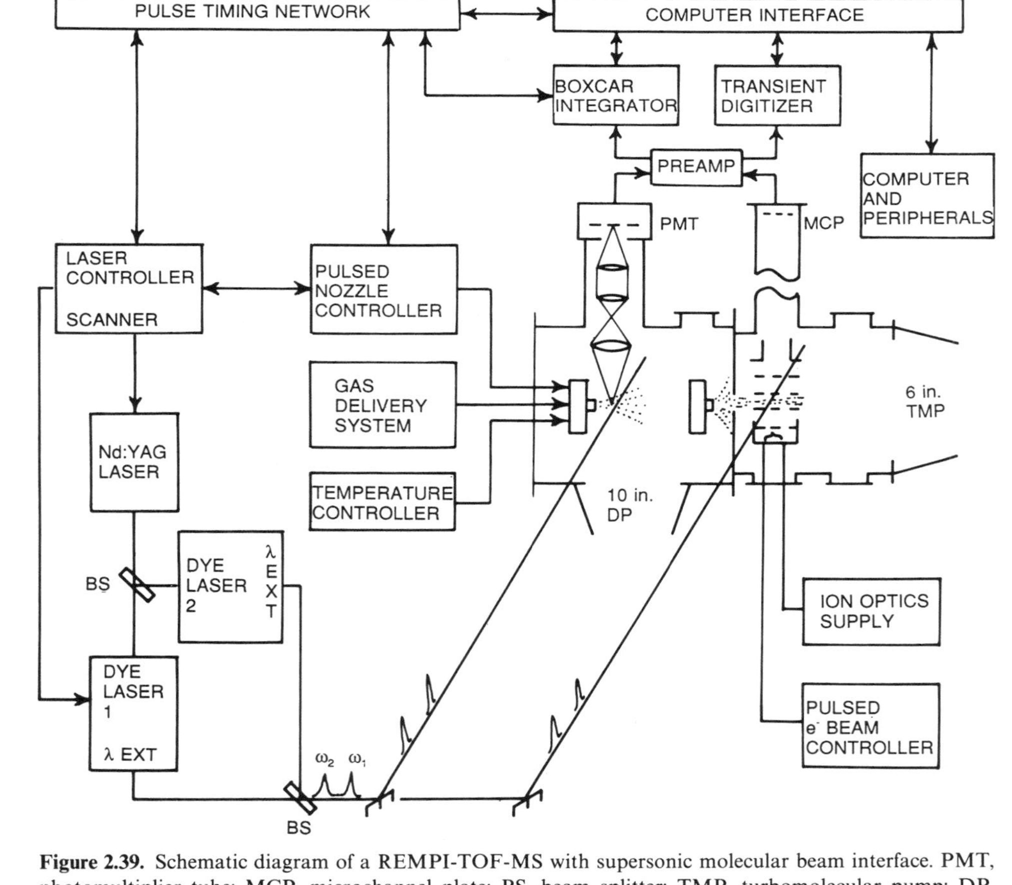

Figure 2.39. Schematic diagram of a REMPI-TOF-MS with supersonic molecular beam interface. PMT, photomultiplier tube; MCP, microchannel plate; BS, beam splitter; TMP, turbomolecular pump; DP, diffusion pump. Reprinted from Syage (1990) with permission of Oxford University Press, Inc.

Indeed, the attainable mass resolution in TOF-MS depends on the space and energy spread of the ions in the accelerating region. When the molecules are forced to expand through an orifice into the vacuum, the Boltzmann velocity distribution becomes strongly peaked around an average velocity. Thus, the molecules are now traveling in the same direction with a minimal energy spread to alleviate mass resolution limitations (Lubman and Jordan, 1985).

Figure 2.39 shows the schematic diagram of the instrument developed by Syage (1990). The sample is introduced in an evacuated steel cylinder, which is then backfilled with the carrier gas to high pressure. The gas sample flow is fed to an in-house-constructed pulsed nozzle capable of high repetition rates up to 100 Hz at temperatures up to 200 °C. Two differentially pumped chambers over a 1 mm diameter skimmer handle the gas load. The expansion into a supersonic beam takes place in the first region. The main chamber includes the excitation and MS analysis systems. Laser ionization is performed by a Nd:YAG pumped tunable dye laser system with a wavelength range from 195 to 440 nm. Typical pulse energies range between 0.1 and 15 mJ depending on the selected wavelength. The output is then focused to interact with the molecular beam pulses between the grids in front of the 1 m drift tube. Time-resolved microchannel plate detection is applied in combination with a 200 MHz transient digitizer.

Figure 2.40 illustrates the application of three REMPI schemes to access different excited electronic states and deposit various amounts of energy into the parent and fragment ions of diisopropyl methylphosphonate (DIMP) (Syage et al., 1987). No molecular ions are observed, even though ions are formed with as little as 13,000 cm^{-1} internal energy. The ionization through the weak and structureless electronic state at 266 nm yields dominant fragments at m/z 123 and 97, which are also observed during electron ionization in the gas phase. However, the use of a second wavelength for ionization and ion absorption allows the generation of additional low alkyl fragments. Interestingly, ionization through the transition at 401 nm removes previously detected ions from the mass spectrum and gives only a prominent fragment at m/z 43, accompanied by m/z 41. This quite different result is attributed to excitation to a resonance state where decomposition competes with ionization, followed by nonresonant ionization of the neutral isopropyl fragment.

These phenomena are further exploited by the development of, e.g., double-resonance dissociation. A first laser excitation populates a single vibrational level of the ion, and a second scanning laser dissociates the ion by resonance-enhanced multiphoton dissociation. The signal of the parent ions decreases in favor of the corresponding fragment peak intensity (Syage and Wessel, 1987). The practical feasibility of this method for environmental analysis is evidenced by the detection limits of 300 ppt for a variety of

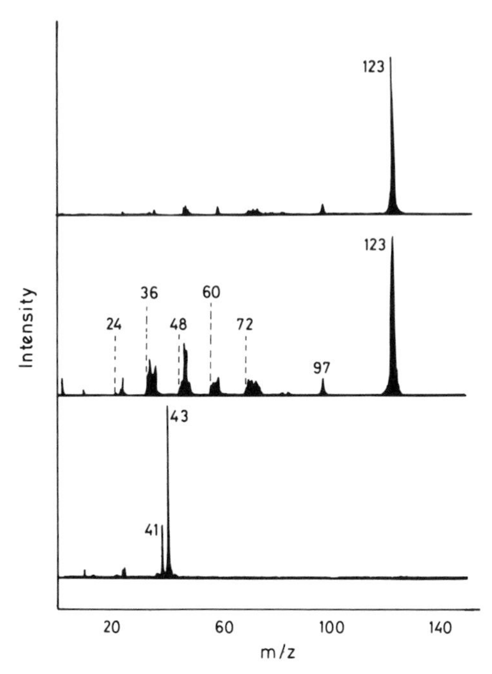

Figure 2.40. Positive ion mass spectra of diisopropyl methylphosphonate, recorded by REMPI-TOF-MS at different wavelengths. Reprinted from Syage et al. (1987) with permission of the Optical Society of America.

organophosphonate and sulfide molecules in humid air (Syage et al., 1987). Apart from the inherent sensitivity, this example demonstrates how the REMPI process provides an additional degree of freedom to the MS analysis. Specifically, the structural information can be manipulated through the fine tuning of the energy deposited in the parents and the selective excitation of molecular vs. fragment ions.

Similar experiments have been performed by Lubman's group at the University of Michigan (see Pang et al., 1988). A series of publications has also shown the extent of separability attainable for structural isomers and identical molecules with different isotopes (Tembreull and Lubman, 1984; Tembreull et al., 1988; Lubman et al., 1985). Subsequent experiments employ pulsed laser desorption in conjunction with the molecular beam and TOF techniques and will be considered later (in Section 2.7).

REMPI under Atmospheric Pressure. The concept of laser ionization under atmospheric pressure in conjunction with ion mobility spectrometry has furthered the development of a sensitive and rapid method to monitor trace contaminants in the environment (Kolaitis and Lubman, 1986). Indeed, a ppb concentration at atmospheric pressure means detection of 3×10^{10} molecules cm^{-3}, which represents a density comparable to or better than that normally probed in molecular beam experiments. The ion mobility drift technique allows separation of ions, produced at atmospheric pressure, in a rapid and continuous manner before transfer into the mass analyzer.

Figure 2.41 illustrates the apparatus, modified for laser ionization (Kolaitis and Lubman, 1986). The sample is mixed with the carrier gas and fed into the ion mobility drift tube. The ion source and ion–molecule reaction regions actually refer to the original mode of operation, where a ^{63}Ni (β) radioisotope source produces the ions. Laser ionization with 266 nm pulses from a frequency-quadrupled Nd:YAG takes place in the middle of the drift region. No resonance experiments have been attempted. The ion mobility depends on the collision-controlled behavior of charged species at atmospheric pressure under the influence of a low electric field. Nitrogen, air, argon, methane in argon, and carbon dioxide can be used as the drift gas. The separated ions are then injected through a pinhole orifice into a quadrupole MS. The mass filter can be scanned rapidly up to m/z 1200 with unit resolution. Intense molecular ions are observed in the quasi-absence of fragments, in spite of a power density level on the order of 30 $MW \cdot cm^{-2}$. Note that the degree of optical selectivity, as reached in the previous experiments on species in vacuum, cannot be attained under atmospheric pressure.

2.6. LASER DESORPTION AND NONLASER POSTIONIZATION

This section will be confined to the more recent approaches. In fact, as already discussed, the earliest experiments used lasers for evaporation in conjunction with the then-current methods for ionization. (These experiments were covered in Section 2.2).

Examples selected for consideration in the subsection concern the combination of laser ablation for sample introduction of solids in ICP-MS and the implementation of lasers to increase the yield and potentially the spatial resolution in glow discharge (GD) MS. As for organic applications (Section 2.6.2, below), attention will be paid to earlier experiments using laser-induced desorption in conjunction with CI as well as to the recent approach of laser ablation/electron ionization in FTMS.

2.6.1. Inorganic Applications

Most instruments routinely employed for trace element analysis use solutions for sample introduction into the atomization or excitation sources. The

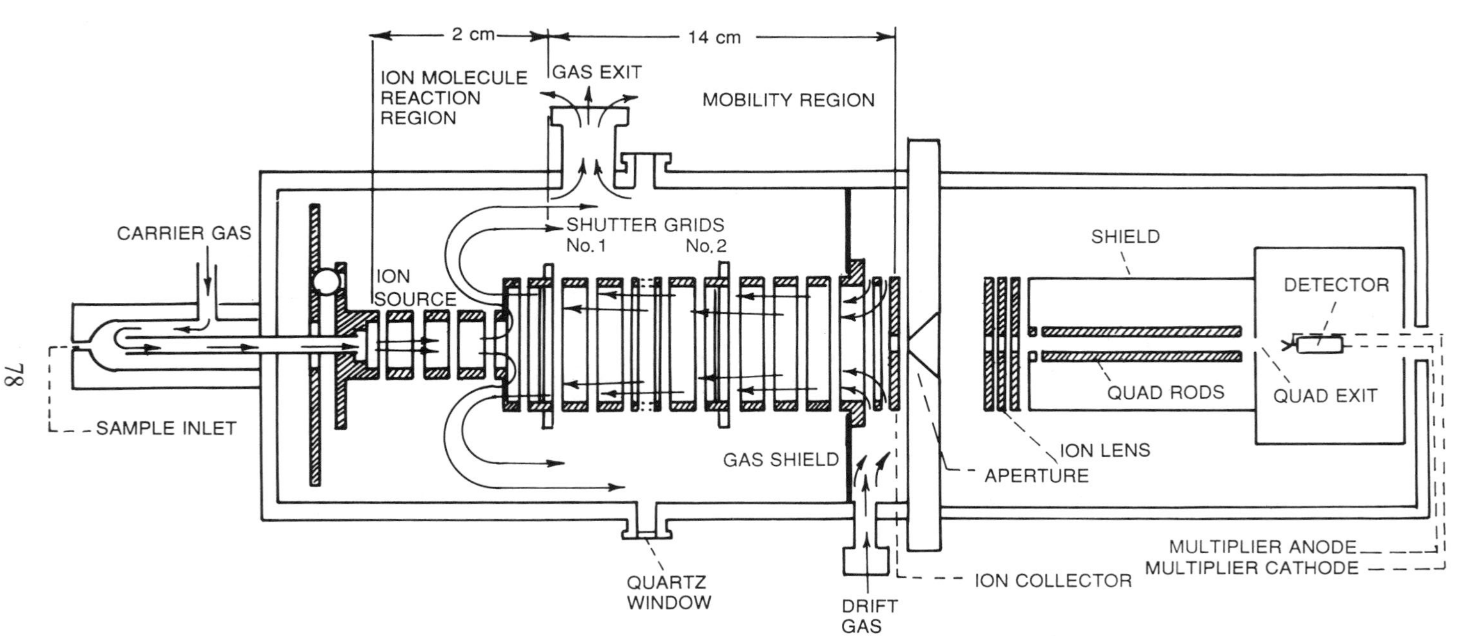

Figure 2.41. Cross section of the atmospheric pressure ionization MS–ion mobility drift tube apparatus. Reprinted from Kolaitis and Lubman (1986) with permission of the American Chemical Society.

development of alternative techniques for solid samples is particularly noteworthy because this complements the analytical capabilities of well-established approaches with other particularly interesting features. In general, solid probing implies the potential of compositional analysis with modest-to-high spatial resolution. Specifically when laser-based approaches are used, there is the additional advantage of applicability with virtually no restriction on the physical properties of the starting material, such as electrical conductivity. Of course, sample preparation is largely reduced, however, at the risk of incomplete and/or fractional volatilization of the sample during the laser interaction.

ICP-MS is now widely utilized as an extremely sensitive and versatile technique for multielemental trace analysis with a high sample throughput at a relatively modest operating cost. A suitable interface to handle solid samples is based on laser ablation. The phenomenological description of this process is complex and beyond the scope of this chapter. Although a free-running laser permits ablation of higher amounts of material and hence increases the sensitivity, *Q*-switched sources are preferentially used. The latter reduces the problem of fractional volatilization, produce more particles in a size range ($< 10\,\mu m$) that is adequate for subsequent entrainment and transport by gas flow (Arrowsmith, 1990).

The feasibility of this approach was first demonstrated by Gray (1985) using a low-repetition-rate ruby laser in the free-running as well as the *Q*-switched mode. Later, the analytical applicability was significantly improved by the use of higher-repetition-rate sources (Arrowsmith, 1987); reproducibility and precision were considerably enhanced.

Figure 2.42 illustrates a laser ablation cell for ICP-MS (Arrowsmith, 1987). The sample is mounted on an adjustable flange under a small ablation cell. The latter is connected to the gas flow supply on the one side and the transfer line to the ICP torch on the other. The entire unit is then enclosed in a gas-tight second volume filled with the plasma support gas, just above atmospheric pressure. Hence, clean gas flows in and around the cell, whereas ablated material, not entrained by the other gas stream, can escape from the cell. The inlet gas flows through an annular device to form a symmetric gas sheath around the base of the cell. The sample is irradiated by a Nd:YAG laser operated at 1064 nm. The beam is focused to about $40\,\mu m$ using a single lens, whereas the use of an additional aperture reduces the spot to about $20\,\mu m$. However, the irradiated area must be optimized as regards sensitivity. A coaxially aligned He–Ne laser acts as beam spot locator for optical viewing of the sample surface via a TV camera and monitor with an overall magnification of $300\times$.

Figure 2.43 gives an example of the crude depth profiling potential that can be achieved with consecutive laser ablation on the same spot (Arrowsmith,

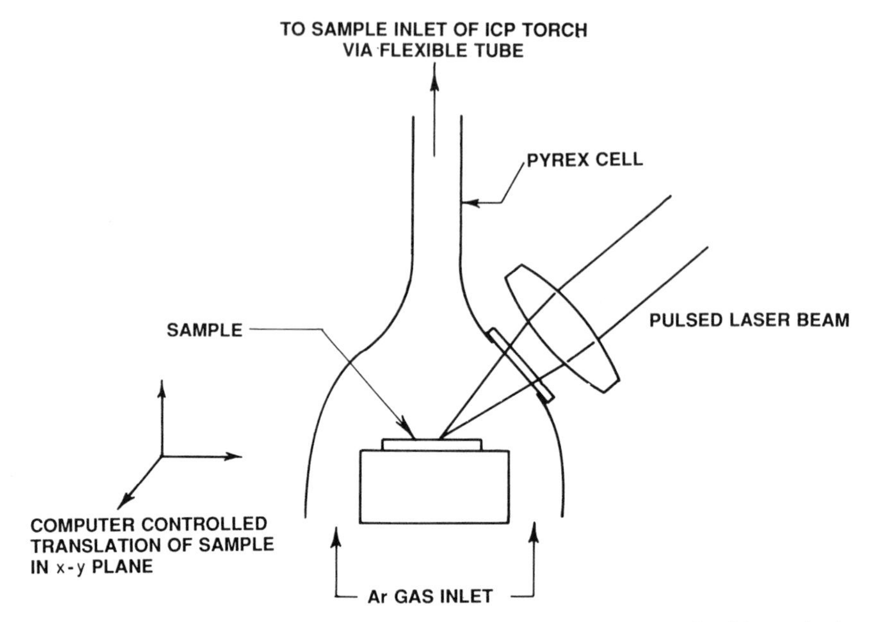

Figure 2.42. Schematic diagram of the laser ablation system for analysis of solid samples by ICP-MS. Reprinted from Arrowsmith (1987) with permission of the American Chemical Society.

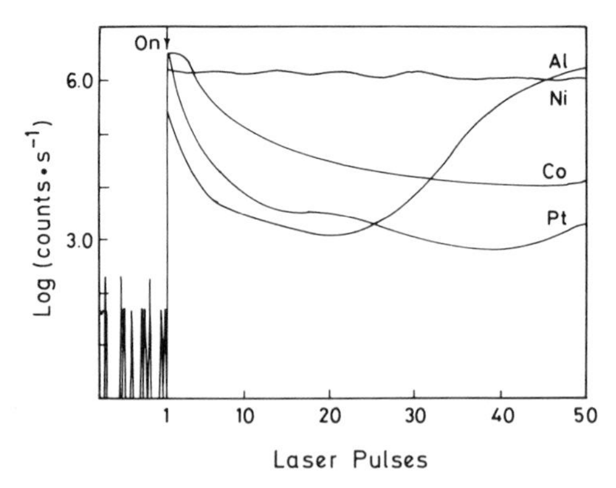

Figure 2.43. Elemental depth profiles obtained by ICP-MS and successive ablation of a multilayer thin film material at a pulse repetition rate of 1 Hz. Reprinted from Arrowsmith (1990) with permission of Oxford University Press, Inc.

1990). The sample is a multilayer film with a Co–Pt layer close to the surface. The depth ablated per shot is estimated at about 5 nm. Nevertheless, successive laser shots make the crater wider and the Co and Pt signal does not return to the baseline. Data acquisition is performed on a Sciex 250 Elan ICP-MS with digital control of the quadrupole scan. Transient signals from single laser shots are measured in the selected ion monitoring mode. The dead time between data points is about 5 ms. Hence, a reasonably high duty cycle and sample rate per mass can be achieved, depending on the number of selected masses. Moreover, a retarding potential can be applied under computer control to the quadrupole to attenuate the signal during the time of data acquisition of a particular mass. This feature can be used to bring an element within the dynamic range that would normally cause detector saturation. More recent examples of local analysis as well as other aspects of laser solid sampling in ICP have recently been reviewed (Denoyer et al., 1991; Gray, 1989).

A second development is the use of laser irradiation in the source of GD-MS. The GD basically consists of a sample cathode and an anode sealed into a low-pressure rare gas environment, normally argon at about 1 torr. The anode is normally the ion source housing, including the ion exit orifice. The voltage across the electrodes, causes breakdown of the gas and formation of a plasma. Accelerated positive argon ions strike the cathode and sputter sample atoms. These atoms are then ionized by collisions with electrons and energetic metastable atoms in the plasma. The implementation of laser ablation to perform or assist the sputtering process provides several advantages. Whereas normal GD requires a conducting sample, laser ablation can be carried out on insulators. Additionally, sampling location can be selected, thus providing potential for local analysis. Finally, the delivered photon energy can be controlled (Hess and Harrison, 1990). So far, however, these projects need further research before the stage of applicability in practical analysis is reached.

2.6.2. Organic Applications

Cotter and his group have explored the combination of LD with CI as a direct exposure technique on a double-focusing Dupont 21-491 MS (Cotter, 1980). Instrumental modifications are minimal. A TEA CO_2 laser (GenTec, Dalton, Quebec, Canada), providing 40 ns pulses of 0.7 J at 10.6 μm with a repetition rate of 1.2 Hz is directed toward the sample coating on a Vespel tip in the standard introduction probe for solids. The intensity of the mass spectra drastically decreases during consecutive analyses of the same spot. A timing circuitry allows a variable delay between the laser firing and actual data collection by the 100 MHz transient recorder. Fast scanning of the electrostatic sector permits coverage of a mass range of 10 amu within 2 ms. Nevertheless, the main signals of most products last up to several minutes, so there is ample time to record a part of the spectrum after each laser interaction. Figure 2.44 shows the analysis of a labile steroid glucuronide by LD-CI and direct exposure CI (Cotter, 1980). The conventional introduction

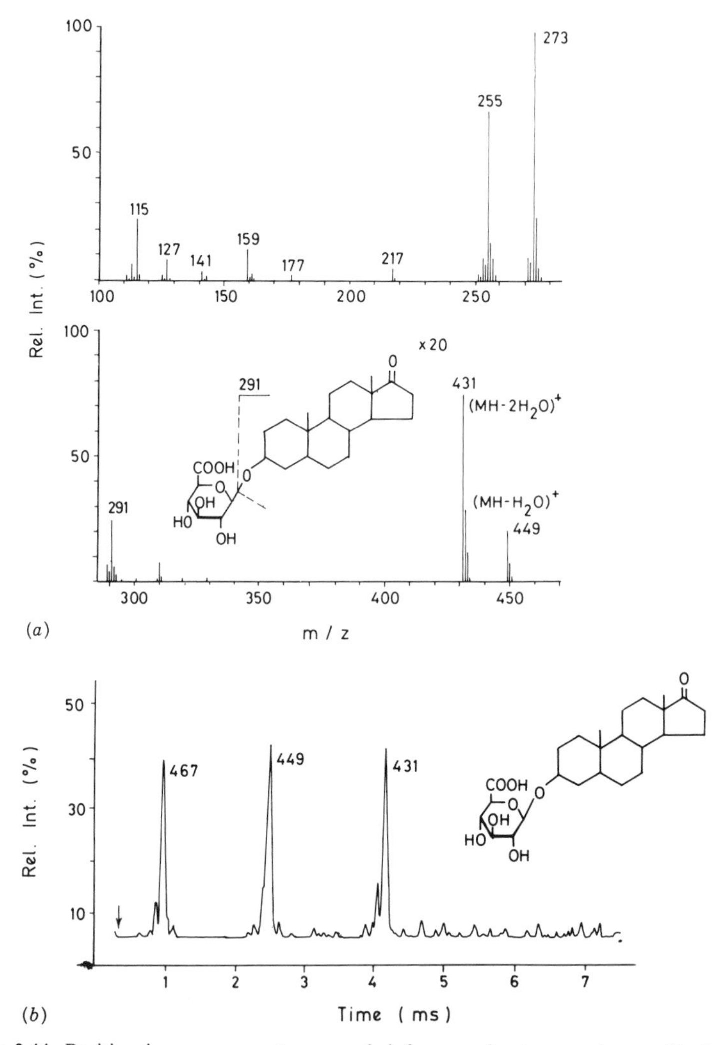

Figure 2.44. Positive ion mass spectra recorded from androsterone glucuronide by direct exposure chemical ionization (a) and laser desorption chemical ionization (b). Only the parent region of the latter spectrum is shown. Reprinted from Cotter (1980) with permission of the American Chemical Society.

probe based on conductive heating only yields pyrolysis products. Direct exposure CI allows detection of the dehydrated molecules in the parent region of the mass spectrum, whereas laser irradiation of the solid sample yields intact molecules in the gas phase, which are then converted into $(M+H)^+$ ions by the reagent gas.

One of the difficulties evidenced in these early experiments is related to the large difference in duty cycle between the laser and the magnetic sector instrument. The irradiation-produced burst of neutrals lasts for less than a millisecond, but a sector cannot be scanned nearly that fast. Thus, many laser pulses are needed to acquire a complete mass spectrum. As an ion storage type of analyzer, FTMS is certainly better suited for such experiments. Early experiments involving the laser desorption of thymidine in the presence of CI reagent gases methane and ammonia have been unsuccessful (McCrery and Gross, 1985). However, more recently, it has been shown that a high density of reagent ions allows effective LD-CI of this and other compounds. To avoid space charge broadening of the FTMS signals, ejection of the reagent ions is required after the reaction (Amster et al., 1989). The method provides an extra measure of control over the degree of fragmentation and the relative abundance of the molecular ion by the use of different reagent gases.

Modification of an ion quadrupole storage trap to perform laser desorption in the cavity in combination with postionization with electron impact has been reported as well (Heller et al., 1989). The experiments are carried out on a Finnigan ion trap detector system (San Jose, California). The popularity of these versatile and low-cost analyzers suggests that they will have many future applications.

Whereas the previous work is more or less focused on qualitative or structural identification, the research of McIver's group at the University of California, Irvine, has yielded a particularly appealing method for surface analysis of both organic and inorganic constituents (Land et al., 1990). Basically, laser-induced thermal desorption is combined with EI or CI of the generated neutrals and mass analysis is done by FTMS. Conceptually, the method ultimately exploits a variety of advantageous features. First of all, laser heating of solids offers the possibility of performing high-resolution layer-by-layer characterization and in addition allows the release of labile molecules without thermal destruction. Postionization in the gas phase by means of EI or CI permits identification of the species by using the vast amount of reference data. FTMS provides a unique set of performance characteristics with respect to mass resolution and accurate m/z assignment, especially when sensitivity is considered. Moreover, the instrument excellently suits the analysis of transient bursts generated by the pulsed laser irradiation. Finally, recording of the mass spectrum does not destroy the ions, which

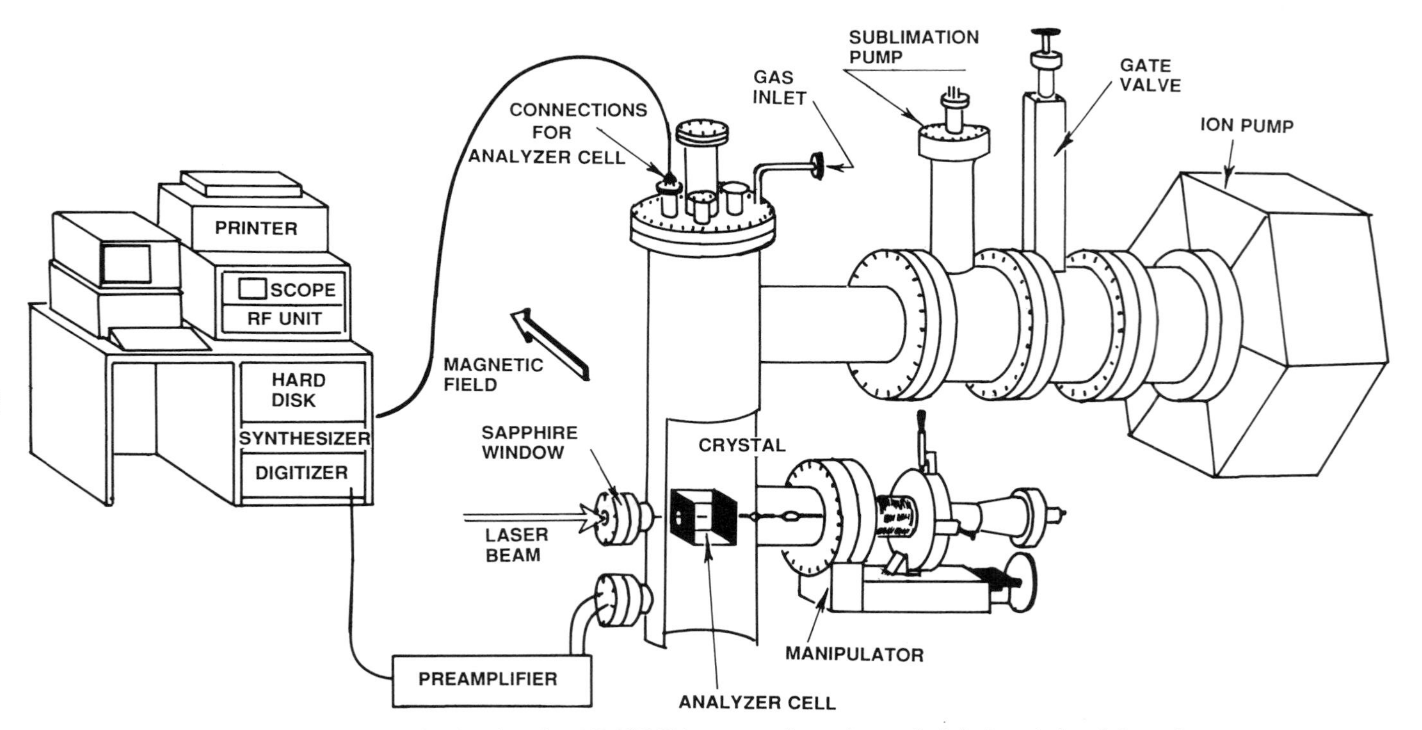

Figure 2.45. Perspective drawing of an LD-FTMS instrument for surface analysis by laser-induced thermal desorption and postionization. Reprinted from Sherman et al. (1985) with permission of Elsevier Science Publishers.

remain available for additional experiments such as CID or laser photodissociation that can more precisely determine their structure.

Figure 2.45 shows a perspective drawing of the instrument (Sherman et al., 1987). The main chamber is evacuated to a base pressure of 3×10^{-11} torr after bake-out. The sample is mounted on a movable 6 mm diameter stage with heating and cooling capabilities in front of a hole in the 6 cm cubic FTMS cell, placed in a 1.1 T vertically oriented magnetic field. Molecular adsorbates can be prepared in situ on a platinum single crystal by backfilling the vacuum chamber. The gas pressure and/or exposure time provides flexibility to prepare different coverages. A 248 nm excimer or 1.06 μm Nd:YAG laser beam impinges perpendicularly on the specimen. The irradiated area is about 0.2 mm^2, and the power density in the focal spot typically lies between 2 and 20 $MW \cdot cm^{-2}$. A rhenium filament produces a 1–10 μA electron beam at a distance of about 3.1 cm above the sample in the same direction as the magnetic field. Figure 2.46 shows the sequence of events in an LD-EI experiment (Land et al., 1990). The laser beam rapidly heats the near-surface region of the sample. The neutrals expand into the vacuum and are then exposed to the electron beam in the middle of the cell. The resulting ions are trapped on small orbits. After excitation a coherently moving ion packet induces image currents on the receiver plates. The transient is amplified, digitized, and stored in the computer, where a fast Fourier transform analysis allows recovery of the individual cyclotron frequencies. The transient signal lasts for several tenths of a second at pressures below 10^{-8} torr, and a mass resolution of over 10,000 can be routinely achieved.

To illustrate the method's capabilities, we have selected two examples. The first concerns the analysis of the surface of a magnetic hard disk (Land et al.,

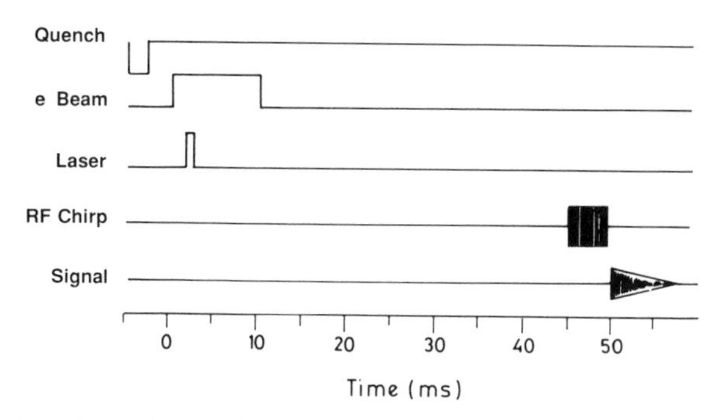

Figure 2.46. Chronology of pulses in a laser-induced thermal desorption FTMS experiment using postionization with electron impact. Reprinted from Land et al. (1990) with permission of Oxford University Press, Inc.

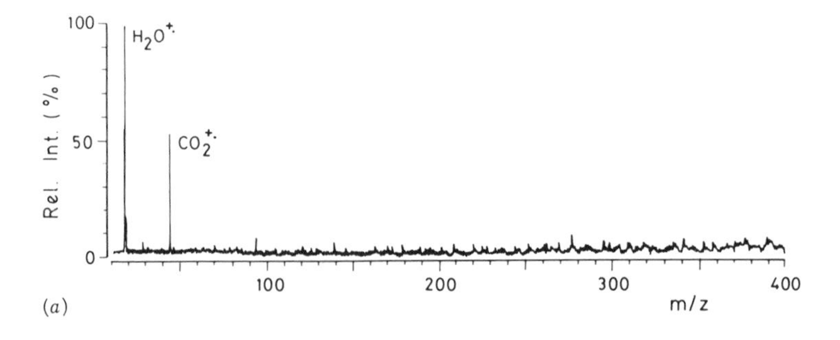

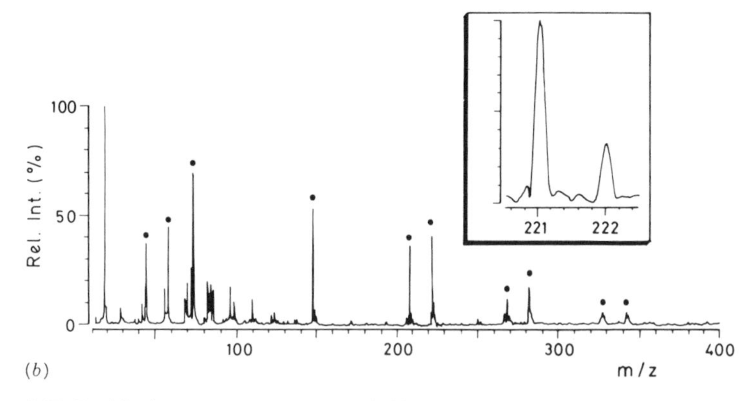

Figure 2.47. Positive ion mass spectrum recorded by a laser-induced thermal desorption FTMS from the surface of a computer hard disc using laser power between 1 and 20 MW·cm^{-2} (b) and less than 1 MW·cm^{-2} (a). Reprinted from Land et al. (1987) with permission of the American Chemical Society.

1987). The material consists of a 2 mm Al base onto which a 1 μm P-doped Ni is evaporated. Next comes the magnetic layer, a 50 nm thick film deposited by evaporation of a Cr-containing Co alloy, which is then covered with 30 nm of sputter-deposited carbon. Adsorption of molecular species on this carbon is of great importance for disk performance. Figure 2.47 shows a series of representative spectra (Land et al., 1987). Only CO_2 and H_2O are seen in the spectrum under the lowest power densities utilized around 1 MW·cm^{-2}. The background contribution from the chamber can be easily checked by EI without LD and accounts for most of the observed water signal but only for 20% of the one for CO_2. At higher laser power density a multitude of peaks up to *m/z* 341 is found. The accurate mass assignment of FTMS as

Table 2.3. Structural Assignment for Major Peaks in the Positive Ion Mass Spectrum, Recorded from a Magnetic Disc by Thermal Laser Desorption and FTMS

Possible Ion Structure	Measured m/z	Calculated m/z
SiO^+	43.979	43.972
$HC_2{=}\overset{+}{Si}{-}CH_3$	57.065	57.016
$H_3C{-}\overset{+}{Si}(CH_3)_2$	73.043	73.047
$(H_3C)_3Si{-}\overset{+}{O}{=}Si(CH_3)_2$	147.07	147.07
$H{-}Si(CH_3)_2{-}O{-}Si(CH_3)_2{-}\overset{+}{O}{=}Si(CH_3)_2$	207.04	207.07
$H_3C{-}Si(CH_3)_2{-}O{-}Si(CH_3)_2{-}\overset{+}{O}{=}Si(CH_3)_2$	221.10	221.08
$H{-}Si(CH_3)_2[{-}O{-}Si(CH_3)_2{-}]_2\overset{+}{O}{=}Si(CH_3)H$	266.99	267.07
$H{-}Si(CH_3)_2[{-}O{-}Si(CH_3)_2{-}]_2\overset{+}{O}{=}Si(CH_3)_2$	281.05	281.09
$H{-}Si(CH_3)_2[{-}O{-}Si(CH_3)_2{-}]_3\overset{+}{O}{=}SiH_2$	326.97	327.08
$H{-}Si(CH_3)_2[{-}O{-}Si(CH_3)_2{-}]_3\overset{+}{O}{=}Si(CH_3)H$	340.96	341.09

Source: Reprinted from Land et al. (1987) with permission of the American Chemical Society.

well as proposed ion structures are summarized in Table 2.3 (Land et al., 1987). This information already points to the presence of poly(dimethylsiloxanes) on the disk. Additional support is provided by the relative peak intensities.

It should be noted that, unlike most ion transport analyzers, the relative peak intensities in FTMS are not affected by temporary fluctuations in the concentration of the analyte in the source. The measurement is performed on the accumulated ion bunch between the quench and excitation pulses. Furthermore, a series of signals between *m/z* 30 and 110 are readily assigned to fragments of aliphatic alcohols. These data evidence the great potential of FTMS, particularly in combination with LD-EI, to achieve identification of unknown components in the demanding situation of a heterogeneous

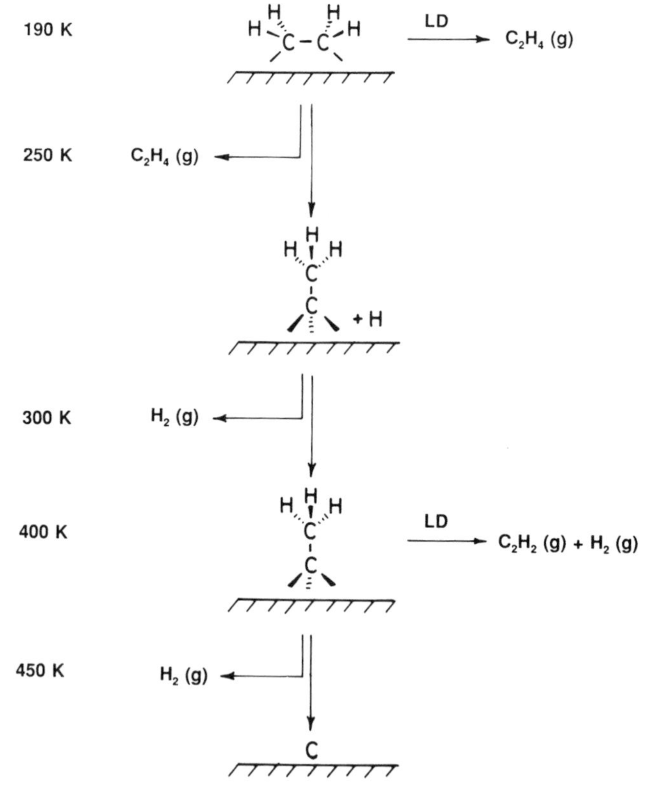

Figure 2.48. Sequence of reactions in the dehydrogenation of ethylene on platinum: on the left are the results under slow heating conditions; on the right are the reactions induced by fast laser heating. Reprinted from Sherman et al. (1987) with permission of Elsevier Science Publishers.

material. Indeed, all the necessary information should be obtainable from only one laser-generated ion burst. As with the TOF laser microprobes, penetration through the upper layers and irradiation at high power density allows us to characterize the elemental constituents as well.

A second example nicely illustrates the extremely subtle and detailed information that can be gained from the application of LD-EI-FTMS in the field of catalysts. Some important reactions occur when unsaturated hydrocarbons come into contact with platinum. For instance, ethylene exists intact on Pt at low temperature but rearranges at higher temperature to a C—CH_3 ethylidene intermediate and surface hydrogen. Above 450 K, ethylidene is converted to surface carbon and desorbs H_2. This conversion has been monitored with the LD-EI-FTMS technique. Ethylene releases ethane, while the ethylidene intermediate generates acetylene and hydrogen. The sequence and results are illustrated in Figures 2.48 and 2.49, respectively (Sherman et al., 1987). After annealing is performed up to 450 K, a prominent hydrogen signal is observed whereas the major contribution at *m/z* 28 comes from CO^+ and no longer from ethylene. Instead, an intense acetylene signal

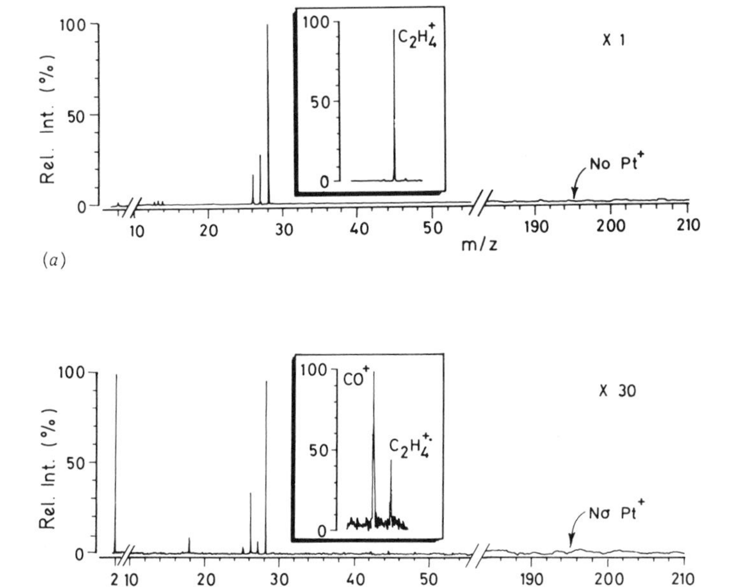

Figure 2.49. Laser-induced thermal desorption FT mass spectrum in the positive ion detection mode from ethylene on a platinum surface (a) before and (b) after annealing to 400 K. Reprinted from Sherman et al. (1987) with permission of Elsevier Science Publishers.

is observed. Note that these spectra are recorded from a one-monolayer adsorbate, probed over an area of about $0.5\,mm^2$ (Land et al., 1990).

2.7. LASER DESORPTION AND IONIZATION IN A TWO-STEP PROCESS

2.7.1. Inorganic Applications

Some of the previously described methods involving the postionization of sputtered neutrals are or can be equipped with laser sources to perform the initial sampling of the solid material. This holds true in particular for the SALI technique. However, additional information is readily available through the aforementioned references. Hence, these techniques will not be treated in further detail because it is more profitable to present a few different approaches.

The use of laser ablation coupled with RIMS detection of sputtered neutrals has a number of interesting advantages and applications. First of all, the duty cycle of sampling and detection is optimized. Indeed, the combination of short-pulsing low-repetition-rate lasers with a continuous supply of analyte in the gas phase by thermal evaporation or continuous beam excitation of solids implies a substantial loss of analyte. In principle, taking a beam radius of 2.5 mm and an atomic velocity of $5 \times 10^4\,cm \cdot s^{-1}$, the sample turnover could be as high as 100 kHz (Nogar et al., 1985). Laser ionization at 10–100 Hz means an efficiency of 0.1% only. The pulsed production of neutrals allows ample progress in terms of the required sample quantity. This argument not only applies to laser ablation as the first step but also to a pulsed primary ion beam, as used in SALI or SIRIS. A second advantage is the obtainable spatial resolution for the sample probing step. Again, tight focusing is feasible for a laser as well as primary ion beams, but it is experimentally much easier in the first case. Last, photon-induced production of neutrals is entirely compatible with insulating materials, whereas the use of primary ions requires compensation of the charging effects.

It is important to separate both temporally and spatially the ablation and ionization events. Ideally, the desorption laser has a short pulse duration of less than 100 ns to maximize the spatial overlap between the ablation plume and the ionizing laser field and hence to maximize the sensitivity. In addition, the short ablation pulse can be effectively treated as an impulse gate function in any TOF analysis. Variable wavelength is a desirable feature since the desorption of adsorbed species and the bulk ablation of substrate material

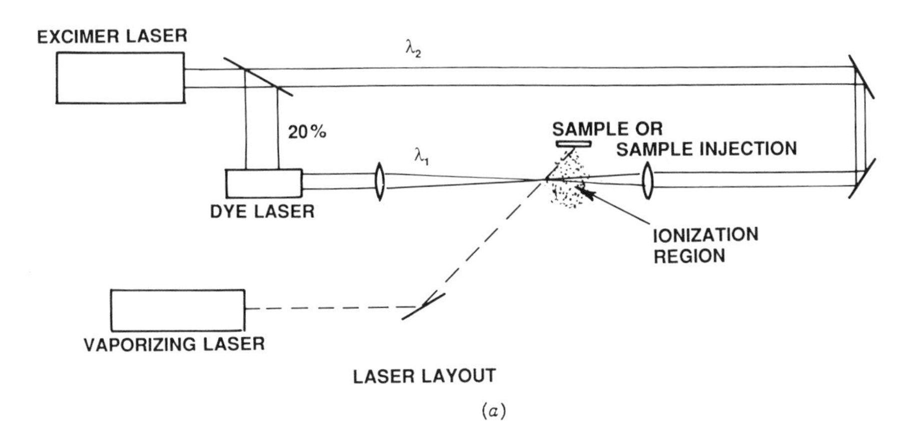

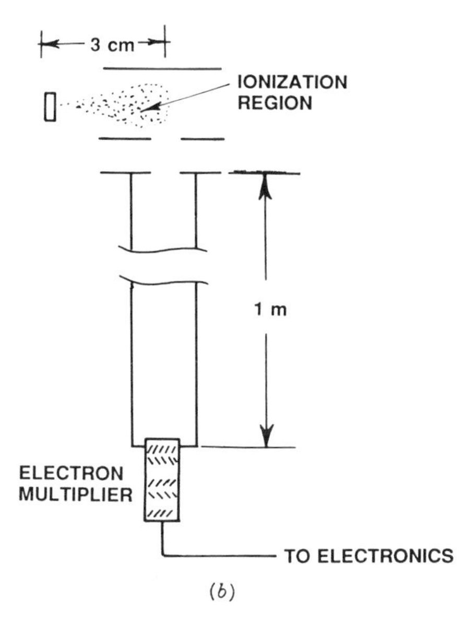

Figure 2.50. Schematic diagram of the laser layout needed to achieve resonance ionization TOF-MS. (a) Laser system; (b) TOF system. Reprinted from Williams et al. (1984) with permission of the American Chemical Society.

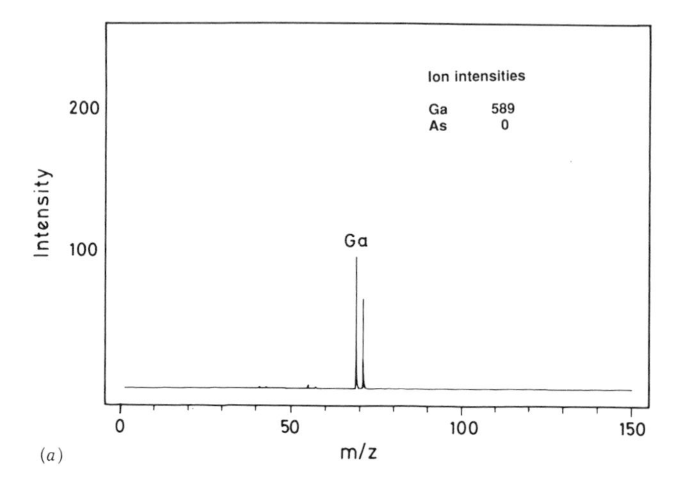

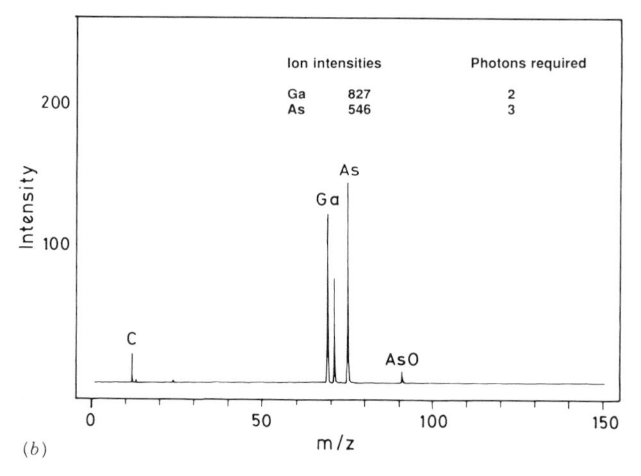

Figure 2.51. Positive ion mass spectrum from high-purity GaAs, recorded by TOF-LMMS without (a) and with (b) postionization. Reprinted from Schueler and Odom (1987) with permission of the American Institute of Physics.

are sensitive to the incident laser frequency and absorption bands. Many of these features are appropriate for the ionizing laser as well. Nevertheless, broad tunability is now considered essential to ensure a wide range of detectable species through various ionization schemes.

The first application on laser ablation and RIMS was reported by Mayo et al. (1982). Beekman et al. (1980) proposed LD-REMPI-TOF-MS for the isotopic analysis of solid samples. The isotopic shifts for atoms with Z between 10 and 100 are less than the Doppler line widths at temperatures where the vapor pressure is about 1 torr. Under these conditions, the RIS processes all isotopes of an element simultaneously. If the RIS is saturated for all isotopes of the element under investigation, then the ionized atoms will be an accurate isotopic representation of the sample. The instrument is schematically represented in Figure 2.50 (Williams et al., 1984). A Nd:YAG is used for ablation and an excimer for the ionization. Part of the output is coupled to a tunable dye laser, which pumps the neutrals to an excited state, and the rest achieves the ionization. Subsequent work in this field has recently been reviewed (Nogar and Estler, 1990). However, the method does not yet compete successfully with other techniques such as RIMS on the sputtered neutrals generated by primary ion beams.

Nevertheless, Odom and Schueler (1990) have exploited the idea on a commercial TOF-LMMS. Hereby, most of the possible advantages, which are mentioned before, have been combined in a really practical analytical tool for local analysis. However, the ionizing laser is not tunable, which prevents the application of RIS conditions. The instrumentation is basically a LIMA® 2A commercial microprobe, which was extensively discussed in Section 2.3. The frequency-quadrupled Nd:YAG 266 nm laser beam, which is focused through the Cassegrain system, ablates the sample with a lateral resolution of about 2 μm. A similar laser is used for the ionization. This beam passes parallel to the sample at a distance of about 0.6 mm. The focal diameter is 100 μm, and the photon flux density is typically 10^{28} $cm^{-3} \cdot s^{-1}$. The delay between the ablating and ionizing laser can be adjusted up to 2.5 μs. Figure 2.51 illustrates the mass spectra recorded for GaAs with and without application of the postionizing beam (Schueler and Odom, 1987). The large difference in ionization potential between Ga and As explains the absence of the latter element in the mass spectrum when no postionization is applied. Precise timing between the two lasers is critical. Figure 2.52 shows the use of postionization to remove the disturbing signals from organic species on a sample of $MoSi_{2.25}$ (Odom and Schueler, 1990). Such contaminants on the surface are typically introduced by various chemical processing steps and/or inadvertent handling. Their presence in the mass spectrum frequently prevents the use of isotope patterns. Postionization of the laser-ablated neutrals generates new interfering contributions, however, such as the signals from $^{92}Mo^{2+}$, which overlap with the ^{46}SiO cluster ions.

2.7.2. Organic Applications

One of the major challenges an analytical chemist has to deal with still is the identification and structural characterization of a thermolabile, polar,

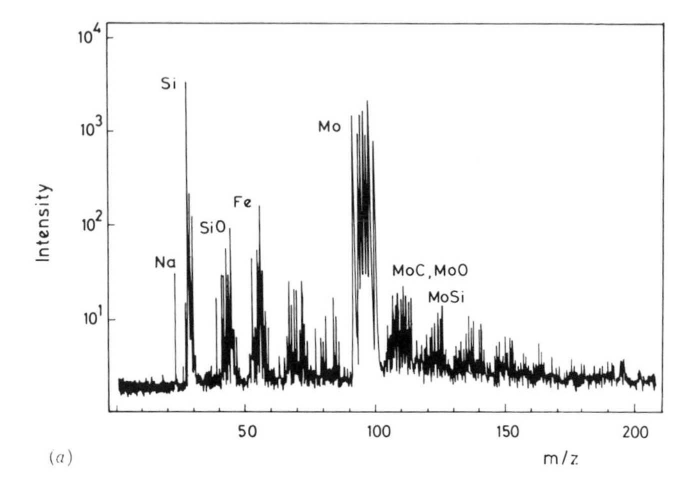

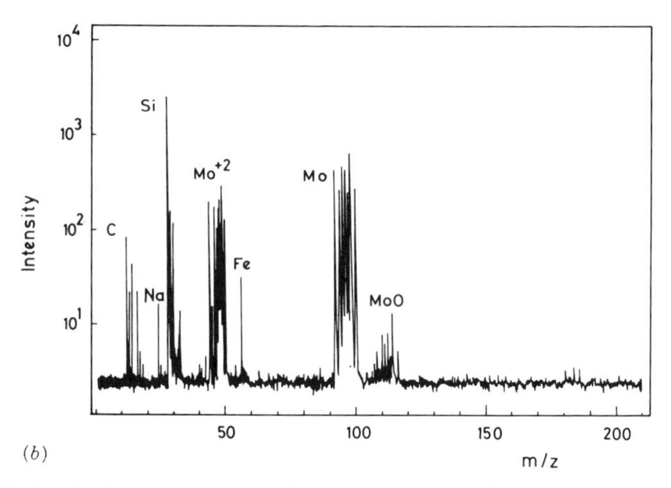

Figure 2.52. Positive ion mass spectrum obtained from an organically contaminated surface of $MoSi_{2.25}$ by laser DI (a) and additional postionization (b). Reprinted from Odom and Schueler (1990) with permission of Oxford University Press, Inc.

and/or high-molecular-weight organic compound available in extremely small quantities and often accompanied by impurities or other constituents. Ideally, the MS should be able to yield information on the molecular weight as well as on the functionalities present in the molecule. Hence, a soft ionization technique alone does not provide a satisfactory solution. Since the generation of fragments always implies reduction of the parent peak intensity, a certain control over the degree of fragmentation is required, unless a combination of several techniques is employed. Moreover, the MS should allow measurements in the high mass range with high mass resolution and accurate mass determination capabilities. The need for sensitivity is obvious, and a high degree of selectivity is desirable to eliminate purification steps. Such a long-awaited technique is not yet available, but recent investigation and development of pulsed laser desorption combined with MUPI and TOF-MS represents significant progress in that direction.

These instruments employ ultrafast heating of solid samples by laser irradiation, allowing neutral molecules to be desorbed without thermal decomposition even from labile molecules. The subsequent cooling by a pulsed supersonic jet has already been discussed. The REMPI process provides the ability to tune the amount of fragmentation by simply controlling the photon intensity. Finally, the TOF-MS analysis is performed on ion bunches, of which the velocity distribution and the time spread is largely optimized. As a result, a rather high mass resolution is attainable even without an ion reflector.

Figure 2.53 illustrates the main operational subunits of the commercial Bruker RETOF instrument (Bremen, Germany) (Grotemeyer et al., 1986a). A small pulsed CO_2 laser delivering 100 mJ per pulse of 27 μs is initially used for ablation of solid samples. No particular attempt at fine focusing the desorbing laser is made so far. The emitted neutrals collide with the supersonic jet of Ar atoms produced through a pulsed nozzle. The pulse length is about 200 μs, and the repetition rate is 2–10 Hz. A skimmer separates the desorption subunit and the ionization chamber. The repeller electrode removes the ions formed during the initial irradiation of the solid sample. The ionization is then achieved by an excimer pumped dye laser producing 2 J pulses of 5 ns. Frequency doubling gives 150 μJ in 5 ns pulses between 300 and 260 nm. The wavelength is matched to the intermediate energy level of the analyte molecules. Ionization then results from the absorption of an additional photon. High ionization efficiencies can be obtained using low laser intensity levels. Roughly 5% of the neutrals can be ionized, with ionization cross sections of 10^{-18}–10^{-17} cm^2 for the two absorption steps and a 1 kW, 5 ns laser pulse. As a result, further photon absorption by neutrals or ions is avoided to eliminate an extensive degree of fragmentation. Selection of an appropriate wavelength allows one to keep the excess energy in the molecules above the ionization threshold strictly limited to maintain the parent peak intensity. Daughters are introduced in the mass spectrum simply by increase of the laser energy. Mass analysis is performed by a TOF system with a 1 m drift tube and ion reflector. A 200 MHz transient recorder is used to digitize

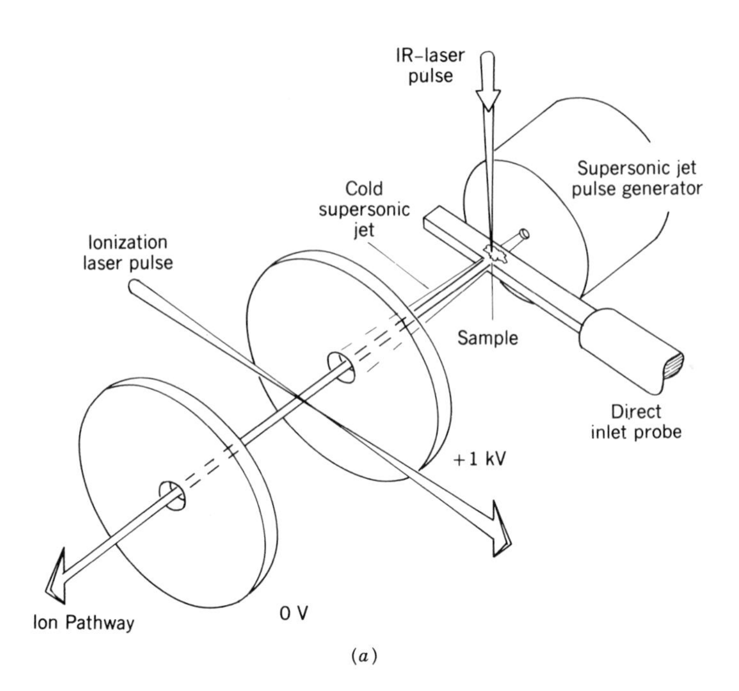
IR–laser
pulse
Supersonic jet
pulse generator
Cold
supersonic
jet
Ionization
laser pulse
Sample
Direct
inlet probe
+1 kV
Ion Pathway
0 V

(a)

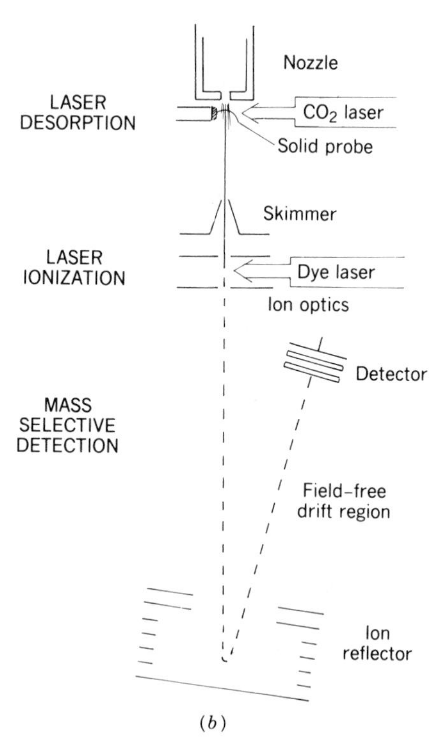
Nozzle
LASER
DESORPTION
CO2 laser
Solid probe
Skimmer
LASER
IONIZATION
Dye laser
Ion optics
Detector
MASS
SELECTIVE
DETECTION
Field–free
drift region
Ion
reflector

(b)

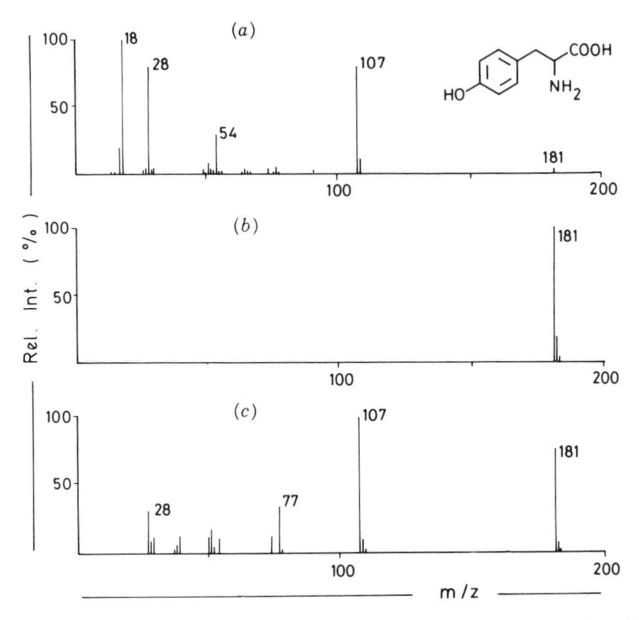

Figure 2.54. Comparison of EI (a) and MUPI [(b) low flux density; (c) high flux density] positive ion mass spectra of the amino acid tyrosine. Reprinted from Grotemeyer and Schlag (1988) with permission of John Wiley & Sons, Ltd.

and store the signals. More recently, a 10 ps laser has been used to improve the ionization step. Unit resolution in the mass range up to 5000 has been demonstrated (Grotemeyer and Schlag, 1987).

From 1986 onward, a series of publications has described the application of the method to a variety of compounds such as organic complexes as well as selected biomolecules such as chlorophyls, a range of amino acids, and protected nucleosides and oligonucleotides (Grotemeyer et al., 1986b, 1987; Grotemeyer and Schlag, 1988; Beavis et al., 1988). Figure 2.54 provides a good illustration of the tunable degree of fragmentation attainable for a simple molecule by LD-MUPI (Grotemeyer and Schlag, 1988). The data from conventional EI-MS are included as well. It should be noted, however, that the observed difference between the parent vs. fragment ion production in both methods is not only affected by the ionization technique but also by the precise method of sample introduction. Thermal decomposition on a

Figure 2.53. Scheme of the LD-REMPI RETOF mass spectrometer. (a) Source; (b) diagram of entire system. Reprinted from Grotemeyer et al. (1986a) with permission of John Wiley & Sons, Ltd.

direct probe may occur. Additional data also show that, depending on the structure actually investigated, fragmentation cannot be avoided completely. This holds true in particular for compounds such as oligonucleotides (Lindner et al., 1990). Incomplete cooling and resonance exchange between given moieties can occur besides the excess amount of energy deposited in the molecule above the ionization threshold as a result of the selected wavelength. However, the REMPI approach on separately desorbed neutrals remains an effective tool for the investigation of several important biomolecules.

The application of separate desorption and ionization steps with jet cooling largely optimizes the critical features of the ion bunch before TOF-MS separation. Hence, the use of an ion reflector is not entirely mandatory and home-built instruments use linear drift tubes (Engelke et al., 1987; Lubman, 1988; Lubman and Li, 1990). Intense further research in this field is being carried out by Lubman's group at the University of Michigan. (Extensive additional information will be found in his Part C of Chapter 3 of this volume.)

2.8. LASERS AT OTHER STAGES OF THE MS EXPERIMENT

The use of lasers in mass spectrometry is not limited to the initial step of ion formation. The unique features of lasers as versatile and experimentally convenient sources of excitation with high monochromaticity, intensity, and directionality in combination with continuous progress in laser technology and the wide availability of lasers have resulted in the development of numerous additional applications. The purpose of this section is to discuss a few examples of multistage MS experiments involving photodissociation. Additionally, the use of lasers in the ionization process, insofar as the interaction does not occur directly with the analyte, will be presented as well. These topics are particularly relevant in view of the increasing use of FTMS and ion trap analyzers. More and more laser applications at different MS stages are foreseeable in the near future.

The field of multistage MS has grown impressively during the last few years. The technique exploits two or more mass-separating steps so that the mass-selecting ions exiting from one MS can be modified by dissociation or reaction prior to the mass analysis of the resulting ions. The most currently used form, namely, tandem MS, has two distinct but related fields of application: the first is structural elucidation; the second is analysis of mixtures where the first mass analysis takes over the role of the commonly used GC or LC preseparation steps. The coupling to modern ionization techniques, as in fast atom bombardment (FAB), necessitates the use of sequential techniques. The first step eliminates or minimizes the chemical interference from the matrix and isolates the parent ion. Structurally specific information is then gained from the subsequent fragmentation.

The most commonly used method to further characterize the ions from the first mass selection is collison-induced dissociation (CID). Grazing collisions of kilovolt ions convert part of the ion's translational energy into internal energy, resulting in unimolecular fragmentation like that found in normal EI mass spectra. Low-energy collisions can be more efficient in depositing internal energy into the molecule. Alternatively, photodissociation exploits the fine tuning of the additional excess of energy imparted to the ions under study. This approach becomes particularly interesting when large complex molecules are involved (Beynon et al., 1982; Fukuda and Campana, 1985). More sophisticated techniques exploit the neutralization reaction (Wesdemiotis and McLafferty, 1987; Terlouw, 1989). Finally, a quite popular technique exploits ion–molecule reactions to yield information concerning specific functionalities and chemically reactive sites in the analyte ions. The recent proliferation of FTMS and other ion storage MSs has contributed to the exponentially growing interest in multistage experiments. Nevertheless, the other types of MS analyzers, specifically magnetic sector, TOF, and quadrupole, are also used. Within the context of this section attention will be focused primarily on laser-induced photofragmentation of mass-selected ions. CID experiments on laser-desorbed or non-laser-generated ions in the ion source have been mentioned previously.

As regards MS/MS experiments, distinction should be made between instruments that separate ions in space and in time. Time-dependent ion transport analyzers are obviously less popular for photodissociation studies, because the interaction time between the ion and photon beam is inevitably limited and in most cases too short to achieve efficient excitation and fragmentation. Hence, the majority of the research in this field is performed on ion storage instruments, but some experiments still are conducted on magnetic sector and TOF-type systems.

2.8.1. Multistage Experiments in Magnetic Sector MS

Several double-focusing magnetic MSs have been modified for laser-induced photofragmentation of gas phase ions. Both normal and reverse geometry instruments have been used. An AEI MS 902 (Kratos, Manchester, UK)–based instrument has been developed at Texas A&M University to study photodecomposition in the first field-free region between the ion source and the electrostatic analyzer (Krailler et al., 1985). Photofragmentation of mass-analyzed ions has also been performed on a reverse geometry VG ZAB-2F instrument (Jarrold et al., 1983, 1984). The latter setup provides lower sensitivity in comparison with the previous one, is better suited to measure

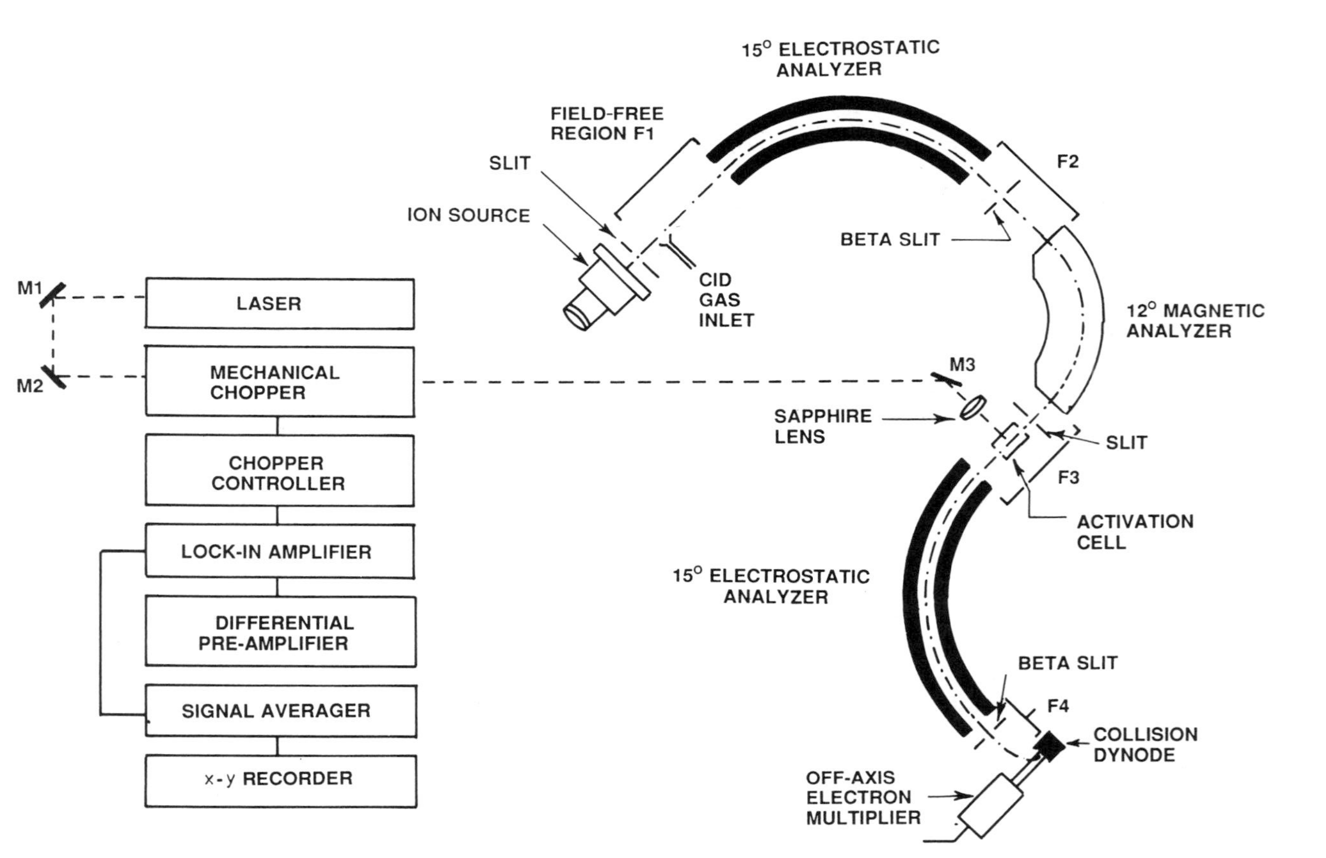

Figure 2.55. Scheme of laser ion beam photodissociation MS/MS instrumentation. Reprinted from Tecklenburg et al. (1987) with permission of the American Chemical Society.

photokinetic energy distributions, and can also provide information on product angular distributions (Krailler et al., 1985). More recently, a Varian MAT 731 Mattauch-Herzog instrument has been adapted to permit the use of soft ionization techniques such as field ionization (FI), field desorption (FD), and Cs^+ ion bombardment, in combination with subsequent photofragmentation (Welch et al., 1986).

Figure 2.55 shows the schematic diagram of a Kratos MS-50 TA triple mass analyzer, based on the conventional combination of electrostatic and magnetic sectors (Tecklenburg and Russell, 1987). Gas phase ions have been produced in the source either by electron ionization and later also by FAB (Tecklenburg et al., 1987). The beam of an argon ion laser is used directly or is coupled to a dye laser, covering the wavelength range between 550 and 675 nm. A mechanical chopper allows photofragmentation to be performed either in the first field-free region between the acceleration plates and the first electrostatic sector or in the collision cell between the magnetic and the final electrostatic analyzer. However, in the first case, the beam passes through the ion source and hence photon absorption is not confined to the field-free region. The use of a chopped laser beam allows separation of the photon-induced signal from the metastable and collision-induced products by means of a lock-in amplifier. Multiple scans, typically up to 32, have to be accumulated. The specially designed biased activation cell with adjustable voltages on the entrance and exit lenses permits a significant improvement of the signal-to-noise (S/N) ratio and hence drastic shortening of the data acquisition time (Tecklenburg et al., 1989). The formation of odd-electron fragments from protonated parents upon UV and visible photon excitation has been demonstrated for dinitrophenyl derivatives of some amino acids (Tecklenburg et al., 1987).

2.8.2. Multistage Experiments in TOF-MS

Several photofragmentation studies using TOF systems have been reported recently, most of which are performed on inorganic cluster ions. A tandem TOF system has been constructed to study the photodissociation of $(CO_2)_n^+$ clusters (Johnson et al., 1984). A special technique has been applied to generate large clusters without internal excitation by nucleation onto the monomer during the free jet expansion. The ion clusters are allowed to drift in a 10 cm field-free region and are subsequently introduced into the TOF-MS by a drawout pulse, which provides the time origin of the mass spectra. A Nd:YAG dye laser beam is introduced coaxially with the drift tube to photodissociate the clusters during their free flight. The reflecting field separates the parent from the daughter ions and produces a so-called secondary TOF mass spectrum. The decay of ions within the drift region produces an energy deficiency proportional to the mass ratio of daughters over parents. If this difference is small enough, then the reflector action compensates the flight time and sharp daughter ion peaks are still detected. Otherwise, the daughters follow a shorter path and arrive at the detector before their parents. As a result, the daughter signals before each parent peak become a secondary TOF spectrum, which is extending to shorter times the more the mass difference between parents and daughters increases. This of course implies that the acceptance area of the detector is sufficient because parent and daughter ions follow spatially resolved trajectories in the second drift tube. This approach has previously been exploited for the study of metastable decay of organic ions (Boesl et al., 1982). The instrument makes profitable use of the spatial focusing that can be achieved by the drawout pulse technique.

The cluster ion generation gives a broad spatial distribution with a beam cross section of about 25 cm^2 after the first field-free region. Time-lag focusing enables the parents to be obtained with a resolution of 120, in spite of the large energy spread, around 500 eV, for the ions when introduced into the MS. The mass resolution of the secondary daughter ion spectrum is typically 50. Note that the photodissociation in this experiment is not performed on one given parent ion. In fact, the initial ion bunch contains clusters of several sizes and all these species are decomposed during their flight. Hence, each individually registered mass spectrum contains the information from numerous parent/daughter systems.

The photodissociation of mass-selected parents in tandem TOF systems has been achieved by the use of gating pulse techniques. Figure 2.56 illustrates the setup developed at Rice University, Houston, Texas (O'Brien et al., 1988). A short extraction pulse introduces a sample from the cluster ion beam into the first 3 m drift tube. The static deflectors reduce the molecular beam velocity, and the einzel lenses serve focusing. The pulsed mass gate at the end of the first TOF consists of three electrodes. Rejection takes place when the outer two are grounded, the inner one being kept at a potential higher than the energy of the ions arriving from the field-free region. If the central grid is grounded as well, then ions are injected into the photodecomposition region. A decelerating field at the entrance of this chamber is mandatory to increase the interaction and subsequent reaction time. Nd:YAG or excimer laser excitation is applied depending on the application and wavelength required. The ions continue to drift for several centimeters into the extraction region of the second TOF-MS, which is mounted perpendicularly to the first one. The tube length is now 1 m. The turning radius—and hence the time spent on that trajectory—depends on the ion mass. The extraction pulse duration must be optimized in addition to the standard time-lag focusing procedures (Zheng et al., 1986). A pair of microchannel plates are used for detection. Some other

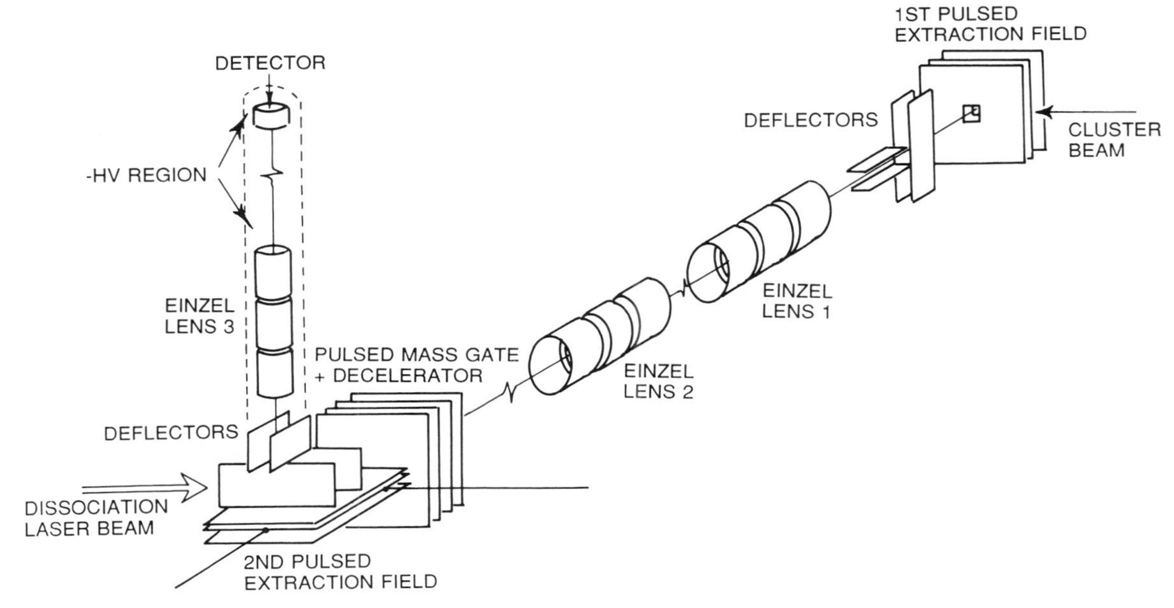

Figure 2.56. Overview of the tandem TOF-MS for photofragmentation studies. Reprinted from O'Brien et al. (1988) with permission of the American Institute of Physics.

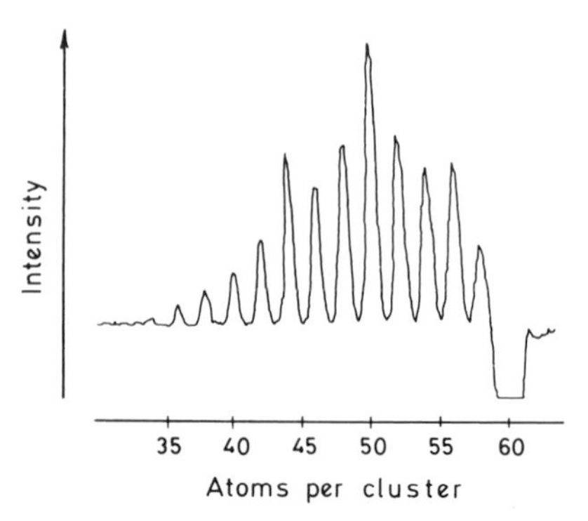

Figure 2.57. Positive daughter ion mass spectrum of C_{60}^+, mass selected and fragmented with $15\,mJ\,cm^{-2}$ ArF excimer light. Reprinted from O'Brien et al. (1988) with permission of the American Institute of Physics.

groups actively involved in this field have constructed largely similar instruments according to the same functional principles (Geusic et al., 1986, 1987; Cox et al., 1986).

Figure 2.57 shows the fragments from a mass-selected C_{60}^+ cluster, irradiated by a 193 nm ArF excimer beam at $15\,mJ \cdot cm^{-2}$. The negative peak comes from the subtraction of the signal intensities, recorded during alternating experimental cycles with the laser on and off (O'Brien et al., 1988). Additional work has been performed on semiconductor clusters such as GaAs, Si, and Ge anions, and several metal cluster ions from iron, nickel, and niobium (Liu et al., 1986; Brucat et al., 1986).

To overcome the critical timing problems encountered in the aforementioned systems, a reflectron-based design has been elaborated (LaiHing et al., 1989; Wiley et al., 1990). A pulsed mass gate is inserted after the first 1.3 m flight tube before the reflecting region. The 12° reflector follows the current construction scheme. Additional windows are provided for the excimer beam. The laser irradiates the selected ion packet at the peak of its trajectory, where the average forward velocity of the ions is minimal, which reduces the timing problem of laser firing. The mass spectra are then recorded after subsequent analysis in a 1 m drift tube. Up to 800 alternate cycles with the laser on and off are accumulated. Resonance-enhanced photodissociation studies have been performed with this system, complemented by an excimer pumped dye laser (Cheng et al., 1990).

2.8.3. Multistage Experiments in Quadrupole Ion Trap MS

To our knowledge, there are virtually no applications reported on the use of triple quadrupole instruments for study of laser-induced photofragmentation. One of the reasons probably is the advent of ion storage techniques. Quadrupole ion trap systems have encountered tremendous interest in many applications as versatile, low-cost, and user-friendly mass analyzers, capable of real multistage experiments up to 11 consecutive daughter generations from only one ionization event (Louris et al., 1990). Recently developed scan programs for combined application of dc and RF fields allow the ejection of all ions except the parents of interest from the ion trap, and hence unambiguous MS^n data can be obtained. An extremely elegant way to perform photodissociation in an ion trap analyzer is illustrated in Figure

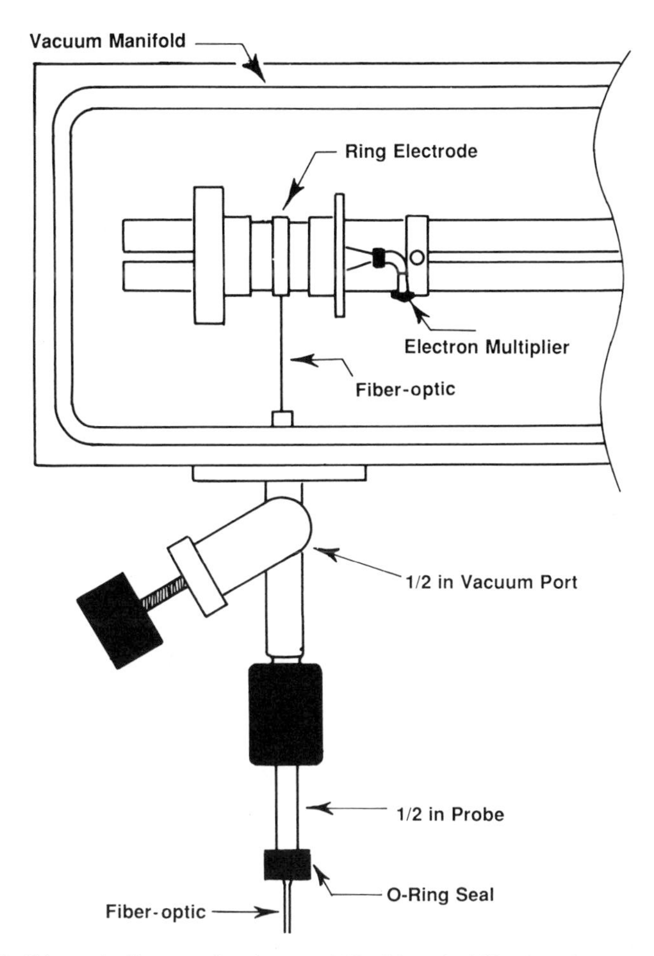

Figure 2.58. Schematic diagram of an ion trap MS with optical fiber interface. Reprinted from Louris et al. (1990) with permission of Elsevier Science Publishers.

2.58 (Louris et al., 1987a). A 1.3 mm aperture is drilled in the ring electrode, and the laser beam is introduced through a plastic-clad fused silica core optical fiber. The light exits as a strongly divergent beam and strikes the opposite side of the ring electrode, which is a convex, highly polished surface. Hence practically the entire volume of the ion trap is illuminated either directly or by reflection. Several laser pulses, delivering 5–20 mJ each, can be fired consecutively at 10 Hz to perform efficient dissociation of the stored ions.

Figure 2.59 shows a comparison between photodissociation and CID spectra for the molecular ion of *n*-butylbenzene (Louris et al., 1987a). The example is a well-studied test sample to verify the conditions and energy deposition in ion dissociation experiments (Louris et al., 1987b). Depending on the wavelength and the power density of the exciting radiation, photodissociation efficiencies for the molecular ions of up to 98% can be obtained whereas CID experiments yield a 90% figure. The ratio of the odd-electron fragments at m/z 92 vs. the even-electron systems at m/z 91 shows that photodissociation permits the deposition of more energy into the parents than does collisions with the noble gas as used for CID. Moreover, the degree of fragmentation can be additionally varied by selecting the photon energy and beam intensity. It should be noted that photodissociation, as opposed to CID, also allows an accurate control of actually deposited excess energy, as opposed to CID. This feature has been the driving force behind the application of photoionization in fundamentally oriented MS studies, even before the

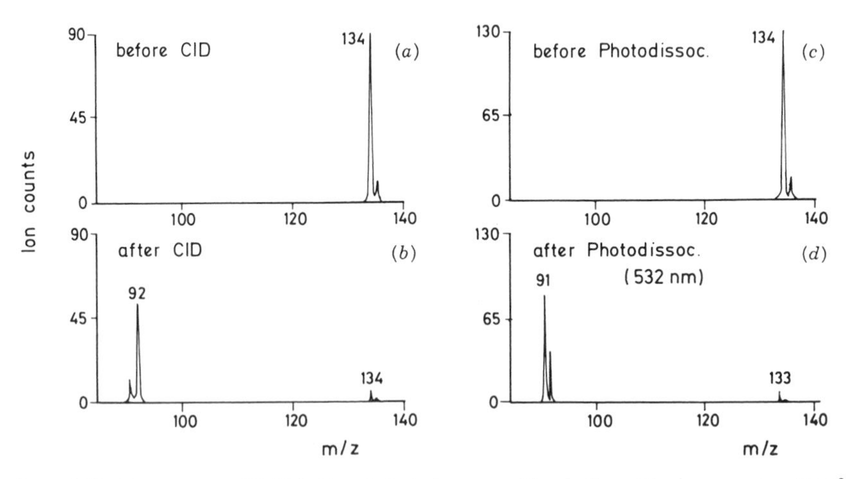

Figure 2.59. Comparison of CID (a,b) and photodecomposition (c,d) positive ion mass spectra of *n*-butylbenzene. Reprinted from Louris et al. (1990) with permission of Elsevier Science Publishers.

advent of lasers made the experiments more accessible and convenient to perform.

2.8.4. Multistage Experiments in FTMS

The wide applicability of laser irradiation to a variety of tasks during a multistage experiment is perhaps most convincingly evidenced by reports in the current FTMS literature. The large number of recently published papers reflects the excellent match between both technologies. As it is not possible to give a more comprehensive review within the scope of this contribution, our attention will be confined to applications concerning photofragmentation, generation of selective reagent gases for chemical ionization, and synthesis of labile short-lived compounds within the cell of the MS analyzer.

There are fundamental reasons why photodecomposition studies are the most intensively performed on FTMS instruments. Indeed, the analyzer cell must be operated at extremely low pressure, less than 10^{-8} torr, to avoid collisions between the ions and neutral gas molecules. This makes the instrument in principle less ideal for CID experiments, even though in practice the problem can be adequately solved by elaborating suitable compromises between the conflicting aspects. Nevertheless, photon-induced excitation of mass-selected parents offers, apart from the fine tuning of the actually deposited internal energy, a particularly appealing solution for two specific problems in FTMS: first, the vacuum conditions are not affected, as opposed to CID; secondly, sufficient excitation of high-mass parents can be more easily achieved by photons than by collisions, since the kinetic energy of these ions is rather low.

Photodecomposition experiments have been conducted in both single and dual cell instruments with ionization inside the cell region as well as with external sources. To complement the previously mentioned instruments, the use of a tandem quadrupole FTMS has been selected to stress the benefits of photodissociation for obtaining sequence information on picomole amounts

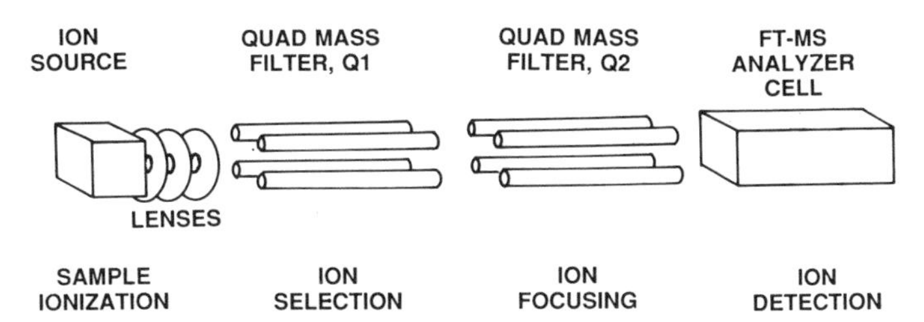

Figure 2.60. Scheme of the tandem quadrupole FTMS. Reprinted from Hunt et al. (1985a) with permission of the American Chemical Society.

of peptides. Figure 2.60 gives the schematic diagram of the instrument (Hunt et al., 1985a,b). Ion generation of large biomolecules is performed by bombardment of the sample in a suitable liquid matrix with 10 keV Cs^+ ions on the gold-plated tip of a stainless steel probe. The source, lens system, and first quadrupole mass filter are standard Finnigan MAT components. The second set of quadrupole rods primarily serves to guide the ions over a distance of 86 cm through the fringing field of the 7 T superconducting magnet into the cell. Differential pumping maintains a vacuum better than 10^{-9} torr in the guiding quadrupole and ion cyclotron resonance (ICR) cell. The quadrupole drivers have been modified to obtain a high mass cutoff above 10 kDa, with a low mass value to be selected anywhere below 3 kDa. The 193 nm ArF excimer laser beam for photodissociation is brought in from the back of the magnet and enters the ICR cell via an aperture in the trapping plate, which is covered with a 90% nickel screen.

Figure 2.61 illustrates the photodissociation mass spectrum, generated from the protonated parent ions of a 15-residue tryptic peptide with relative molecular mass of 1771 (Hunt et al., 1987). A sweep-out pulse has been employed to eject most of the ions below m/z 1600. A single 193 nm pulse has been applied to generate the daughter ions. Up to 50 cycles have been accumulated to produce the represented mass spectrum. Yet, the entire data set is obtained from the same 10 pmol sample. The acquisition takes less than 20 s. The mass assignment is within 0.2 amu of the expected values. The results have permitted unambiguous determination of the peptide sequence, except for one residue. Additional examples demonstrate the method's powerful capabilities to generate fragment ions indicative of the structure (Shabanowitz and Hunt, 1990). Sensitivity is certainly high considering the amount of information generated. Indeed, these data actually concern a two-stage MS experiment with high front- and back-end resolution, which is unattainable on many hybrid instruments.

Multistage FTMS and detection of granddaughter ion mass spectra by photodissociation has been performed on porphyrins, metalloporphyrins, and alkaloids (Nuwaysir and Wilkins, 1989). An extensive review of the recent literature in this field lies beyond the scope of this chapter.

A major analytical goal of CI-MS analysis of mixtures is the selective ionization of constituents with minimal fragmentation and maximal chemical discrimination. In fact, the chemistry of the gas phase ion–molecule interactions may provide an elegant way to substitute or at least assist the chromatographic preseparation step in conventional MS analysis (Kondrat and Cooks, 1978). The concept also holds the potential for an added dimension in the application of tandem MS of complex samples. The use of metal ions as reagents in CI has been relatively little investigated but is now emerging as a new and developing area of current research. The role of FTMS used in conjunction with lasers is undeniable here.

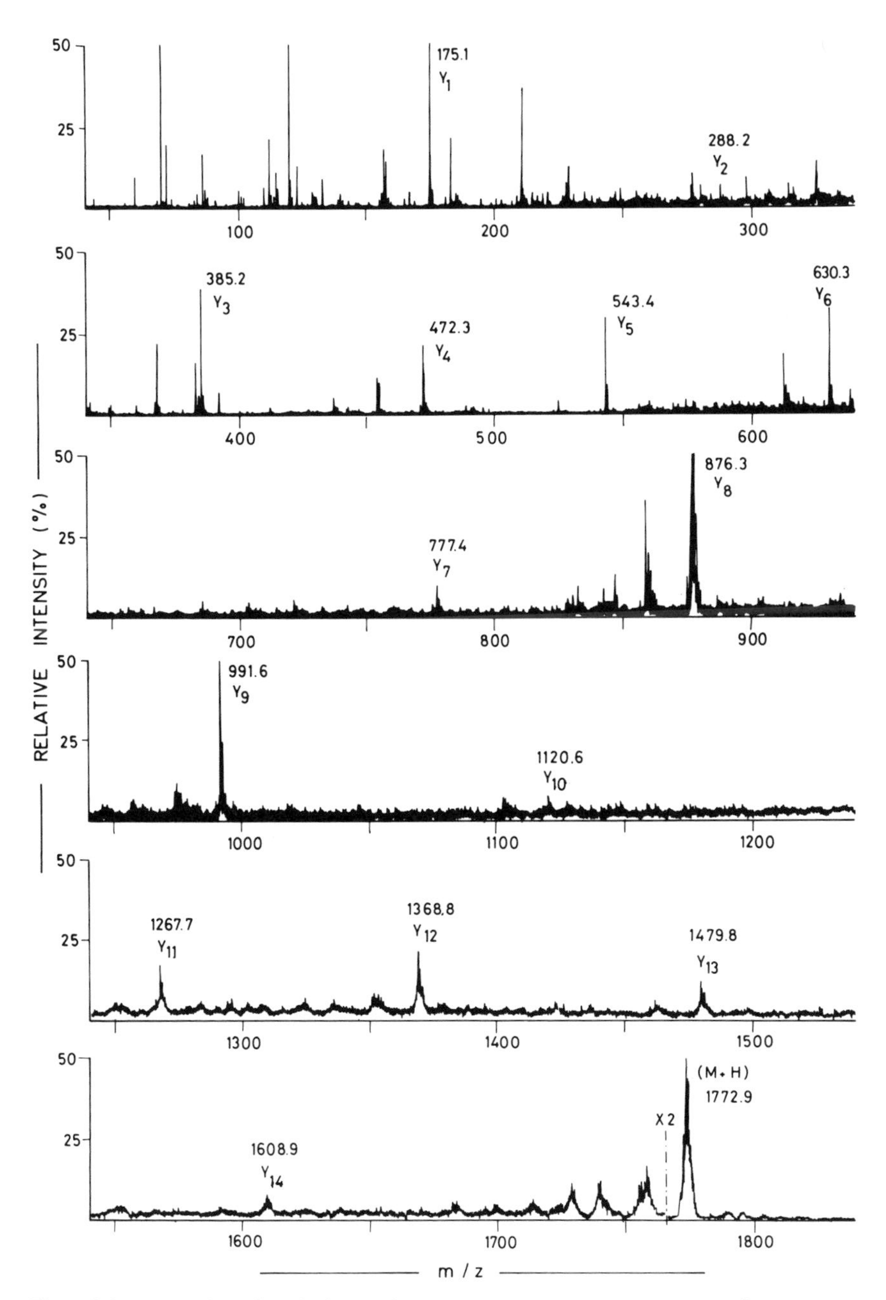

Figure 2.61. Laser photodissociation positive ion mass spectrum of the $(M+H)^+$ ion from 10 pmol of a tryptic peptide. Reprinted from Hunt et al. (1987) with permission of the Royal Society of Chemistry.

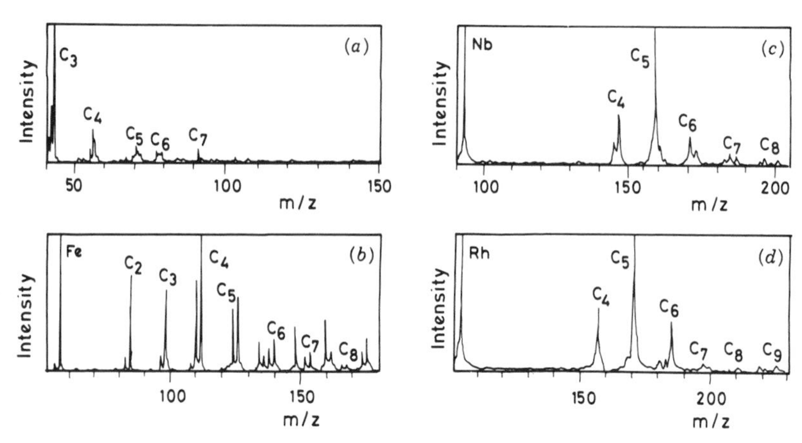

Figure 2.62. Positive ion mass spectra from a gasoline sample recorded by FTMS with EI (a) and CI using Fe^+ (b), Nb^+ (c), and Rh^+ (d) as reagent. Reprinted from Freiser (1985) with permission of Elsevier Science Publishers.

The generation of metal ions to perform selective CI is particularly easy in FTMS by laser irradiation of metal foils or by decomposition of volatile complexes such as metal carbonyl compounds. Usually, the metal ion is then combined with an organic ligand by an ion–neutral reaction in the cell. Since the seminal paper by Allison et al. (1979), various studies have exploited the reactivity of bare metal ions to locate double and triple bonds, an important pursuit in petroleum, lipid, and pheromone chemistry (see references in Gross, 1989). Metal ions offer a choice between soft ionization, providing molecular weight information, and hard ionization, yielding fragment ions characteristic of structural features of the analyte. Various transition and other metal ions have been evaluated as potential CI reagents (Forbes et al., 1987; Chowdhury and Wilkins, 1987, 1988; Chowdhury et al., 1989). The analysis of simple mixtures of alkene and alkyne isomers by GC-FTMS and Fe^+ CI has demonstrated the capability to differentiate 3-heptene from 1- and 2-heptene (Peake et al., 1987). The adduct of the former compound with Fe^+ eliminates ethylene, whereas propylene is lost for the two other isomers.

Figure 2.62 illustrates the differentiation capabilities of a metal ion CI in comparison to EI for the light components in gasoline, introduced by a simple leak valve (Freiser, 1985). The EI data display the alkyl series from C_3 to C_7, as well as signals referring to benzene and toluene. Application of FTMS with Fe^+ CI allows detection of the alkyl constituents up to C_9, as well as the two aromatic molecules, which are now clearly identified by the much higher intensity of the molecular ions. The data obtained by the use of Nb^+ and Rh^+ CI allow confirmation of the components' structure, because

$$RhC_5H_6^+ + c\text{-}C_5H_{10} \longrightarrow RhC_{10}H_{12}^+ + 2H_2$$

$$RhC_5H_6^+ \xrightarrow{h\nu} [RhC_5H_6^+]^* \xrightarrow{c\text{-}C_5H_{10}} RhC_{10}H_{10}^+ + 3H_2$$

$$RhC_5H_6^+ \xrightarrow[-H^{\cdot}]{h\nu} RhC_5H_5^+ \xrightarrow{h\nu} \text{other products}$$

Figure 2.63. Reaction schemes of $RhC_5H_6^+$ with cyclopentane under photoexcitation. Reprinted from Gord et al. (1989) with permission of the American Chemical Society.

neither Nb^+ nor Rh^+ tends to cleave C—C bonds. However, these metal ions do not provide information about the aromatic analogues in the mixture. The absence of an NbO^+ signal indicates that oxygen-containing compounds are not present in the mixture in significant amounts (Freiser, 1985).

Numerous examples exist in the current literature of the use of lasers in the synthesis or decomposition of organometallic compounds within the FTMS cell. A recent nice example of photoexcitation of an ion–molecule complex shows a combination of both (Gord et al., 1989). Laser irradiation of an Rh foil in an external source is applied to produce Rh^+. The initial starting product $RhC_5H_6^+$ has been obtained by reaction with cyclopentane. Thermalized $RhC_5H_6^+$ reacts further to yield $RhC_{10}H_{12}^+$, but under photoexcitation the species forms with cyclopentane the rhodocenium ion $RhC_{10}H_{10}^+$ or undergoes decomposition into $RhC_5H_5^+$ and other products. The schemes are summarized in Figure 2.63. Both reactions of the photoexcited $RhC_5H_6^+$ ions are competitive, which implies that the photodissociation proceeds slowly, either because the photon energy is near to the dissociation energy or because multiple photon absorption is required to produce dissociation. Figure 2.64 illustrates the positive ion mass spectra with and without laser irradiation. The production of $RhC_{10}H_{12}^+$ is observed as expected in the absence of an additional photon excitation. Laser irradiation of the ions in the cell allows detection of the photon-induced formation of $RhC_{10}H_{10}^+$ and $RhC_5H_5^+$. The former product could have been a photofragment of $RhC_{10}H_{12}^+$. Hence, continuous ejection has been employed to verify that the product was not produced from $RhC_{10}H_{12}^+$. The time scale used here is sufficiently short to prevent any significant photon absorption at the laser power used. This represents one of the very few examples of photon-induced reactivity during ion–molecule reactions in the gas phase.

In the field of organometallics, numerous examples are found of the synthesis of short-lived species inside the FTMS cell. Most frequently, laser desorption is used to generate the starting material, for instance, the bare

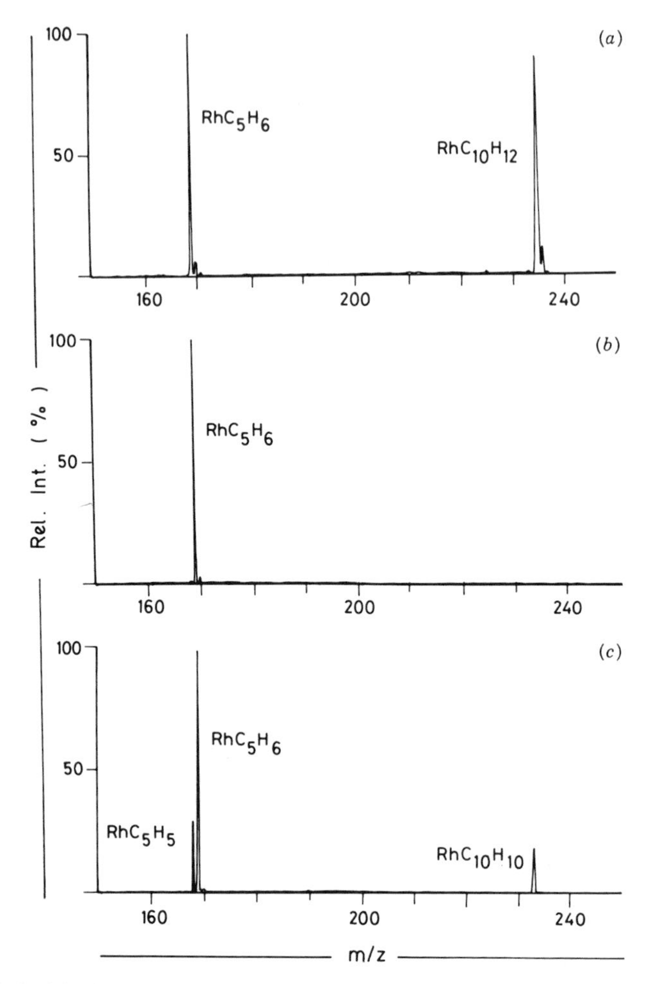

Figure 2.64. Positive ion mass spectra recorded by FTMS from $RhC_5H_6^+$ and cyclopentane (a) without irradiation, and under continuous ejection of $RhC_{10}H_{12}^+$ (b) without and (c) with irradiation. Reprinted from Gord et al. (1989) with permission of the American Chemical Society.

metal ions. Introduction of the ligands in the gas phase systematically allows building up of the complex via ion–molecule reactions. Striking experiments have involved up to 20 steps in the formation of large cluster ions (Forbes et al., 1988). A nice illustration of the effective potential of FTMS to assist synthesis, analysis, and fundamental physicochemistry is provided by the formation and decomposition of Cr^{2+} ions (Houriet and Vulpius, 1988). Figure 2.64 illustrates the mass spectra of the species in the cell at different

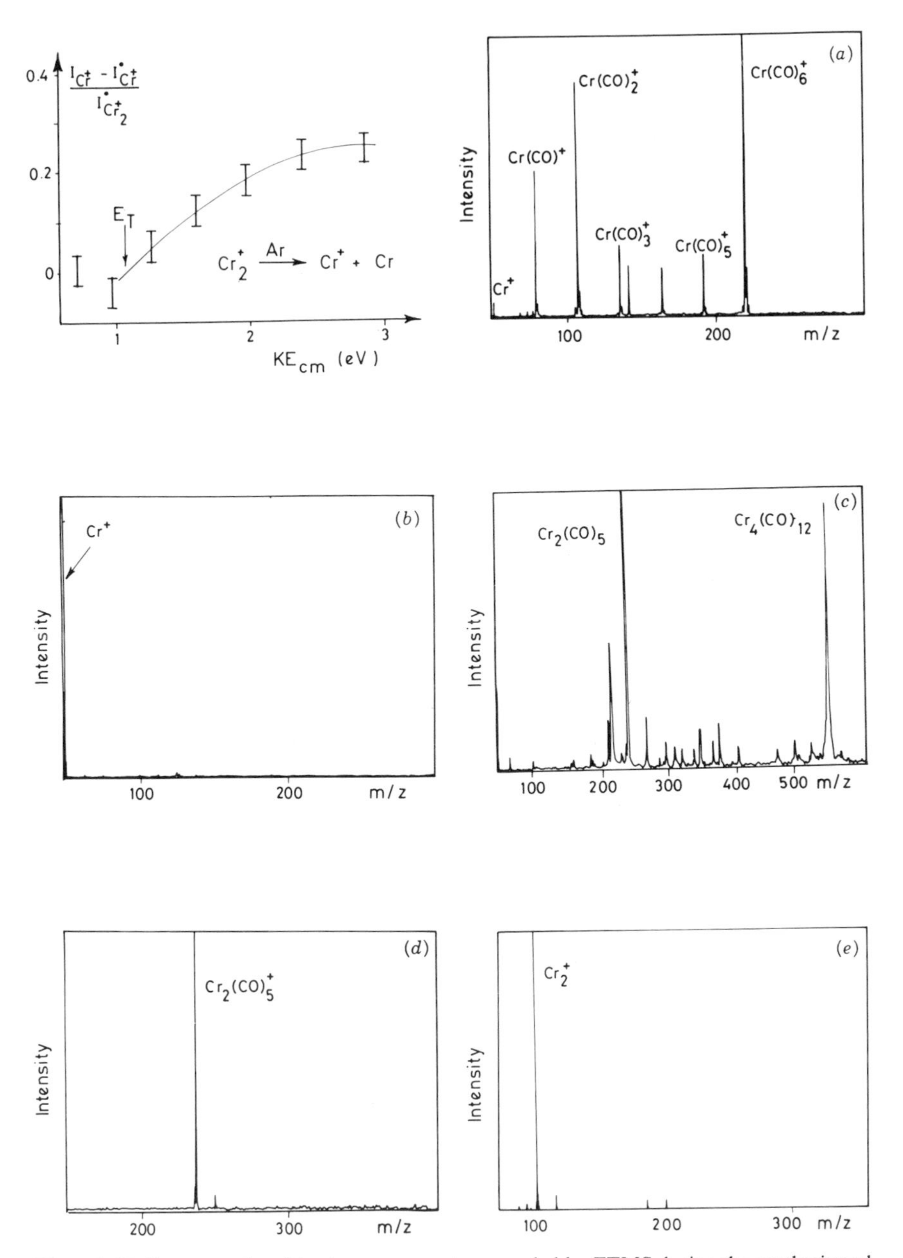

Figure 2.65. Sequence of positive ion mass spectra recorded by FTMS during the synthesis and collision-induced decomposition of Cr_2^+ and energy dependence of the latter reaction (a)–(e), different stages of the multistage experiment. Reprinted from Houriet and Vulpius (1988) with permission of Elsevier Science Publishers.

stages of the experiment. The sequence starts with the generation of Cr^+ ions by EI of $Cr(CO)_6$ and subsequent notch ejection to isolate Cr^+ in the cell. The reaction with $Cr(CO)_6$ then proceeds to form various products due to consecutive ion–molecule reactions. Time and pressure can be adjusted to optimize the formation of $Cr_2(CO)_5^+$ ions, which are then again isolated by notch ejection. The final step involves the photodecomposition of the metal ligand ions by irradiation with an argon ion laser pulse at 415.5 nm. The specific cleavage of the metal–carbonyl bonds occurs, accompanied by the production of some Cr^+ as well. The mass spectrum is then cleaned out by a third notch ejection, leaving only Cr_2^+ in the cell. The preparation and purification of Cr_2^+ is then completed, and the species are available for study of the affinity for various ligands, determination of the bond dissociation energy, etc. The results of the latter experiment are illustrated in the Figure 2.65e. The amount of dissociation is monitored as a function of the collision energy with the argon bath gas. It is evident that similar studies on isolated intermediates and short-lived species are required to complete our understanding of isolated vs. bulk metal reactivity.

The capabilities of FTMS for multistage experiments to generate, synthesize, purify, and characterize a variety of chemical species, stable and labile, remain one of the unique features of the method. Unlike ion transport MS systems, the basic FTMS hardware does not need extension or rebuilding for such experiments. Within given limits, it is the lifetime of the ions and not the instrumental hardware which determines the feasible number of preparation steps or of subsequent generations daughter ions. The ion trap technology gives access to subsequent daughter ion studies but does not as yet provide comparable possibilities to attempt in situ formation and purification of labile species.

ACKNOWLEDGMENTS

Luc Van Vaeck and Wim Van Roy are indebted to the National Science Foundation, Belgium (N.F.W.O), as research associate and research assistant, respectively.

LIST OF ACRONYMS

AC	alternating current
CEMA	channeltron electron multiplier array
CI	chemical ionization
CID	collision-induced dissociation
CW	continuous wave
DC	direct current
DCI	direct chemical ionization
DI	desorption ionization
DIMP	diisopropyl methylphosphonate
EAR	energy and angle refocusing
EI	electron ionization
FAB	fast atom bombardement
FD	field desorption
FI	field ionization
FID	flame ionization detector
FTMS	Fourier transform mass spectrometry
GC	gas chromatography
GDMS	glow discharge mass spectrometry
ICP	inductively coupled plasma
ICR	ion cyclotron resonance
IR	infrared
LAMMA®	laser microprobe mass analyzer
LAMMS, LIMS, LMMS	laser microprobe mass spectrometry
LC	liquid chromatography
LD	laser desorption
LIF	laser-induced fluorescence
LIMA®	laser ionization mass analyzer
LPMS	laser probe mass spectrograph
LTE	local thermal equilibrium
MCA	multichannel analyzer
MIKES	mass-analyzed ion-kinetic energy spectrometer
MS^n	*n* times MS
MS/MS	mass spectrometry/mass spectrometry
MUPI	multiphoton ionization
PAH	polycyclic aromatic hydrocarbons
REMPI	resonance-enhanced multiphoton ionization
RF	radio frequency
RIMS	resonance ionization mass spectrometry
RIS	resonance ionization spectrometry
SALI	surface analysis by laser ionization
SARISA	surface analysis by resonance ionization of sputtered atoms
SIM	selective ion monitoring
SIMS	secondary ion mass spectrometry
SIRIS	sputter-initiated resonance ionization spectrometry
TECD	total electron capture detector
TLC	thin-layer chromatography
TOF	time-of-flight
TR	transient recorder
UV	ultraviolet

REFERENCES

Adams, F., and Mauney, T. (1986). *Adv. Mass Spectrom.* **9A**, 507–524.

Adams, F., and Van Vaeck, L., eds. (1986). *Proceedings of Third International Laser Microprobe Mass Spectrometry Workshop.* Antwerp, Belgium.

Allison, J., Freas, R. B., and Ridge, D. P. (1979). *J. Am. Chem. Soc.* **101**, 1332–1333.

Amster, I. J., Land, D. P., Hemminger, J. C., and McIver, R. T., Jr. (1989). *Anal. Chem.* **61**, 184–185.

Anderson, J. B., Andres, R. P., and Fenn, J. B. (1966). *Adv. Chem. Phys.* **10**, 275.

Arrowsmith, P. (1987). *Anal. Chem.* **59**, 1437–1444.

Arrowsmith, P. (1990). In *Lasers and Mass Spectrometry* (D. M. Lubman, ed.), pp. 180–204. Oxford University Press, Oxford.

Bahr, U., and Hillenkamp, F. (1985). *Adv. Mass Spectrom.* **10B**, 853–854.

Ban, V. S., and Knox, B. E. (1969). *Int. J. Mass Spectrom. Ion Phys.* **3**, 131–141.

Beavis, R., Lindner, J., Grotemeyer, J., Atkinson, I. M., Keene, F. R., and Knight, W. (1988). *J. Am. Chem. Soc.* **110**, 7534–7535.

Becker, C. H. (1986). *Scanning Electron Microsc.* **4**, 1267–1276.

Becker, C. H. (1990). In *Lasers and Mass Spectrometry* (D. M. Lubman, ed.), pp. 84–102. Oxford University Press, Oxford.

Becker, C. H., and Gillen, K. T. (1984). *Anal. Chem.* **56**, 1671–1674.

Beekman, D. W., Calcott, T., Kramer, S. D., Arakawa, E. T., Hurst, G. S., and Nussbaum, E. (1980). *Int. J. Mass Spectrom. Ion Phys.* **34**, 89–97.

Berkowitz, J., and Chupka, W. A. (1964). *J. Chem. Phys.* **40**, 2735–2736.

Bernal, E., Levine, G. L. P., and Ready, J. F. (1966). *Rev. Sci. Instrum.* **37**(7), 938–941.

Beynon, J. H., Cooks, R. G., Amy, J. W., Baitinger, W. E., and Ridley, T. Y. (1973). *Anal. Chem.* **45**, 1023A–1031A.

Beynon, J. H., Brenton, A. G., and Harris, F. M. (1982). *Int. J. Mass Spectrom. Ion Phys.* **45**, 5–34.

Bingham, R. A., and Salter, P. L. (1976a). *Int. J. Mass Spectrom. Ion Processes* **21**, 133–144.

Bingham, R. A., and Salter, P. L. (1976b). *Anal. Chem.* **48**, 1735–1740.

Boesl, U., Neussner, H. J., Weinkauf, R., and Schlag, E. W. (1982). *J. Phys. Chem.* **86**, 4857–4863.

Brenna, J. T. (1989). In *Microbeam Analysis—1989* (P. E. Russell, ed.), pp. 306–310. San Francisco Press, San Francisco.

Brenna, J. T., Creasy, W. R., McBain, W., and Soria, C. (1988). *Rev. Sci. Instrum.* **59**, 873–879.

Briukhanov, A. S., Boriskin, A. I., Bikovski, Y. A., Eriomenko, V. M., and Yariomenko, V. M. (1983). *Int. J. Mass Spectrom. Ion Phys.* **47**, 35–38.

Brown, R. S., and Wilkins, C. L. (1987). *ACS Symp. Ser.* **359**, 127–139.

Brucat, P. J., Zheng, L.-S., Pettiette, C. L., Yang, S., and Smalley, R. E. (1986). *J. Chem. Phys.* **84**, 3078–3088.

Bruynseels, F., and Van Grieken, R. (1986). *Int. J. Mass Spectrom. Ion Processes* **74**, 161–177.

Bruynseels, F., Storms, H., Tavares, T., and Van Grieken, R. (1985). *J. Environ. Anal. Chem.* **13**, 1–14.

Buchanan, M. V., ed. (1987). *Fourier Transform Mass Spectrometry: Evolution, Innovation, and Applications*, ACS Symp. Ser. No. 359. Am. Chem. Soc., Washington, DC.

Busch, K. L., Hsu, B. H., Wood, K. V., and Cooks, R. G. (1984). *J. Org. Chem.* **49**, 764–769.

Casal, B., Ruiz-Hitzky, E., Van Vaeck, L., and Adams, F. C. (1988). *J. Inclusion Phenom.* **6**, 107–118.

Chamel, A., and Eloy, J. F. (1983). *Scanning Electron Microsc.* **2**, 841–851.

Chen, C. H., Payne, M. G., Hurst, G. S., Kramer, S. D., Allman, S. L., and Phillips, R. C. (1990). In *Lasers and Mass Spectrometry* (D. M. Lubman, ed.), pp. 3–36. Oxford University Press, Oxford.

Cheng, P. Y., Wiley, K. F., Salcido, J. E., and Duncan, M. A. (1990). *Int. J. Mass Spectrom. Ion Processes* **102**, 67–80.

Chowdhury, A. K., and Wilkins, C. L. (1987). *J. Am. Chem. Soc.* **109**, 5336–5343.

Chowdhury, A. K., and Wilkins, C. L. (1988). *Int. J. Mass Spectrom. Ion Processes* **82**, 163–176.

Chowdhury, A. K., Cooper, J. R., and Wilkins, C. L. (1989). *Anal. Chem.* **61**, 86–88.

Clarke, N. S., Davey, A. R., and Ruckmann, J. C. (1984). In *Microbeam Analysis—1984* (A. D. Romig and J. I. Goldstein, eds.), pp. 31–40. San Francisco Press, San Francisco.

Cody, R. B., and Kissinger, J. A. (1987). *ACS Symp. Ser.* **359**, 59–80.

Cody, R. B., Bjarnason, A., and Weil, D. A. (1990). In *Lasers and Mass Spectrometry* (D. M. Lubman, ed.), pp. 316–339. Oxford University Press, Oxford.

Colby, S. M., Stewart, M., and Reilly, J. P. (1990). *Anal. Chem.* **62**, 2400–2403.

Conzemius, R. J., and Capellen, J. M. (1980). *Int. J. Mass. Spectrom. Ion Phys.* **34**, 197–271.

Conzemius, R. J., Schmidt, F. A., and Svec, H. J. (1981). *Anal. Chem.* **53**, 1899–1902.

Conzemius, R. J., Simons, D. S., Shankai, Z., and Byrd, G. D. (1983). In *Microbeam Analysis—1983* (R. Gooley, ed.), pp. 301–331. San Francisco Press, San Francisco.

Conzemius, R. J., Zhao, S., Houk, R. S., and Svec, H. J. (1984). *Int. J. Mass Spectrom. Ion Processes* **61**, 277–292.

Cotter, R. J. (1980). *Anal. Chem.* **52**, 1767–1770.

Cotter, R. J. (1989). *Biomed. Environ. Mass Spectrom.* **18**, 513–532.

Cotter, R. J., and Tabet, J. C. (1983). *Int. J. Mass Spectrom. Ion. Phys.* **53**, 151–166.

Cox, D. M., Trevor, D. J., Reichmann, K. C., and Kaldor, A. (1986). *J. Am. Chem. Soc.* **108**, 2457–2458.

Daniel, W. M., Delorenza, D. J., and Wilson, H. R. (1988). In *Microbeam Analysis—1988* (D. E. Newbury, ed.), pp. 264–268. San Francisco Press, San Francisco.

Davis, D. V., Cooks, R. G., Meyer, B. N., and McLaughlin, J. L. (1983). *Anal. Chem.* **55**, 1302–1305.

Denoyer, E. R., Fredeen, K. J., and Hager, J. W. (1991). *Anal. Chem.* **63**, 445A–457A.

De Waele, J. K., and Adams, F. C. (1988). *Scanning Microsc.* **2**, 209–228.

De Waele, J. K., Swenters, I. M., and Adams, F. C. (1985). *Spectrochim. Acta* **40B**, 795–800.

De Waele, J. K., Wouters, H., Van Vaeck, L., Adams, F., and Ruiz-Hitzky, E. (1987). *Anal. Chim. Acta* **195**, 331–336.

Dietze, H. J., and Becker, S. (1985). *Fresenius' Z. Anal. Chem.* **302**, 490–492.

Dietze, H. J., Becker, S., Opausky, I., Matus, L., Nyary, I., and Frecska, J. (1983). *Mikrochim. Acta* **3**, 263–270.

Dingle, T. (1981). *Vacuum* **31**, 571–577.

Dingle, T., Griffiths, B. W., Ruckman, J. C., and Evans, C. A. (1982). In *Microbeam Analysis—1982* (K.F.J. Heinrich, ed.), pp. 365–368. San Francisco Press, San Francisco.

Dobson, R. L. M., D'Silva, A. P., Weeks, S. J., and Fassel, V. A. (1986). *Anal. Chem.* **58**, 2129–2137.

Donohue, D. L., Young, J. P, and Smith, D. H. (1983). *Int. J. Mass Spectrom. Ion Phys.* **43**, 293–307.

Eloy, J. F. (1969). *Méthodes Phys. Anal.* **5**, 157–161.

Eloy, J. F. (1971). *Int. J. Mass Spectrom. Ion Processes* **6**, 101–115.

Eloy, J. F. (1985). *Scanning Electron Microsc.* **2**, 563–576.

Eloy, J. F. (1986). *Scanning Electron Microsc.* **4**, 1243–1253.

Eloy, J. F., and Dumas, J. L. (1966). *Méthodes Phys. Anal.* **2**, 251–256.

EMAL Technical Information Bulletin (1982), Moscow V/O Sojazzugranibor.

Emary, W. B., Wood, K. V., and Cooks, R. G. (1987). *Anal. Chem.* **59**, 1069–1071.

Engelke, F., Hahn, J. H., Henke, W., and Zare, R. N. (1987). *Anal. Chem..* **59**, 909–912.

Fassett, J. D., and Kingston, H. M. (1985). *Anal. Chem.* **57**, 2474–2478.

Fassett, J. D., Travis, J. C., Moore, L. J., and Lyttle, F. E. (1983). *Anal. Chem.* **55**, 765–770.

Fenner, N. C., and Daly, N. R. (1966). *Rev. Sci. Instrum.* **37**, 1068–1070.

Fenner, N. C., and Daly, N. R. (1968). *J. Mater. Sci.* **3**, 259–261.

Fletcher, R. A., and Simons, D. S. (1985). In *Microbeam Analysis—1985* (J. T. Armstrong, ed.), pp. 319–321. San Francisco Press, San Francisco.

Forbes, R. A., Tews, E. C., Huang, Y., Freiser, B. S., and Perone, S.P. (1987). *Anal. Chem.* **59**, 1937–1944.

Forbes, R. A., Laukien, F. H., and Wronka, J. (1988). *Int. J. Mass Spectrom. Ion Processes* **83**, 23–44.

Freiser, B. S. (1985). *Anal. Chim. Acta* **178**, 135–158.

Fukuda, E. K., and Campana, J. E. (1985). *Anal. Chem.* **57**, 949–952.

Geusic, M. E., McIlrath, T. J., Jarrold, M. F., Bloomfield, L. A., and Brown, W. L. (1986). *J. Chem. Phys.* **84**, 2421–2422.

Geusic, M. E., Jarrold, M. F., McIlrath, T. J., Freeman, R. R., and Brown, W. L. (1987). *J. Chem. Phys.* **86**, 3862–3869.

Glish, G. L., Goeringer, D. E., Asano, K. G., and McLuckey, S. A. (1989). *Int. J. Mass Spectrom. Ion Processes* **94**, 15–24.

Gord, J. R., Buckner, S. W., and Freiser, B. S. (1989). *J. Am. Chem. Soc.* **111**, 3753–3754.

Gray, A. (1985). *Analyst.* **110**, 551–556.

Gray, A. L. (1989). *Adv. Mass Spectrom.* **11B**, 1674–1693.

Griffiths, I. W., and Heesterman, P. J. L. (1990). *Int. J. Mass Spectrom. Ion Processes* **99**, 79–98.

Gross, M. L. (1989). *Adv. Mass Spectrom.* **11A**, 792–811.

Grotemeyer, J., and Schlag, E. W. (1987). *Org. Mass Spectrom.* **22**, 758–760.

Grotemeyer, J., and Schlag, E. W. (1988). *Biomed. Environ. Mass Spectrom.* **16**, 143–149.

Grotemeyer, J., Boesl, U., Walter, K., and Schlag, E. W. (1986a). *Org. Mass Spectrom.* **21**, 645–653.

Grotemeyer, J., Boesl, U., Walter, K., and Schlag, E. W. (1986b). *J. Am. Chem. Soc.* **108**, 4233–4236.

Grotemeyer, J., Walter, K., Boesl, U., and Schlag, E. W. (1987). *Int. J. Mass Spectrom. Ion Processes* **79**, 69–83.

Güçer, S., Van Vaeck, L., and Adams, F. (1989). *Spectrochim. Acta* **44B**, 1021–1029.

Hardin, E. D., and Vestal, M. L. (1981). *Anal. Chem.* **53**, 1492–1497.

Hardin, E. D., Fan, T. P., Blakley, C. R., and Vestal, M. L. (1984). *Anal. Chem.* **56**, 2–7.

Harding-Barlow, I., Snetsinger, K. G., and Keil, K. (1973). In *Microprobe Analysis* (C. A. Anderson ed.), pp. 423–455. Wiley, New York.

Hayes, J. M., and Small, G. J. (1983). *Anal. Chem.* **55**, 565A–574A.

Heimbrook, L. A., Moyers, K. W., and Hillenius, S. J. (1989). In *Microbeam Analysis—1989* (P. E. Russell, ed.), pp. 335–336. San Francisco Press, San Francisco.

Heinen, H. J., Meier, S., Vogt, H., and Wechsung, R. (1983). *Int. J. Mass Spectrom. Ion Phys.* **47**, 19–22.

Heinen, H. J., Holm, R., and Storp, S. (1984). *Fresenius' Z. Anal. Chem.* **319**, 606–610.

Heller, D. N., Lys, I., Cotter, R. J., and Uy, O. M. (1989). *Anal. Chem.* **61**, 1063–1086.

Hercules, D. M., Novak, F. P., Visnawadham, S. K., and Wilk, Z. A. (1987). *Anal. Chim. Acta* **195**, 61–71.

Heresh, F., Schmid, E. R., and Huber, J. F. K. (1980). *Anal. Chem.* **52**, 1803–1807.

Hess, K. R., and Harrison, W. W. (1990). In *Lasers and Mass Spectrometry* (D. M. Lubman, ed.), pp. 205–222. Oxford University Press, Oxford.

Hillenkamp, F., and Kaufmann, R. eds. (1981). *Fresenius' Z. Anal. Chem.* **308**, 195–320.

Holm, R., and Holtkamp, D. (1989). In *Microbeam Analysis—1989* (P. E. Russell, ed.), pp. 325–329. San Francisco Press, San Francisco.

Holm, R., Kampf, G., Kirchner, D., Heinen, H. J., and Meier, S. (1984). *Anal. Chem.* **56**, 690–692.

Honig, R. E. (1958). *J. Appl. Phys.* **29**, 549–555.

Honig, R. E., and Woolston, J. R. (1963). *Appl. Phys. Lett.* **2**, 138–139.

Houriet, R., and Vulpius, T. (1988). *Chem. Phys. Lett.* **154**, 454–457.

Hunt, D. F., Shabanowitz, J., McIver, R. T., Jr., Hunter, R. L., and Syka, J. E. P. (1985a). *Anal. Chem.* **57**, 765–768.

Hunt, D. F., Shabanowitz, J., Yates, J. R., III, McIver, R. T., Jr., Hunter, R. L., Syka, J. E. P., and Amy, J. (1985b). *Anal. Chem.* **57**, 2728–2733.

Hunt, D. F., Shabanowitz, J., and Yates, J. R., III (1987). *J. Chem. Soc., Chem. Commun.*, pp. 548–550.

Hurst, G. S. (1981). *Anal. Chem.* **53**, 1448A–1456A.

Hurst, G. S., Payne, M. G., Kramer, S. D., and Young, J. P. (1979). *Rev. Mod. Phys.* **51**, 767–819.

Ijames, C. F., and Wilkins, C. L. (1988). *J. Am. Chem. Soc.* **110**, 2687–2688.

Jacob, W. A., De Nollin, S., Hertsens, R. C., and De Smet, M. (1984). *J. Trace Microprobe Tech.* **2**, 161–184.

Jansen, J. A., and Witmer, A. W. (1982). *Spectrochim. Acta* **37B**(6), 483–491.

Jarrold, M. F., Illies, A. J., and Bowers, M. T. (1983). *J. Chem. Phys.* **79**, 6086–6096.

Jarrold, M. F., Illies, A. J., and Bowers, M. T. (1984). *J. Chem. Phys.* **81**, 214–221.

Jochum, K. P., Matus, L., and Seufert, H. M. (1988). *Fresenius' Z. Anal. Chem.* **331**, 136–139.

Johnson, M. A., Alexander, M. L., and Lineberger, W. C. (1984). *Chem. Phys. Lett.* **112**, 285–290.

Karas, M., and Hillenkamp, F. (1989). In *Microbeam Analysis—1989* (P. E. Russell, ed.), pp. 353–354. San Francisco Press, San Francisco.

Karas, M., Bachmann, D., and Hillenkamp, F. (1985). *Anal. Chem.* **57**, 2935–2939.

Karas, M., Bachmann, D., Bahr, U., and Hillenkamp, F. (1987). *Int. J. Mass Spectrom. Ion Processes* **78**, 53–68.

Karas, M., Bahr, U., Ingendoh, A., and Hillenkamp, F. (1989). *Angew. Chem., Int. Ed. Engl.* **28**, 760–761.

Kaufmann, R. (1982). In *Microbeam Analysis—1982* (K. J. F. Heinrich, ed.), pp. 341–358. San Francisco Press, San Francisco.

Kaufmann, R., Hillenkamp, F., Wechsung, R., Heinen, H. J., and Schurmann, M. (1979). *Scanning Electron Microsc.* **2**, 279–290.

Kaufmann, R., Wieser, P., and Wurster, R. (1980). *Scanning Electron Microsc.* **2**, 607–622.

Knox, B. E. (1968). *Mater. Res. Bull.* **3**, 329–336.

Knox, B. E. (1971). *Dyn. Mass Spectrom.* **2**, 61–91.

Kohler, V. L., Harris, A., and Wallach, E. R. (1986). In *Microbeam Analysis—1986* (A. D. Romig and W. F. Chalmers, eds.), pp. 264–268. San Francisco Press, San Francisco.

Kolaitis, L., and Lubman, D. M. (1986). *Anal. Chem.* **56**, 1993–2001.

Kondrat, R. W., and Cooks, R. G. (1978). *Anal. Chem.* **50**, 81A–92A.

Kovalev, I. D., Maksimov, G. A., Suchkov, A. I., and Larin, N. V. (1978). *Int. J. Mass Spectrom. Ion Phys.* **27**, 101–137.

Krailler, R. E., Russell, D. H., Jarrold, M. F., and Bowers, M. T. (1985). *J. Am. Chem. Soc.* **107**, 2346–2354.

Kramer, S. D., Young, J. P., Hurst, G. S., Payne, M. G. (1979). *Opt. Commun.* **30**, 47–50.

Krier, G., Verdun, F., and Muller, J. F. (1985). *Fresenius' Z. Anal. Chem.* **322**, 379–382.

Kubis, A. J., Somayula, K. V., Sharkey, A. G., and Hercules, D. M. (1989). *Anal. Chem.* **61**, 2516–2523.

LaiHing, K., Cheng. P. Y., Tayler, T. G., Wiley, K. F., Peschke, M., and Duncan, M. A. (1989). *Anal. Chem.* **61**, 1460–1465.

Land, D. P., Tai, T. L., Lindquist, J. M., Hemminger, J. C., and McIver, R. T., Jr. (1987). *Anal. Chem.* **59**, 2924–2927.

Land, D. P., Wang, D. T. S., Tai, T. L., Sherman, M. G., Hemminger, J. C., and McIver, R. T., Jr. (1990). In *Lasers and Mass Spectrometry* (D. M. Lubman, ed.), pp. 157–178. Oxford University Press, Oxford.

Liebl, H. (1983). *Int. J. Mass Spectrom. Ion Phys.* **46**, 511–514.

LIMA Technical Documentation (1985). Cambridge Mass Spectrometry, Cambridge, UK.

Lincoln, K. A. (1965). *Anal. Chem.* **37**(4), 541–543.

Lincoln, K. A. (1969). *Int. J. Mass Spectrom. Ion Phys.* **2**, 75–83.

Lincoln, K. A. (1974). *Int. J. Mass Spectrom. Ion Processes* **13**, 45–53.

Lindner B., and Seydel, U. (1985). *Anal. Chem.* **57**, 895–899.

Lindner, B., and Seydel, U. (1989). In *Microbeam Analysis—1989* (P. E. Russell, ed.), pp. 286–292. San Francisco Press, San Francisco.

Lindner, U., Grotemeyer, J., and Schlag, E. W. (1990). *Int. J. Mass Spectrom. Ion Processes* **100**, 267–285.

Liu, Y., Zhang, Q.-L., Tittel, F. K., Curl, R. F., and Smalley, R. E. (1986). *J. Chem. Phys.* **85**, 7434–7441.

Louris, J. N., Brodbelt, J. S., and Cooks, R. G. (1987a). *Int. J. Mass Spectrom. Ion Processes* **75**, 345–352.

Louris, J. N., Cooks, R. G., Syka, J. E. P., Kelley, P. E., Stafford, G. C., Jr., and Todd, J. F. J. (1987b). *Anal. Chem.* **59**, 1677–1685.

Louris, J. N., Brodbelt-Lustig, J. S., Cooks, R. G., Glish, G. L., Van Berkel, G. J., and McLuckey, S. A. (1990). *Int. J. Mass Spectrom. Ion Processes* **96**, 117–137.

Lubman, D. M. (1987). *Anal. Chem.* **59**, 31A–40A.

Lubman, D. M. (1988). *Mass Spectrom. Rev.* **7**, 559–592.

Lubman, D. M., and Jordan, R. M. (1985). *Rev. Sci. Instrum.* **56**, 373–376.

Lubman, D. M., and Li, L. (1990). In *Lasers and Mass Spectrometry* (D. M. Lubman, ed.), pp. 353–382. Oxford University Press, Oxford.

Lubman, D. M., Tembreull, R., and Sin, C. H. (1985). *Anal. Chem.* **57**, 1084–1087.

Mamyrin, B. A., and Schmikk, D. V. (1979). *Sov. Phys.—JETP (Engl. Transl.)* **49**, 762–764.

Mamyrin, B. A., Karataev, V. I., Schmikk, D. V., and Zagulin, V. A. (1973). *Sov. Phys.—JETP (Engl. Transl.)* **37**, 45–48.

Marshall, A. G. (1989). *Adv. Mass Spectrom.* **11A**, 651–669.

Marshall, A. G., and Verdun, F. R. (1990). *Fourier Transforms in NMR, optical and MS: A User's handbook*, Elsevier. New York.

Mathey, A., Van Vaeck, L., and Steglich, W. (1987). *Anal. Chim. Acta* **195**, 89–96.

Mathey, A., Van Vaeck, L., and Ricci, P. (1989). In *Microbeam Analysis—1989* (P. E. Russell, ed.), pp. 350–352. San Francisco Press, San Francisco.

Mattews, L. E., Baxter, C. S., and Leake, J. A. (1989). In *Microbeam Analysis—1989* (P. E. Russell, ed.), pp. 264–268. San Francisco Press, San Francisco.

Matus, L., Seufert, M., and Jochum, K. P. (1988). *Int. J. Mass Spectrom. Ion Processes* **84**, 101–111.

Mauney, T. (1985). In *Microbeam Analysis—1985* (J. T. Armstrong, ed.), pp. 299–304. San Francisco Press, San Francisco.

Mauney, T., and Adams, F. (1984). *Int. J. Mass Spectrom. Ion Processes* **59**, 103–119.

Mayo, S., Lucatorto, T. B., and Luther, G. G. (1982). *Anal. Chem.* **54**, 553–556.

McCrery, D. A., and Gross, M. L. (1985). *Anal. Chim. Acta* **178**, 105–116.

McCrery, D. A., Ledford, E. B., and Gross, M. L. (1982). *Anal. Chem.* **54**, 1435–1437.

McLuckey, S. A., Schoen, A. E., and Cooks, R. G. (1982). *J. Am. Chem. Soc.* **104**, 848–850.

Michiels, E., Van Vaeck, L., and Gijbels, R. (1984). *Scanning Electron Microsc.* **3**, 1111–1128.

Moesta, P., Seydel, U., Lindner, W., and Grisebach, H. (1982). *Z. Naturforsch.* **37C**, 748–751.

Moore, W. J., Fassett, J. D., and Travis, J. C. (1984). *Anal. Chem.* **56**, 2770–2775.

Morelli, J. J. (1990). In *Lasers and Mass Spectrometry* (D. M. Lubman, ed.), pp. 138–156. Oxford University Press, Oxford.

Muller, J. F., Pelletier, M., Krier, G., Weil, D., and Campana, J. (1989). In *Microbeam Analysis—1989* (P. E. Russell, ed.), pp. 311–316. San Francisco Press, San Francisco.

Musselman, I. H., Simons, D. S., and Linton, R. W. (1988). In *Microbeam Analysis—1988* (D. E. Newbury, ed.), pp. 356–364. San Francisco Press, San Francisco.

Nogar, N. S., and Estler, R. C. (1990). In *Lasers and Mass Spectrometry* (D. M. Lubman, ed.), pp. 65–83. Oxford University Press, Oxford.

Nogar, N. S., Estler, R. C., and Miller, C. M. (1985). *Anal. Chem.* **57**, 2441–2444.

Nuwaysir, L. M., and Wilkins, C. L. (1989). *Anal. Chem.* **61**, 689–694.

Nuwaysir, L. M., and Wilkins, C. L. (1990). In *Lasers and Mass Spectrometry* (D. M. Lubman, ed.), pp. 291–315. Oxford University Press, Oxford.

O'Brien, S. C., Heath, J. R., Curl, R. F., and Smalley, R. E. (1988). *J. Chem. Phys.* **88**, 220–230.

Odom, R. W., and Schueler, B. (1990). In: *Lasers and Mass Spectrometry* (D. M. Lubman, ed.), pp. 103–137. Oxford University Press, Oxford.

Opsal, R. B., and Reilly, J. P. (1988). *Anal. Chem.* **60**, 1060–1065.

Opsal, R. B., Owens, K. G., and Reilly, J. P. (1985). *Anal. Chem.* **57**, 1884–1889.

Pallix, J. P., Becker, C. H., and Newman, N. (1988a). *J. Vac. Sci. Technol.* [2]**A6**, 1049–1052.

Pallix, J. B., Becker, C. H., Missert, N., Naito, M., Hammond, R. H., and Wright, P. (1988b). In *Secondary Ion Mass Spectrometry SIMS VI* (A. Benninghoven, A. M. Huber, and H. E. Werner, eds.), pp. 817–818. Wiley, New York.

Pallix, J. B., Becker, C. H., Missert, N., Char, K., and Hammond, R. H. (1988c). *AIP Conf. Proc.* **165**, 413–420.

Pallix, J B., Schühle, U., Becker, C. H., and Huestis, D. L. (1989). *Anal. Chem.* **61**, 805–811.

Pang, H. M., Sin, C. H., and Lubman, D. M. (1988). *Appl. Spectrosc.* **42**, 1200–1206.

Parks, J. E. (1990). In *Lasers and Mass Spectrometry* (D. M. Lubman, ed.), pp. 37–64. Oxford University Press, Oxford.

Paul, W., and Steinwedel, H. (1953). *Z. Naturforsch.* **A8**, 448–450.

Payne, M. G., Chen, C. H., Hurst, G. S., Kramer, S. D., Garrett, W. R., and Pindzola, M. (1982). *Chem. Phys. Lett.* **79**, 142–148.

Peake, A., Huang, S. K., and Gross, M. L. (1987). *Anal. Chem.* **59**, 1557–1563.

Pelletier, M., Krier, G., Muller, J. F., Weil, D., and Johnston, M. (1988). *Rapid Commun. Mass Spectrom.* **2**, 146–150.

Pelletier, M., Krier, G., Muller, J. F., Campana, J., and Weil, D. (1989). In *Microbeam Analysis—1989* (P. E. Russell, ed.), pp. 339–349. San Francisco Press, San Francisco.

Pellin, M. J., Young, C. E., Calaway, W. F., Burnett, J. W., Jorgensen, B., Schweitzer, E. L., and Gruen, D. M. (1987). *Nucl. Instrum. Methods Phys. Res.* **B18**, 446–451.

Pierce, J. L., Busch, K. L., Cooks, R. G., and Walton, R. A. (1982). *Inorg. Chem.* **21**, 2597–2602.

Posthumus, M. A., Kistemaker, P. G., Meuzelaar, H. L. C., and Ten Noever de Brauw, M. C. (1978). *Anal. Chem.* **50**, 985–991.

Ruckmann, J. C., Davey, A. R., and Clarke, N. S. (1984). *Vacuum* **34**, 911.

Russell, D. H. (1986). *Mass Spectrom. Rev.* **5**, 167–189.

Russell, P. E., ed. (1989). *Microbeam Analysis—1989*, pp. 261–376. San Francisco Press, San Francisco.

Sack, T. M., McCrery, D. A., and Gross, M. L. (1985). *Anal. Chem.* **57**, 1290–1295.

Schmidt, P. F. (1989). In *Microbeam Analysis—1989* (P. E. Russell, ed.), pp. 383–386. San Francisco Press, San Francisco.

Schmidt, P. F., and Brinkmann, B. (1989). In *Microbeam Analysis—1989* (P. E. Russell, ed.), pp. 330–331. San Francisco Press, San Francisco.

Schueler, B., and Odom, R. W., (1987). *J. Appl. Phys.* **61**, 4652–4661.

Schühle, U., Pallix, J. B., and Becker, C. H. (1988). *J. Am. Chem. Soc.* **10**, 2323–2324.

Seydel, U., and Lindner, B., eds. (1983). *Proceedings of the Second LAMMA Symposium.* Borstel, Germany.

Shabanowitz, J., and Hunt, D. F. (1990). In *Lasers and Mass Spectrometry* (D. M. Lubman, ed.), pp. 340–352. Oxford University Press, Oxford.

Shaw, R. W., Young, J. P., and Smith, D. H. (1988). *Anal. Chem.* **60**, 282–283.

Sherman, M. G., Kingsley, J. R., Hemminger, J. C., and McIver, R. T., Jr. (1985). *Anal. Chim. Acta* **178**, 79–89.

Sherman, M. G., Land, D. P., Hemminger, J. C., and McIver, R. T., Jr. (1987). *Chem. Phys. Lett.* **137**, 298–300.

Siegel, M. W., and Vasile, M. J. (1981). *Rev. Sci. Instrum.* **52**, 1603–1615.

Simons, D. S. (1983). *Int. J. Mass Spectrom. Ion Phys.* **55**, 15–30.

Smalley, R. E., Wharton, L., and Levy, D. H. (1977). *Acc. Chem. Res.* **10**, 139–145.

Smith, D. H., Young, J. P., and Shaw, R. W. (1989). *Mass Spectrom. Rev.* **8**, 345–378.

Spengler, B., Karas, M., Bahr, U., and Hillenkamp, F. (1987). *J. Phys. Chem.* **91**, 6502–6506.

Stafford, G. C., Kelley, P. E., Syka, J. E. P., Reynolds, W. E., and Todd, J. F. J. (1989). *Int. J. Mass Spectrom. Ion Processes* **60**, 85–94.

Stefani, R. (1988). *Analusis* **16**, 147–156.

Stoll, R., and Röllgen, F. W. (1979). *Org. Mass Spectrom.* **14**, 642–645.

Stoll, R., and Röllgen, F. W. (1982). *Z. Naturforsch.* **37A**, 9–14.

Syage, J. A. (1990). In *Lasers and Mass Spectrometry* (D. M. Lubman, ed.), pp. 468–489. Oxford University Press, Oxford.

Syage, J. A., and Wessel, J. E. (1987). *J. Chem. Phys.* **86**, 3313–3320.

Syage, J. A., Pollard, J. E., and Cohen, R. B. (1987). *Appl. Opt.* **26**, 3516–3520.

Tabet, J. C., and Cotter, R. J. (1983). *Int. J. Mass Spectrom. Ion Processes* **54**, 151–158.

Tabet, J. C., and Cotter, R. J. (1984). *Anal. Chem.* **56**, 1662–1667.

Tanaka, K., Waki, H., Ido, Y., Akita, S., Yoshida, Y., and Yoshida, K. (1988). *Rapid Commun. Mass Spectrom.* **2**, 151–153.

Tecklenburg, R. E., Jr., and Russell, D. H. (1987). *J. Am. Chem. Soc.* **109**, 7654–7662.

Tecklenburg, R. E., Jr., Miller, M. N., and Russell, D. H. (1987). *J. Am. Chem. Soc.* **111**, 1161–1171.

Tecklenburg, R. E., Jr., Sellers-Hann, L., and Russell, D. H. (1989). *Int. J. Mass Spectrom. Ion Processes* **87**, 111–120.

Tembreull, R., and Lubman, D. M. (1984). *Anal. Chem.* **56**, 1962–1967.

Tembreull, R., Sin, C. H., Li, P., Pang, H. M., and Lubman, D. M. (1988). *Anal. Chem.* **57**, 1186–1192.

Terlouw, J. K. (1989). *Adv. Mass Spectrom.* **11B**, 984–1010.

Todd, J. F. J. (1981). In *Dynamic Mass Spectrometry VI* (D. Price and J. F. J. Todd, eds.), pp. 44–70. Heyden, London.

Tourmann, J. L., and Kaufmann, R. L. (1989). In *Microbeam Analysis—1989* (P. E. Russell, ed.), pp. 359–363. San Francisco Press, San Francisco.

Tuithoff, H. H., and Boerboom, A. J. H. (1976). *Int. J. Mass Spectrom. Ion Phys.* **20**, 107–121.

Tuithoff, H. H., Boerboom, A. J. H., and Meuzelaar, H. L. C. (1975). *Int. J. Mass Spectrom Ion Phys.* **17**, 299–307.

Van Breemaen, R. B., Snow, M., and Cotter, R. J. (1983). *Int. J. Mass Spectrom. Ion Phys.* **49**, 35–50.

Vandeputte, D. F., Ameloot, P. C., Cleymaet, R., Coomans, D., and Van Grieken, R. E. (1990a). *Biol. Trace Elem. Res.* **23**, 133–144.

Vandeputte, D. F., Jacob, W. A., and Van Grieken, R. E. (1990b). *Biomed. Environ. Mass Spectrom.* **10**, 33–37.

Van der Peyl, G. J. Q., van der Zande, W. J., and Kistemaker, P. G. (1984). *Int. J. Mass Spectrom. Ion Phys.* **62**, 51–71.

Van Vaeck, L., and Gijbels, R. (1989). In *Microbeam Analysis—1989* (P. E. Russell, ed.), pp. xvii–xxv. San Francisco Press, San Francisco.

Van Vaeck, L., and Gijbels, R. (1990a). *Fresenius' Z. Anal. Chem.* **337**, 743–754.

Van Vaeck, L., and Gijbels, R. (1990b). *Fresenius' Z. Anal. Chem.* **337**, 755–765.

Van Vaeck, L., Claereboudt, J., De Nollin, S., Jacob, W., Adams, F., Gijbels, R., and Cautreels, W. (1985). *Adv. Mass Spectrom.* **9B**, 1985–1986.

Van Vaeck, L., Claereboudt, J., Van Espen, P., Adams, F., Gijbels, R., and Cautreels, W. (1986a). *Adv. Mass Spectrom.* **9B**, 957–958.

Van Vaeck, L., Claereboudt, J., Veldeman, E., Vermeulen, M., and Gijbels, R. (1986b). *Bull. Soc. Chim. Belge* **95**, 351–372.

Van Vaeck, L., Van Espen, P., Jacob, W., Gijbels, R., and Cautreels, W. (1988a). *Biomed. Environ. Mass Spectrom.* **16**, 113–119.

Van Vaeck, L., Van Espen, P., Gijbels, R., and Lauwers, W. (1988b). *Biomed. Environ. Mass Spectrom.* **16**, 121–130.

Van Vaeck, L., Van Espen, P., Adams, F., Gijbels, R., Lauwers, W., and Esmans, E. (1989a). *Biomed. Environ. Mass Spectrom.* **18**, 581–591.

Van Vaeck, L., Bennett, J., Van Espen, P., Gijbels, R., Adams, F., and Lauwers, W. (1989b). *Org. Mass Spectrom.* **24**, 782–796.

Van Vaeck, L., Bennett, J., Van Espen, P., Gijbels, R., Adams, F., and Lauwers, W. (1989c). *Org. Mass Spectrom.* **24**, 797–806.

Van Vaeck, L., Bennett, J., Lauwers, W., Vertes, A., and Gijbels, R. (1990). *Mikrochim. Acta* **3**, 283–303.

Vastola, F. J., and Pirone, A. J. (1968). *Adv. Mass Spectrom.* **4**, 107–111.

Vastola, F. J., Mumma, R. O., and Pirone, A. J. (1970). *Org. Mass Spectrom.* **3**, 101–104.

Verbueken, A. H., Van Grieken, R. E., De Broe, M. E., and Wedeen, R. P. (1987). *Anal. Chim. Acta* **195**, 97–115.

Verbueken, A. H., Bruynseels, F. J., Van Grieken, R., and Adams, F. (1988). In *Inorganic Mass Spectrometry* (F. Adams, R. Gijbels, and R. Van Grieken, eds.), pp. 173–254. Wiley, New York.

Verdun, F. R., Krier, G., and Muller, J. F. (1987). *Anal. Chem.* **59**, 1383–1387.

Vogt, H., Heinen, H. J., Meier, S., and Wechsung, R. (1981). *Fresenius' Z. Anal. Chem.* **308**, 195–200.

Wanczek, K. P. (1984). *Int. J. Mass Spectrom. Ion Processes* **60**, 11b–60b.

Welch, M. J., Sams, R., and White, E. V. (1986). *Anal. Chem.* **58**, 890–894.

Wesdemiotis, C., and McLafferty, F. W. (1987). *Chem. Rev.* **87**, 485–500.

Wiley, K. F., Cheng, P. Y., Tayler, T. G., Bishop, M. B., and Duncan, M. A. (1990). *J. Phys. Chem.* **94**, 1544–1549.

Wiley, W. C., and McLaren, I. H. (1955). *Rev. Sci. Instrum.* **26**, 1150.

Wilk, Z. A., and Hercules, D. M. (1987). *Anal. Chem.* **59**, 1819–1825.

Wilkerson, C. W., Jr., and Reilly, J. P. (1990). *Anal. Chem.* **62**, 1802–1808.

Wilkerson, C. W., Jr., Colby, S. M., and Reilly, J. P. (1989). *Anal. Chem.* **61**, 2669–2673.

Wilkins, C. L., Weil, D. A., Yang, C. L. C., and Ijames, C. F. (1985). *Anal. Chem.* **57**, 520–524.

Wilkins, C. L., Chowdhury, A. K., Nuwaysir, L. M., and Coates, M. L. (1989). *Mass Spectrom. Rev.* **8**, 67–92.

Williams, M. W., Beekman, D. W., Swan, J. B., and Arakawa, E. T. (1984). *Anal. Chem.* **56**, 1348–1350.

Wouters, L., Van Grieken, R., Linton, R., and Bauer, C. (1988). *Anal. Chem.* **60**, 2218–2220.

Wouters, L., Artaxo, P., and Van Grieken, R. (1990). *Int. J. Environ. Anal. Chem.* **38**, 427–438.

Wright, L. G., Cooks, R. G., and Wood, K. V. (1985). *Biomed. Mass Spectrom.* **12**, 159.

Young, J. P., Hurst, G. S., Kramer, S. D., and Payne, M. G. (1979). *Anal. Chem.* **51**, 1050A–1060A.

Young, J. P., Shaw, R. W., and Smith, D. H. (1989a). *Anal. Chem.* **61**, 1271A–1279A.

Young, J. P., Shaw, R. W., and Smith, D. H. (1989b). *Appl. Spectrosc.* **7**, 1164–1168.

Zakett, D., Schoen, A. E., and Cooks, R. G. (1981). *J. Am. Chem. Soc.* **103**, 1295–1297.

Zheng, L.-S., Karner, C. M., Brucat, P. J., Yang, S. H., Pettiette, C. L., Craycraft, M. J., and Smalley, R. E. (1986). *J. Chem. Phys.* **85**, 1681–1688.

CHAPTER

3

METHODS UTILIZING LOW AND MEDIUM LASER IRRADIANCE

A. LASER-INDUCED THERMAL DESORPTION AND MATRIX-ASSISTED METHODS

AKOS VERTES

Department of Chemistry
George Washington University
Washington, D.C.

RENAAT GIJBELS

Department of Chemistry
University of Antwerp (UIA)
Antwerp, Belgium

3A.1. INTRODUCTION

Producing ions from large molecules is of marked importance in mass spectrometry. In this chapter we survey different laser desorption methods in view of their virtues and drawbacks in volatilization and ion generation. Laser-induced thermal desorption and matrix-assisted laser desorption are assessed with special emphasis on the recent breakthrough in the field ($m/z > 100{,}000$ ions produced by matrix-assisted laser desorption). Efforts to understand and describe laser desorption and ionization are also discussed. We emphasize the role of restricted energy transfer pathways as a possible explanation for the volatilization of nondegraded large molecules.

With the growing importance of biomedical investigations in organic analysis, the emphasis has been shifting to the detection and structure determination of ever larger and more complex molecules. Gas phase analytical methods in general—and mass spectrometry was not an exception—exhibited

Laser Ionization Mass Analysis, Edited by Akos Vertes, Renaat Gijbels, and Fred Adams.
Chemical Analysis Series, Vol. 124.
ISBN 0-471-53673-3

increasing difficulties with these larger and larger molecular masses because of the involatility and instability of such materials.

The early answer to this challenge came in the 1970s, in the form of so-called soft ionization techniques. Field desorption, chemical ionization, plasma desorption, secondary ion, electrohydrodynamic, and laser desorption ion sources were developed to cope with nonvolatile compounds. A thorough overview of these methods with respect to their high mass capabilities was published by Daves 1979. There seemed to be two distinct strategies to follow: (a) optimization of volatilization conditions in terms of sample dispersion and heating rate; or (b) direct ionization of molecules from a surface by external electrostatic field, particle bombardment, or laser radiation. At this stage of development, however, analytical possibilities were limited to the mass range $m/z < 10{,}000$.

During the 1980s four techniques were able to overcome this limit: ^{252}Cf plasma desorption mass spectrometry (Sundqvist and Macfarlane, 1985) became available in the $m/z < 45{,}000$ mass range; fast atom bombardment (FAB)—an offspring of secondary ionization with special sample preparation—reached the $m/z \approx 24{,}000$ region (Barber and Green, 1987); the two forerunners, however, became electrospray ionization (Fenn et al., 1990) and matrix-assisted laser desorption (MALD) (Karas et al., 1989a), with $m/z = 133{,}000$ and $m/z = 250{,}000$ high mass records, respectively. The latest developments indicate capabilities for both techniques even beyond these limits (Nohmi and Fenn, 1990; Williams and Nelson, 1990).

The primary information stemming from MALD and electrospray measurements is the mass-to-charge ratio (m/z) of the molecular ions. It is the quick and accurate determination of molecular weight that makes these methods especially attractive. However, one should not overemphasize the value of molecular weight determination by these methods. Several other long-established techniques are available in the same or even in a broader molecular weight range. Ultracentrifugation, light scattering, gel permeation chromatography, and gel electrophoresis cover the molecular weight region from ~ 1000 to $\sim 5{,}000{,}000$ (Siggia, 1968). The cost and availability of the necessary instrumentation compares favorably to mass spectrometric equipment. When assessing the different molecular weight determination methods it is necessary to keep in mind that mass spectrometry directly provides mass-to-charge ratio data, whereas the separation techniques measure particle mobilities and may incorporate systematic errors.

An important aspect of the analytical value of mass spectrometric methods is their remarkably high sensitivities and low detection limits. Recent reports on electrospray ionization put sensitivities in the low picomole per microliter region and detection limits in the low femtomole region for several $556 <$

$m/z < 66{,}000$ peptides (Van Berkel et al., 1990). Both sensitivity and detection limits, however, are highly compound dependent. The detection limit increases with increasing molecular weight, because the ion signal is distributed over more charged states (Loo et al., 1990).

Similar, or even better results were obtained with laser desorption. The sensitivity was in the low picomole per microliter region, whereas the detection limits were in the subpicomole (Karas et al., 1990a: $10{,}000 < m/z < 100{,}000$ peptides) or in the subfemtomole region (Hahn et al., 1987: for porphyrin derivatives).

Naturally, these numbers should be evaluated with the achievements of other techniques in mind. As an example, we cite here some results on the analytical performance of open tubular liquid chromatography. As little as 4 fmol protein was hydrolyzed and derivatized, resulting in approximately 25 nL volume analyte solution. In this mixture quantitative determination of 14 amino acid constituents was carried out with about 5% error (Oates and Jorgenson, 1990). Chromatographic and immunoassay analyses are serious competitors in terms of sensitivity, with usually much lower instrumentation costs and consequently with better availability.

Efforts to volatilize large molecules and detect minute quantities with the aid of lasers and mass spectrometry will be the topic of this section. Laser desorption mass spectrometry of nonvolatile organic molecules was surveyed by Shibanov (1986), but vigorous development in the field since then has changed the landscape considerably.

We will focus our attention on what the different laser desorption and ionization methods have to offer over the capabilities of the more widespread analytical tools. Special emphasis will be given to the mass range and mass accuracy, to detection limits and sensitivity, to the freedom from matrix interferences, and to the availability of structural information. Two types of experiments will be examined in separate sections: laser-induced thermal desorption in Section 3A.2, and matrix-assisted laser desorption in Section 3A.4. Between these discussions, Section 3A.3 presents some theoretical considerations on the laser irradiance threshold for plume formation. The growing number of applications have made it necessary to discuss the MALD analysis of biological molecules separately. This is the subject of Section 3A.5.

In Section 3A.6 we survey and comment on the existing ideas and models that seek to explain laser desorption of large molecules. The concepts range from rapid heating (Beuhler et al., 1974) to mechanical approaches (Williams, 1990), to restricted energy transfer (Vertes and Gijbels, 1991), to expansion cooling of the laser-generated plume (Vertes, 1991b). The short final section is devoted to our present understanding of ion formation processes in MALD.

3A.2. LASER-INDUCED THERMAL DESORPTION

The idea of laser-induced thermal desorption (LITD) stems from early studies of the influence of laser radiation on small molecules adsorbed on solid surfaces. If the substrate absorbs the laser radiation, it heats up on a time scale comparable to the laser pulse length. The resulting temperature rise leads to the detachment of the adsorbed molecules.

It was soon realized that Q-switched laser pulses (whose durations are on the nanosecond time scale) may lead to subthermal velocity distributions of the desorbed particles (Wedler and Ruhmann, 1982). Further studies suggested that the desorption process is adiabatic, especially in the case of monolayers (Simpson and Hardy, 1986).

The possibility of desorbing cold molecules from hot surfaces seemed very promising and triggered studies on larger molecules. The conflict of volatilization vs. thermal lability of these compounds was thought to be overcome (van der Peyl et al., 1982a). Calculated surface temperatures showed strong correlation with the generated ion currents. The effort to implement the method for higher mass compounds, however, was apparently not completely successful: $m/z \approx 10{,}000$ seemed to be an upper limit of ion production by LITD (Ijames and Wilkins, 1988).

Clearly, the evolution of substrate surface temperature plays a decisive role in the course of events in LITD (Vertes et al., 1989b). Temporal and spatial temperature distributions, $T(\mathbf{r}, t)$, in the substrate can be described by the heat conduction equations:

$$C(T)\frac{\partial T(\mathbf{r}, t)}{\partial t} = -\operatorname{div} J(\mathbf{r}, t) + I(\mathbf{r}, t) \tag{1}$$

and

$$J(\mathbf{r}, t) = K(T)\nabla T(\mathbf{r}, t) \tag{2}$$

where $I(\mathbf{r}, t)$ is the source term describing the laser heating and $J(\mathbf{r}, t)$ is the heat flow; $C(T)$ and $K(T)$ are the specific heat and the heat conductivity, respectively, generally exhibiting considerable temperature dependence. The source term can be expressed in terms of laser irradiance, $I_0(\mathbf{r}, t)$, corrected for surface reflection, R:

$$I(\mathbf{r}, t) = \alpha(1 - R)I_0(\mathbf{r}, t)\exp[-\alpha z) \tag{3}$$

where the light propagates along the z-axis perpendicular to the substrate surface; α is the light absorption coefficient of the substrate at the laser wavelength.

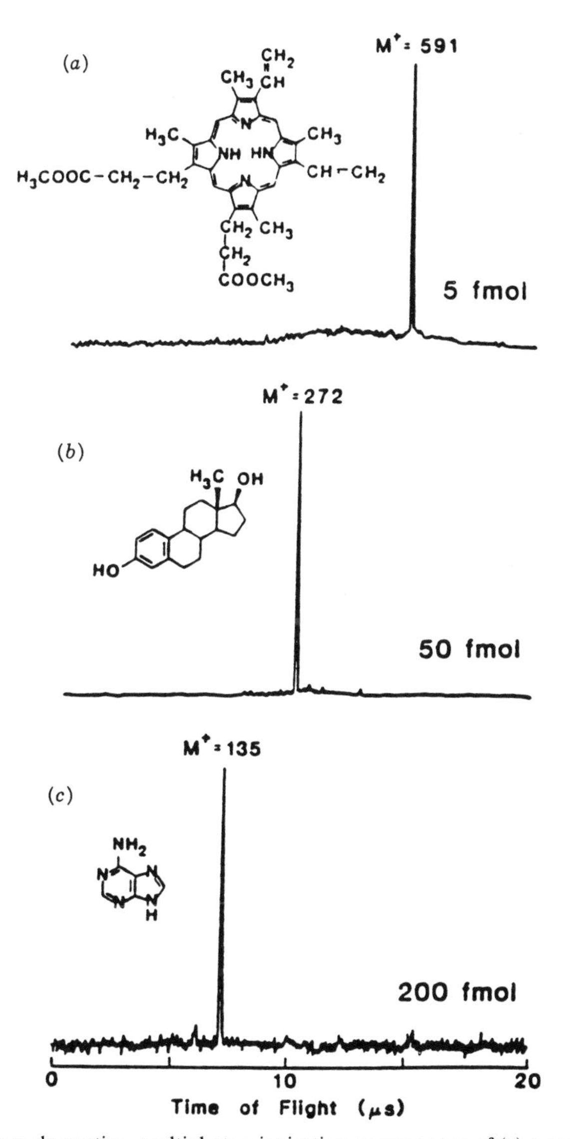

Figure 3A.1. Laser desorption–multiphoton ionization mass spectra of (a) protoporphyrin IX dimethyl ester, (b) β-estradiol, and (c) adenine. The amount of each sample desorbed by a CO_2 laser pulse is 5, 50, and 200 fmol, respectively. Operating conditions: CO_2 laser fluence, $\sim 400\,mJ/cm^2$; Nd:YAG laser fluence, $\sim 3\,mJ/cm^2$; duty cycle = 10 Hz, and signal averaging time = 20 s. From Hahn et al. (1987) with permission of the American Chemical Society.

The solution of the linear problem (K and C being constant) for general laser intensity distributions and for the special case of a Gaussian beam was found by Lax (1977). The non linear problem of temperature-dependent heat conductivity was also solved analytically (Lax, 1978).

Experimental results also became available to trace surface temperature rise as the consequence of a laser pulse. The resistance change of a vapor-deposited platinum stripe on an insulating substrate provided time-resolved information on surface heating (Zenobi et al., 1988). Comparison of the measured temperatures with analytical solutions of Eqs. (1)–(3) in the linear, one-dimensional case and with numerical solutions in the nonlinear case concluded in favor of the latter (Philippoz et al., 1989).

Both experiments and calculations indicated extremely high heating rates, generally in excess of 10^8 K/s. In a study of surface decomposition, the kinetics of reaction and desorption were considered to be competing first-order processes (Deckert and George, 1987). It was possible to show the existence of a crossover point of the product yield curves, indicating the takeover of desorption.

Analytical applications of LITD excel in particularly low detection limits. In the case of protoporphyrin IX dimethyl ester, subfemtomolar quantities of the analyte provided detectable signals ($S/N=2$) (Hahn et al., 1987). Figure 3A.1 shows molecular ion signals for three compounds utilizing femtomolar quantities of sample. The molecules were desorbed by a CO_2 laser pulse, and subsequently ionized by a frequency-quadrupoled Nd:YAG laser. Ion detection was carried out by a time-of-flight mass spectrometer. Linear dependence of the parent ion signal intensity with adsorbate concentration was found over a range of 5 orders of magnitude (Zare et al., 1988), offering remarkable quantitation possibilities.

3A.3. THE IRRADIANCE THRESHOLD FOR PLUME FORMATION

As laser irradiance increases, laser desorption—characterized alone by energy transfer across the steady surface and by occasional detachment of particles—gives rise to plume formation. To see all forms of laser–solid interaction it is important to realize that even for a given material there are several different regimes depending on the laser irradiance. The simplest case involves the following possibilities:

a. Surface heating with thermal desorption
b. Surface melting with surface evaporation
c. Volume evaporation
d. Formation of an optically thick plume

e. Plasma absorption in the plume
f. Optical breakdown (in transparent insulators)

The separation of these regimes according to the principal determining factors (i.e., laser irradiance, I_0, and the solid optical absorption coefficient, α) are shown in Figure 3A.2. The three different thresholds—here volatilization threshold, I_0^{volat}, plasma ignition threshold, I_0^{plasm}, and optical breakdown threshold, I_0^{break}—are marked by dashed vertical lines. Although their values are well defined, variations can be expected in a given experiment according to the choice of laser or target material.

In an earlier study we dealt with the transition from normal absorption of the laser-generated plume to plasma absorption characterized by 10^8–10^9 W/cm^2 threshold irradiance (Vertes et al., 1989a). For this high-irradiance transition between regimes d and e, the threshold was established on the basis of investigating normal absorption vs. plasma absorption.

Matrix-assisted laser desorption is a much milder process typically requiring about 10^6 W/cm^2 irradiance and low-melting-point or low-sublimation-temperature material. Recently, it has been shown by elaborate irradiance threshold measurements that the underlying process is a collective effect, similar in principle to phase transitions (Ens et al., 1991). Furthermore, threshold irradiance values change little by changing the mass of the guest mole-

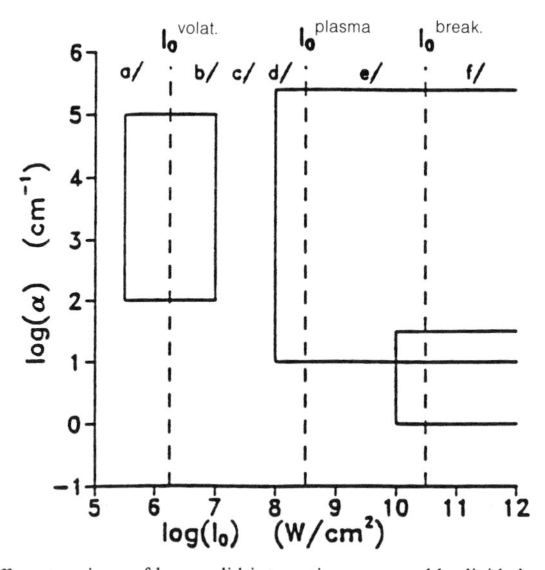

Figure 3A.2. Different regimes of laser solid interaction are roughly divided according to laser irradiance and optical absorption coefficient. From Vertes (1991b) with permission of San Francisco Press Inc. See explanation in text.

cules over an order of magnitude (Hedin et al., 1991). Therefore, we assume that the threshold of our concern marks the transition between regimes a and b or between processes a and c. To estimate the required irradiance we set the following condition: in order to reach volatilization due to laser heating the elevated surface temperature, T_{surf}, must exceed the melting or sublimation temperature of the matrix, T_{subl}:

$$T_{surf} \geqslant T_{subl} \tag{4}$$

The rise of surface temperature, $\Delta T_{surf}(t)$, at time t under the influence of a uniform penetrating light source with I_0 irradiance is (von Allmen, 1987, p. 55):

$$\Delta T_{surf}(t) = \frac{I_0}{K}\left\{\frac{\delta}{\pi^{1/2}} - \frac{1}{\alpha}\left[1 - \exp\left(\frac{\alpha\delta}{2}\right)^2 \operatorname{erfc}\left(\frac{\alpha\delta}{2}\right)\right]\right\} \tag{5}$$

Table 3A.1. Melting Temperatures and Wavelengths and Molar Extinction Coefficients of UV Absorption Maxima for Most Successful Matrices Used in Laser Desorption of Peptides and Proteins

Laser Desorption Matrix	T_m (°C)	λ_{max} (nm)	ε_{max} (L/mol·cm)
Nicotinic acid	236	262	2750
(3-pyridinecarboxilic acid)		217	8610
Pyrazinoic acid	225	267	7820
(2-pyrazinecarboxilic acid)		208	7960
Vanillic acid	214	259	11900
(4-hydroxy-3-methoxybenzoic acid)		217	22900
Sinapinic acid	192	240	10000
(*trans*-3,5-dimethoxy-4-hydroxy-cinnamic acid)		320	15849
Caffeic acid	225	324	n.a.[a]
(*trans*-3,4-dihydroxycinnamic acid)		297	
		235	
Ferrulic acid	171	320	n.a.
(*trans*-4-hydroxy-3-methoxy-cinnamic acid)		235	
2,5-Dihydroxybenzoic acid	205	335	n.a.
		237	

Source: From Grasselli and Ritchey (1975).
[a]n.a. = not available.

where $\delta = 2(\kappa t)^{1/2}$ and the thermal diffusivity $\kappa = KV_M/c_p$; here K, V_M, and c_p are the thermal conductivity, the molar volume, and the specific heat of the material, respectively. Although analytical formulas are shown to overestimate the temperature jump in pulsed surface heating, at the irradiances of our concern ($< 5 \times 10^7$ W/cm^2) the error is negligible (Philippoz et al., 1989).

If we substitute the values for material parameters of nicotinic acid ($\alpha_{265} = 4 \times 10^4$ cm^{-1}; $c_p = 150$ J/mol·K; $V_M = 83.5$ cm^3/mol; $K = 2 \times 10^{-3}$ W/cm·K), the surface temperature rise is $\Delta T_{surf}(t) = 202$ K by the end of a frequency-quadrupoled Nd:YAG laser pulse ($t = 10$ ns; $I_0 = 10^6$ W/cm^2). This value, if superimposed on room temperature, compares extremely well with the sublimation temperature of the matrix $T_{subl} = 236$ °C. Detailed investigations show the threshold irradiance of ion generation with MALD for both the nicotinic acid matrix (Beavis and Chait, 1989a) and the sinapinic acid matrix (Ens et al., 1991) to be around 10^6 W/cm^2. The most successful matrices, their melting points, and light absorption characteristics are listed in Table 3A.1. Other quantities influencing Eq. (5), such as K, V_M, and c_p, differ little for these materials. Based on these data, Eq. (5) can be evaluated for the listed matrices. The calculated threshold irradiances show insignificant variation in accordance with experimental observations.

3A.4. MATRIX-ASSISTED LASER DESORPTION

Matrix-assisted laser desorption (MALD) as a soft volatilization and ionization technique was introduced a few years ago (Tanaka et al., 1988; Karas and Hillenkamp, 1988). Since then it has acquired the reputation of being a strong candidate for the analysis of high-molecular-weight biopolymers. MALD stayed in competition with electrospray ionization for the availability of a similar mass range and for their comparable detection limits. In other respects, e.g., the problem of interferences and coupling to separation techniques, they are considered to be complementary methods. However, in the atmosphere of excitement caused by the vast array of potential new applications, not too much attention has been paid to answer basic questions concerning the mechanism of the volatilization and ionization phenomena.

In laser desorption experiments the sample should have a high absorption coefficient at the laser wavelength. For commonly used ultraviolet (UV) lasers, this condition is not met for a large number of important biomolecules. Early on, the idea of mixing these samples with a good absorber came as a possible enhancement method (Karas et al., 1987). Indeed, resonantly absorbing substances exhibited about an order of magnitude lower threshold irradiances than the analytes themselves. The lower the irradiance is, the softer the ionization method can be considered.

Based on this observation, the MALD method was introduced, essentially as a special sample preparation technique. In the first version of the experiment a dilute solution of the analyte ($\sim 10^{-5}$ M) was mixed with an equal amount of a 5×10^{-2} M aqueous solution of nicotinic acid. Solutions of the large (guest) molecules and of the matrix (host) material were mixed, providing 1:1000 to 1:10,000 molar ratios. A droplet of this mixture was air-dried on a metallic substrate and introduced into the mass spectrometer. Moderate irradiance (10^6–10^7 W/cm^2) frequency-quadrupoled Nd:YAG laser pulses (266 nm, 10 ns) were used to desorb and ionize the sample. The generated ions were typically analyzed by a time-of-flight mass spectrometer. The astonishing finding was that extremely large molecules (molecular weight exceeding 100,000 Da) could be transferred to the gas phase and ionized by this method.

Achieving softer ionization conditions for a broader class of materials was a remarkable advance in itself. Further developments came with the

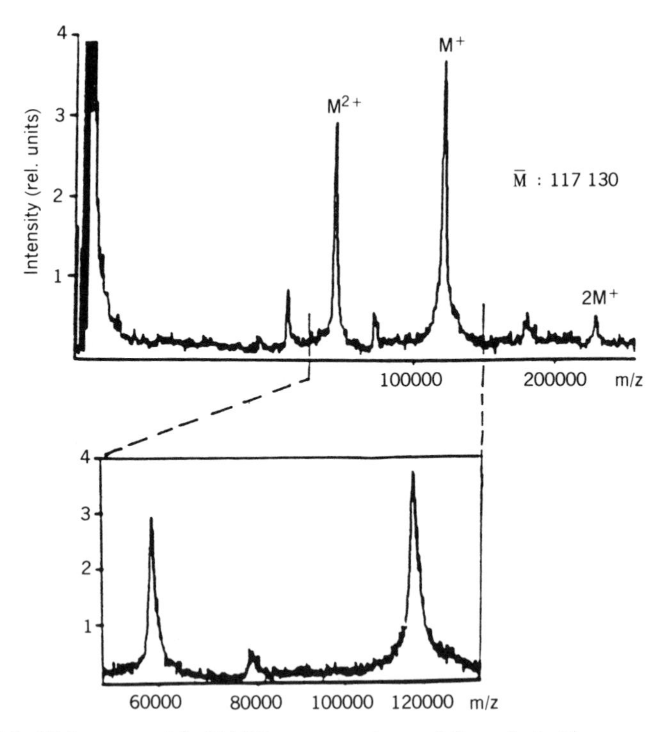

Figure 3A.3. High-mass region MALD mass spectrum of β-D-galactosidase enzyme. Signal averaging improved the signal-to-noise ratio (30 shots). From Karas et al. (1989b) with permission of VCH Verlagsgesellschaft mbH.

introduction of about 20 keV postacceleration at the end of the flight tube. The purpose of this modification was to increase the detection sensitivity for large mass ions. These particles otherwise hit the conversion dynode of the electron multiplier detector with relatively low velocity, resulting in poor conversion. Laser desorption of large ($m/z > 10{,}000$) (Karas and Hillenkamp, 1988) and very large ($m/z > 100{,}000$) (Tanaka et al., 1988; Karas et al., 1989a) ions generated great interest among mass spectroscopists who had been struggling with the volatilization problem for more than a decade.

Owing to the extended mass range, several important classes of molecules (proteins, certain polymers) became available for mass spectrometric analysis and investigation. A typical β-D-galactosidase MALD spectrum is shown in Figure 3A.3. Better preparation procedures and the introduction of new matrices yielded even higher upper mass limits, reaching $m/z \approx 250{,}000$ Da (Karas et al., 1989b). These findings had a considerable impact on the pace of laser desorption investigations. In the 1988–1992 period the number of related articles showed exponential growth as a consequence of the discovery of MALD.

A growing number of groups joined the investigation (Beavis and Chait, 1989a; Nelson et al., 1989; Salehpour et al., 1989; Ens et al., 1990; Frey and Holle, 1990; Hettich and Buchanan, 1990; Nuwaysir and Wilkins, 1990; Spengler and Cotter, 1990; Vertes et al., 1990b; Zhao et al., 1990a). Detection limits ranging from 5 pmol to 50 fmol and mass accuracy of about 0.1% were established (Karas et al., 1989c). Both characteristics, however, showed strong sample dependence. Beavis and Chait (1989b) introduced cinnamic acid derivatives as new matrix materials showing similar or superior features to nicotinic acid. Negative ion spectra of proteins were also observed (Beavis and Chait 1989a; Salehpour et al., 1989). High-mass glycoproteins and hydrophobic proteins not accessible to plasma desorption mass spectrometry showed encouraging response in MALD experiments (Salehpour et al., 1990).

Mass accuracy of the MALD method was squeezed to $\pm 0.01\%$ by using internal calibration of the mass scale (Beavis and Chait, 1990a). In most cases, the required sample amount (0.5–1.0 μL of sample solution containing several picomoles or several hundred femtomoles of analyte) compared favorably with the sample requirements of other mass spectroscopic techniques.

Frequency-tripled Nd:YAG laser ($\lambda = 355$ nm) was successfully tried to demonstrate that MALD is feasible with more affordable nitrogen lasers ($\lambda = 337$ nm) (Beavis and Chait, 1989c). Recently, possibilities have been broadened even further by the experiments on MALD using infrared (IR) radiation of Q-switched Er:YAG laser ($\lambda = 2.94\ \mu$m) (Overberg et al., 1990) and of CO_2 laser ($\lambda = 10.6\ \mu$m) (Hillenkamp, 1990).

Some early efforts were directed toward broadening the range of applicable matrices and laser wavelengths and resulted in a rough outline of required matrix material properties (Beavis and Chait, 1990c). A promising candidate for a successful matrix should exhibit the following features:

a. Strong light absorption at the laser wavelength
b. Low volatilization temperature (volatilization shall take place preferably in the form of sublimation)
c. Common solvent with the analyte
d. Ability to separate and surround the large molecules in a solid solution without forming covalent bonds

With these principles in mind several groups set out to search for appropriate and better matrices. In a thorough study Beavis and Chait (1990c) tested about 50 different matrix materials. When UV lasers were used the best results in peptide volatilization and ionization could be obtained by applying nicotinic acid, 2,5-dihydroxybenzoic acid, pyrazynoic acid, vanillic acid, ferrulic acid, sinapinic acid, caffeic acid, and certain other cinnamic acid derivatives. The important optical and thermal properties of some UV matrices are listed in Table 3A.1. Nitrobenzyl alcohol on fibrous paper substrate turned out to be a useful matrix as well (Zhao et al., 1990b). In addition to the matrices used in the UV experiments carboxylic acids, glycerol, urea and tris buffer proved also to be applicable in the IR experiments. Low analyte specificity and moderate volatility in the vacuum system complemented the list of desirable features expected from prospective matrices.

Experiments aimed at further elucidating the physical background of the MALD phenomena were launched in several laboratories. Initial kinetic energy distributions of MALD-generated ions were measured utilizing pulsed ion extraction in a Wiley–McLaren-type time-of-flight instrument (Spengler and Cotter, 1990). Departing ion energies showed about 1 eV mean value. The question of substrate participation was analyzed in a study where laser and ion optics were situated on opposite sides of suspended sample crystals (Vertes et al., 1990b). Successful matrix-assisted experiments on various low-mass peptides supported the idea of negligible substrate participation with strongly absorbing matrices.

Several groups were able to establish laser irradiance thresholds of ion generation (e.g., see Hedin et al., 1991; Ens et al., 1990). Threshold phenomena had been reported many times earlier in the general context of laser ionization. Although the absolute numbers for irradiance varied from study to study or even from sample to sample, the values were generally somewhere around several times 10^6 W/cm^2. These uncertainties were attributed to poorly defined light intensity measurements and to uncontrolled variations in sample morphology. Ens et al. (1990) used single ion counting to determine the irradiance threshold of insulin ion generation and the irradiance dependence

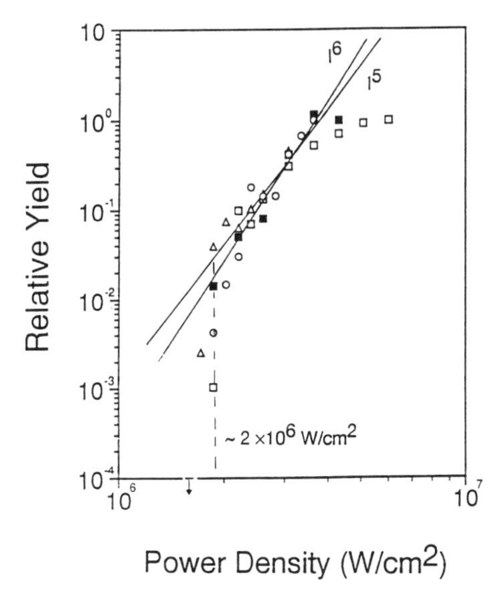

Figure 3A.4. Irradiance dependence of the insulin molecular ion yield clearly exhibits threshold behavior. From Ens et al. (1990) with permission of John Wiley & Sons, Ltd. The various symbols correspond to different detector openings.

of ion yield (see Figure 3A.4). With the analysis of pulse height distributions it also was inferred that MALD is the result of collective processes. In a different experiment it was shown that mass spectrometry is not the only method to separate and detect the large molecules (Williams, 1990).

Velocity distributions were measured for high-mass polypeptide molecular ions (Beavis and Chait, 1991b) and for neutral molecules of gramicidin S (Huth-Fehre and Becker, 1991) generated by MALD. There were several important observations in these studies. The polypeptide molecular ions showed similar *velocity* distributions, exhibiting 750 m/s average velocity independent of molecular mass in the 1000–15,600 Da range. This result was interpreted as an indication of jet-expansion-type processes during the desorption (Beavis and Chait, 1991b). Velocity distribution measurements on neutral molecules of ferrulic acid matrix and gramicidin S, a 1141.5 Da peptide, indicated common velocity maximum between 300 and 400 m/s (Huth-Fehre and Becker, 1991). This value is in good agreement with the drift velocities established by the cool plume model (see Section 3A.6.2, below). It is also remarkable that the matrix and the guest molecules showed very similar velocity distributions.

Detection mechanisms of large molecules were investigated in order to improve detection efficiency (Spengler et al., 1990c; Kaufmann et al., 1991). With increasing molecular weight of the 5–40 keV primary ions, secondary electron emission at the conversion dynode was found to be diminishing and to give rise to the generation of small secondary ions. Tandem time-of-flight measurements revealed the nature of these secondary ions. The negative secondary ion spectra consisted of H^-, C_2H^-, and $C_2H_2^-$ ions, whereas the positive secondary ions were identified as H^+, Na^+, K^+, and ions at $m/z = 28$, 41, 43, 45, 73 and minor components up to $m/z = 326$ (Kaufmann et al., 1992). The formation of these secondary ions was attributed to sputtering of surface contaminants rather than collision-induced dissociation of the primary ions.

Preliminary results were reported on the metastable decay of MALD-generated ions: 354.5 nm laser desorption of insulin molecules resulted in molecular ions decaying in the field-free flight tube, losing probably a relatively small neutral entity (Beavis and Chait, 1991a). Beavis and Chait also concluded that metastable decay is an indication of high internal (vibrational) temperatures of the desorbed ions. Metastable decay of peptide and protein ions in MALD was also observed in a time-of-flight instrument with an ion reflector (Spengler et al., 1991; Spengler and Kaufmann, 1992). The decay channels were sample specific but generally included the loss of ammonia and segments of amino acid side chains. This postsource decay showed potential to extract structural information on peptides and proteins.

Desorbed neutrals and molecular ions were investigated using tunable UV laser light for desorption, supersonic jet cooling, and resonance-enhanced multiphoton ionization (REMPI) (Frey and Holle, 1990). It is instructive to compare the effect of matrix assistance and supersonic cooling on the fragmentation patterns. Without cooling and without matrix no molecular ion signals were observed for gramicidin D (MW = 1881), whereas the introduction of either jet cooling or matrix assistance alone was sufficient to suppress the fragmentation. Preliminary estimates of neutral velocity distributions indicate about 300 K translational temperature, independent of laser wavelength ($\lambda = 10.6\ \mu m$ or 266 nm).

Pilot experiments to utilize MALD in Fourier transform mass spectrometers (FTMS) have been described (Hettich and Buchanan, 1990). Low-mass ions ($m/z < 2000$) exhibited strong matrix enhancement under FTMS conditions, but efforts to detect molecular ions in the higher mass region were unsuccessful. The origin of this deficiency was linked to the detection limitations of the FTMS instrument. Another attempt to combine MALD with FTMS concluded that the nicotinic acid matrix material sublimed from the probe tip within 15 min under the usual vacuum of FTMS measurements (10^{-8} torr) (Nuwaysir and Wilkins, 1990). Possible alternatives for matrix material were tested, with encouraging results in the case of sinapinic acid and sucrose. High-mass capabilities of MALD-FTMS have been demonstrated only recently by Castoro et al. (1992). With proper adjustment of the ion source they were able to detect ions with masses as high as 34,000 Da. Some spectra showed much higher mass resolution than those obtained by time-of-flight instruments.

Recently, a different scheme of laser volatilization has been introduced (Nelson et al., 1989; Becker et al., 1990). Thin films of the aqueous analyte solution were frozen onto cooled metal probe tips. The laser radiation ablated the ice film. The volatilized material was either collected on a solid surface and subsequently analyzed using gel electrophoresis or directly analyzed by time-of-flight mass analysis with or without postionization. Electrophoresis results on DNA digest showed very-high-mass particles ($m/z \approx 6$ MDa) in the plume, whereas TOF detection revealed $m/z \approx 18{,}500$ ions (Nelson et al., 1990). These results were rationalized in terms of sudden heating of the metal surface followed by fast heat transfer to the ice layer and by explosive boiling (Nelson and Williams, 1990).

3A.5. MALD IN THE ANALYSIS OF BIOLOGICAL MOLECULES

Applications of MALD-MS are just beginning to mushroom. The most important achievements are expected in areas where the unique features of the method can be utilized. The unparalleled mass accuracy in the high-mass region have even made it possible to revise the calibration curves of certain commercial gel electrophoresis molecular weight marker kits (Kratzin et al., 1989).

3A.5.1. Peptides and Polypeptides

Although many of the peptides can be readily analyzed by fast atom bombardment (FAB), there are special classes of these materials where MALD has demonstrated a competitive advantage. Most significantly, ion production from high-molecular-weight polypeptides by MALD does not show the cutoff around 5000 Da quite customary in the case of FAB measurements. Bovine trypsin inhibitor (6500 Da), melittin (2847 Da), and Met-Lys bradykinin (1320 Da) were studied with MALD to assign correct molecular weights to molecular weight marker kits (Kratzin et al., 1989).

An important feature of MALD is its relatively uniform sensitivity to different peptides. FAB tends to provide unstable or low ion currents for hydrophilic peptides, whereas MALD spectra show no deterioration for this important class of compounds. Two neuropeptides (substance P and bombesin), six analogues of melanocyte-stimulating hormone core, and collagenase enzyme substrates were investigated using transmission geometry MALD. Strong molecular ion peaks were present in the positive ion spectra, indepen-

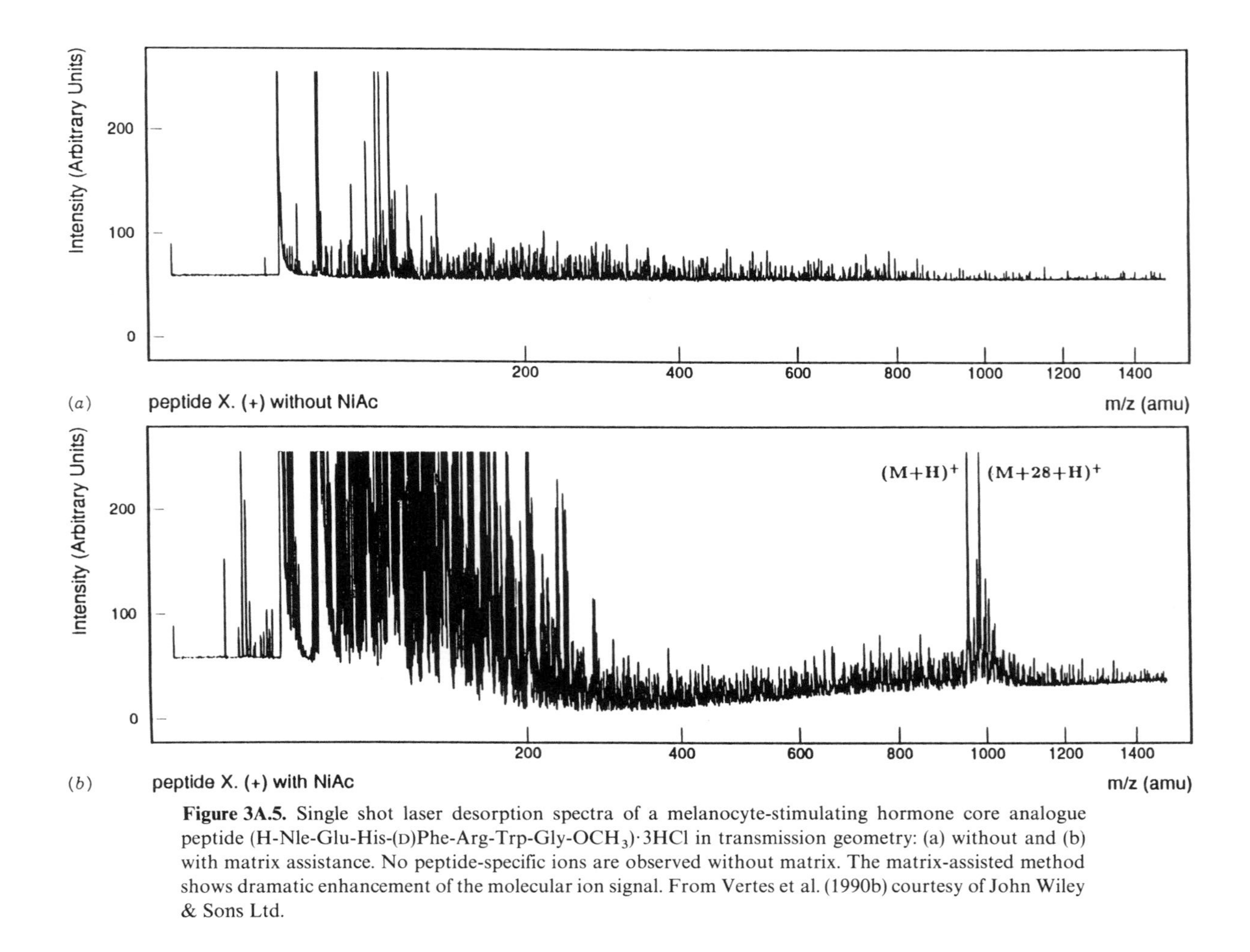

Figure 3A.5. Single shot laser desorption spectra of a melanocyte-stimulating hormone core analogue peptide (H-Nle-Glu-His-(D)Phe-Arg-Trp-Gly-OCH_3)·3HCl in transmission geometry: (a) without and (b) with matrix assistance. No peptide-specific ions are observed without matrix. The matrix-assisted method shows dramatic enhancement of the molecular ion signal. From Vertes et al. (1990b) courtesy of John Wiley & Sons Ltd.

dent of the chemical nature of the peptide (Vertes et al., 1990b). In Figure 3A.5 the dramatic enhancing effect of the matrix is demonstrated for (H-Nle-Glu-His-(D)Phe-Arg-Trp-Gly-OCH_3)·3HCl.

MALD shows detection limits in the femtomol and subfemtomole range for peptides (Strobel et al., 1991). This is clearly an asset in many areas of biochemical research (neuropeptides, etc.) where often only minute traces of the analyte are available.

A combination of enzymatic digestion and MALD can contribute to protein sequence determination. This technique is based on the capability of MALD to produce interference-free spectra of complex mixtures. Digestion of the protein by the appropriate enzyme provides a mixture of peptide fragments that can be analyzed at different stages of degradation by MALD (Schaer et al., 1991a,b). In another study, peptide sequencing is envisioned utilizing metastable decay in the field-free region of a reflectron-type time-of-flight spectrometer (Spengler et al., 1992).

3A.5.2. Proteins

Most of the measurements involving MALD have been carried out on proteins. From the time of the effect's discovery, most of the initial exploratory-type experiments have used protein samples. Here, however, we shall only focus on protein studies emphasizing applications.

Protein molecular mass determinations are particularly straightforward with MALD. The most intensive peaks in the high-mass region of the spectra are often related to the protonated molecules. Doubly and multiply charged molecular ions, as well as cluster ions, are also formed under certain conditions. The purity and molecular mass of proteins were investigated early on in several laboratories (Gabius et al., 1989; Karas et al., 1989c; Salehpour et al., 1989; Beavis and Chait, 1990d).

It is very rare to observe fragment ions using the matrix-assisted technique. This feature makes it easy to find the molecular mass of proteins. To recover the primary structure, however, knowledge of fragmentation patterns is necessary. As we mentioned earlier, several groups suggested getting around this problem by utilizing enzymatic digestion (Schaer et al., 1991b; Juhasz et al., 1992; Chait et al., 1992). Both digestion with carboxypeptidases Y and B and tryptic digestion are used in combination with MALD. By calculating the mass differences between peptides successively degraded by carboxypeptidase, it is possible to identify the released amino acids. With tryptic digestion, an amino acid sequence can be confirmed within minutes, providing unprecedented ease and rapidity of analysis.

Further development is expected in enhancing sequencing capabilities by MALD. One option is to observe metastable decay of laser-desorbed ions.

These processes take place in the field-free drift region of the time-of-flight mass spectrometer. In a preliminary study several useful fragmentation pathways of relatively small peptides were observed (Spengler et al., 1992). Another possible way to increase fragmentation is to use less effective matrices or matrix mixtures (Wilkins et al., 1992).

MALD also is useful to detect the presence or absence of certain functional groups on large proteins. Because of its superior mass accuracy MALD is

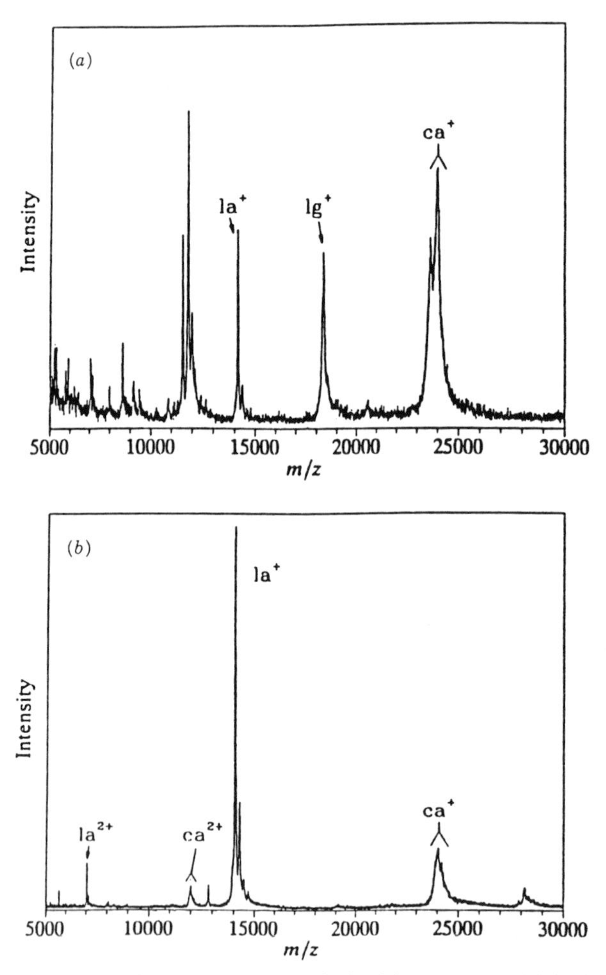

Figure 3A.6. Mass spectra of complex mixtures obtained from commercial bovine milk (a) and human breast milk (b). Several protein peaks are labeled [lactalbumin (la), lactoglobulin (lg), and casein (ca)], with the protonation states of the ions also indicated. From Beavis and Chait (1990d) with permission of the authors.

able to distinguish between untreated monoclonal antibodies and their derivatives conjugated with chelators and drugs (Siegel et al., 1991). Similarly, metal binding properties of protein surfaces can be studied (Yip and Huthchens, 1992). Mass accuracy for very large protein molecules ($\sim 150{,}000$ Da) is around $\pm 0.1\%$. Therefore, the presence of conjugated units is detected if their mass exceeds 150 Da. In the case of smaller protein substrates the mass accuracy can reach 0.01%, providing increased potential for functional group identification.

Mixtures of proteins are also readily analyzed by MALD. Proteins produce spectra even in complex mixtures of organic and inorganic constituents such as unpurified biological fluids. This feature was emphasized in an experiment where the mass spectra of commercial bovine milk and human breast milk were taken (Beavis and Chait, 1990b,d) (see Figure 3A.6). Physiological salt concentrations—usually prohibitive for other MS methods—are not detrimental to MALD spectra (Karas et al., 1990b). A comparative study of MALD and electrospray MS showed clear advantages in favor of the MALD technique, particularly for mixtures of more than two components (Allen et al., 1991). Off-line coupling of MALD with separation techniques such as high-performance liquid chromatography (HPLC), thin-layer chromatography (TLC), and gel electrophoresis further enhances the possibilities of mixture analysis.

One of the most promising applications of MALD is related to identifying protein expression errors and to investigating posttranslational modifications such as glycosylation, phosphorylation, sulfation, proteolytic processing, and disulfide bond formation among others. The method is based on comparing the molecular weight of the protein as measured by MALD to its value calculated from the related DNA sequence, or to its value without modification. The difference between these values can reveal protein expression errors (Keough et al., 1992) or even the nature of a small blocking group (Hillenkamp et al., 1991). A number of possible posttranslational modifications and the associated mass change are listed in Table 3A.2. Measuring the mass and determining the sequence of a protein produced by translation can also help in confirming the sequence of the original DNA.

Monitoring and quality control of protein synthesis and identification of by-products are also among the demonstrated capabilities of MALD (Boernsen et al., 1991). Comparison of reversed-phase HPLC, tricine–sodium dodecyl sulfate–poly(acrylamide) gel electrophoresis (SDS-PAGE) and MALD in the detection and identification of impurities in commercial β-galactosidase (MW $\sim 100{,}000$ Da) showed unambiguous advantages of MALD. Both other methods require more sample preparation and longer analysis time, and also provide less accurate mass information on the unknown species. In purification during synthetic peptide production and in

Table 3A.2. Posttranslational Modifications of Proteins and Resulting Mass Changes

Posttranslational Modification	Mass Change (Da)
Acetylation	42.04
Biotinylation (amide bond to lysine)	226.29
Carboxylation of Asp and Glu	44.01
C-terminal amide formed from Gly	−0.98
Cysteinylation	119.14
Deamidation of Asn and Gln	0.98
Deoxyhexoses (Fuc)	146.14
Disulfide bond formation	−2.02
Farnesylation	204.36
Formylation	28.01
Hexosamines (GlcN, GalN)	161.16
Hexoses (Glc, Gal, Man)	162.14
Hydroxylation	16.00
Lipoic acid (amide bond to lysine)	188.30
Methylation	14.03
Myristoylation	210.36
N-Acetylhexosamines (GlcNAc, GalNAc)	203.19
Oxidation of Met	16.00
Palmitoylation	238.41
Pentoses (Xyl, Ara)	132.12
Phosphorylation	79.98
Proteolysis of a single peptide bond	18.02
Pyridoxal phosphate (Schiff base formed to lysine)	231.14
Pyroglutamic acid formed from Gln	−17.03
Sialic acid (NeuNAc)	291.26
Stearoylation	266.47
Sulfation	80.06

Source: From Finnigan MAT (1991).

protein separation from biological products, HPLC is the method of choice. Therefore, it is not surprising that when used to assess the homogeneity of the purified product HPLC is often not able to resolve components which co-purify with the compound of interest. MALD, based on a completely different physical principle, is usually successful in these hard-to-handle situations.

Quantitation of proteins with MALD is under investigation too (Nelson et al., 1992). Preliminary results show linear response over 1 decade concentration range, using internal protein standards and the standard addition method. The estimated error of concentration determination is 10–20%.

3A.5.3. Nucleotides

There is limited experience with MALD of nucleotides. Preliminary studies show that volatilization of deoxyribonucleic acids (DNA) is possible both from frozen aqueous solutions (Nelson et al., 1989) and from rhodamine 6G dye matrix (Romano and Levis, 1991). In the first study the excimer pumped dye laser operated at 581 nm, where the DNA and the water matrix are transparent. The laser energy was therefore deposited into the copper substrate. In the second investigation, a frequency-doubled Nd:YAG laser was used at 532 nm, very close to the absorption maximum of rhodamine 6G (526 nm). After collecting the ablated material on a target both groups used PAGE for separation and autoradiography for detection. When the autoradiograms of the ablated and collected material are compared to the band distributions from the starting digest, it appears plausible that DNA molecules up to 410 kDa (622 base pairs) survived the vaporization intact or with very little fragmentation.

Mass spectrometric detection of the volatilized species is much more difficult. Molecular ions of double-stranded oligomeric nucleic acids were produced from frozen aqueous matrix only up to masses ~ 18,500 Da (Nelson et al., 1990). Smaller oligonucleotides in the mass range below 2000 Da

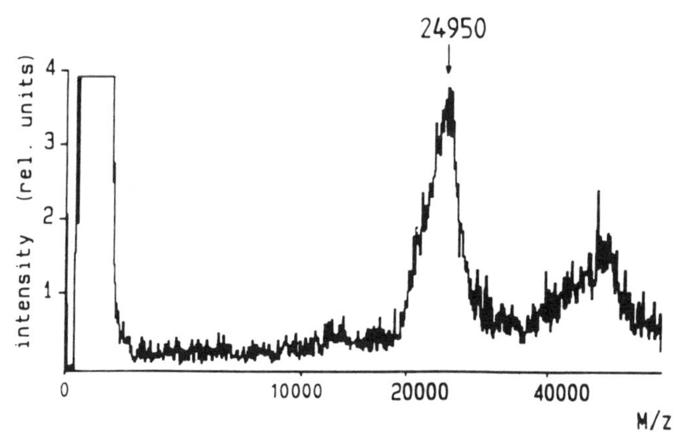

Figure 3A.7. Laser desorption–ionization mass spectrum of yeast t-RNA of molecular weight 24,952 Da; 0.1 μg/μL aqueous solution mixed with 5×10^{-2} mol/L nicotinic acid. Average of 20 single spectra. From Hillenkamp et al. (1990) courtesy of Elsevier Science Publishers B.V.

desorbed from nicotinic acid matrix also exhibited significant molecular ion signals (Spengler et al., 1990b). Yeast t-RNA of mass 24,952 Da was desorbed using 266 nm laser radiation (Hillenkamp et al., 1990). So far, this is among the largest nucleotide ions observed in a mass spectrometer. In Figure 3A.7 one can notice the corresponding broad and asymmetric molecular ion peak. The lower edge tailing of this peak is believed to be the consequence of gradual nucleobase cleavage from the RNA ion. Indeed, the related nucleobase ion signals were detected in the low-mass region of the spectrum.

Difficulties with mass spectrometric detection of intact nucleotide ions can be attributed to their strong UV absorption and to their reduced ability to accommodate positive charge. To overcome these obstacles future experiments will need to focus on finding matrix-laser combinations that ensure selective energy deposition into the matrix and utilize negative ions and neutrals for increased detection efficiency.

3A.5.4. Carbohydrates

Application of matrices for carbohydrate analysis provides less impressive spectrum enhancement over conventional techniques than it does for proteins. The standard method of measuring oligosaccharides by mass spectrometry is derivatization followed by FAB analysis. Permethylation or peracetylation of carbohydrates often improves the sensitivity of FAB analysis by about an order of magnitude. The accessible molecular weight range for this method, however, does not exceed 3500 Da.

FAB or liquid secondary ion mass spectrometry (SIMS) determination of carbohydrates, especially in their underivatized form, also is hindered by low ion yields. To overcome this difficulty matrix assistance was introduced long before the discovery of MALD (Harada et al., 1982): α- and γ-cyclodextrin and some smaller oligosaccharides were measured using diethanolamine matrix to produce abundant amine adduct ion peaks. Other FAB matrices and matrix mixtures were also assessed (Voyksner et al., 1989). With this method quasi-molecular ions up to 3700 Da are detected in the positive ion FAB spectra of native oligosaccharides.

In some cases direct laser desorption offers a practical alternative. Extensive fragmentation has recently been observed in IR laser desorption of neutral oligosaccharides (Martin et al., 1989; Spengler et al., 1990a). Fragmentation patterns are useful for structure elucidation and provide complementary information to FAB experiments. The major fragmentation pathway has been identified as a retro-aldol-type pericyclic hydrogen rearrangement leading to fragmentation of the sugar ring (Spengler et al., 1990a). The presence of this process depends on the opening of the hemiacetal saccharide ring to the linear saccharide form. Cyclic acetals usually do not exhibit this

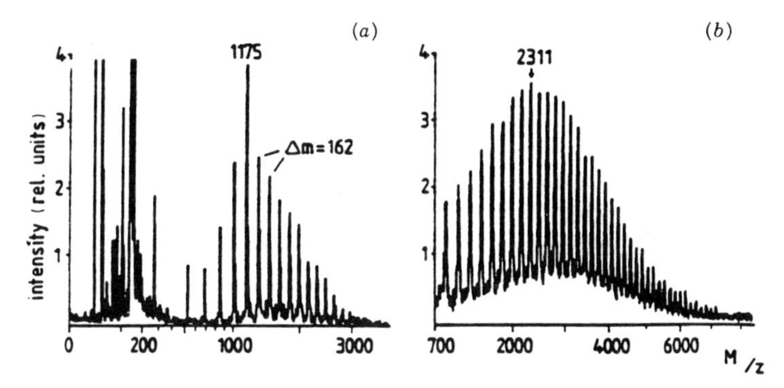

Figure 3A.8. Matrix LDI spectrum of maltodextrins (a) and dextrans (b). The various oligoglucan species differ in mass by 162 Da as indicated in the figure. *Conditions:* matrix, dihydroxybenzoic acid; sample, 100 ng (a), 200 ng (b). The spectra are averages of 20 (a) and 60 (b) single spectra. From Stahl et al. (1991) with permission of the American Chemical Society.

reaction; therefore the presence of the corresponding derivatives can be detected.

MALD analysis of glucans, oligofructans, mannose-containing oligosaccharides (Stahl et al., 1991), dextran hydrolyzates, high-mannose oligosaccharides (Mock et al., 1991), and cyclodextrins (Boernsen et al., 1990) has been shown to have significant advantages. Among them are the high throughput of the measurement and the direct information on molecular masses. Typical analysis time is around 5–15 min, which compares favorably with the duration of gel filtration and HPLC determinations. Mass spectrum evaluation is made simple by the lack of fragmentation and by the dominance of alkalinated quasi-molecular ions. Typical spectra of maltodextrins and dextrans are displayed in Figure 3A.8. Carbohydrate-related ions can be observed up to 7000 Da exhibiting clear improvement over earlier techniques. Another advantage of MALD is the capability of measuring complex carbohydrate mixtures. Figure 3A.8b shows as many as 39 dextran components. Furthermore, as we discussed earlier, the technique is relatively insensitive to the presence of inorganic salts. These features make it possible to simplify sample preparation (for example, no desalting is needed) and save some of the separation steps as well.

3A.5.5. Other Compounds

An obvious and very important extension of the MALD technique would be its application to polymers of industrial origin. This idea was already present in one of the first publications describing MALD (Tanaka et al., 1988). Poly(propylene glycol) with an average molecular weight of 4 kDa

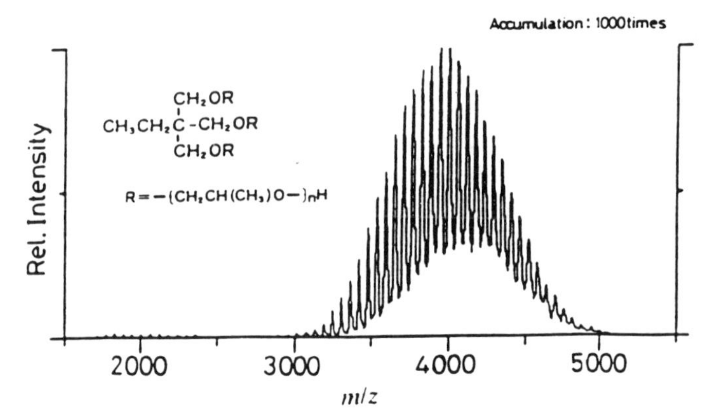

Figure 3A.9. Laser ionization mass spectrum of poly(propylene glycol). Average molecular weight is 4 kDa (PPG4K). From Tanaka et al. (1988) courtesy of Heyden & Sons, Ltd.

(PPG4K) and the 20 kDa fraction of poly(ethylene glycol) (PEG20K) were investigated. The matrix was glycerol suspension of ultrafine cobalt powder (300 Å diameter particles). The PPG4K spectrum is shown in Figure 3A.9. More recent investigations include the measurement of PEG 8K and PEG10K fractions using a sinapinic acid matrix and Fourier transform mass spectrometry (FTMS) detection (Castoro et al., 1992).

The larger and more important group of polymers showing no solubility in water has not yet been measured successfully with MALD. There are several ways to address this task. One possibility is to search for matrices showing good solubility in the organic solvents of these polymers and low sublimation temperatures. Some compounds exhibiting these features are polyaromatic hydrocarbons (PAH). These compounds show strong light absorption at the wavelength of the frequency-quadrupled Nd:YAG laser (266 nm) and readily form radical cations even at very low irradiances. Another approach could be to enhance ion generation. This can be accomplished by using strong ionizing agents in the matrix or by the introduction of a second ionizing laser pulse across the plume.

Highly thermolabile compounds constitute another class of materials that can benefit from the gentle ionization of MALD. An interesting example is vitamin B_{12}. Vitamin B_{12} is itself a strong absorber at 266 nm; therefore its quasi-molecular ions cannot be observed without a matrix. Introduction of nicotinic acid matrix, however, yields a strong $[M+H-CN]^+$ ($m/z = 1239$) ion signal at threshold irradiance (Karas et al., 1987). Increasing the laser irradiance leads to partial fragmentation. At 10 times the threshold irradiance the degree of fragmentation is comparable to the situation without a matrix. Aryltriphenylphosphonium halides represent a group of compounds with

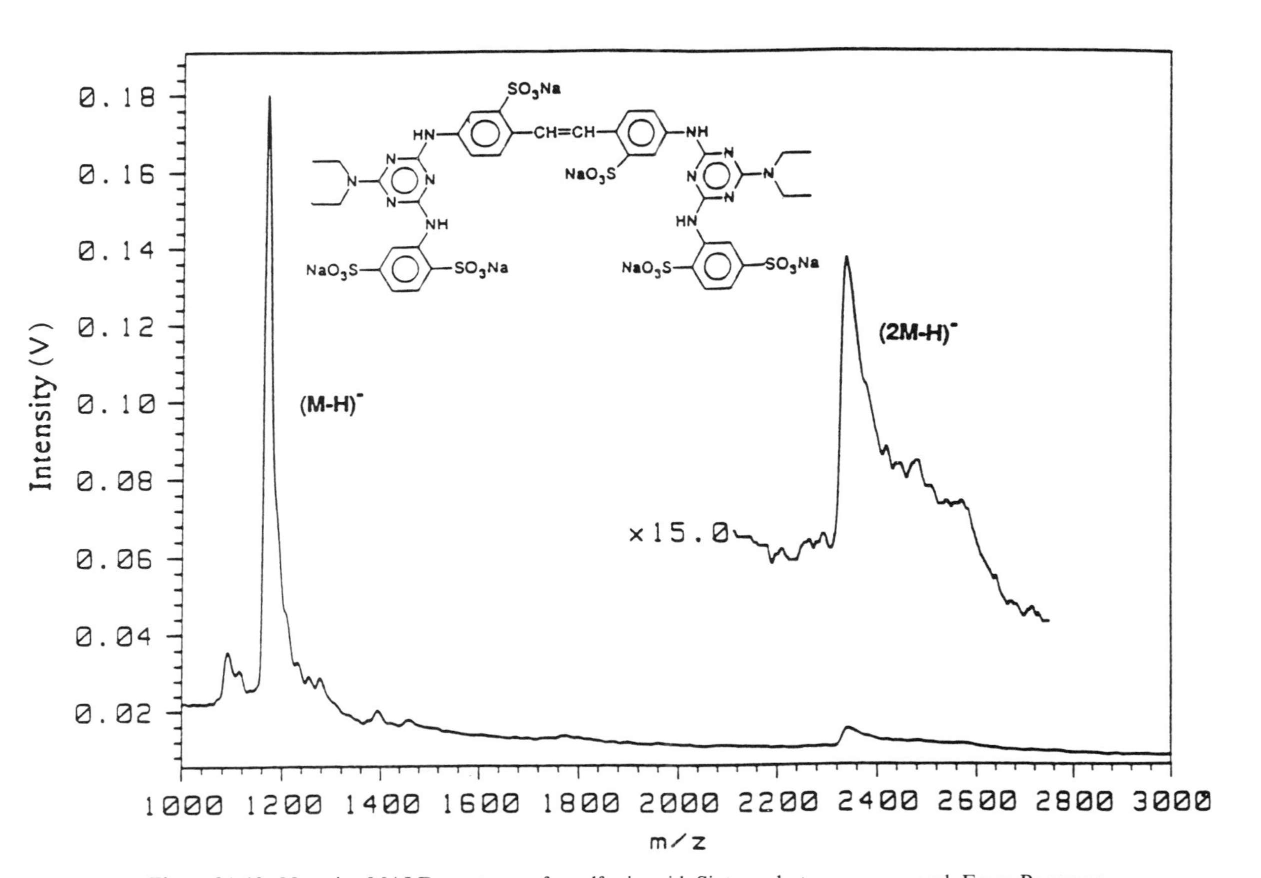

Figure 3A.10. Negative MALD spectrum of a sulfonic acid. Sixteen shots were averaged. From Boernsen et al. (1990).

extreme thermal instability. 1-Naphthylmethyltriphenylphosphonium chloride, for example, can decompose under the slightest thermal load and provide $m/z = 402$ cations. In the presence of nicotinic acid matrix, however, intact molecular ions can be observed (Claereboudt et al., 1991). Sulfonic acids also were measured in the negative ion mode using an Nd:YAG pumped dye laser at 330 nm and sinapinic acid matrix (Boernsen et al., 1990). The molecular ion and the dimer ion peaks at $m/z = 1173.8$ and $m/z = 2350.9$, respectively, are shown in Figure 3A.10. At this point we can conclude that compounds very different in chemical nature are amenable to analysis with MALD.

Rapid development in the diversification of MALD applications is expected in the near future, although the inertia of the standard biochemical approach is immense. Molecular weight determination of large biomolecules is routinely done as part of the gel electrophoresis separation step. It is difficult to compete in price and in ease of operation with the conventional method, especially since the introduction of precast gels has considerably shortened the analysis time. MALD will have to offer substantial superiority in certain features in order to get widespread recognition. The price of the instrumentation will still be prohibitive for many laboratories, but centralized handling of the tough problems in specialized facilities already seems feasible.

3A.6. PROPOSED MECHANISMS AND MODELS

Theoretical investigations date back to efforts devoted to the understanding of plasma desorption and other high-energy-particle–induced desorption techniques (Johnson, 1987). A basic question arises for all these methods: How can one account for the transfer of large molecules to the gas phase without fragmentation or degradation? The intriguing similarity between MALD and some other soft ionization methods is that they start with sudden energy deposition and they yield large molecules in the gas phase. Energy deposition and redistribution processes have emerged as a key factor in the description of MALD mechanisms (Vertes, 1991a; Vertes and Gijbels, 1991). Attempts have also been made to account for the energy transfer to the large molecules during their volatilization (Vertes, 1991b; Vertes and Gijbels, 1991).

A successful model must answer two major questions in connection with the laser-induced volatilization of large molecules

a. What is the nature of the laser-induced process leading to the transition from the solid phase to the gas phase?

b. How can large molecules escape fragmentation in an environment abruptly energized by the laser pulse?

These questions are equally relevant both in LITD and in MALD experiments. The rough options to answer question *a* are to invoke thermal or electronic processes. Closer inspection of the problem has produced several suggested mechanisms falling within these categories. Question *b* is specifically emphasized in the MALD situation where the matrix ions may undergo extensive fragmentation whereas the embedded large molecules desorb intact.

Extremely fast energy deposition—as is the case with laser heating—generates a strongly nonequilibrium population of energy levels in the solid. There are numerous different regimes for the relaxation of these energy distributions. The models suggested for the different regimes of laser desorption all incorporate some hypothesis as to the nature of this energy redistribution process.

3A.6.1. Phase Explosion

Spinodal Decomposition. It is known from thermodynamics that condensed phases can only be superheated up to a point where homogeneous vapor nucleation becomes dominant over heterogeneous nucleation (evaporation) and the whole phase is suddenly transformed into vapor (von Allmen, 1987). A rough estimate of the corresponding temperature is $0.9 \times T_c$, where T_c is the critical temperature. This phenomenon, sometimes called *phase explosion* or *spinodal decomposition*, was referred to in the explanation of fast atom bombardment (FAB) ionization experiments (Vestal, 1983; Sunner et al., 1988). Recently, preliminary molecular dynamics simulations have shown the spinodal decomposition mechanism to be feasible in particle and laser desorption experiments (Shiea and Sunner, 1990). The molecular dynamics simulation was able to visualize strongly nonequilibrium phase transitions induced by high-energy projectiles in two-dimensional Lennard–Jones fluids. Extension of the calculations to three dimensions and to laser excitation in MALD experiments is expected.

Phonon Avalanche. In laser-induced explosive desorption (Domen and Chuang, 1987) the phonon modes of the lattice are pumped by a system of excited anharmonic oscillators, i.e., by highly excited vibrational states of the molecules. Fain and Lin (1989a,b) treated UV laser–induced nonselective desorption theoretically. In their model the adsorbate (CH_2I_2 on Al_2O_3, Ag, or Al) is excited to higher electronic states. The excitation energy is then converted into internal vibrations of the adsorbates and subsequently to phonons. These authors showed that a system of excited anharmonic oscillators can become unstable under the influence of some external force field. If the energy dissipation from the phonon modes is slower than their energy gain the number of phonons shows exponential divergence; in other words, a phonon avalanche can be observed. The main characteristics of the model are the existence of a surface-coverage-dependent threshold laser fluence for the avalanche and the molecular nonselectivity of the desorption above the threshold. On the other hand, for desorption below the threshold their model predicted selectivity. Similar treatment of IR laser–induced desorption exhibited neither molecular selectivity nor threshold behavior and no phonon avalanche was reported (Fain et al., 1989).

3A.6.2. The Cool Plume Model

Hydrodynamic description of laser-generated plume expansion has contributed to the insight of laser–solid interaction (Vertes et al., 1988a, 1990a). Shock wave development under moderate and high irradiance conditions has been demonstrated, with kinetic energy distributions of desorbed and ablated ions successfully accounted for (Vertes et al., 1989c). A refined version of the one-dimensional model was able to handle phase transitions, surface recession, and heat conduction processes in the solid, as well as electron–neutral inverse bremsstrahlung light absorption, multiple ionization, and radiation cooling in the plume (Balazs et al., 1991).

Here we outline a scenario whereby the laser energy deposited into the solid matrix leads to heating and phase transition. The generated plume, in turn, undergoes gas dynamic expansion and exhibits cooling. The entrained large molecules are therefore also stabilized in the expansion. We also discuss hydrodynamic calculations of the plume expansion.

In general, we are interested in the density, temperature, and velocity distributions of the laser-generated plume as they develop in time. In contrast to the laser plasma generation experiments, we expect moderate temperatures during volatilization of low sublimation point matrices.

In the course of laser–solid interaction two distinct phases can be recognized. The first phase covers the period when the solid surface does not reach the phase transition temperature. In this regime material transport can be neglected and the description only accounts for generating a hot spot on the solid surface. The temperature distribution is governed by the relation between the laser heating and cooling of the spot by heat conduction:

$$\frac{\partial(\rho e)}{\partial t} = -\frac{\partial}{\partial z}\left[\kappa \frac{\partial(\rho e)}{\partial z}\right] + \alpha_{\text{solid}} I \tag{6}$$

where ρe and α_{solid} stand for the energy density and the absorption coefficient of the solid material. We have already seen a special solution of this equation for the case of a uniform penetrating source [Eq. (5)]. Because we now allow both for phase transitions in the solid and for different laser pulse profiles, a numerical solution of this equation will be sought. This solution will be used afterward as a boundary condition for the description of plume expansion.

The second phase starts when the surface of the solid is heated above the phase transition temperature. At this stage the vapor pressure of the material becomes significant and material transport across the surface cannot be neglected. To deal with the expansion problem we have to solve a simplified set of hydrodynamic equations expressing the conservation of mass, momentum, and energy:

$$\frac{\partial[\rho]}{\partial t} = -\frac{\partial[\rho v]}{\partial z} \tag{7}$$

$$\frac{\partial[\rho v]}{\partial t} = -\frac{\partial[p + \rho v^2]}{\partial z} \tag{8}$$

$$\frac{\partial\left[\rho\left(e + \frac{v^2}{2}\right)\right]}{\partial t} = -\frac{\partial\left[\rho v\left(e + \frac{p}{\rho} + \frac{v^2}{2}\right)\right]}{\partial z} + \alpha_{\text{plume}} I \tag{9}$$

where ρ, v, p, and α_{plume} denote the density, hydrodynamic velocity, pressure, and absorption coefficient of the plume, respectively. We note that all the transport equations [Eqs. (7)–(9)] are written in one-dimensional form. Therefore, only processes along the z coordinate (perpendicular to the surface) are accounted for; i.e., radial transport is neglected in the model. Assuming a Gaussian beam profile, this approximation is valid in the center of the beam where radial gradients are vanishing. However, the relevance of the results is not altered by this restriction, because at volatilization threshold (where most of the experimental work is done) only the center of the spot is hot enough to contribute substantially to the volatilization process.

Coupling between Eqs. (7)–(9) and Eq. (6) is provided by the Clausius–Clapeyron equation for vapor pressure. Under laser desorption conditions (i.e., near threshold irradiance), the plume remains optically thin; thus, the laser absorption term in Eq. (9) can be neglected. The plume is heated only by the transfer of warm material across the interface and cooled by the expansion process. Owing to the relatively low temperatures, thermal ionization and radiative cooling are not significant factors either. Solutions of Eqs. (6)–(9) were found by a computer code developed and reported earlier (Balazs et al., 1991).

The nicotinic acid–vacuum interface was investigated under the influence of a 10 ns frequency-quadrupled Nd:YAG laser pulse (Vertes et al., 1991; Vertes, 1991b). The temporal profile of the $10^7\ \text{W/cm}^2$ irradiance pulse is

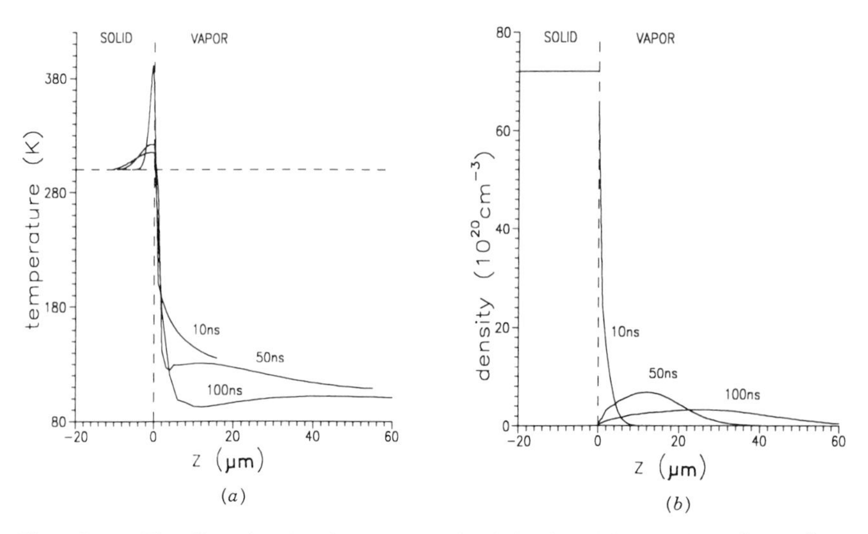

Figure 3A.11. The effect of a 10 ns frequency-quadrupled Nd:YAG laser pulse (10^7 W/cm^2) on the matrix–vacuum interface. Spatial temperature (a) and density (b) distributions are shown at different times. Surface position is marked by the vertical dashed line. In graph (a) the starting temperature is shown by the horizontal dashed line. From Vertes et al. (1991).

approximated by a square wave. Spatial distribution of the number density and temperature across the surface are shown in Figure 3A.11. Three different time stages are depicted in order to visualize postpulse behavior.

The spatial density profile at the end of the laser pulse (10 ns) showed monotonous decay (Figure 3A.11b). Subsequent cooling of the surface lowered the rate of evaporation precipitously. Consequently, the density immediately above the target dropped quickly and the vapor detached from the surface to produce a drifting and expanding plume packet (25 ns, 50 ns). We estimated the kinetic energy of particles stemming from plume translation. The drift velocity of the plume center of mass is $v_{\mathrm{drift}} = 340$ m/s at 70 ns. Drift velocity measurements of neutrals desorbed from ferrulic acid matrix due to a 10^7 W/cm^2, 266 nm laser pulse show velocity maxima between 300 and 400 m/s (Huth-Fehre and Becker, 1991). This is surprisingly good agreement. If this result is converted to kinetic energy ($E_{\mathrm{kin}} \approx 60$ meV), it is also a value close to the measured energies of molecules in the plume: $E_{\mathrm{kin}} = 40$ meV (2×10^6 W/cm^2, 248 nm, tryptophan target) (Spengler et al., 1988).

It is worthwhile to note that the plume density was relatively high; at the 50 ns time stage the maximum density still exceeded one-tenth of the solid density. With further expansion of the vapor cloud the density dropped quickly. The high initial plume density has important repercussions on the possibility of gas phase processes. It seems feasible that certain reactions are induced in this dense cloud of particles. Most important among them can be protonation, alkalination, and adduct ion formation of the guest molecules. Indeed, there is clear experimental evidence that the appearance of sodium- and/or potassium-containing quasi-molecular ions is bound to the presence of sodium and/or potassium ion signals and to the generation of a dense plume (Claereboudt et al., 1991).

Integrating the density distributions along the z-axis provides a measure of the total desorbed amount from unit surface. In Figure 3A.12 the laser irradiance dependence of this quantity is depicted. Two interesting features of this figure are the existence of a threshold irradiance and the power law relationship. Comparing Figures 3A.4 and 3A.12 positive correlation can be established between the cool plume model and measurements. Measured laser irradiance thresholds for ion production are around 2×10^6 W/cm^2, whereas calculated plume formation thresholds are close to 3×10^6 W/cm^2. The measured ion yield, Y_i, shows power law dependence on the laser irradiance: $Y_i \propto I^6$. Integrated plume density distributions, Y_n, from the model exhibit power law dependence on laser irradiance as well, but the exponent is somewhat higher: $Y_n \propto I^{7.6}$. Parametric studies of matrix properties also show strong correlation between phase transition temperature and plume density.

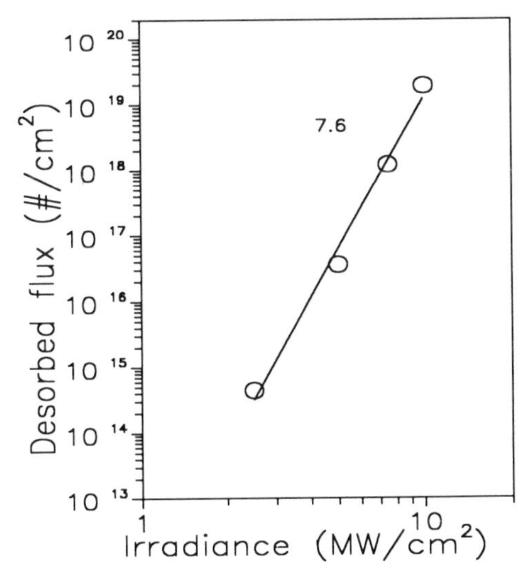

Figure 3A.12. Calculated desorbed flux of matrix molecules as a function of laser irradiance (nicotinic acid matrix; 4 × ω Nd:YAG laser). The cool plume model predicts power law with 7.6 exponent.

Another interesting feature of the plume is the spatial and temporal variation of the temperature. The surface of nicotinic acid heats up to the phase transition temperature, but quick decay of plume temperature is observed with increasing distance from the surface (Fig. 3A.11a). The actual value of the temperature drops well below room temperature due to expansion cooling. The obvious consequence of such cooling would be a stabilizing effect for the entrained large molecules. The situation is reminiscent of two-laser experiments in which jet cooling of large molecules is introduced between the desorption and ionization step.

Experimental verification of low or moderate plume temperatures, with thermally labile molecules being used as "molecular thermometers," was reported earlier (Claereboudt et al., 1991). Measuring laser-induced thermal decomposition processes of thermally extremely labile substances in native phase and in the presence of nicotinic acid matrix indicated the participation of a cooling mechanism if the matrix was present. Aryltriphenylphosphonium halide guest molecules were used as molecular thermometers in typical MALD experiments. The mass spectra revealed no thermal decomposition and low internal energies of the guest molecules, thereby supporting the cool plume model.

3A.6.3. Desorption Induced by Electronic Transitions

Low-energy electrons or UV photons may excite the surface adsorbate complex to a repulsive antibonding electronic state. A possible relaxation of this excitation is the departure of the adsorbate from the surface. The desorption induced by electronic transition (DIET) mechanism has been discovered and rediscovered several times (Menzel and Gomer, 1964; Antoniewicz, 1980). A relatively new development in utilizing the DIET model is its application to bulk etching of organic polymers by far-UV radiation (Garrison and Srinivasan, 1984, 1985). In the framework of this DIET model it is possible to rationalize polymer ablation without melting or any other thermal effect. In the irradiated volume of the polymer the monomer units are thought to be instantaneously excited to the repulsive state and leave the bulk of the solid with coherent motion.

Investigation of desorption processes induced by low-energy electron impact has already led to the idea of repulsive state participation in the mechanism of detachment (Menzel and Gomer, 1964; Antoniewicz, 1980). If by electron impact or by the absorption of UV photons an adsorbed molecule is excited electronically, the resulting state can be an antibonding state, a higher laying excited state, or an ionized state. Because of the Franck–Condon principle these excited states are not in their equilibrium geometry. The adsorbates in these configurations are frequently in the repulsive range of the interaction potential.

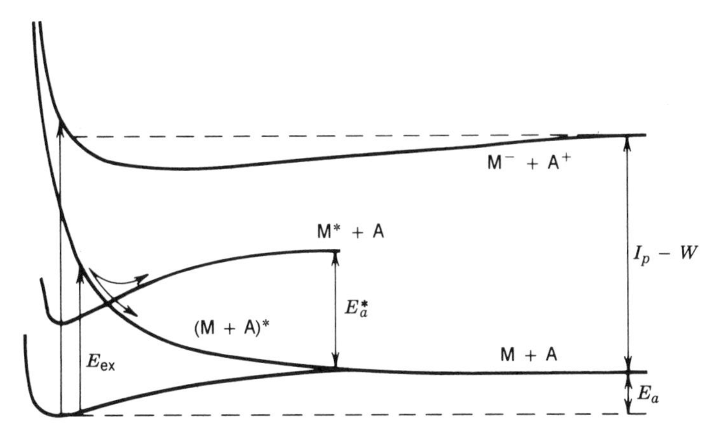

Figure 3A.13. Development of laser-induced desorption from a metallic surface (M) via repulsive states. The adsorbate (A)–surface interaction can be electronically excited to a repulsive state. Decay to M + A desorbed state or intersystem crossing to the hypersurface of the M* + A system can be observed. Higher excitation energy or two-photon processes may provide enough energy for ion desorption: $M^- + A^+$.

In Figure 3A.13 we schematically depict how the system will behave in the case of electronic excitation. If the antibonding state is reached, the adsorbed particle will experience a monotonous repulsive field and leave the surface with $(E_{ex} - E_a)$ kinetic energy, where E_{ex} and E_a are the excitation and adsorption energies, respectively. If the antibonding potential surface is intersected by the potential surface of a bonding excited state, intersystem crossing may occur and the adsorbate can be retarded. Ion desorption is observed in the case of excitation to even higher electronic states, generally attained by multiphoton processes.

It is the peculiarity of this completely nonthermal mechanism that in case of coherent electronic excitation the energy conversion leads to coherent translational motion of the desorbed particles. As a result, melting of the surface layer is avoided and extremely well-defined pits are produced. The removed particles exhibit velocity distributions oriented strongly toward the surface normal (Garrison and Srinivasan, 1985).

Microscopic modeling of laser ablation by molecular dynamic simulation supports the feasibility of this mechanism, especially for far-UV radiation (Garrison and Srinivasan, 1984, 1985). Ab initio density functional calculations and effective medium theory reveal finer details of the adsorbate–surface interaction and its perturbation by electronic excitation (Avouris et al., 1988). It is shown that the desorption of ions from metal surfaces is largely enhanced by the screening of the image charge in the metal. There are similarities between UV laser etching and the UV-MALD experiment. The recently demonstrated possibility of IR-MALD, however, seems to be in contradiction to the basic assumptions of the DIET mechanism. Also, in the framework of this model, multicharged ions remain strongly bounded to the surface, a prediction not supported by experiments with large adsorbed molecules (Hillenkamp, 1989).

Energy distribution measurements of desorbed ions and neutrals show a marked difference in the translational energy of the two species (Spengler et al., 1988; Beavis and Chait, 1991b; Huth-Fehre and Becker, 1991). Neutrals have kinetic energies in the order of 0.1 eV, whereas the ions exhibit at least 10 times higher values. If the desorption is attributed to electronic transitions, it is straightforward to rationalize this difference. The neutrals can be desorbed by single-photon excitation to a repulsive state, whereas the ions are the result of a two-photon process leading to the ionic state (Figure 3A.13). The amount of kinetic energy of desorbing particles is a function of the potential surface shape and the excitation energy. Excitation to the steep repulsing part of the ionic state interaction potential may result in departing energetic ions.

3A.6.4. The Pressure Pulse Mechanism

Disintegration via Mechanical Stress and Shock. Another way that disintegration occurs is by thermally induced mechanical stress (Beavis et al., 1988) and shock (Lindner and Seydel, 1985). The energy absorbed from the laser pulse causes inhomogeneous heating of the sample. Thermal expansion of the illuminated region produces mechanical stress. Even at moderate laser irradiances the thermally induced stress, σ, may exceed the critical stress value, σ^*, where cracks are formed and mechanical fragmentation occurs. In principle, crack formation can be observed if the strain energy exceeds the energy of new surfaces created by the cracks. It is also possible to prove that upon mechanical fragmentation a large number of quite small fragments are formed (Vertes and Levine, 1990); therefore, many of the embedded guest molecules can be released into the gas phase.

The condition that must be met for intact large molecules to be released by thermally induced stress and crack formation is

$$\tau(\sigma = \sigma^*) \ll \tau(T_G \gg T_0) \qquad (10)$$

where $\tau(T_G \gg T_0)$ is the time needed for the guest temperature to depart substantially from the initial temperature of the system, T_0.

In Figure 3A.14 the time history of crack formation is shown for a sample containing large molecules embedded in alkali halide matrix. Because of the long pulse duration of the applied CO_2 laser (10 μs), energy exchange has

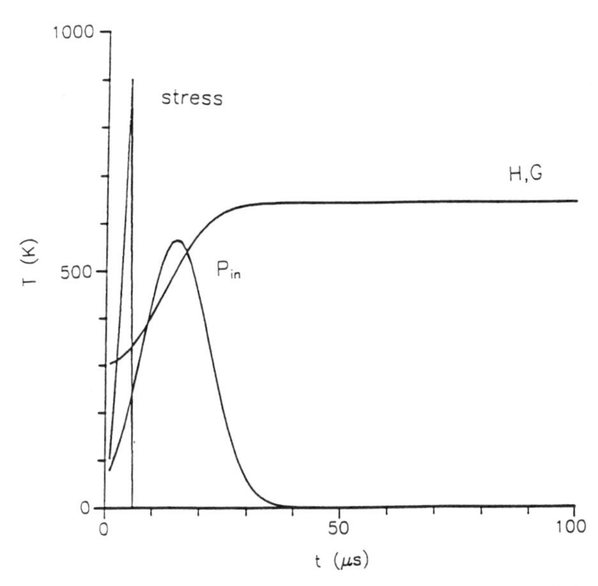

Figure 3A.14. Temporal evolution of host (*H*) and guest (*G*) temperature and thermally induced stress for a CO_2 laser pulse. The pulse profile vs. time is labeled P_{in}. The stress curve is $1000(\sigma/\sigma^*)$, where σ^* is the critical stress. At $\sigma=\sigma^*$ the matrix fragments, releasing some part of the guest molecules. From Vertes and Levine (1990) with permission of Elsevier Science Publishers B.V.

already become complete between host and guest molecules by the beginning of the excitation. The energy deposition by the laser pulse, P_{in}, generates rapidly rising thermal stress, which reaches the critical value before the overall temperature can rise excessively.

The material parameters that determine the instance of crack formation are the light absorption coefficient, the volume thermal expansion coefficient, the bulk modulus, and the critical stress value. In order to achieve earlier crack formation and guest molecules liberated with lower internal energy content, modifications are needed to increase the volume thermal expansion coefficient and/or the bulk modulus of the sample or to decrease the critical stress value. In terms of this model it is feasible that the introduction of fine metal powder into the sample improves the efficiency of volatilization (Beavis et al., 1988).

Another mechanism of releasing large molecules by mechanical effects is shock-wave-induced desorption. It has been proved that laser pulses at elevated irradiances generate shock waves in the ablated plume and compression waves in the solid target (Vertes et al., 1988a, 1989c). If a slab-like sample is thick enough not to be perforated by the laser pulse but thin enough to experience the effect of the compression wave at the back surface, intact large molecules can be detached from the sample (Lindner and Seydel, 1985). The description of this phenomenon is complex and beyond the scope of the present discussion.

Pressure Pulse Model. Originally the pressure pulse model was developed for fast heavy-ion-induced sputtering (Johnson and Sundqvist, 1991; Johnson et al., 1991). In this model the penetrating primary particles deposit their kinetic energy into the solid and set up a pressure gradient. Upon exceeding a threshold value this pressure gradient drives the expansion of the surface layer leading to desorption. A similar scenario is envisaged for laser desorption. As the laser light penetrates the target, energy is deposited and pressure gradient builds up. This model also predicts the existence of a fluence threshold. The value of the threshold fluence depends on the amount of deposited energy converted into expansion, on the cohesive energy of the solid, and on energy transfer rates in the solid. According to this description the fluence threshold should change with varying laser pulse length. Therefore, a critical test of the model may be performed by using picosecond and nanosecond laser pulses to desorb large molecules from the same matrix (Johnson and Sundqvist, 1991).

3A.6.5. Energy Redistribution Processes

Heating of the matrix lattice by the laser pulse gives no satisfactory explanation of all the findings of UV-MALD experiments. In contrast to the presence of intact guest molecules in the plume, host fragments can be abundant in the mass spectrum. This observation points to the importance of exploring the possible energy transfer pathways in the system. In the UV-MALD experiments, primary energy deposition leads to electronic excitation of the host molecules ($\pi \rightarrow \pi^*$ transition in the case of nicotinic acid matrix) (Irinyi and Gijbels, 1992). Quick internal conversion processes lead to vibrationally highly excited ground states. Part of the host molecules will decompose from these vibrational states, but another part transfers its energy to the lattice. Thus, the lattice is heated and the phase transition temperature can eventually be reached.

Vibrational Energy Transfer Experiments. Intramolecular and intermolecular vibrational energy transfer in condensed phases has been studied by picosecond laser spectroscopic methods (Seilmeier and Kaiser, 1988). These investigations were aimed at elucidating vibrational energy redistribution and equilibration after ultrashort light pulse excitation of 10 ps duration and less. Large organic molecules (usually dyes) dissolved in low-molecular-weight organic liquids exhibit the following features:

a. If excited above 1000 cm^{-1}, they often redistribute the excess energy over the vibrational manifold of the electronic ground state with a relaxation time shorter than 1 ps. It has also been shown that the transient internal temperature of the excited molecule is a meaningful term.
b. Intermolecular energy transfer is also very fast and has a relaxation time of about 10 ps, largely dependent on the degree of excitation and on the solvent molecules.
c. Excitation to high vibrational states of the S_0 ground state may exhibit fast relaxation to the bottom of the first electronically excited state (S_1). Vibrational redistribution of the energy in S_0 and in S_1 has relaxation times on the order of 0.5 ps.

Similar studies were carried out on suface adsorbates (Heilweil et al., 1989).

From our point of view such investigations should have a slightly different emphasis. First of all, in LITD or in MALD the energy is dumped into the solid substrate or into the matrix, not into the large molecule. Therefore, reverse flow of energy is generated and should be studied. Second, the amount of deposited energy is deliberately set to levels where desorption and/or phase disintegration occur; therefore, highly anharmonic displacements are induced. We cited the foregoing studies because they indicate the possibility of time-resolved energy transfer experiments of a similar nature. Scarce reports of vibrational temperature measurements on molecules after laser evaporation from a cryogenic matrix show vibrational cooling down to $T = 170 \pm 30$ K (aniline in CO_2 matrix; Nd:YAG laser; 3×10^8 W/cm^2) (Elokhin et al., 1990). Under somewhat different conditions, similar temperatures were also indicated by the *cool plume model* (see Section 3A.6.2).

Desorption vs. Fragmentation in LITD. There is a long history of investigating desorption and fragmentation kinetics at different heating rates (Beuhler et al., 1974; Deckert and George, 1987). The microscopic dynamics of energy transfer from a rapidly heated surface to the adsorbate species has been treated by stochastic trajectory modeling on a computer (Lucchese and Tully, 1984; Lim and Tully, 1986), by classical molecular dynamics simulation (Holme and Levine, 1989), and by evaluation of the survival probability of adsorbates without degradation (Muckerman and Uzer, 1989; Zare and Levine, 1987). All these studies arrived at similar conclusions: if the surface heating is rapid enough the desorbing species may have considerably lower temperature than the surface itself. In the case of the frequency mismatch between the physisorption bond and the chemical bonds of adsorbate, an energy transfer bottleneck is formed (Zare and Levine, 1987). The bottleneck model in its original form was suitable to explain LITD but not MALD experiments.

The Homogeneous Bottleneck Model. What prevents guest molecules from heating up in an environment where host molecules and lattice vibrations are both highly excited? The homogeneous bottleneck model (HBM), proposed for the description of MALD of large molecules, suggests that there is an obstacle in the energy transfer toward the embedded guest molecules. This so-called energy transfer bottleneck is caused by mismatch between the guest–host interaction frequency and the internal vibrational frequencies of the guest molecule. A simple kinetic model of the energy transfer processes shows that at an appropriately high sublimation rate the guest molecules will be liberated internally cold (Vertes et al., 1990a; Vertes and Levine, 1990).

In order to follow the path of the deposited energy in the target we partition its energy density, ρe, in the following way:

$$\rho e = (1 - x)H + L + xG + B \qquad (11)$$

where H, L, G, and B denote the energy density content of the host, the lattice, and the guest and energy density used for bond breaking, respectively; x is the volume fraction of the guest molecules.

We take the area of the volume element heated by the laser to be determined by the laser beam cross section and its thickness by the inverse of the host absorption coefficient, α_{0H}^{-1}. The energy density is increased in this volume by laser heating:

$$\frac{d(\rho e)_{\text{heat}}}{dt} = \frac{\alpha_{0H} + \alpha_{0G}}{\pi^{1/2}} I_0 \exp\left[-\frac{(t - t_0)^2}{\tau_p^2} \right] \qquad (12)$$

Here, I_0 stands for the laser irradiance; t_0 and τ_p describe the center and the half-width of a Gaussian laser pulse; and α_{0H} and α_{0G} are the effective absorption coefficients of the host and guest (weighted by .their concentration).

In the power density regime we are discussing, there are two main mechanisms to cool the excited volume: phase transformation and heat conduction. Inspecting the enthalpies of the possible phase transition processes, we conclude that the two most effective cooling phase transitions are evaporation and sublimation.

The cooling rate, expressed by the phase transition enthalpy, ΔH_{phtr}, and temperature, T_{phtr} is written as

$$\frac{d(\rho e)_{\text{cool}}}{dt} = \frac{\alpha_{0H} p_0 \Delta H_{\text{phtr}}}{(2\pi MRT)^{1/2}} \exp\left[\frac{\Delta H_{\text{phtr}}(T_L - T_{\text{phtr}})}{RT_L T_{\text{phtr}}}\right] \tag{13}$$

where p_0, is the ambient pressure; T_L is the lattice temperature expressed by $T_L = V_M L/C_p$; V_M and C_p are the molar volume and the lattice specific heat of the host with molecular weight M.

To follow the energy redistribution processes we introduce the kinetic equations where the energy exchange terms are proportional to the energy differences (Vertes et al., 1990a):

$$\frac{dH}{dt} = \frac{\alpha_{0H}}{\alpha_{0H} + \alpha_{0G}} \frac{d(\rho e)_{\text{heat}}}{dt} - \alpha_{HL}(H - L) - \alpha_{HG}(H - G) - \alpha_{HB} H \tag{14}$$

$$\frac{dL}{dt} = \alpha_{HL}(H - L) - \alpha_{LG}(L - G) - \frac{d(\rho e)_{\text{cool}}}{dt} \tag{15}$$

$$\frac{dG}{dt} = \frac{\alpha_{0G}}{\alpha_{0H} + \alpha_{0G}} \frac{d(\rho e)_{\text{heat}}}{dt} + \alpha_{HG}(H - G) + \alpha_{LG}(L - G) - \alpha_{GB} G \tag{16}$$

$$\frac{dB}{dt} = \alpha_{HB} H + \alpha_{GB} G \tag{17}$$

Here, α_{HL}, α_{HG}, α_{HB}, α_{LG}, and α_{GB} are the host–lattice, host–guest, host-bond-breaking, lattice–guest, and guest-bond-breaking energy transfer coefficients, respectively. It is assumed that all the processes except bond breaking are reversible.

The physical picture of laser volatilization underlying these equations is the following. The laser radiation electronically excites mostly the host and, with a lower cross section, also the guest molecules. With very fast internal conversion processes (on the picosecond time scale) the electronic excitation leads to internal vibrational excitation. At a given rate (α_{HL}) these internal vibrations are transferred to lattice vibration and are also channeled directly to guest vibrations (α_{HG}). The lattice is cooled by the phase transformation and transfers energy to the guest molecules (α_{LG}). The guest heating rate is determined by direct light absorption and energy transfer from the lattice and from the host molecules. Both the host and the guest molecules are subject to irreversible fragmentation, which consumes some part of the energy. An energy transfer bottleneck is an extremely low value of one or more of the transfer coefficients.

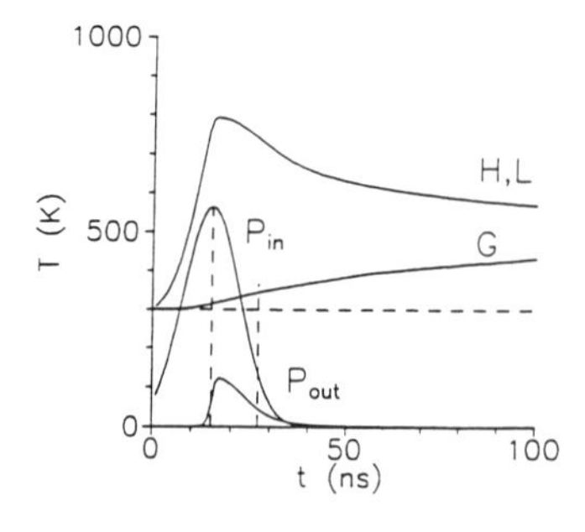

Figure 3A.15. Energy pathways for UV laser irradiation of nicotinic acid matrix containing 10^{-4} volume fraction of an $M_G = 10^5$ protein; P_{in} and P_{out} are the laser power input and the power output carried by sublimation per unit matrix area, respectively. See text for further details. From Vertes et al. (1990a) courtesy of John Wiley & Sons, Ltd.

In Figure 3A.15 the time development of host (H) and guest (G) internal temperatures are shown. The laser power input, P_{in}, heats up the host molecules. The lattice temperature is proved to be identical with the host temperature. Here P_{out} shows the power drained from the system by phase transition, in this case by sublimation. Inspecting Figure 3A.15 we note that near the maximum desorption rate the guest molecules are still close to their initial temperature. Thus, in this model, the possibility of volatilizing internally cold large molecules has been demonstrated. Recent experiments show that the matrix-assisted method is not bound to desorption from thin layers adsorbed on a substrate. In accordance with the predictions of the HBM theory, it can be carried out without substrate, from suspended crystals (Vertes et al., 1990b).

The criterion for liberating internally cold molecules from a strongly absorbing matrix can be expressed in a very general form:

$$\tau(T_L = T_{subl}) \ll \tau(T_G \approx T_H) \tag{18}$$

where $\tau(T_L = T_{subl})$ is the time required to reach sublimation temperature, T_{subl}, by the lattice temperature, T_L, and $\tau(T_G \approx T_H)$ is the approximate time needed to reach the equality of T_H and T_G, the host and the guest temperature.

The key material factors in the model appear to be the low heat of sublimation, subcritical concentration of the guest molecules, and the irradiance input in a short time compared to the sublimation induction period (Vertes et al., 1990a). The model is homogeneous in that the energy density is taken to be uniform within the "hot region" of the matrix. The two competing effects are the rates of energy transfer from the matrix to the guest molecules and the desorption by sublimation. It is the bottleneck for energy transfer to the embedded guest molecules that makes their energy content lag behind that of the matrix. This is particularly so for an initially cold sample. When a sufficiently high rate of sublimation can be achieved, the guest molecules (or adduct ions) will desorb internally cold and will thus not fragment.

This simple competitive kinetic model is able to reproduce several experimental findings: the existence of a laser irradiance threshold and its estimated value; the suitability of matrices with low phase transition temperatures; the low volume concentration requirement for the large molecules; and the need for a short laser pulse, i.e., for fast heating rates. It also predicts that sample cooling should extend the possibilities of the method. A generalization of the idea involved in MALD suggests that new desorption techniques can be successful if the liberation of large molecules by phase disintegration precedes their destruction by fragmentation (Vertes, 1991a).

There are unanswered questions in all the models discussed above. None of the models accounts for ion formation—a prerequisite for mass analysis and detection in mass spectrometry. A further shortcoming of these theories is that none of them explains the large differences in efficiency between matrices with similar phase transition temperatures and optical characteristics (Beavis and Chait, 1990c).

3A.7. ION FORMATION

The origin of the ions in the system is a subject of continuing debate. It is clear from postionization experiments that the degree of ionization in the plume is very low. In light of the calculated temperature values, it is also obvious that thermal ionization is not a feasible ion production mechanism. Rather, protonation and other adduct ion formation mechanisms shall be regarded as the primary source of ions. These processes can be discussed in terms of preformed ion volatilization and gas phase ion formation. Guest particles that are already in their ionized form in the solid state (preformed ions) can give rise to gas phase ions extremely easily. Molecular ions of these substances in MALD experiments were produced even without ion production from the matrix (Claereboudt et al., 1991). These findings support the suggestion that the role of the matrix is to embed and separate the guest particles in the solid phase and to entrain them in the course of volatilization.

Ion formation mechanisms can be classified according to the ion formation site and according to the process involved. Certain molecular solids are built up of ion associates. For example, nicotinic acid—a frequently used matrix in MALD experiments—is built up of molecular dimers held together by charge disproportionation. Defects and impurities are also sources of ionized particles. Surface states and adsorption itself are frequently promoting ion formation by preforming the gas phase ion. However, gas phase processes can also be effective in producing the ionized species. Electron impact ionization and ion–molecule reactions have high cross sections, and the ions generated in the gas phase are directly collected by the accelerating field of the spectrometer.

Thermal ion formation is not likely in the case of large molecules because the required temperatures would certainly destruct these molecules. Photoprocesses are very inefficient for IR radiation but are worthy of some interest in UV experiments. Indeed, two photons of the most often used frequency-quadrupled Nd:YAG laser provide enough energy to ionize many organic molecules.

Inspecting MALD spectra tells us that the most favored channel of molecular ion formation is cationization. Protonated and/or alkalinated molecular ions are usually abundant in positive ion spectra. Not much is known about the formation of these ions. It seems likely that protonated ions are already formed on the surface or even in the solid phase by proton hopping (Kammer, 1990) and alkalinated ions are mostly the products of gas phase ion–molecule reactions (van der Peyl et al., 1982b).

In the case of negligible dipole moments ion–molecule reactions are described by the Gioumousis–Stevenson theory (see Gioumousis and Stevenson, 1958). A simple rate constant expression can be derived for the case of the cationization of a very large molecule by a relatively light ion ($M_2 \ll M_1$):

$$k = 2\pi e \left(\frac{\alpha_{\text{pol}}}{M_2} \right)^{1/2} \tag{19}$$

where α_{pol} and M_2 are the polarizability of the large molecule and the mass of the light ion, respectively. Evaluation of Eq. (19) indicates that protonation is six times faster than the reaction with K^+ if the concentrations are similar. Calculations of the desorption profiles and ion–molecule reaction rates show feasible ion production rates for small molecules also (Vertes et al., 1988b, 1989d). However, Eq. (19) predicts moderate deterioration of gas phase cationization with increasing ion mass.

There are other types of ions abundant in the high-mass region. Adduct ions are formed by the combination of the high-mass molecules with fragment ions of the matrix. Their origin is probably also a gas phase ion–molecule process. Multiply charged species are easy to form from large molecules, as is learned from electrospray ionization experiments. Because of the possibility of independent ionization at different sites of the molecule and because of distant charge locations, the increasing size can be accompanied by increasing number of charges. Ionized clusters quite often appear in the spectra. It is unlikely that these clusters are formed in the gas phase, for guest molecule concentration is already very low in the matrix. Most experiments are performed on peptides and on proteins. These compounds are amphoteric in character and readily associate in solution. In principle the associates can survive sample preparation and desorb as a single particle.

Because only a small fraction of the desorbed species is ionized, increasing ionization efficiency might well produce significant improvement in detection limits and an even broader range of applications. Clearly, more work is needed to enhance our understanding of laser-induced ion formation from large molecules.

ACKNOWLEDGMENTS

One of the authors (A. V.) is indebted to the National Science Foundation (Grant # CTS-9212389) and to the George Washington University Facilitating Fund for their support.

REFERENCES

Allen, M. H., Grindstaff, D. J., Vestal, M. L., and Nelson, R. W. (1991). *Biochem. Soc. Trans.* **19**, 954–957.

Antoniewicz, P. R. (1980). *Phys. Rev. B* **21**, 3811–3815.

Avouris, P., Kawai, R., Lang, N. D., and Newns, D. M. (1988). *J. Chem. Phys.* **89**, 2388–2396.

Balazs, L., Gijbels, R., and Vertes, A. (1991). *Anal. Chem.* **63**, 314–320.

Barber, M., and Green, B. N. (1987). *Rapid Commun. Mass Spectrom.* **1**, 80–83.

Beavis, R. C., and Chait, B. T. (1989a). *Rapid Commun. Mass Spectrom.* **3**, 233–237.

Beavis, R. C., and Chait, B. T. (1989b). *Rapid Commun. Mass Spectrom.* **3**, 432–435.

Beavis, R. C., and Chait, B. T. (1989c). *Rapid Commun. Mass Spectrom.* **3**, 436–439.

Beavis, R. C., and Chait, B. T. (1990a). *Anal. Chem.* **62**, 1836–1840.

Beavis, R. C., and Chait, B. T. (1990b). *Proc. 38th ASMS Conf. Mass Spectrom. Allied Top., Tucson, Arizona, 1990*, pp. 26–27.

Beavis, R. C., and Chait, B. T. (1990c). *Proc. 38th ASMS Conf. Mass Spectrom. Allied Top., Tucson, Arizona, 1990*, pp. 152–153.

Beavis, R. C., and Chait, B. T. (1990d). *Proc. Natl. Acad. Sci.* USA **87**, 6873–6877.

Beavis, R. C., and Chait, B. T. (1991a). In *Methods and Mechanisms for Producing Ions from Large Molecules* (K. G. Standing and W. Ens, eds.), pp. 227–234. Plenum, New York.

Beavis, R. C., and Chait, B. T. (1991b). *Chem. Phys. Lett.* **181**, 479–484.

Beavis, R. C., Lindner, J., Grotemeyer, J., and Schlag, E. W. (1988). *Z. Naturforsch.* **43a**, 1083.

Becker, C. H., Jusinski, L. E., and Moro, L. (1990). *Int. J. Mass Spectrom. Ion Processes* **95**, R1–R4.

Beuhler, R. J., Flanigan, E., Greene, L. J., and Friedman, L. (1974). *J. Am. Chem. Soc.* **96**, 3990–3999.

Boernsen, K. O., Schaer, M., and Widmer, H. M. (1990). *Chimia* **44**, 412–416.

Boernsen, K. O., Schaer, M., Gassmann, E., and Steiner, V. (1991). *Biol. Mass Spectrom.* **20**, 471–478.

Castoro, J. A., Koster, C., and Wilkins, C. (1992). *Rapid Commun. Mass Spectrom.* (in press).

Chait, B. T., Wang, R., Beavis, R. C., and Kent, S. B. H. (1992). *Proc. 40th ASMS Conf. Mass Spectrom. Allied Top., Washington, D. C., 1992*, pp. 1939–1940.

Claereboudt, J., Claeys, M., Gijbels, R., and Vertes, A. (1991). *Proc. 39th ASMS Conf. Mass Spectrom. Allied Top., Nashville, Tennessee, 1991*, pp. 322–323.

Daves, G. D., Jr. (1979). *Mass Spectrom.* **12**, 359–365.

Deckert, A. A., and George, S. M. (1987). *Surf. Sci. Lett.* **182**, 215.

Domen, K., and Chuang, T. J. (1987), *Phys. Rev. Lett.* **59**, 1484–1487.

Elokhin, V. A., Krutchinsky, A. N., and Ryabov, S. E. (1990). *Chem. Phys. Lett.* **170**, 193–196.

Ens, W., Mao, Y., Mayer, F., and Standing, K. G. (1990). *Proc. 38th ASMS Conf. Mass Spectrom. Allied Top., Tucson, Arizona*, pp. 24–25.

Ens, W., Mao, Y., Mayer, F., and Standing, K. G. (1991). *Rapid Commun. Mass Spectrom.* **5**, 117–123.

Fain, B., and Lin, S. H. (1989a). *Chem. Phys. Lett.* **157**, 233–238.

Fain, B., and Lin, S. H. (1989b). *J. Chem. Phys.* **91**, 2726–2734.

Fain, B., Lin, S. H., and Gortel, Z. W. (1989). *Surf. Sci.* **213**, 531–555.

Fenn, J. B., Mann, M., Meng, C. K., Wong, S. F., and Whitehouse, C. M. (1990). *Mass Spectrom. Rev.* **9**, 37–70.

Finnigan MAT (1991). *Analytical Biochemistry*, LASERMAT Appl. Data Sheet No. 5. San Jose, CA.

Frey, R., and Holle, A. (1990). *Proc. 38th ASMS Conf. Mass Spectrom. Allied Top., Tucson, Arizona, 1990*, pp. 212–213.

Gabius, H. J., Bardosi, A., Gabius, S., Hellmann, K. P., Karas, M., and Kratzin, H. (1989). *Biochem. Biophys. Res. Commun.* **163**, 506–512.

Garrison, B. J., and Srinivasan, R. (1984). *Appl. Phys. Lett.* **44**, 849–851.

Garrison, B. J., and Srinivasan, R. (1985). *J. Appl. Phys.* **57**, 2909–2914.

Gioumousis, G., and Stevenson, D. P. (1958). *J. Chem. Phys.* **29**, 294.

Grasselli, J. G., and Ritchey, W. M., eds. (1975). *Atlas of Spectral Data and Physical Constants for Organic Compounds.* CRC Press, Cleveland, OH.

Hahn, J. H., Zenobi, R., and Zare, R. N. (1987). *J. Am. Chem. Soc.* **109**, 2842–2843.

Harada, K., Suzuki, M., and Kambara, H. (1982). *Org. Mass Spectrom.* **17**, 386–391.

Hedin, A., Westman, A., Hakansson, P., Sundqvist, B. U. R., and Mann, M. (1991). In *Methods and Mechanisms for* Producing Ions from Large Molecules (K. G. Standing and W. Ens, eds.), pp. 211–219. Plenum, New York.

Heilweil, E. J., Casassa, M. P., Cavanagh, R. R., and Stephenson, J. C. (1989). *Annu. Rev. Phys. Chem.* **40**, 143–171.

Hettich, R. L., and Buchanan, M. V. (1990). *Proc. 38th ASMS Conf. Mass Spectrom. Allied Top., Tucson, Arizona, 1990*, 156–157.

Hillenkamp, F. (1989). *Adv. Mass Spectrom.* **11A**, 354.

Hillenkamp, F. (1990). *Proc. 38th ASMS Conf. Mass Spectrom. Allied Top., Tucson, Arizona, 1990*, pp. 8–9.

Hillenkamp, F., Karas, M., Ingendoh, A., and Stahl, B. (1990). In *Biological Mass Spectrometry* (A. L. Burlingame and J. A. McCloskey, eds.), pp. 49–60. Elsevier, Amsterdam.

Hillenkamp, F., Karas, M., Beavis, R. C., and Chait, B. T. (1991). *Anal Chem.* **63**, 1193A–1203A.

Holme, T. A., and Levine, R. D. (1989). *Surf. Sci.* **216**, 587–614.

Huth-Fehre, T., and Becker, C. H. (1991). *Rapid Commun. Mass Spectrom.* **5**, 378–382.

Ijames, C. F., and Wilkins, C. L. (1988). *J. Am. Chem. Soc.* **110**, 2687–2688.

Irinyi, G., and Gijbels, R. (1992). *Proc. 4th Sanibel Conf. Mass Spectrom. ASMS, Sanibel Island, Florida, 1992*, p. 59.

Johnson, R. E. (1987). *Int J. Mass Spectrom. Ion Processes* **78**, 357–392.

Johnson, R. E., and Sundqvist, B. U. R. (1991). *Rapid Commun. Mass Spectrom.* **5**, 574–578.

Johnson, R. E., Banerjee, S., Hedin, A., Fenyo, D., and Sundqvist, B. U. R. (1991). In *Methods and Mechanisms for Producing Ions from Large Molecules* (K. G. Standing and W. Ens, eds.), pp. 89–99. Plenum, New York.

Juhasz, P., Papayannopoulo, I. A., Zeng, C., Papov, V., and Biemann, K. (1992). *Proc. 40th ASMS Conf. Mass Spectrom. Allied Top., Washington, D.C., 1992*, pp. 1913–1914.

Kammer, H. F. (1990). In *Mass Spectrometry of Large Non-Volatile Molecules for Marine Organic Chemistry* (E. R. Hilf and W. Tuszynski, eds.), pp. 61–72. World Scientific, Singapore.

Karas, M., and Hillenkamp, F. (1988). *Anal. Chem.* **60**, 2299–2301.

Karas, M., Bachmann, D., Bahr, U., and Hillenkamp, F. (1987). *Int. J. Mass Spectrom. Ion Processes* **78**, 53–68.

Karas, M., Bahr, U., and Hillenkamp, F. (1989a). *Int. J. Mass Spectrom. Ion Processes* **92**, 231–242.

Karas, M., Bahr, U., Ingendoh, A., and Hillenkamp, F. (1989b). *Angew. Chem., Int. Ed. Engl.* **28**, 760–761.

Karas, M., Ingendoh, A., Bahr, U., and Hillenkamp, F. (1989c). *Biomed. Environ. Mass Spectrom.* **18**, 841–843.

Karas, M., Bahr, U., Ingendoh, A., Nordhoff, E., Stahl, B., Strupat, K., and Hillenkamp, F. (1990a). *Anal. Chim. Acta* **241**, 175–185.

Karas, M., Bahr, U., Ingendoh, A., Stahl, B., Strupat, K., Nordhoff, E., and Hillenkamp, F. (1990b). *Proc. 2nd Int. Symp. Appl. Mass Spectrum. Health Sci., Barcelona, Spain.*

Kaufmann, R., Spengler, B., and Kirsch, D. (1991). In *Methods and Mechanisms for Producing Ions from Large Molecules* (K. G. Standing and W. Ens, eds.), pp. 235–245. Plenum, New York.

Kaufmann, R., Kirsch, D., Rood, H., and Spengler, B. (1992). *Rapid Commun. Mass Spectrom.* **6**, 98–104.

Keough, T., Lacey, M. P., Oppenheimer, C. L., and Thaman, D. A. (1992). *Proc. 40th ASMS Conf. Mass Spectrom. Allied Top., Washington, D.C., 1992*, pp. 1907–1908.

Kratzin, H. D., Wiltfang, J., Karas, M., Neuhoff, V., and Hischmann, N. (1989). *Anal. Biochem.* **183**, 1–8.

Lax, M. (1977). *J. Appl. Phys.* **48**, 3919–3924.

Lax, M. (1978). *Appl. Phys. Lett.* **33**, 786–788.

Lim, C., and Tully, J. C. (1986). *J. Chem. Phys.* **85**, 7423–7433.

Lindner, B., and Seydel, U. (1985). *Anal. Chem.* **57**, 895–899.

Loo, J. A., Edmonds, C. G., Smith, R. D., Lacey, M. P., and Keough, T. (1990). *Biomed. Environ. Mass Spectrom.* **19**, 286–294.

Lucchese, R. R., and Tully, J. C. (1984). *J. Chem. Phys.* **81**, 6313–6319.

Martin, W. B., Silly, L., Murphy, C. M., Raley, T. J., Jr., Cotter, R. J., and Bean, M. F. (1989). *Int. J. Mass Spectrom. Ion Processes* **92**, 243–265.

Menzel, D., and Gomer, R. (1964). *J. Chem. Phys.* **41**, 3311–3328.

Mock, K. K., Davey, M., and Cottrell, J. S. (1991). *Biochem. Biophys. Res. Commun.* **177**, 644–651.

Muckerman, J. T., and Uzer, T. (1989). *J. Chem. Phys.* **90**, 1968–1973.

Nelson, R. W., and Williams, P. (1990). *Proc. 38th ASMS Conf. Mass Spectrom. Allied Top, Tucson, Arizona, 1990*, pp. 168–169.

Nelson, R. W., Rainbow, M. J., Lohr, D. E., and Williams, P. (1989). *Science* **246**, 1585–1587.

Nelson, R. W., Thomas, R. M., and Williams, P. (1990). *Rapid Commun. Mass Spectrom.* **4**, 348–351.

Nelson, R. W., McLean, M. A., and Vestal, M. L. (1992). *Proc. 40th ASMS Conf. Mass Spectrom. Allied Top., Washington, D.C., 1992*, pp. 1919–1920.

Nohmi, T., and Fenn, J. B. (1990). *Proc. 38th ASMS Conf. Mass Spectrom. Allied Top. Tucson, Arizona, 1990*, pp. 10–11.

Nuwaysir, L. M., and Wilkins, C. L. (1990). *Proc. 38th ASMS Conf. Mass Spectrom. Allied Top., Tucson, Arizona, 1990*, pp. 844–845.

Oates, M. D., and Jorgenson, J. W. (1990). *Anal. Chem.* **62**, 1577–1580.

Overberg, A., Karas, M., Bahr, U., Kaufmann, R., and Hillenkamp, F. (1990). *Rapid Commun. Mass Spectrom.* **4**, 293–296.

Philippoz, J. M., Zenobi, R., and Zare, R. N. (1989). *Chem. Phys. Lett.* **158**, 12–17.

Romano, L. J., and Levis, R. J. (1991). *J. Am. Chem. Soc.* **113**, 9665–9667.

Salehpour, M., Perera, I., Kjellberg, J., Hedin, A., Islamian, M. A., Hakansson, P., and Sundqvist, B. U. R. (1989). *Rapid Commun. Mass Spectrom.* **3**, 259–263.

Salehpour, M., Perera, I. K., Kjellberg, J., Hedin, A., Islamian, M. A., Hakansson, P., Sundqvist, B. U. R., Roepstorff, P., Mann, M., and Klarskov, K. (1990). In *Ion Formation from Organic Solids (IFOS V)* (A. Hedin, B. U. R. Sundqvist, and A. Benninghoven, eds.), pp. 119–124. Wiley, Chichester.

Schaer, M., Boernsen, K. O., and Gassmann, E. (1991a). *Rapid Commun. Mass Spectrom.* **5**, 319–326.

Schaer, M., Boernsen, K. O., Gassmann, E., and Widmer, H. M. (1991b). *Chimia* **45**, 123–126.

Seilmeier, A., and Kaiser, W. (1988). In *Ultrashort Laser Pulses and Applications* (W. Kaiser, ed.), pp. 279–317. Springer-Verlag, Berlin.

Shibanov, A. N. (1986). In *Laser Analytical Spectrochemistry* (V. S. Letokhov, ed.), pp. 353–404. Adam Hilger, Bristol.

Shiea, J., and Sunner, J. (1990). *Proc. 38th ASMS Conf. Mass Spectrom. Allied Top., Tucson, Arizona, 1990*, pp. 166–167.

Siegel, M. M., Hollander, I. J., Hamann, P. R., James, J. P., Hinman, L., Smith, B. J., Farnsworth, A. P. H., Phipps, A., King, D. J., Karas, M., Ingendoh, A., and Hillenkamp, F. (1991). *Anal. Chem.* **63**, 2470–2481.

Siggia, S. (1968). *Survey of Analytical Chemistry*, pp. 141–175. McGraw-Hill, New York.

Simpson, C. J. S. M., and Hardy, J. P. (1986). *Chem. Phys. Lett.* **130**, 175–180.

Spengler, B., and Cotter, R. J. (1990). *Anal. Chem.* **62**, 793–796.

Spengler, B., and Kaufmann, R. (1992). *Analusis* **20**, 91–101.

Spengler, B., Bahr, U., Karas, M., and Hillenkamp, F. (1988). *Anal. Instrum.* **17**, 173–193.

Spengler, B., Dolce, J. W., and Cotter, R. J. (1990a). *Anal. Chem.* **62**, 1731–1737.

Spengler, B., Pan, Y., Cotter, R. J., and Kan, L. S. (1990b). *Rapid Commun. Mass Spectrom.* **4**, 99–102.

Spengler, B., Kirsch, D., Kaufmann, R., Karas, M., Hillenkamp, F., and Giessmann, U. (1990c). *Rapid Commun. Mass Spectrom.* **4**, 301–305.

Spengler, B., Kirsch, D., and Kaufmann, R. (1991). *Rapid Commun. Mass Spectrom.* **5**, 198–202.

Spengler, B., Kirsch, D., Kaufmann, R., and Jaeger, E. (1992). *Rapid Commun. Mass Spectrom.* **6**, 105–108.

Stahl, B., Steup, M., Karas, M., and Hillenkamp, F. (1991). *Anal. Chem.* **63**, 1463–1466.

Strobel, F. H., Solouki, T., White, M. A., and Russell, D. H. (1991). *J. Am. Soc. Mass Spectrom.* **2**, 91–94.

Sundqvist, B., and Macfarlane, R. D. (1985). *Mass Spectrom. Rev.* **4**, 421–460.

Sunner, J., Ikonomou, M. G., and Kebarle, P. (1988). *Int. J. Mass Spectrom. Ion Processes* **82**, 221–237.

Tanaka, K., Waki, H., Ido, Y., Akita, S., Yoshida, Y., and Yoshida, T. (1988). *Rapid Commun. Mass Spectrom.* **2**, 151–153.

Van Berkel, G. J., Glish, G. L., and McLuckey, S. A. (1990). *Anal. Chem.* **62**, 1284–1295.

van der Peyl, G. J. Q., Haverkamp, J., and Kistemaker, P. G. (1982a). *Int. J. Mass Spectrom. Ion Phys.* **42**, 125–141.

van der Peyl, G. J. Q., Isa, K., Haverkamp, J., and Kistemaker, P. G. (1982b). *Nucl. Intrum. Methods* **198**, 125–130.

Vertes, A. (1991a). In *Methods and Mechanisms for Producing Ions from Large Molecules* (K. G. Standing and W. Ens, eds.), pp. 275–286. Plenum, New York.

Vertes, A. (1991b). In *Microbeam Analysis—1991* (D. G. Howitt, ed.), pp. 25–30. San Francisco Press, San Francisco.

Vertes, A., and Gijbels, R. (1991). *Scanning Microsc.* **5**, 317–328.

Vertes, A., and Levine, R. D. (1990). *Chem. Phys. Lett.* **171**, 284–290.

Vertes, A., Juhasz, P., DeWolf, M., and Gijbels, R. (1988a). *Scanning Microsc.* **2**, 1853–1877.

Vertes, A., Juhasz, P., Jani, P., and Czitrovszky, A. (1988b). *Int. J. Mass Spectrom. Ion Processes* **83**, 45–70.

Vertes, A., DeWolf, M., Juhasz, P., and Gijbels, R. (1989a). *Anal. Chem.* **61**, 1029–1035.

Vertes, A., Juhasz, P., Balazs, L., and Gijbels, R. (1989b). In *Microbeam Analysis—1989* (P. E. Russell, ed.), pp. 273–276. San Francisco Press, San Francisco.

Vertes, A., Juhasz, P., DeWolf, M., and Gijbels, R. (1989c). *Int. J. Mass Spectrom. Ion Processes* **94**, 63–85.

Vertes, A., Juhasz, P., and Gijbels, R. (1989d). *Fresenius' Z. Anal. Chem.* **334**, 682.

Vertes, A., Gijbels, R., and Levine, R. D. (1990a). *Rapid Commun. Mass Spectrom.* **4**, 228–233.

Vertes, A., Balazs, L., and Gijbels, R. (1990b). *Rapid Commun. Mass Spectrom.* **4**, 263–266.

Vertes, A., Gijbels, R., and Adams, F. (1990c). *Mass Spectrom. Rev.* **9**, 71–113.

Vertes, A., Irinyi, Gy., Balazs, L., and Gijbels, R. (1991). *Proc. 39th ASMS Conf. Mass Spectrom. Allied Top., Nashville, Tennessee, 1991*, pp. 927–928.

Vestal, M. (1983). *Mass Spectrom. Rev.* **2**, 447–480.

Von Allmen, M. (1987). *Laser-beam Interactions with Materials: Physical Principles and Applications.* Springer-Verlag, Berlin.

Voyksner, R. D., Williams, F. P., Smith, C. S., Koble, D. L., and Seltzman, H. H. (1989). *Biomed. Environ. Mass Spectrom.* **18**, 1071–1078.

Wedler, G., and Ruhmann, H. (1982). *Surf. Sci.* **121**, 464–486.

Wilkins, C. L., Castoro, J. A., and Koester, C. (1992). *Proc. 40th ASMS Conf. Mass Spectrom. Allied Top., Washington, D. C., 1992*, pp. 1923–1924.

Williams, P. (1990). In *Ion Formation from Organic Solids (IFOS V)* (A. Hedin, B. U. R. Sundqvist, and A. Benninghoven, eds.), pp. 131–135. Wiley, Chichester.

Williams, P., and Nelson, R. W. (1990). *Proc. 38th ASMS Conf. Mass Spectrom. Allied Top, Tucson, Arizona, 1990*, pp. 22–23.

Yip, T. T., and Hutchens, T. W. (1992). *Proc. 40th ASMS Conf. Mass Spectrom. Allied Top., Washington, D. C., 1992*, pp. 1915–1916.

Zare, R. N., and Levine, R. D. (1987). *Chem. Phys. Lett.* **136**, 593–599.

Zare, R. N., Hahn, J. H., and Zenobi, R. (1988). *Bull. Chem. Soc. Jpn.* **61**, 87–92.

Zenobi, R., Hahn, J. H., and Zare, R. N. (1988). *Chem. Phys. Lett.* **150**, 361–365.

Zhao, S., Somayajula, K. V., Sharkey, A. G., and Hercules, D. M. (1990a). *Proc. 38th ASMS Conf. Mass Spectrom. Allied Top., Tucson, Arizona, 1990*, pp. 154–155.

Zhao, S., Somayajula, K. V., Sharkey, A. G., and Hercules, D. M. (1990b). *Fresenius, Z. Anal. Chem.* **338**, 588–592.

CHAPTER

3

METHODS UTILIZING LOW AND MEDIUM LASER IRRADIANCE

B. STRUCTURAL CHARACTERIZATION OF ORGANIC MOLECULES BY LASER MASS SPECTROMETRY

L. VAN VAECK, W. VAN ROY, R. GIJBELS, and F. ADAMS

Department of Chemistry
University of Antwerp (UIA)
Antwerp, Belgium

3B.1. INTRODUCTION

Since its first use in organic analysis, mass spectrometry (MS) has relied largely on electron impact (EI) ionization, usually at the energy of 70 eV. The extensive fragmentation often prevents the detection of molecular ions. The advent of positive and negative chemical ionization has enabled us to overcome this problem. Unless the measurement of gases is involved, a suitable sample introduction technique is required to achieve volatilization of the analyte. Thermal evaporation on a direct insertion probe has been popular for the qualitative characterization of solid samples. However, application is confined to relatively volatile, low-molecular-weight (MW), and/or thermally stable samples. The same problem exists for the hyphenated techniques, developed later. The on-line coupling of gas chromatography (GC) and MS has largely contributed to the development of organic quantitative and trace analysis. As a result, the combination of thermal evaporation and gas phase ionization has achieved a key position because of the excellent reproducibility and easiness of achieving comparable experimental conditions on a variety of instruments. All the required information about structure and functionalities of the molecule can thus be obtained in a readily interpretable form, as long as the molecule under study does not undergo decomposition in the introduction system.

Laser Ionization Mass Analysis, Edited by Akos Vertes, Renaat Gijbels, and Fred Adams. Chemical Analysis Series, Vol. 124.
ISBN 0-471-53673-3

The desire to study thermolabile, polar, and high-mass molecules from solid samples has prompted development of the so-called desorption techniques. Various techniques have been elaborated, ranging widely in complexity from rapid heating of a metal substrate, in the absence or presence of a strong electric field, up to MeV particle impact. Flash desorption, field desorption (FD), and field ionization (FI) have preceded the variety of "beam" techniques such as secondary ion MS (SIMS), plasma desorption (PD), heavy ion bombardment MS, fast atom bombardment (FAB) MS, and finally also laser desorption (LD). Sample energization is thus performed by a beam of ions, neutrals, or photons with energies ranging from a few electron volts to more than 100 MeV. The recent evolution has overcome the volatility barrier. It appears that there are no definite upper mass limits on samples for which MW information may be obtained by MS. Most of these apparently dissimilar techniques produce relatively stable even-electron parent ions similar to these present in solution rather than the odd-electron ones generated by gas phase electron interaction. Depending on the method, fragmentation is either extensive or negligible.

Organic chemists are interested in laser-based techniques for several reasons apart from the capability to deal with the characterization of thermolabile and involatile molecules. First of all, laser-based techniques are directly compatible with the measurement of insulating samples. Additionally, commercially available laser sources allow a large degree of experimental freedom. The range of conditions covered by the laser experiments in the literature is quite large, with power densities from ca. 10 up to $10^{10}\ W \cdot cm^{-2}$ and wavelengths from the ultraviolet (UV) to the far infrared (IR). Finally, unlike beams of charged particles, focusing of the laser beam down to the diffraction limit is experimentally easy and permits local and microanalysis. In fact, microprobes have long been unavailable in the field of organic MS, as opposed to the range of instruments which can be used for elemental applications.

Conceptually, LD-MS on organic compounds dates from 1970, when the detection of cationized molecules without fragmentation or decomposition was reported for sodium hexylsulfonates (Vastola et al., 1970). However, the 1978 report by Posthumus et al. was perhaps most responsible for the subsequent interest in using LD for the analysis of polar, thermolabile, and relatively high-MW organic compounds. For two or three years, several laboratories conducted experiments on a variety of instrumental configurations (Hillenkamp, 1983). The use of LD-MS was confined to these home-built setups. In fact, the development of the method may also have suffered from competition with FAB-MS, which is more easily implemented on existing MS systems. Subsequently, the introduction of commercially available laser microprobes, originally intended for elemental determinations, yielded a real renaissance of the LD-MS approach for organic compounds. These instruments not only offer significant assets for structure elucidation, they also permit characterization, down to the micrometer level, of micro-objects and the localization of organic target compounds at solid surfaces or even within complex matrices. The number of publications on the application of time-of-flight (TOF)–based laser microprobe MS (LMMS) instruments apparently has decreased of late, with the exception of reports on the remarkable innovation of matrix-assisted desorption, permitting the detection of giant ions with a MW up to 230 kDa so far. Nevertheless, LD-MS is currently gaining wide acceptance as the method that excellently fulfills the requirements for combination with Fourier transform (FT) MS, both on a bulk and on a micro scale. An extensive description of such laser-based instruments is given in Chapter 2 of this volume.

The purpose of this chapter is to survey the literature data on LD-MS of organic compounds. Next, Section 3B.2 will deal with the background of the major processes in ion generation and fragmentation without attempting a thorough phenomenological description. Indeed, in spite of the substantial progress made during the last few years, a unifying description of the photon–solid interaction and the precise parameters that finally determine the nature and characteristics of the ion generation process is not yet available. Some concepts have been propounded by several authors on the basis of reported mass spectra and will be described. These empirical approaches certainly need refinement as to their physical accuracy; however, they provide interesting keys to the explication of the detected signals in terms of phenomena that are familiar to practical researchers in organic MS.

Section 3B.3, the major portion of this review, will focus on characteristic features of recorded mass spectra. The data are organized according to functional groups, polarity, and molecular weight. Main groups include polycyclic aromatic hydrocarbons (PAHs) and related analogues, small mono- and polyfunctional molecules, a series of salts and organic acids, nucleosides and nucleotides, sugars and polysaccharides, amino acids and peptides, natural products, alkaloids and steroids, biomolecules with low and medium MW, organic complexes, polymers and UV stabilizers, dyes and food additives, and several drugs and metabolites. The discussion is structured according to this classification and allows us to assess the difference or similarity between results obtained under largely varying experimental conditions. In addition, particular attention will be devoted to the use of fragments for structural analysis, on the one hand, and to possible information about desorption–ionization (DI) processes, on the other. Practical analytical applications will be briefly mentioned in the pertinent subsections along with the corresponding mass spectra.

Section 3B.4 will focus on diverse examples that will enable us to weigh the possibilities and limitations. These examples pertain primarily to TOF-

LMMS, but some recent data are reported for LD-FTMS as well. We venture to predict, however, that the rapid pace of development of instruments will make a real FTMS microprobe available for use in the very near future. Here, again, characterization of molecules by means of structurally relevant information and not by means of the constituting elements will be deemed essential. The question of surface sensitivity will be addressed, as well as the possibilities of detecting targets embedded in a heterogeneous matrix.

The reader is referred to the preceding review by Vertes and Gijbels (Part A of this chapter) for further information on high-mass analysis of biomolecules via matrix-assisted desorption. Also, the application of laser ionization in the gas phase lies beyond the scope of the present contribution which deals exclusively with analysis of organic compounds in the solid and/or liquid state. We also would like to reiterate that complete coverage of this field is not feasible within the framework of this chapter. Hence, the selection made may show some personal bias. As current users of TOF-LMMS, we could not resist the temptation to present a few of our own results.

3B.2. SOME GENERAL CONCEPTS IN LD-MS

This section will briefly summarize some mechanistic concepts to lay the foundation for the later discussion of the mass spectra.

Numerous experiments have been performed to elucidate the nature of the laser desorption process. Difficulties arise as a result of the many parameters that can be manipulated. Power density, wavelength, pulse duration, substrate, matrix, sample thickness, electronic absorption of the analyte, and the orientation of the incoming laser beam to the sample can all affect the appearence of the mass spectrum. Surveys of the experimental parameters employed in several laboratories are available (Hillenkamp, 1983; van der Peyl et al., 1982; Van Vaeck and Gijbels, 1989; Conzemius and Capellen, 1980). As is readily conceivable with the wide variety of parameters in separate experiments different desorption mechanisms have been proposed.

Laser desorption is distinguished from SIMS, FAB, and PD-MS, all methods based on particle bombardment. LD mass spectra often closely resemble particle bombardment mass spectra (Forest et al., 1989; Zakett et al., 1981; Schueler and Krueger, 1980; Balasanmugan et al., 1981; Schiebel and Schulten, 1986). However, it is important to realize that a photon beam may be energetic yet lack appreciable momentum; hence the primary excitation process in LD must be different from that of either SIMS or PD-MS.

As in the case for SIMS and FAB, kinetic energy (KE) distributions of ions in LD have been used to gain mechanistic insights (Cotter and Tabet, 1983;

van der Peyl et al., 1984; Hardin and Vestal, 1981; Mauney and Adams, 1984a). An adequate description of the ion emission must account for the following experimental observations:

1. Higher KE particles (tens of electron volts) are emitted immediately after a laser pulse in contrast to later times (several microseconds later). The ions emitted in the early stages are usually atomic or smaller molecular fragments. They appear to represent nonthermal components of desorption because the KEs correspond to an unrealistically high temperature.
2. Protonated or cationized molecules have relatively low KEs.
3. Many more neutral particles than ions are emitted, and the release of neutrals occurs for longer times.

There are unifying models that adequately cover the entire range of LD-MS experiments. It is now generally accepted that the mass spectral information must be related to one or more ion formation mechanisms from the following principal classes: thermal, selvedge, shock-wave-driven, and resonant processes. It appears that the thermal type of ion generation predominates under low-power-density conditions, when desorption proceeds as a function of substrate temperature. The second type of ionization processes involves interaction in the gas phase of neutrals and co-desorbed ions, most often thermionically emitted alkali ions. However, an active chemical process between the constituents may be involved, bearing some resemblance to observations in FAB and molecular SIMS (Cooks and Busch, 1983). The third series of processes refers to the nonthermal explosive ablation of different species, i.e., neutral and cationized molecules as well as large clusters and microparticulates. Little fragmentation is observed, and sample probe or substrate does not participate in the desorption process. Finally, resonant desorption occurs when the analyte itself undergoes an electronic transition under laser irradiation. Fragmentation is always extensive near the threshold, and there appears to be no effect of the sample probe.

In the following subsections we shall describe some of the experimental data and relevant aspects of the aforementioned major ion formation modes. Anticipating the systematic discussion of the mass spectra later on, we have preferred to make an explicit distinction between ion formation in LD-MS and LMMS. The latter technique specifically works with a laser-irradiated spot close to the diffraction limit of 0.5 μm at 266 nm, whereas the former instruments typically use spots from 50 μm upward. For reasons that are not yet clear, the detection and role of the odd-electron molecular ions in LMMS differ enough from those in LD-MS to warrant a separate discussion.

3B.2.1. Thermal Ionization and Generation of Neutrals in LD-MS

It should be noted that the thermolability of, say, organic salts is a relative concept. Quaternary ammonium and phosphonium compounds have been shown to produce intact preformed ions in the gas phase when resistive heating is applied (Stoll and Röllgen, 1980; Cotter and Yergey, 1981a,b). The detection of undegraded species in MS depends on a variety of experimental factors, specifically:

- The heating rate
- The transfer efficiency between the introduction system and the ion source, when applicable, or the analyzer itself
- The time domain for ion formation and detection

The direct DI methods of solids also permit the use of samples inside the ion source. As a result, efficient measurements are ensured and, even when only a very limited number of molecules escapes thermal decomposition, there is a significant chance for detection.

It is important to realize that the intact deliberation of thermolabile compounds from the solid state remains entirely compatible with the extremely high temperatures attained during laser irradiation. Apart from the creation of so-called hot spots and cooler regions further away, there is the fundamental effect of ultrafast heating, which favors intact release without decomposition of heat-sensitive molecules. This apparent contradiction was, in fact, first exploited in the early 1970s and later on in the flash desorption technique, developed as an alternative to the conventional resistive heating for thermolabile compounds (Beuhler et al., 1974; Daves et al., 1980).

The phenomenon can be readily rationalized assuming that an Arrhenius-like equation applies to desorption of the intact neutrals and to generation of the thermal degradation products. This model may be a simplification of the real situation within an LD-MS experiment but still remains highly relevant. A nonvolatile compound means that the activation energy for desorption exceeds that for decomposition. Otherwise, the substance can be handled by resistive heating techniques. The Arrhenius plots of the rates of the two processes vs. $1/T$ are schematically presented in Figure 3B.1 (Daves, 1979). If the point of intersection occurs at an experimentally realistic temperature, then the rate of desorption becomes favored above that value in comparison with that governing the decomposition reaction. Otherwise stated, the exponential factors in the rate expressions become comparable for desorption and degradation at elevated temperatures above the intersection value. Use of the fast-heating approach means that the time spent in the low-temperature

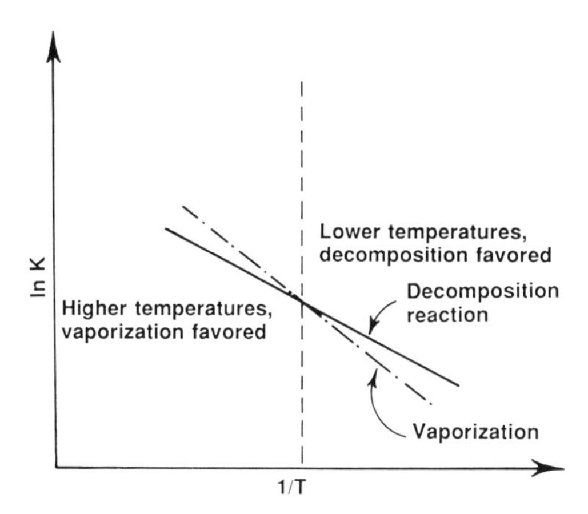

Figure 3B.1. Relationship between the temperature dependencies of evaporation and decomposition of an involatile thermally labile compound. Reprinted from Daves (1979) with permission of the American Chemical Society.

range, where decomposition prevails, is minimized and the relative contribution of the intact molecules in the gas phase is increased.

One of the earliest observations of mechanistic significance in LD-MS is that the emission of ions and neutral species continues for several microseconds after a laser pulse is terminated (Tabet and Cotter, 1984; Vastola and Pirone, 1968; Zakett et al., 1981; Van Breemen et al., 1973; Cotter, 1980). The thermal conductivity of the substrate and electronic absorption bands of the sample also greatly affect the efficiency of the desorption process. Less irradiation power is required to form ions from nonabsorbing samples on metals, which have high thermal conductivities, than from nonabsorbing samples on substrates that have poor heat transfer properties (van der Peyl et al., 1982). When the solid exhibits strong absorption at the wavelength used for irradiation, lattice excitation drastically lowers the threshold power density for the desorption of neutrals (Karas et al., 1985; Spengler et al., 1987).

The generation of gas phase ions from organic salts represents one of the most easily described cases of LD and other DI techniques (Emary et al., 1990). If the sample has an absorption band of an appropriate energy, electronic excitation can occur. For thin layers of nonabsorbing organics, the substrate absorbs most of the energy, transferring it back to the sample as thermal energy. In either case, desorption occurs if the translational energy made available is sufficient to overcome the lattice binding energies. Cation–anion bond breaking (C^+A^-) is the key step that when accomplished results in C^+ and A^- being liberated as charged particles, making this a very efficient ion formation process. Neutral C^+A^- species are also emitted as ionic clusters

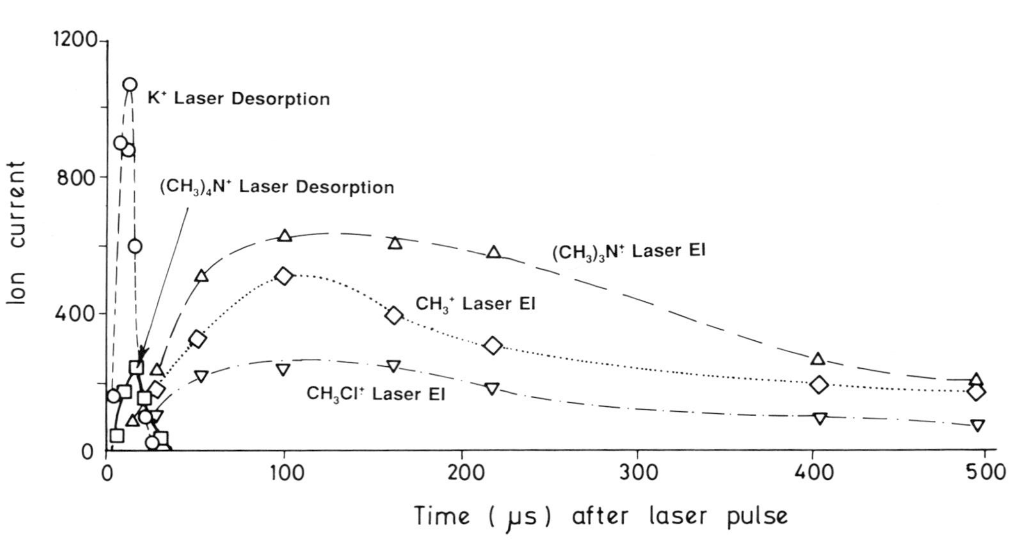

Figure 3B.2. Intensity of ions from tetramethylammonium chloride during the first 500 μs following laser pulse irradiation of the solid sample. Reprinted from Van Breemen et al. (1973) with permission of Elsevier Science Publishers.

resulting from nonsymmetrical combinations of anions and cations. These steps are largely reminiscent of the processes in SIMS (Campana, 1987).

Particularly relevant are the time profiles observed for the intact cation from a tetraalkylammonium salt in comparison to the ones for thermal decomposition products that have been postionized by EI. Figure 3B.2 illustrates the ion signal intensities, monitored from tetramethylammonium chloride at intervals up to 500 μs after a 40 ns laser pulse (Van Breemen et al., 1973). The intensities of Na^+, K^+, and the intact organic cations reach a maximum peak at 5–15 μs after the pulse, and these ions are not detected anymore after 50 μs. In contrast, the neutral species trimethylamine and methylchloride, ionized by EI, can be detected even after 500 μs.

Note that ion formation in LD-MS for nonionic compounds occurs by gas phase cationization and hence is considered in the next subsection.

3B.2.2. Gas Phase Ion–Molecule Reactions in LD-MS

For thin sample films, with laser power densities of approximately 10^8 W$\cdot$cm^{-2} and pulse durations between nanoseconds and microseconds, it is generally accepted that laser desorption occurs via fast heating of the substrate, resulting in thermionic emission from the hot center of the laser spot and vaporization of intact neutral sample molecules further away at lower temperatures. Gas phase reactions in the resultant plasma of ions and neutrals present above the surface produce cationized or protonated molecular ions (van der Peyl et al., 1982, 1983). This concept is consistently supported by a variety of studies of organic neutrals and alkali cations focusing on the time dependency of adduct formation, kinetic energy distribution measurements, and experiments with separate sources.

Ejection of sucrose ions has been observed to occur for about 300 μs after a 10 ns pulse at power density of 10^8 W$\cdot$cm^{-2} (Zakett et al., 1981). Emission of very large ions can also continue for long times after the laser pulse. For example, the $(M+H)^+$ ions of trehalose octapalmitate (MW 2246) was observed 10 μs after a 40 ns laser pulse (Tabet and Cotter, 1984). Such results are indeed suggestive of thermal emission processes. Calculations indicate a slow falloff (microsecond regime) of the surface temperature after it is elevated by a laser pulse (van der Peyl et al., 1982). This is associated with the much larger volume of the hot spot as compared with that in particle desorption methods. The long-lived temperature elevation is considered to be responsible for the longevity of ion and neutral emission.

The kinetic energy distributions of K^+ from KCl have been found to be about 15 eV wide 10 μs after the laser pulse but only about 0.3 eV wide 43 μs after the laser pulse (Cotter and Tabet, 1983). The low energy spread at long delays is consistent with an equilibrium thermal model, but energies of 10 eV or more suggest a nonequilibrium process. The distributions of Na^+ and cationized sucrose show widths greater than 1 eV, with the former being slightly broader than the latter (van der Peyl et al., 1982, 1984). These distributions are much wider than that obtained for the thermionic Na^+ emitted from a 1000 K substrate. The ion "temperature" required to produce the observed $(M+Na)^+$ distribution was calculated to be between 3000 and 3300 K. High temperatures at the center of the laser spot may be responsible for alkali emission, whereas cooler areas further away may be the source of the organics. Alternatively, the extremely high temperature may suggest the existence of a nonequilibrium process, since survival of such a hot molecule under equilibrium conditions is unlikely. Both thermal and nonthermal processes may occur in a given LD experiment (Hillenkamp, 1983).

Experiments in which the organic molecules are physically separated from the alkali sources have demonstrated that gas phase cationization is indeed possible, though other ionization processes are not ruled out. Gas phase cationization of sucrose by a distinct source of K^+ has been studied for joule heating conditions, with continuous wave (CW) laser and pulsed low-power laser irradiation (Stoll and Röllgen, 1982; van der Peyl et al., 1981; Chiarelli and Gross, 1987). Interestingly, in a commercial TOF-LMMS instrument, the detection of cationized molecules from the simultaneous irradiation of physically separated sucrose and LiCl particles has been reported (Wieser and Wurster, 1983).

It should be remembered that the production of cationized molecules is not limited to LD but also occurs on resistive heating of, say, glucose (Cotter and Yergey, 1981b), on the condition that suitable experimental conditions are applied to avoid predominance of thermal degradation.

Laser irradiation of solids creates a selvedge region where the chance for intermolecular interactions and reactions is significant, as opposed to free vacuum conditions where essentially only unimolecular decay occurs. In this region, a more active chemistry can be observed in addition to the simple alkali–ion attachment. The analysis of PAHs in the presence of nitroaromatic compounds is possible with an improved detection limit by means of increased production of $(M+O-H)^-$ ions, which are formed by interaction of desorbed neutral analyte molecules with the nitro-anion, available from the laser-initiated decomposition of analyte (Viswanadham et al., 1988). This so-called solid state chemical ionization approach can be exploited by mixing chlorine-containing compounds with the analyte to promote the detection of glycosides as $(M+Cl)^+$ adducts. Sodium nitrate and sodium peroxide permit use of ion–neutral chemistry with O^- and O_2^- (Balasanmugan et al., 1989).

3B.2.3. Shock Wave or Nonthermal Desorption in LD-MS

Laser desorption at $10^{10}\,W \cdot cm^{-2}$ proceeds without interaction from the substrate (sample probe) and thus is not driven by temperature. The lack of peak broadening in a TOF-MS indicates that the desorption times are much shorter than at lower power densities. A theoretical calculation indicates that the ion formation must occur within 11 μm of the probe surface in the TOF experiment (Chiarelli and Gross, 1987). The selvedge or intermediate region between the solid state and gas phase is not as diffuse and the material ejected during desorption must be largely condensed (microparticulates). Thus there is little probability for gas phase cationization. Apart from these general observations, there still remains a lot of debate about the real physical nature of these processes. Several terms are used to describe such conditions: *shock-wave-driven desorption*; *the volcano model*; and, even more simply, ablation and "*true*" *laser desorption* (Lindner and Seydel, 1985; Hillenkamp et al., 1989; Cotter, 1984; Hillenkamp, 1983). The last two terms are rather vaguely defined and essentially refer to the collective nonequilibrium process of ion generation in general.

The differences between thermal and nonthermal desorption are best illustrated by experiments in which the power density and sample thickness are varied (Lindner and Seydel, 1985). Mixtures of saccharides and alkali salts have been measured in the transmission geometry. Extensive fragmentation is observed when a power density of $10^{8}\,W \cdot cm^{-2}$ is applied on 1 μm thick samples, whereas the signal from the cationized molecule predominates under irradiation of a 20 μm layer at $10^{11}\,W \cdot cm^{-2}$. The former parameters are believed to favor the thermal regime, whereas the reduced fragmentation under the latter conditions has prompted the proposal of *shock-wave-driven ion formation*.

The impact of the laser beam as used in this study leads to a temperature rise of up to a few thousand degrees (Kelvin) at the surface within a duration of 10 ns. The resulting heating rate of $10^{11}\,K \cdot s^{-1}$ exceeds the limit of approximately $10^{9}\,K \cdot s^{-1}$ above which an explosive vaporization (or "phase explosion") should occur. This thermal ablation produces a shock wave that traverses the solid under energy dissipation and subsequently leads to the desorption of intact molecules from the opposed surface via vibrational disturbance of the binding potentials. Since the molecules are not coupled to the lattice long enough to absorb sufficient internal energy, thermolabiles are ejected intactly and do not undergo fragmentation in a later phase. The sample comes off one monolayer at a time, and little substrate interactions occur. The mechanism of phase explosion is also used in other DI techniques, particularly FAB and SIMS (Sunner et al., 1988a,b). Additional experiments with alkali salts on back of thick samples, which were not perforated, have shown that cationization still occurs. Hence, gas phase reactions between neutrals and co-desorbed alkali ions from impurities are responsible for the detected parent ions.

Additional experiments have been conducted with sucrose on a double-split sample probe, providing a physically separated source for alkali ions (Chiarelli and Gross, 1987). The extent of the cationization process has been monitored as a function of the distance and of the power density over a range between 10^6 and 10^{10} W·cm^{-2}. Strong evidence for gas phase cationization has been obtained for a power density of 10^6 W·cm^{-2} up to a distance of 1 mm, but no alkali attachment has been observed when laser irradiation is applied at 10^{10} W·cm^{-2} and the alkali source is only 30 μm away from the sucrose. The lack of cationization in the gas phase under the latter conditions is interpreted as being an indication of the ejection of largely condensed material. The energy deposition in the first several monolayers, and not in the substrate, leads to a short-lived selvedge, preventing alkali attachment to neutrals in the gas phase. Also here cationized molecules still are detected in the absence of a separate alkali source. This observation has been related to the ejection of cationized molecules directly from the condensed phase or from the large clusters ablated from the surface.

The "volcano model" essentially refers to a largely similar process (Hillenkamp et al., 1989). The basic features include:

1. The volume absorption of laser radiation
2. Surface cooling by evaporation or sublimation in conjunction with excess energy accumulation of subsurface layers, probably as little as 100 monolayers deep
3. A subsequent explosive ablation of particulate matter

It is postulated that the dissipation of internal energy occurs through evaporation of low MW components from the primary aggregate. These concepts are documented in the literature of macroscopic laser materials processing. Whether or not this model in any way reflects the real processes remains, at this time, a totally open question. Most experimental indications do not allow us to rule out the occurrence of other competitive mechanisms as well (Hillenkamp et al., 1989).

The nonthermal generation of clusters and smaller to larger molecular aggregates has opened the way to a lot of chemical research. Comparison of LD-MS spectra for a series of amino acids and zwitterionic salts with the corresponding deuterated analogues has yielded indications of the generation of neutral dimeric species, which are decomposed into a positive and negative ion as a result of proton or alkyl group transfer (Parker and Hercules, 1986; Balasanmugan and Hercules, 1984).

3B.2.4. Resonant Desorption in LD-MS

The striking similarity of the main characteristics in a variety of LD-MS results has led to the generally accepted notion that the wavelength used is at least a parameter of some importance (Hercules et al., 1982; Conzemius and Capellen, 1980; Schueler et al., 1981). It should be realised, however, that many, if not most, of the compounds studied only exhibit negligible absorption of the wavelength used for irradiation. Nevertheless, early work in TOF-LMMS has also demonstrated that laser irradiation of organic compounds at 532 nm essentially produces pyrolysis products and carbon clusters whereas structurally relevant information is obtained using the frequency-quadrupled Nd:YAG output at 266 nm (Southon et al., 1984). The observation of radical cations generated from PAHs, when desorbed at 266 nm, has provided support for a resonant desorption mechanism initiated by electronic excitation of the sample (Heinen, 1981). More recent work on amino acids has shown that a match between the irradiating wavelength and the molecular absorption band of the analyte leads to a drastic change of such features as molecular ion production vs. abundance of fragments, threshold irradiance levels, and influence of increased power density (Karas et al., 1985; Hillenkamp et al., 1986, 1987; Spengler et al., 1987).

Above the threshold irradiance under nonresonant conditions, energy transfer between the laser field and the analyte occurs through so-called nonlinear absorption (Hillenkamp et al., 1987). This means any process in which the absorbed power per unit of sample volume ceases to be proportional to the incident flux. Excitation with an absorption wavelength permits direct transfer of energy into the analyte molecules. However, the results also indicate that this classical or linear absorption is not the only process occurring and probably not the predominating energy deposition mechanism (Karas et al., 1985). It is assumed that most of the energy is transferred to the lattice by exciton–phonon coupling. The desorption of neutrals and ions would then result from a short but strong perturbation of the lattice. The limited degree of fragmentation observed in conjunction with the initial ion energy of several electron volts indicates that this collective process of lattice disintegration takes place in nonequilibrium conditions. Additionally, some of the molecules or ions remain in the singlet excited state or get excited by photon absorption after their release from the solid state. Excited amino acids can then act as proton donors for the ground state molecules because of their increased acidity, in analogy with the well-documented situation in solution chemistry. Also direct photofragmentation may occur as a consequence of the electronic singlet state excitation of the molecules. It has been found that the resonant desorption conditions lower the irradiance threshold for detection of aliphatic and aromatic amino acids and make the results less affected by other variables than nonresonant absorption (Karas et al., 1985).

A recent investigation of nucleosides in LD-FTMS has pointed to the role of photochemical phenomena (Chiarelli and Gross, 1989a). Data are compared for purine and pyrimidine analogues irradiated by 1064 and 266 nm at power densities between 10^5 and 10^{10} $W{\cdot}cm^{-2}$. The use of IR instead of UV photons lowers the threshold irradiance level required to detect comparable ion currents. The relative contribution of so-called doubly sodiated molecular and fragment ions, i.e., $(M—H+2Na)^+$ and $(B—H+2Na)^+$, with B denoting the purine or pyrimidine moiety, in comparison to the $(M+Na)^+$ and $(B+Na)^+$ signals not only depends on the wavelength and power density but also on the specific photochemical reactions of the different purine and pyrimidine groups. In particular, the formation of doubly sodiated molecules relates to the lactim–lactam equilibrium in the excited T_1 state. The negligible role of gas phase cationization is associated with the preferential stacking of the molecules in the solid state, which leads to self-association. The direct ejection of large aggregates, consisting of an alkali ion and several neutral nucleosides, is proposed, followed by decomposition into the monocationized molecules. The time scale of the FTMS method used here is in the millisecond range and hence compatible with the time for complete breakdown of the large aggregates.

3B.2.5. Ionization in TOF-LMMS

Hillenkamp (1983) divides the LD mechanism into four separate processes. The first two, i.e. thermal evaporation of ions and gas phase combination of ions and neutrals, are generally accepted. The third, which is not useful for analytical purposes, occurs at high power densities above 10^{10} $W{\cdot}cm^{-2}$ and involves formation of hot plasmas that destroy molecular identity. The fourth process is referred to as “true” laser desorption. This could be called the analogue to such high-energy processes as the collision cascade in SIMS because its existence is indicated by the high kinetic energies of some laser-desorbed ions (Pachuta and Cooks, 1987). Furthermore, cluster ion emission from materials such as graphite does not occur by heating, yet it does occur by LD (Furstenau and Hillenkamp, 1981). In fact, the notion of *true laser desorption* comes close to the concept of phase explosion already mentioned in Sections 3A.6 and 3B.2.3.

One notable and frequent feature observed in most LD mass spectra of organic compounds is the production of molecular ions by cationization or protonation rather than by electron abstraction. In the negative mode deprotonation rather than electron attachment is usually observed. Radical ions are rarely detected. However, studies discussed in later subsections provide evidence that the situation is somewhat different in TOF-LMMS.

Recognizing that one of the obstacles for widespread application of TOF-

LMMS is lack of a unifying insight to guide experimental design, we have followed an entirely empirical approach. Mass spectra of a set of selected compounds, covering a wide range of polarity, functionalities, and expected DI behavior have been compared systematically with those from other MS techniques. We have tried to elaborate a simple conceptual framework that would enable us to rationalize the structural information from the TOF-LMMS data on organic products. It is important to stress that this approach does not aim at a realistic description of the DI process but at a working model adequate to interpret analytical results in microprobe applications. In our experience—and stated in very general terms—monitoring the MW information, if not supplemented by any fragment peaks, is not sufficient in practical problems. Identification of an organic target molecule in a complex environment requires a detailed interpretation of the fragments that characterize the functionalities in the molecule.

Our results often point to a predominant role of thermal processes, on the one hand, and to radical parent ion formation, on the other. Both observations are not generally recognized. We found also that the characteristic features of the recorded mass spectra, and the aforementioned signals in particular, strongly depend on the experimental conditions. Hence, the DI model strictly applies to results recorded according to the so-called threshold procedure (Van Vaeck et al., 1990).

Basically, we attempt interpretation of TOF-LMMS results in terms of energy, pressure, and time. It is readily conceived that laser impact creates an energy gradient along the sample surface and starts a series of processes ranging from complete disintegration in the central area through pyrolysis and thermal decomposition to gradually softer conditions at the periphery. Disintegration denotes the formation of nonspecific elemental cluster ions. Depending on the operating conditions, evidence for one or more processes can be more or less pronounced. So far, we have consistently attempted interpretation by using desorption of the molecules as neutrals and subsequent ionization in the selvedge region as the major pathway for nonionic compounds. A substantial fraction of the preformed ions from salts is released intact from the solid state, but thermal degradation remains important as well. The resulting neutral products are then reionized in the selvedge. Electron interaction, yielding odd-electron molecular ions ($M^{+\cdot}$ or $M^{-\cdot}$) and the subsequently formed fragments, occurs as well as adduct formation by combination with co-desorbed alkali ions. These concepts are summarized in Figure 3B.3 (Van Vaeck et al., 1989b).

The concept of *pressure-related effects* refers to the density of neutrals in the selvedge region. The density of neutrals under threshold conditions depends on the volatility of the product. The term *threshold laser energy* refers to the application of the lowest energy density possible on the minimum

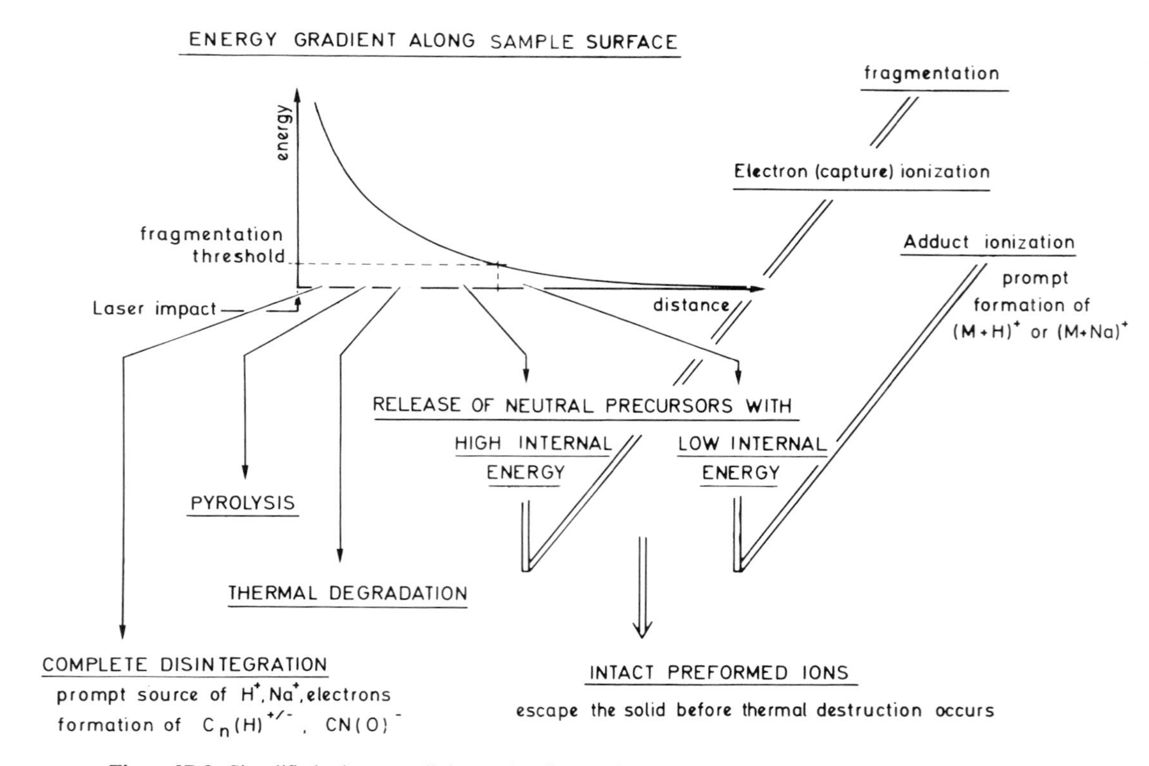

Figure 3B.3. Simplified scheme outlining molecular and fragment ion formation according to the empirical model for one-step DI in TOF-LMMS. Reprinted from Van Vaeck et al. (1989b) with permission of John Wiley & Sons, Ltd.

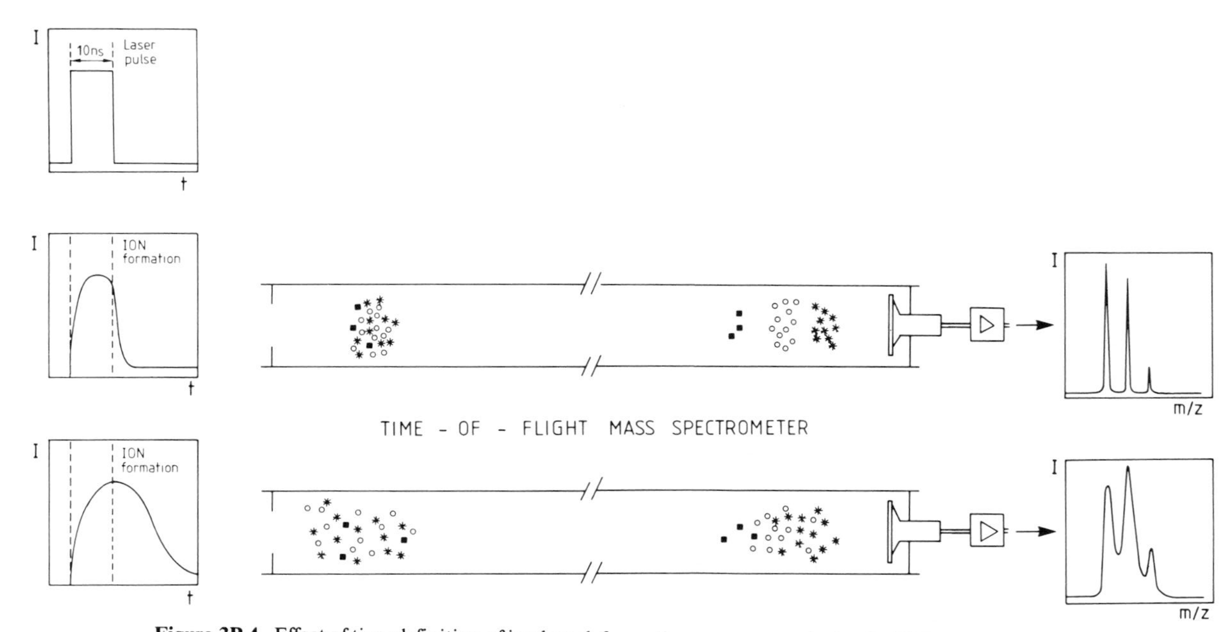

Figure 3B.4. Effect of time definition of ion bunch formation on mass resolution in TOF-LMMS. Reprinted from Van Vaeck et al. (1986a) with permission of the editor, Dr. H. P. Thun.

amount of sample that permits the recording of a mass spectrum with fair signal-to-noise ratio. Low-MW compounds are often most easily desorbed and create a more dense region where second-order interactions are favored. Hence, dimeric clusters are observed and adduct formation prevails. For less volatile compounds, electron ionization–related processes are more important and often adduct formation is not detected anymore. Unimolecular decay occurs in the free vacuum outside the selvedge region. So far, radical precursors have been found to be the most adequate way to account for the observed fragments, which comprise odd- and even-electron species. The fact that the signal due to the radical molecular ions is not detected is related to the quantitative conversion of the parents into fragments in the same way as in conventional EI mass spectra. The assumed density in the selvedge region increases the probability of stabilising third-body collisions upon electron attachment.

Finally, the time domain should be considered. TOF-LMMS exploits only the promptly generated ions, as schematically shown by the effect of the time profile of ion formation on the mass resolution in Figure 3B.4 (Van Vaeck et al., 1986a). As a result, whenever postlaser desorption and ion formation occur, a significant discrimination occurs in the TOF-LMMS instruments as

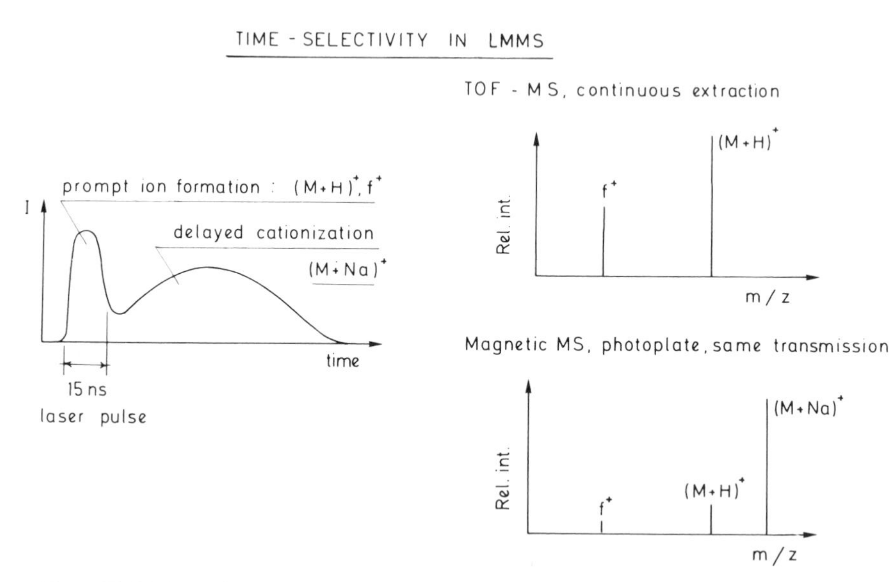

Figure 3B.5. Time selectivity of the TOF-LMMS instrumentation in comparison with that of a magnetic instrument with photoplate detection, permitting inclusion of delayed ion formation in the measured signal. Reprinted from Van Vaeck et al. (1990) with permission of Springer-Verlag.

opposed to "integrating" analyzers such as a magnetic sector or FTMS. The importance of the time domain in ion generation vs. that for mass analysis is schematically shown in Figure 3B.5 (Van Vaeck et al., 1990). This figure can explain why, for instance, cation attachment does not predominate in the TOF-LMMS mass spectra to the same extent as it does in the LD mass spectra. In fact, we usually find that cationization leads to a clearly distinct peak shape in contrast to protonated molecules or fragments. This observation is related to the delayed thermionic flux of alkali-metal ions.

3B.3. SURVEY OF LD MASS SPECTRA FOR DIAGNOSTIC ANALYSIS

3B.3.1. Polycyclic Aromatic Hydrocarbons (PAHs) and Related Analogues

PAHs are thermally stable, and at least the lower analogues can be handled successfully with conventional sample introduction systems. PAHs have been popular in LD-MS research from the very beginning when Vastola and colleagues pioneered photon irradiation of solids with and without EI post-ionization (see Vastola and Pirone, 1968). The reason for this popularity is quite obvious. The conjugated aromatic system ensures a high degree of resonance stabilization, and hence these analogues can cope with a wide range of power densities without destruction of the molecular identity. Additionally, PAHs are apparently extremely useful compounds for probing ion formation mechanisms in LD-MS, specifically the negative ions where electron capture ionization (ECI), solid state chemical ionization (CI), and complete destruction into carbon clusters can be observed. Finally, a lot of interest has been raised by the possible application of recently developed microanalytical techniques in environmental analysis—for instance, in atmospheric particle research.

Early reports dealing with TOF-LMMS data on PAHs offered evidence of the production of prominent signals from $M^{+\cdot}$ with a very low degree of fragmentation and the quasi-absence of low-mass signals, whereas the protonated molecules were found to predominate for the higher analogues (Heinen, 1981; Dutta and Talmi, 1981). The formation of radical ions was considered a remarkable exception, and this notion has tended to persist up to the present. The negative mass spectra contain only a series of nonspecific carbon clusters C_n^-, C_nH^-, and $C_nH_2^-$, but practically no structurally relevant peaks in the molecular ion region have been obtained (Heinen, 1981; Rosmarinowsky et al., 1985).

Two systematic studies on TOF-LMMS of PAHs confirm the detection of $M^{+\cdot}$ as the base peak, whereas the $(M+H)^+$ yields a minor contribution for both the lower and higher analogues (Van Vaeck et al., 1985a; Balasanmugan et al., 1986). Cationization by the inevitably present impurities does not occur. Experiments with deuterated analogues show that the hydrogenated carbon clusters, generated from PAHs in the positive ion mode under application of high-power-density conditions, are involved in $(M+H)^+$ production (Balasanmugan et al., 1986). Molecular information can be obtained in the negative ion detection mode for most analogues by means of prominent $(M-H)^-$, $M^{\cdot-}$, and/or $(M+H)^-$ signals. The relative intensities of these signals depend on the analogue actually studied and the experimental conditions in a transmission or reflection-type geometry irradiation of the sample. Table 3B.1 compares relative peak intensities for the molecular ion region of a few analogues in the LAMMA® 500 (transmission mode analysis) and LAMMA® 1000 (reflection mode) (Leybold-Heraeus, Cologne, Germany) (Van Vaeck et al., 1985b; Balasanmugan et al., 1986). The production of $M^{\cdot-}$ in the reflection-type instrument under threshold conditions is tentatively explained by the higher density in the selvedge region, which increases the third-body collisional stabilization. However, other processes cannot be ruled out. It should be noted that the process of hydride uptake is only rather exceptionally seen, and apart from the PAHs we have no examples in our data base (Van Vaeck and Lauwers, 1990).

Application of a tunable dye laser lowers the threshold irradiation power density of $M^{\cdot+}$ production and decreases the peak intensity of $(M+H)^+$ relative to that of $M^{\cdot+}$ (Muller et al., 1985). The formation of cationized PAHs in the presence of alkali salts has been studied by LD-FTMS (Greenwood et al.,

Table 3B.1. Comparison of Relative Peak Intensities in the Molecular Region of Negative Ion Mass Spectra from Some PAHs Recorded by TOF-LMMS in the Transmission (LAMMA® 500) and Reflection (LAMMA® 1000) Geometry

			$(M-H)^-$		$M^{\cdot-}$		$(M+H)^-$	
Compound	Formula	Mol. Wt.	500	1000	500	1000	500	1000
Pyrene	C_6H_{10}	202	100	100	27 ± 3	11	95 ± 5	45
Triphenylene	$C_{18}H_{12}$	228	36 ± 3	0.0	7 ± 2	100	100	0.0
Benzo[*a*]-pyrene	$C_{20}H_{12}$	252	52 ± 4	0.0	18 ± 3	24	100	0.0
Benzo[*ghi*]-perylene	$C_{22}H_{12}$	276	100	0.0	28 ± 3	75	60 ± 5	0.0
Dibenz[*a,c*]-anthracene	$C_{22}H_{14}$	278	100	0.0	26 ± 3	25	48 ± 5	0.0
Coronene	$C_{28}H_{12}$	300	100	0.0	28 ± 3	100	16 ± 3	0.0

Source: Reprinted from Van Vaeck et al. (1985a) and Balasanmugan et al. (1986) both with permission of the American Chemical Society.

1990). The ionization thresholds for $M^{+\cdot}$ and $(M+Na)^+$ appear to be indistinguishable, confirming the facile nature of the electron detachment process for these analogues. The ratio of cationized molecules over $M^{+\cdot}$ measured in mixtures of 1:1 by mass, exhibits a decreasing trend with the aromatic array size of the molecule. Increase of the power density increases the relative contribution of cationization, but the overall ion yield is not significantly enhanced. Application of IR laser desorption and tunable dye laser ionization of the cooled neutrals in the gas phase allows generation of the $M^{+\cdot}$ with or without the low-mass carbon clusters (Köster et al., 1989). The complete breakdown of organic molecules in the gas phase by multi-photon absorption has been extensively investigated. The occurrence of photofragmentation according to the so-called *ladder mechanism* has been extensively reviewed (Neussner, 1987).

Data in the literature on aza-heterocyclic analogues are more limited. Apart from carbazole, which yields extremely sensitive $(M-H)^-$ anions, the general behavior is analogous to that of the pure PAHs (Van Vaeck et al., 1985a).

Several applications deal with identification of molecular ions from PAHs and the aza analogues in environmental, coal, and oil samples (Mauney and Adams, 1984b; Denoyer et al., 1982; Dutta et al., 1984). However, the presence of large aromatic systems gives rise to the detection of complicated patterns of carbon clusters as well. The generation of these ions and possible use of typical distributions in mass spectra as "fingerprints" for complex materials have been investigated (Fletcher and Currie, 1988). The formation of high-mass carbon clusters up to several thousand daltons has been reported for the laser irradiation of even low-MW analogues, e.g., coronene, on the condition that the local power density is sufficient (Mauney, 1987; Lineman et al., 1989). These observations provide indirect support to the empiric DI approach and specifically the ranges of processes indicated in Figure 3B.3 (Van Vaeck et al., 1989b). Additionally, it becomes clear that the photon–solid interaction and the subsequently created selvedge region is not always the benign environment an analyst desires to perform soft ionization and uncomplicated characterization. A great deal of ion–molecule chemistry is certainly involved, but these ions are not structure specific and hence this topic lies beyond the scope of the present contribution. The high sensitivity of the PAHs in LD-MS facilitates applications such as mapping of thin-layer chromatography (TLC) plates (Kubis et al., 1989).

Related compounds with one or several heteroatoms other than the aza-aromatic skeleton have been studied as well. Whereas anthracene only yields carbon clusters in the negative ion detection mode, incorporation of Group V and VI elements permits detection of intense $(M-H)^-$ in conjunction with structurally meaningful fragments, some of which are odd-electron systems (Van Vaeck et al., 1989b). Dibenzothiophene is found to produce (M − H + S) ions in both positive and negative mass spectra. These cations exhibit minor intensity in comparison with $M^{+\cdot}$ but dominate the anion spectra even under threshold irradiation and at several wavelengths between 225 and 286 nm (Muller et al., 1985).

The presence of functional groups on the aromatic system may change the mass spectral features significantly. Aminopyrene has been studied at different wavelengths and power densities (Muller et al., 1985). Radical molecular ions remain prominent in both positive and negative mass spectra in spite of the additional functional group. The experimental irradiation conditions affect the intensity of the fragments, but the ion current apparently remains mainly carried by odd-electron systems.

Similar observations apply to quinone-type derivatives. These compounds typically exhibit the elimination of CO from $M^{+\cdot}$; hence, as long as unimolecular decay is involved and the even-electron rule is not violated, the role of even-electron precursors remains clearly separated from the decomposition of the radical parents irrespective of the degree of fragmentation. Hence, these analogues are useful test compounds to monitor radical ion formation in terms of experimental parameters, for instance, morphology of the analyzed micro-object (Van Vaeck et al., 1985b, 1990). Note that the negative mass spectra generally show radical molecular anions in the complete absence of subsequent fragments. This observation is assigned to the collisional stabilization in the selvedge (Van Vaeck et al., 1989b). Niessner et al. (1985) have exploited the capabilities of TOF-LMMS for the identification of secondary pollutants from PAHs, generated by heterogeneous gas-particle reactions in simulated atmospheric conditions.

A particularly interesting case concerns the nitro functional group. Positive mass spectra show the expected behavior with production of $M^{+\cdot}$ as a rather weak signal, detectable under irradiation with a selected wavelength, and fragmentation with the subsequent loss of NO and CO or elimination of HNO_2 (Muller et al., 1985). Especially noteworthy are the negative mass spectra, which show the formation of $(M-H+O)^-$ ions, in addition to the $M^{-\cdot}$ and $(M-NO)^-$ signals. The relative intensities of these signals are again influenced by the wavelength and the applied power density (Muller et al., 1985). The generation of high-mass adducts has been extensively studied and related to so-called solid state CI (Balasanmugan et al., 1983; Viswanadham et al., 1988). The term essentially refers to ion–molecule reactions in the selvedge between the nitro-aromatic compound and NO_2^-, generated from the analyte in the same way as the nonspecific carbon clusters provide hydrogen to protonated molecules. The formation of $(M-H+O)^-$ ions in the condensed phase before the exciting beam impinges on the sample can be considered in analogy with the precursor model, which has been proposed in static SIMS (Pachuta and Cooks, 1987; Benninghoven, 1983). However, the influence of spot size and applied laser power density on the relative importance of $(M-H+O)^-$ vs. $(M-NO)^-$ fragments can be used to confirm the less likely nature of the latter possibility. The actual separation of mechanisms, proceeding in the exploding solid phase and the relatively dense environment of the selvedge, is a rather subtle one for which the recorded mass spectra do not provide pertinent information. The addition of nitro-ion sources such as sodium nitrate to unsubstituted PAHs allows a highly sensitive detection of the corresponding $(M-H+O)^-$ signal (Balasanmugan et al., 1989).

3B.3.2. Simple Molecules

This section deals with a variety of small polyfunctional and medium-sized monofunctional nonionic molecules with a MW up to 500. Most data available in the literature describe the results for multifunctional derivatives, hampering a convenient classification. For the sake of clarity, we have opted to use different subsections on nucleosides, amino acids and peptides, sugars and polysaccharides, porphyrins and organic complexes, etc. A separate subsection is devoted to a series of drugs, alkaloids, and polyfunctional polymer additives, covering a MW range between 300 and 1000. Almost inevitably, there is some overlap between the different subsections and the classification is rather arbitrary in nature.

Starting with the low-MW solids, it is found that LD-MS permits detection of cluster-type ions, consisting of several neutral molecules and a charged entity. The latter can be an elemental ion or cluster but also a deprotonated molecule. This feature does not appear anymore in the mass spectra of compounds in the MW range above approximately 300. The generation of these bimolecular recombination ions tends to reduce the diagnostic use and specificity of the corresponding signals, particularly when a mixture of several analyte molecules is measured.

The formation of higher organic clusters has been studied systematically for a series of pyridine derivatives with one and two carboxyl groups or an amide group, as well as the corresponding *N*-oxides (Dang et al., 1984a). Table 3B.2 surveys the structural assignment of the major signals. Deuterium labeling and comparison of several isomers with functional groups on different positions confirm these structural assignments. Note that the protonated dimer itself is not detected in the positive mass spectrum. In contrast, the anions in the parent region refer to the combination of a deprotonated molecule with an intact one. It is found that increase of the laser power on the sample favors detection of these larger cluster ions. Confirming data on nicotinic acid are available (Heinen et al., 1981).

TOF-LMMS data on squaric acid (3,4-dihydroxy-3-cyclobutene-1,2-

Table 3B.2. Survey of TOF-LMMS Data in the Positive and Negative Ion Detection Mode from a Series of Pyridine Derivatives

Compound	Structure	Size of the Largest Cluster Observed		Compound	Structure	Size of the Largest Cluster Observed	
		Positive	Negative			Positive	Negative
Nicotinic acid	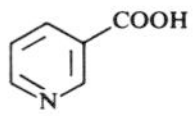 COOH, N	3	2	Nicotinamide	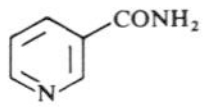 $CONH_2$, N	2	3
Isonicotinic acid	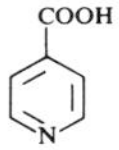 COOH, N	3	2	Isonicotinamide	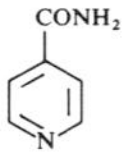 $CONH_2$, N	2	3
2,6-Pyridine-dicarboxylic acid	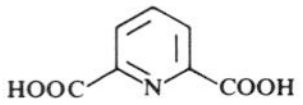 HOOC, N, COOH	2	2	Picolinamide	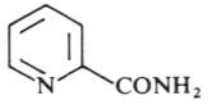 N, $CONH_2$	2	3
Nicotinic acid *N*-oxide	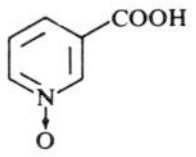 COOH, N, O	2	2	6-Chloro-nicotinamide	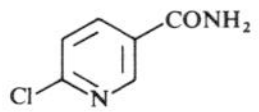 $CONH_2$, Cl, N	3	3
Picolinic acid *N*-oxide	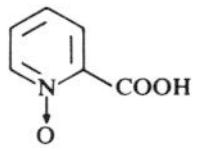 N, COOH, O	2	2	Pyridoxine·HCl	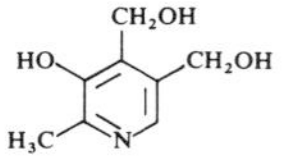 CH_2OH, HO, CH_2OH, H_3C, N	3	2
Isonicotinic acid *N*-oxide	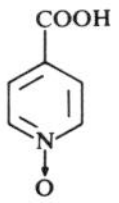 COOH, N, O	2	2	Pyridoxamine·2HCl	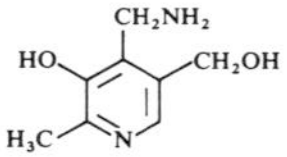 CH_2NH_2, HO, CH_2OH, H_3C, N	2	3

Source: Reprinted from Dang et al. (1984a) with permission of Elsevier Science Publishers.

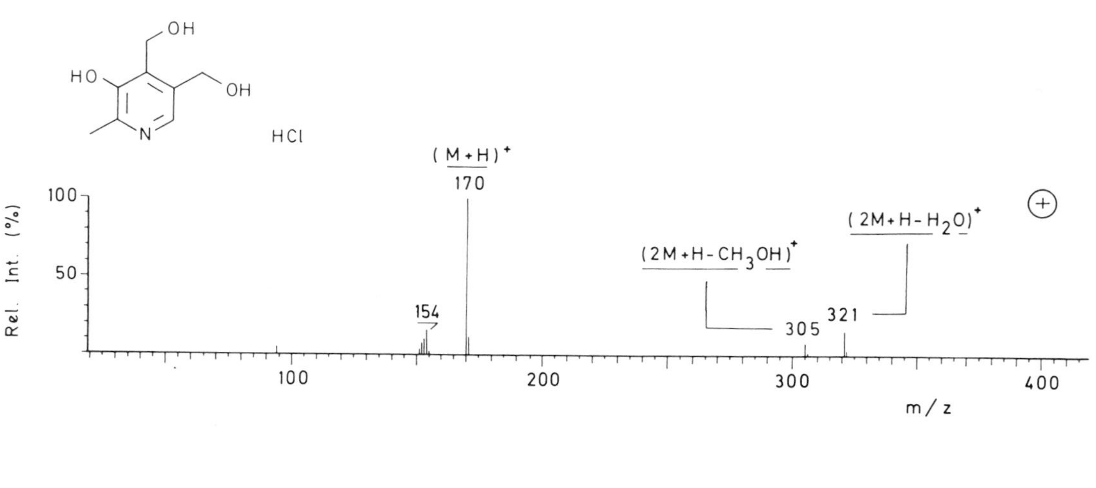

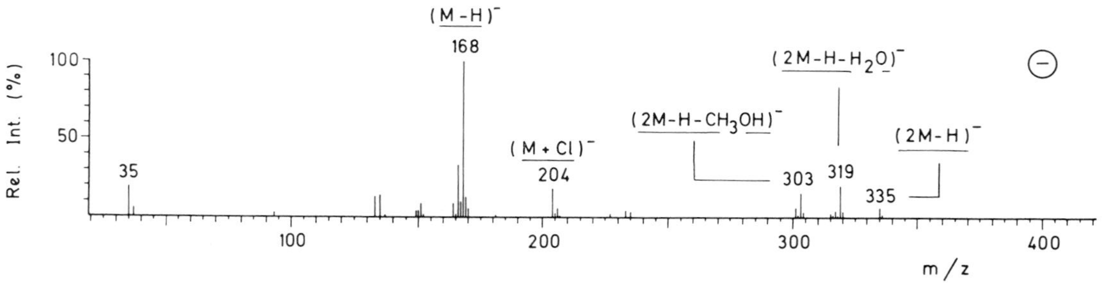

Figure 3B.6. Positive and negative ion mass spectra of pyridoxine hydrochloride recorded by TOF-LMMS. The MW of the free base is 169. Reprinted from Van Vaeck et al. (1989d) with permission of John Wiley & Sons, Ltd.

dione) show low-intensity signals for the cationized and protonated molecules in the positive ion detection mode, whereas the base peak is due to $(M+H-CO)^+$ (Byrd et al., 1986). In contrast to field desorption (FD) no dimeric cations have been observed. Negative mass spectra show $(M-H)^-$ as a base peak, accompanied by an intense signal for the $(2M-H)^-$ organic cluster ion and a significant signal formally corresponding to the loss of H_2 from the intact molecule. As a result, it is again likely that these fragments are odd-electron systems.

The formation of higher cluster ions in pyridoxine hydrochloride is illustrated by the TOF-LMMS data in Figure 3B.6 (Van Vaeck et al., 1989d). Note the detection of a chlorine adduct to the neutral molecule in the negative mass spectrum and the minor signal at m/z 154 in the positive mode. The latter fragments have to be assigned to radical systems that are normally not compatible with the actually observed parents such as the protonated molecules or one of the higher organic cluster ions. The compound has been analyzed under the hydrochloride form and can be considered as a salt in the condensed state. The formation of the radical fragments is consistently rationalized within the framework of the empirical model, but this approach does not exclude other possibilities such as direct photofragmentation. The point of interest concerns the fact that, if our interpretation also includes smaller signals, detection of radical ions is certainly not uncommon (as is generally accepted in the literature).

The formation of protonated dimers is observed in the mass spectra recorded by laser desorption and atmospheric pressure ionization with a Ni β-source for nicotinic acid but not for pyridoxine hydrochloride (Kolaitis and Lubman, 1986). No signals are present between the dimeric and monomeric adduct ions. Although TOF mass analysis is involved, the ion mobility interface between the source and analyzer certainly makes the time domain different from that in LMMS.

It should be noted that a particular technique has been developed to extend the working range of the commercial TOF-LMMS instrument toward volatile liquids (Morelli and Hercules, 1989). So-called *ambient pressure laser mass spectrometry* (APLMS) makes use of a thin polymer film as a "balloon-like" seal, covering the droplets or adsorbed film of the liquid to be investigated on a metal substrate. A specially designed sample stage providing cooling of the analyte has been used for comparison. Molecules such as benzene, glycerol, benzyl alcohol, amyl acetate, etc. have thus been studied. Particularly noteworthy in respect to the previous observations is the detection of higher MW clusters, including the protonated dimer form. Abundant fragments associated with the breakdown of both the mono- and dimeric forms are found as well. Attachment of $M^{+\cdot}$ to one or several neutral molecules has been observed for benzene, toluene, and benzyl alcohol. Di-n-butylpiperidyl phosphoramidate has permitted detection of protonated adducts containing up to four molecular entities. It is suggested that the formation of self-solvated clusters during laser ionization can be more important than is generally accepted for the medium-sized molecules also. However, the instrumentation actually used needs improvement, specifically postionization, to increase the detection capability for high masses.

A series of additional substituents on benzene derivatives containing one or more nitro groups has been included in the aforementioned work on the solid state CI of PAH related molecules (Viswanadham et al., 1988; Balasanmugan et al., 1983, 1989; see Section 3B.3.1, above). It is interesting to note that TOF-LMMS results for nitrosulfonic acids on a surface do not exhibit the formation of $(M-H+O)^-$ ions. The base peak is due to $(M-H)^-$, and intense $(M-NO)^-$ and $(M-NO_2)^-$ fragment signals are observed. No peaks in the molecular ion region of the positive mass spectra are present; only $(M-NO_2)^+$ is detected (Anderson et al., 1985).

A series of papers has dealt with the mass spectra of small biomolecules. Ascorbic acid is a well-studied example. Positive mass spectra recorded by TOF-LMMS and irradiation with 10 ns, 266 nm pulses show sensitive detection of the protonated molecules and abundant fragments corresponding to $(M+H-44)^+$ (Heinen et al., 1980). Cationization is negligible, and no higher organic clusters are observed. Interestingly, the use of a pulsed CO_2 laser at 10.6 μm in conjunction with FTMS yielded a base peak from the $(M+H)^+$ ions in the presence of a quite significant signal from $(2M + H)^+$. The major fragment is also different and formally refers to $(M + H - 46)^+$ (Coad et al., 1987). The formation of the dimeric ions depends on the delay between irradiation and excitation. The increased interval trapping time obviously favors generation of higher species by ion–molecule reactions. The negative mass spectra also show significant differences for both methods. TOF-LMMS essentially detects deprotonated molecules in the presence of low-intensity fragments at *m/z* 115 and 71, but no precise interpretation has been reported (Lohmann et al., 1984). The base peak in the negative mass spectra of the FTMS study refers also to $(M-H)^-$ but abundant signals are detected for the $(M-OH)^-$ and $(2M-H)^-$ ions. The intensity of the dimeric ions again correlates directly with the applied trapping delay (Coad et al., 1987). Finally, irradiation of the sample under atmospheric pressure by mounting the analyte externally to the source on a polymer film essentially removes the fragments from the TOF-LMMS data (Holm et al., 1984). The observed differences can be attributed to the laser irradiation conditions, but it remains clear that the pressure-related effects and the time domain also need attention to explain the observed signals. In fact, these considerations have provided a basis for the empirical DI model as outlined earlier.

Various data from several sources are available on medium-sized, essentially monofunctional molecules, ionized by a well-focused 266 nm laser and subsequently mass analyzed in the LAMMA® or LIMA® instruments. In fact, the data cover a wide range of structural features and no systematic investigation of selected classes has been attempted. The majority of these compounds can be analyzed by conventional techniques as well. The motivation for these studies primarily stems from possible applications in local analysis. The data exhibit quite different features with regard to the importance of radical ion formation, the degree of fragmentation into diagnostically relevant daughters, the generation of nonspecific carbon clusters, and the availability of structural information in the negative mass spectra.

The generation of radical parent ions remains prominent for molecules with a highly aromatic structure such as diphenoxybenzene (Southon et al., 1984). Increase of the laser energy allows the detection of fragments. However, the possibility of controlling the relative abundance of fragment ions over molecular ones is limited to the so-called volatile compounds, i.e., molecules that are easily desorbed, with a minimum of functional groups. A simple halogenated diphenylacetylene derivative no longer shows parent ions, but major fragments are still of the odd-electron type, as illustrated in Figure 3B.7 (Van Vaeck et al., 1986b). Retinoic acid has no aromatic rings, but extensive conjugation over the adjacent double bonds is available. The formation of $M^{+\cdot}$ critically depends on the power density distribution, as evidenced by the dramatic influence of laser beam focusing (Wilk et al., 1988), but extensive fragmentation occurs in any situation. In fact, TOF-LMMS results show that the capability to generate molecular ions with and without daughters is certainly inferior to the two-step laser DI approach (Lubman and Li, 1990; Grotemeyer and Schlag, 1988b).

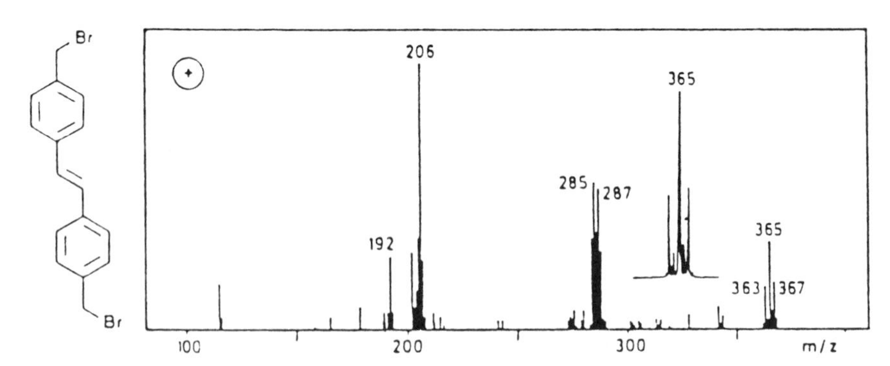

Figure 3B.7. Positive ion mass spectrum of brominated di(methylphenyl)acetylene (MW 366 for two ^{79}Br isotopes) recorded by TOF-LMMS. Reprinted from Van Vaeck et al. (1986b) with permission of John Wiley & Sons, Ltd.

The presence of heteroatoms and polarized bonds in the molecule seems to suppress the signals from radical ions in TOF-LMMS spectra. At the same time, a characteristic distribution of structural information between the positive and negative ion mode mass spectra occurs. Relevant data concern such examples as barbital, pyridoxamine, pyridoxal phosphate, or ciprofloxacin (Heinen et al., 1980; Dang et al., 1984b; Van Vaeck et al., 1989d; Holm and Holtkamp, 1989). Creatinine shows fragment peaks assigned to the rather uncommon loss of CN and CN_2 from the protonated molecules, yielding the base peak, whereas intense signals refer to the attachment of Na^+ and K^+. The lack of high resolution or of additional possibilities to study the actual transitions of the parent ions makes the interpretation often unsatisfactory and explains the current interest in the development of FTMS in conjunction with focused laser irradiation. Abundant cationized molecules are observed. Conventional and external mounting of the sample shows cleaner mass spectra for laser irradiation of creatinine under atmospheric pressure (Heinen and Holm, 1984). Mass spectra for some industrially used disulfides have been recorded (Heinen and Holm, 1984; Holm and Holtkamp, 1989). The cleavage of the sulfur bridge predominates and reduces the peak intensity of the ions in the MW region of the mass spectra. However, sensitivity and specificity of fragments has permitted successful identification of these components in micrometer-scale heterogeneities of rubber samples.

The mass spectra of phthalic acids have been studied, and the metastable decomposition of the parents during the trajectory in the first drift region has been monitored (Rosmarinowsky et al., 1985). The parent peak in the positive mass spectrum corresponds to the protonated anhydride under threshold irradiation, whereas increased power density on the sample permits detection of cationized molecules as well as doubly- and even triply-sodiated species such as $(M-nH+(n+1)Na)^+$. Negative ion data show $(M-H)^-$ and abundant signals for the fragments, owing to the consecutive loss of CO_2 from the deprotonated molecules. The abundance of nonspecific carbon clusters depends on the laser energy applied.

The occurrence of electron attachment to produce $M^{\cdot-}$ still remains an intriguing process. Strong signals evidencing the occurrence of this phenomenon are less frequently observed in comparison to $M^{\cdot+}$ in the positive ion mass spectra recorded by TOF-LMMS. Nevertheless, there are several examples, and in most cases the molecular structure and the functional groups are such that one does not really expect that the ECI would be particularly favored.

Figure 3B.8 shows the positive TOF-LMMS spectra for methyl red, whereas the major fragments in the negative ion detection mode are represented in Figure 3B.9 (Van Vaeck et al., 1989b). The positive mass spectrum shows the features one expects from the method, i.e., soft ionization combined with

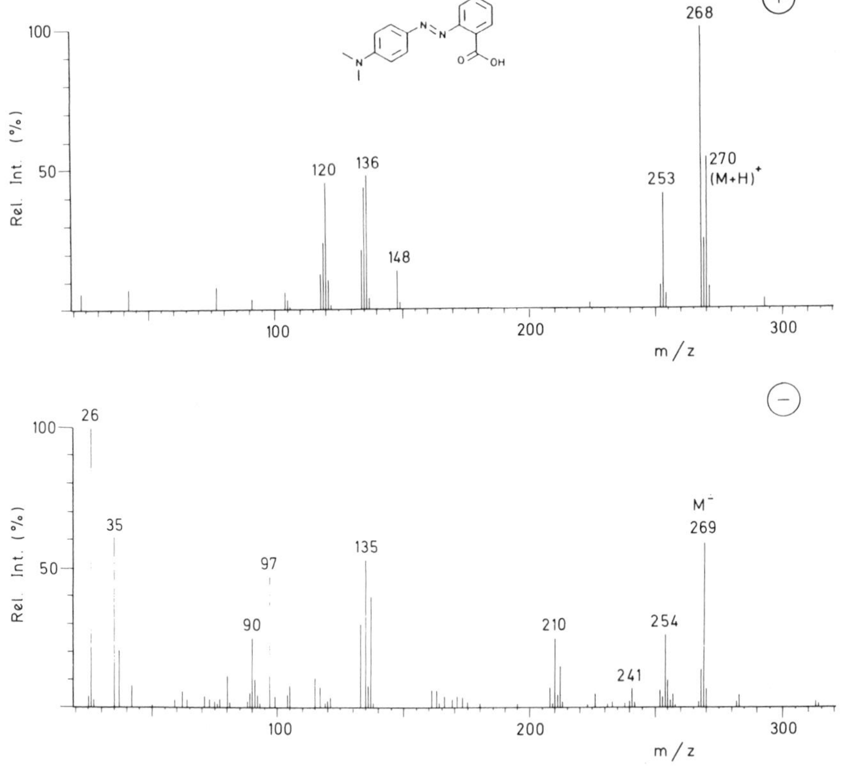

Figure 3B.8. Positive and negative ion mass spectra of methyl red (MW 269) recorded by TOF-LMMS. Reprinted from Van Vaeck et al. (1989b) with permission of John Wiley & Sons, Ltd.

extensive fragmentation. Again, the data largely compare with those in a reference EI-MS registry, except for the parent region (McLafferty and Stauffer, 1989). Instead of a medium-intensity signal for the $M^{+\cdot}$, LMMS shows a base peak from the $(M-H)^+$ ions, accompanied by a substantial contribution from the protonated molecules. Interestingly, the odd-electron molecular ion peak dominates the parent region of the negative ion mass spectrum in TOF-LMMS, in spite of the fact that the molecule contains a carboxylic acid. The acidic hydrogen provides an obvious way to reach the even-electron state,

Figure 3B.9. Structural assignment of the major signals in the negative ion mass spectrum of methyl red recorded by TOF-LMMS. Reprinted from Van Vaeck et al. (1989b) with permission of John Wiley & Sons, Ltd.

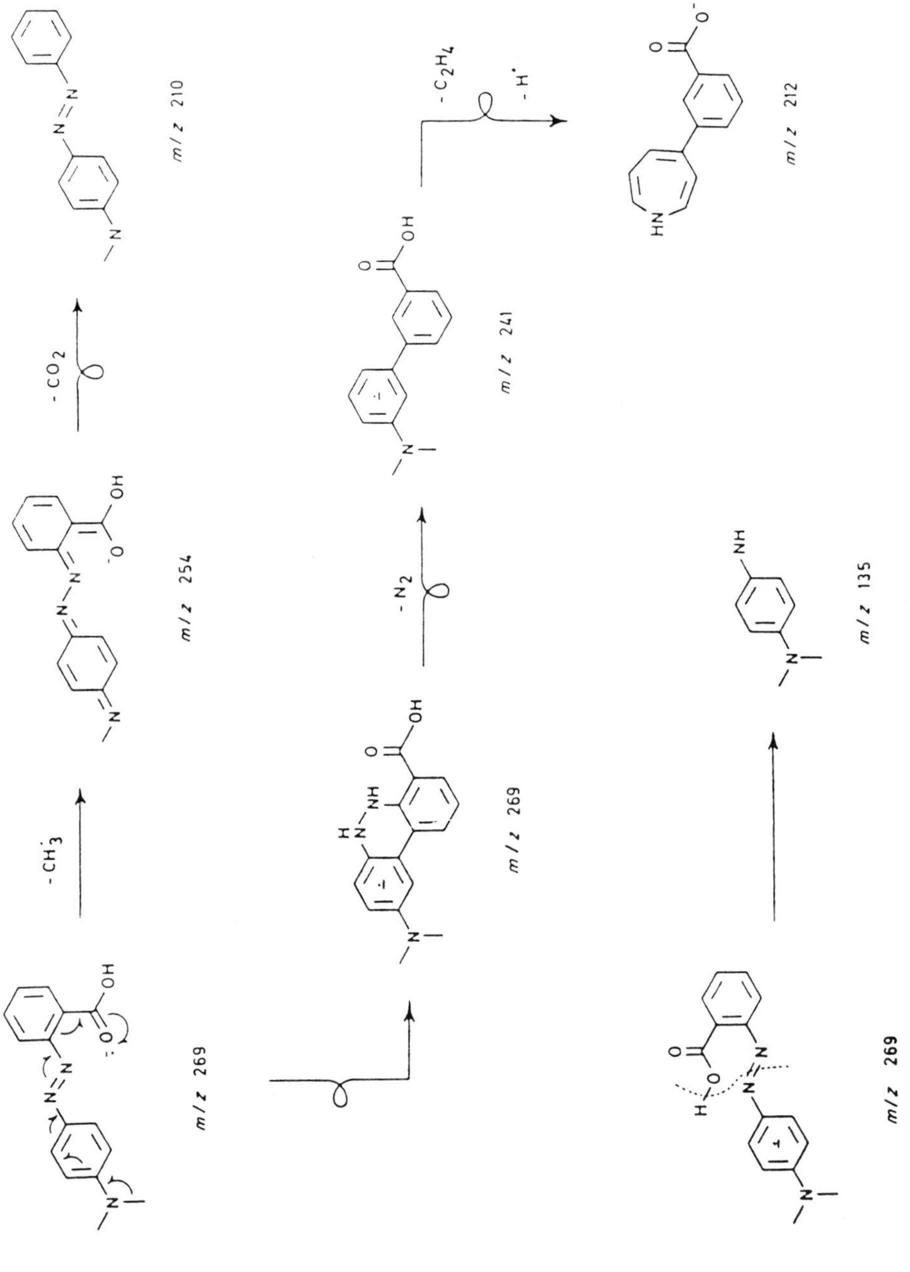

m/z 269

$-CH_3^{\cdot}$

m/z 254

$-CO_2$

m/z 210

m/z 269

$-N_2$

m/z 241

$-C_2H_4$

$-H^{\cdot}$

m/z 212

m/z 269

m/z 135

a pathway which however is not used. Apparently, the third-body collisions in the selvedge region allow for sufficient stabilization of at least a substantial number of the initially generated $M^{+\cdot}$. A roughly five times higher fraction of the total ion current is carried by the fragments detected. TOF-LMMS results do not provide pertinent information about the generation of these daughters by dissociative ECI or alternatively by fragmentation from the $M^{-\cdot}$ as indicated in the pathway of Figure 3B.9. It should be stressed that the proposed mechanisms issue from the desire to rationalize the structural information in the mass spectrum: they do not aim at describing the actual formation in detail. Hence, alternative structures and routes can be elaborated. Those represented maintain consistency with the interpretation of other results in our TOF-LMMS data base. The main point of interest concerns the fact that radical ion formation should not be considered a highly unlikely process outside the class of PAHs.

This statement is further evidenced for instance by the detection of the $M^{+\cdot}$ from styrenesulfonic acid in the study of the degradation of cation exchange resins (Scanlan et al., 1989). In contrast, hydroxynaphthalenesulfonic acid shows only $(M-H)^-$ and subsequent loss of SO_3 (Heinen and Holm, 1984). Both conventional and external mounting of the sample has been applied and yields nearly identical MS information.

TOF-LMMS results on long aliphatic acids show $(M-H)^-$ and signals assigned to the subsequent loss of one and two ethylene molecules. The author interprets the absence of nonspecific carbon clusters as an indication that the elimination occurs before ionization (Heinen, 1981). Threshold conditions remove the higher fragments. Fatty acids have been characterized at the surface of metals in industrial applications by TOF-LMMS on a micrometer scale (Krier and Muller, 1986).

Small organic molecules such as uric acid have been identified from deposits in tissues by means of protonated molecules and structural fragments in positive ion mass spectra and deprotonated molecules in the negative ones (Verbueken et al., 1987). Additional confirmation is obtained from the negative spectra, where deprotonated molecules have been detected.

The ability to produce cationized or protonated molecules in conjunction with structurally specific fragments, which can be readily interpreted using common knowledge from conventional MS, provides a significant asset in local analysis, as shown by the identification of surface contaminants from plasticizers on asbestos fibers by TOF-LMMS (DeWaele et al., 1983a). The volatility of the lower analogues and the negligible sensitivity of the higher ones hinder the analysis of the corresponding pure products. Hence, this example shows the merits of the deductive interpretation as opposed to the comparative approach based on fingerprints of reference samples, and consequently the need to develop a practical model to interpret the signals. The variety of ion formation mechanisms in the selvedge region certainly involves a significant motivation for further elucidation of the processes initiated by laser–solid interaction. However, the incomplete understanding of these mechanisms at this moment tends to hinder practical applications.

A survey of the main signals in the mass spectra of a variety of organic molecules from different classes, including acids, alcohols, aldehydes, amines, esters, ethers, ketones, phenols, is reported in the context of an off-line TLC application (Kubis et al., 1989).

3B.3.3. Medium-Sized Polyfunctional Molecules

This section covers results for a series of polyfunctional UV stabilizers and additives currently used in polymer chemistry, as well as a variety of drugs, alkaloids, and additional biomolecules within a MW range of 300–800. The discussion is confined to nonionic compounds, but some examples of hydrochlorides are included as well. It has been found that data on analytes in the hydrochloride form are not fundamentally different from data on the free bases, apart from the obvious detection of the counterion.

Polymer Additives. Interest in the LD-MS characterization of the primary and secondary antioxidants arises from the capability to selectively monitor these molecules within the complex environment of a polymer sample. However, apart from this application-oriented aspects, the mass spectra have been found to possess interesting features that elucidate some aspects of the basic DI process. Figure 3B.10 surveys the structures of representative commercial additives. To facilitate the discussion, the trade names of these products will be used.

One of the first systematic studies of these compounds by means of TOF-LMMS was performed in our own laboratory (Van Vaeck et al., 1985a). Conventional EI-MS with conventional resistive heating on a direct-probe introduction system was applied for comparison on the same samples. A striking degree of agreement between the results from both methods was observed: the relative peak intensities of the different fragments are comparable for several analogues.

A good example is provided by the mass spectra of Irganox 1076 in Figure 3B.11 (Van Vaeck et al., 1985a). The structural assignment and fragmentation mechanisms in Figure 3B.12 illustrates the detailed characterization of the functional groups and structural units in the molecule (Van Vaeck et al., 1985a). However, the data for Irganox 330 in Figure 3B.13 highlight another relevant feature of the LMMS method as regards its efficacy for structural analysis. In spite of the generally claimed soft ionization capabilities of LMMS, no MW information is available from the mass spectra. Also, the

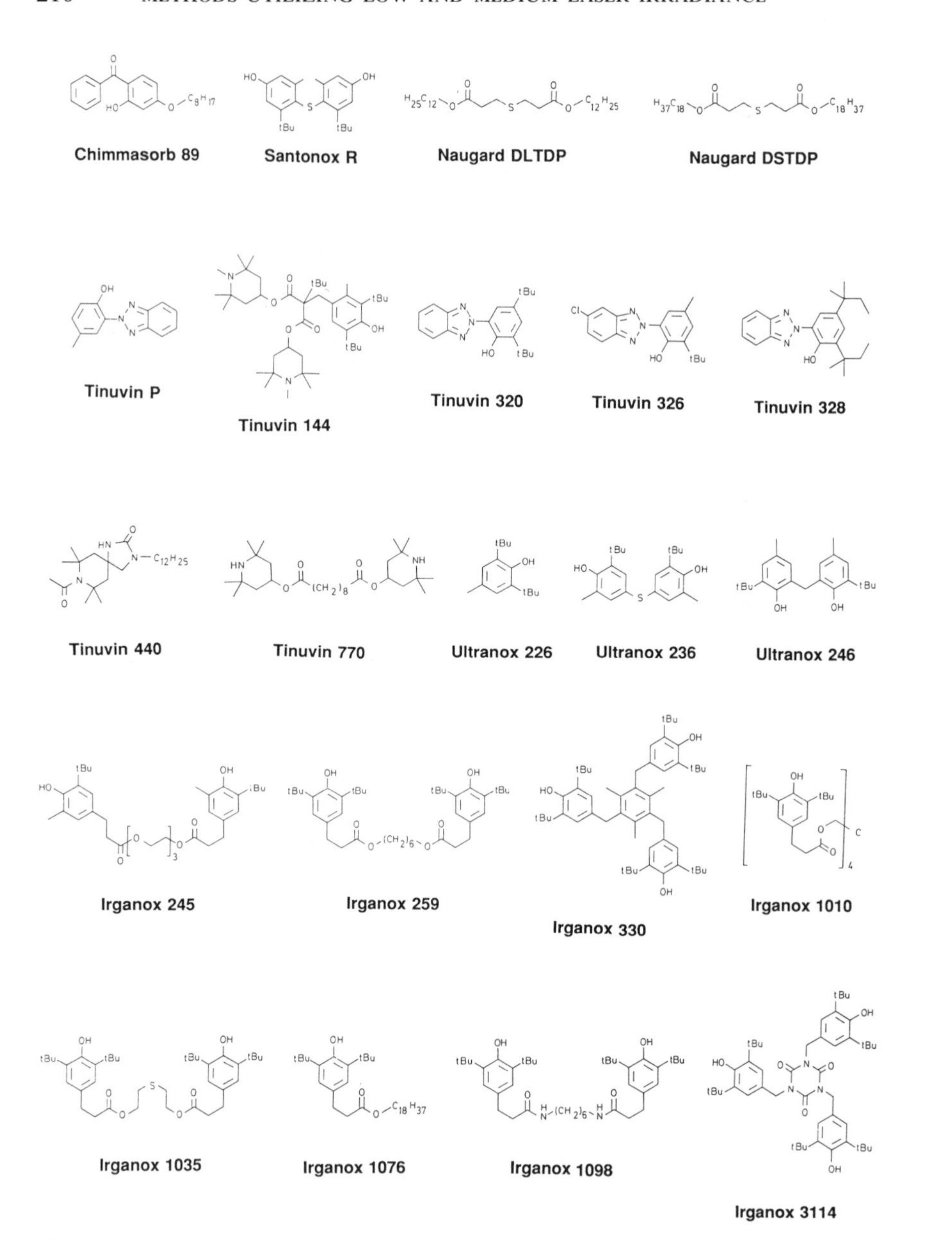

Figure 3B.10. Survey of the structures and trade (product) names of representative polymer additives and UV stabilizers, studied by LD-MS.

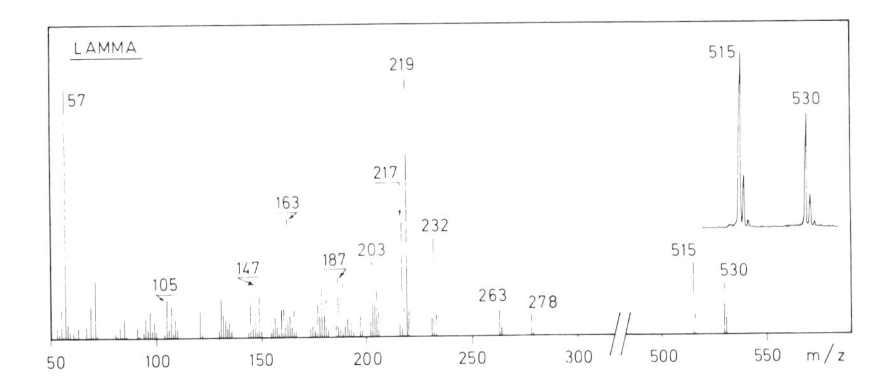

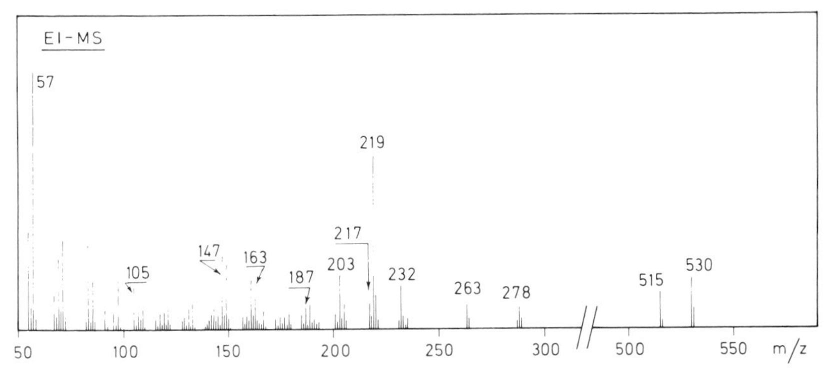

Figure 3B.11. Comparison of positive ion mass spectra of Irganox 1076 (MW 530) recorded by TOF-LMMS and EI-MS. Reprinted from Van Vaeck et al. (1985a) with permission of the American Chemical Society.

negative ion mass spectra do not contain additional information. In contrast, resistive heating of the sample and EI-MS permit detection of $M^{+\cdot}$. This situation is less uncommon than the reader might expect from a survey of the literature. The foregoing apparently remarkable observation results from the fact that LMMS is applied preferentially to so-called problem cases in conventional MS, for which compounds the high heating rate of the solid and presence of the sample inside the ion source represent a definitive advantage. This does not imply that the inherently destructive power of a laser beam represents the most gentle treatment for organic compounds.

Finally, the results and the structural assignment for Chimmasorb 81 given in Figures 3B.14 and 3B.15 illustrate the important contribution of radical species to the total ion current. The generally accepted notion that the detected ions from laser-irradiated solids correspond to even-electron systems is a

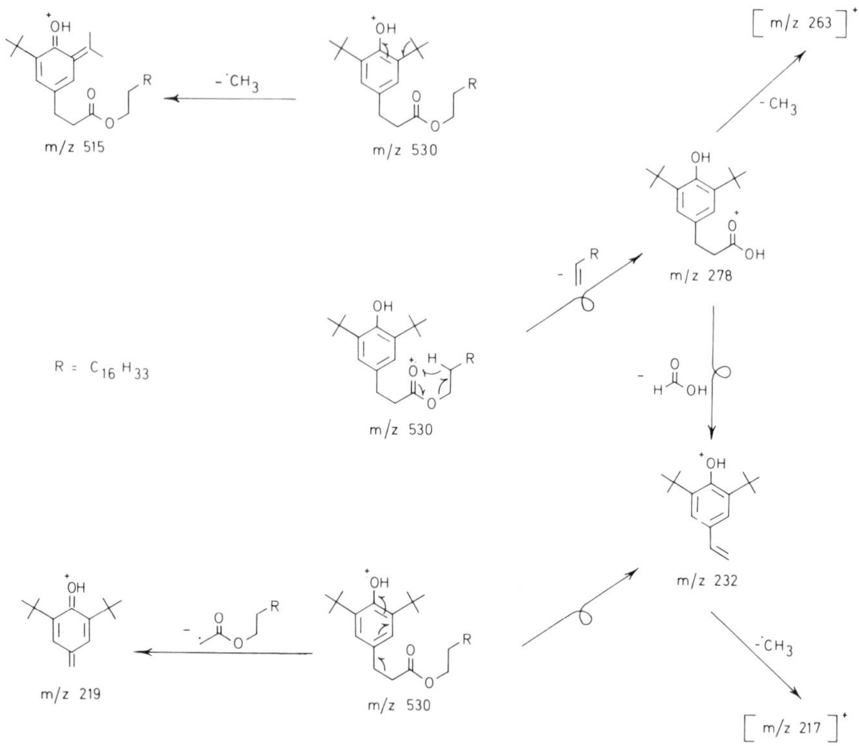

Figure 3B.12. Structural assignment of the major signals in the positive ion mass spectrum of Irganox 1076 recorded by TOF-LMMS. Reprinted from Van Vaeck et al. (1985a) with permission of the American Chemical Society.

misleading simplification when detailed analysis of LMMS data is attempted. However, we have also found that the relative intensity of the odd-electron molecular and fragment ions critically depends on the laser power density. One of the problems involved in interlaboratory comparison of TOF-LMMS remains the rather inadequate description of the so-called threshold level for laser irradiation, which remains after all a relative notion.

An extensive study of the same polymer additives with CO_2 laser irradiation at 10.6 μm on a FTMS dual-cell instrument has been reported (Asamoto

Figure 3B.14. Comparison of positive ion mass spectra of Chimmasorb 81 (MW 326) recorded by TOF-LMMS and EI-MS. Reprinted from Van Vaeck et al. (1985a) with permission of the American Chemical Society.

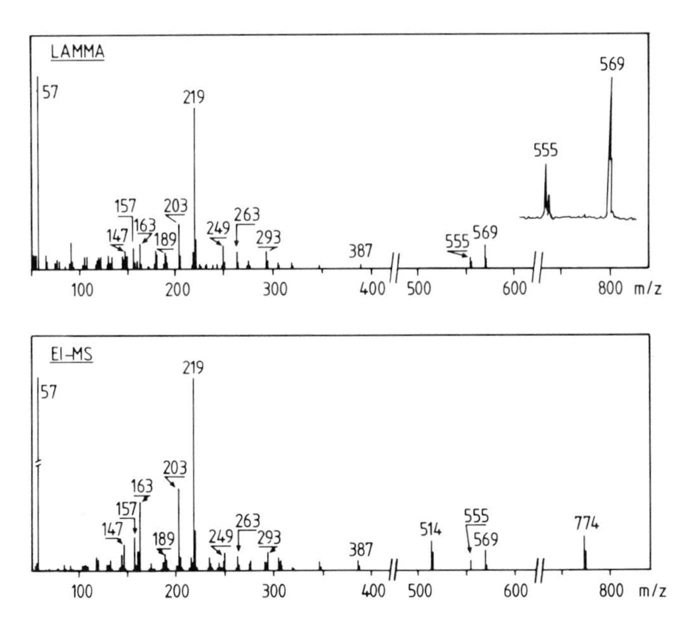

Figure 3B.13. Comparison of positive ion mass spectra of Irganox 330 recorded by TOF-LMMS and EI-MS. Reprinted from Van Vaeck et al. (1985a) with permission of the American Chemical Society.

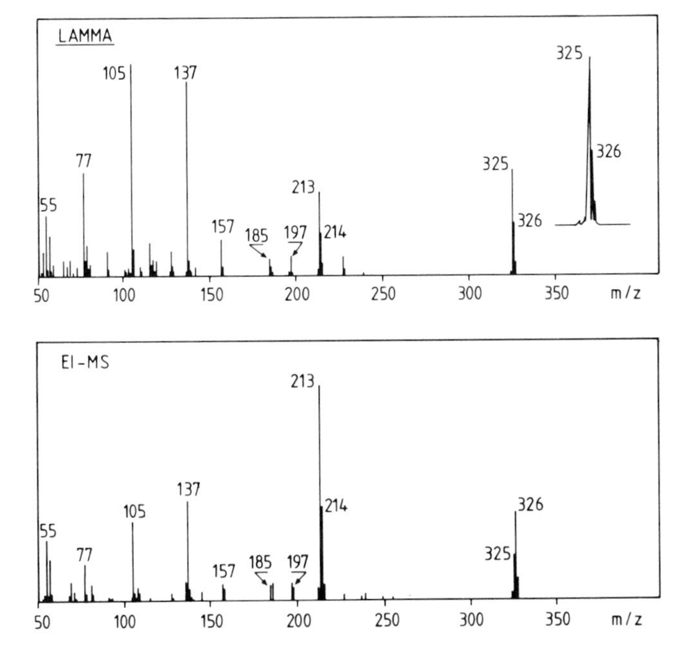

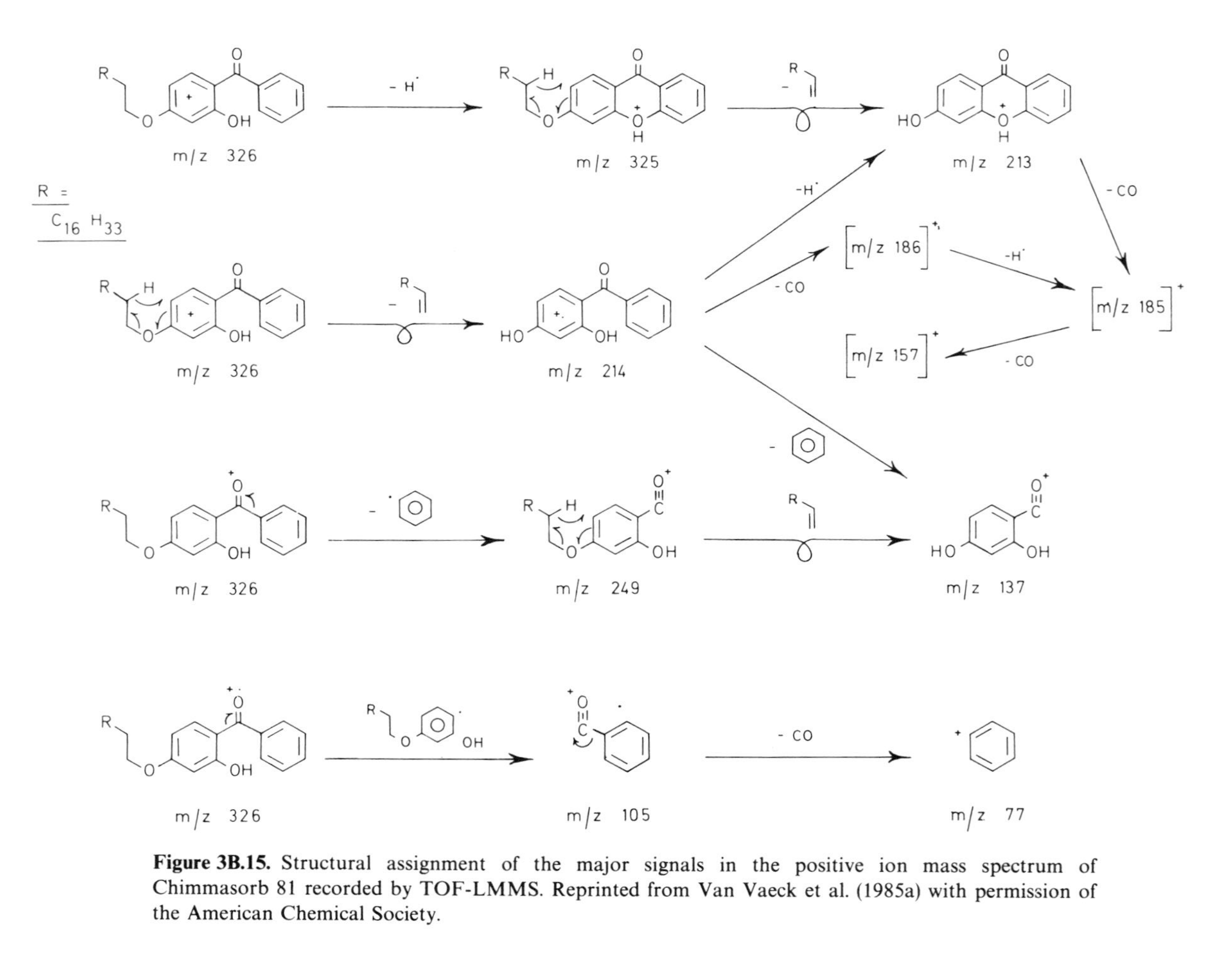

Figure 3B.15. Structural assignment of the major signals in the positive ion mass spectrum of Chimmasorb 81 recorded by TOF-LMMS. Reprinted from Van Vaeck et al. (1985a) with permission of the American Chemical Society.

et al., 1990). The typically applied power density is $1–5 \cdot 10^8$ $W \cdot cm^{-2}$, and the amount of sample evaporated per shot is on the order of 100 ng. The mass spectra show abundant signals from cationized molecules and much lower contributions from the proton attachment. Radical parent ions are observed for analogues from the Ultranox and Tinuvin series. The pattern of fragment ions essentially confirms the trend mentioned previously for the TOF-LMMS results. The corresponding signals are prominent under focused conditions with an ellipsoid spot of about $350 \times 150\,\mu m$. Doubling the irradiated area reduces the relative contribution of the fragments in comparison to the cationized molecules by a factor of roughly 5 to 10, but the relative peak intensities among the different daughter ions are maintained.

The successful characterization of these benzotriazine-based UV stabilizers in this LD-FTMS study contrasts with our own experience on TOF-LMMS analysis of these compounds. The latter method yields only highly irreproducible decomposition patterns containing no structurally specific signals (L. Van Vaeck, unpublished results). This particularly holds true for most of the compounds reported to exhibit strong signals for radical molecular ions in the FTMS study. An unambiguous interpretation of this observation is not yet at hand. Moreover, LD-FTMS has permitted generation of structurally relevant negative mass spectra not observed in TOF-LMMS. Apart from the influence of the laser irradiation conditions, instrumental differences might explain these differences. Both instruments have markedly different capabilities to detect species arising from an extended time domain of the formation process. Also, the residence time of the neutrally desorbed molecules in the selvedge region is different because of the high extraction voltage applied in TOF-LMMS and the low trapping field conditions in FTMS.

The effect of the irradiation wavelength in the IR range on parent and fragment ion production has been studied for a range of secondary antioxidants, hindered phenols and phosphites (Johlman et al., 1990). The use of a CO_2 laser at $10.6\,\mu m$ has been compared with the irradiation by the basic harmonic of an Nd:YAG at $1.06\,\mu m$. The relative importance of the fragments to the total ion current in comparison to the cationized adducts is significantly reduced with the shorter wavelength. However, the mass spectra become more critically affected by a variety of experimental factors such as laser energy/power density and sample preparation. Comparison with FAB mass spectra shows the superior potential of LD to characterize the functional groups and structural moieties in the compounds.

Drugs and Metabolites. Within the framework of the empirical model for DI, a substantial range of drugs and metabolites have been studied with TOF-LMMS. We shall discuss a few relevant compounds in detail. A survey

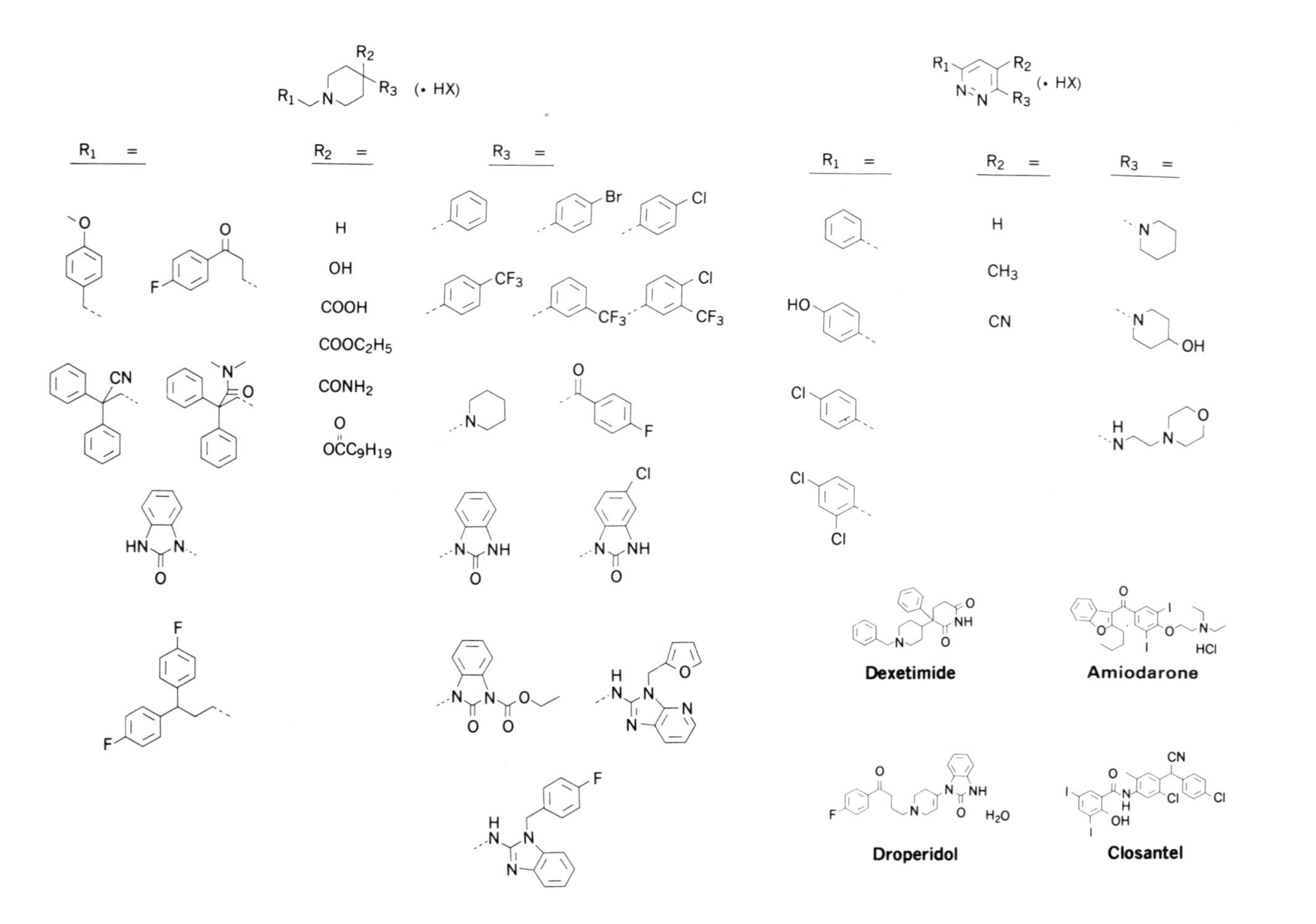

(• HX)
R1 =
R2 =
H
OH
COOH
COOC2H5
CONH2
OCC9H19
R3 =
(• HX)
R1 =
R2 =
H
CH3
CN
R3 =
Dexetimide
Amiodarone
Droperidol
Closantel

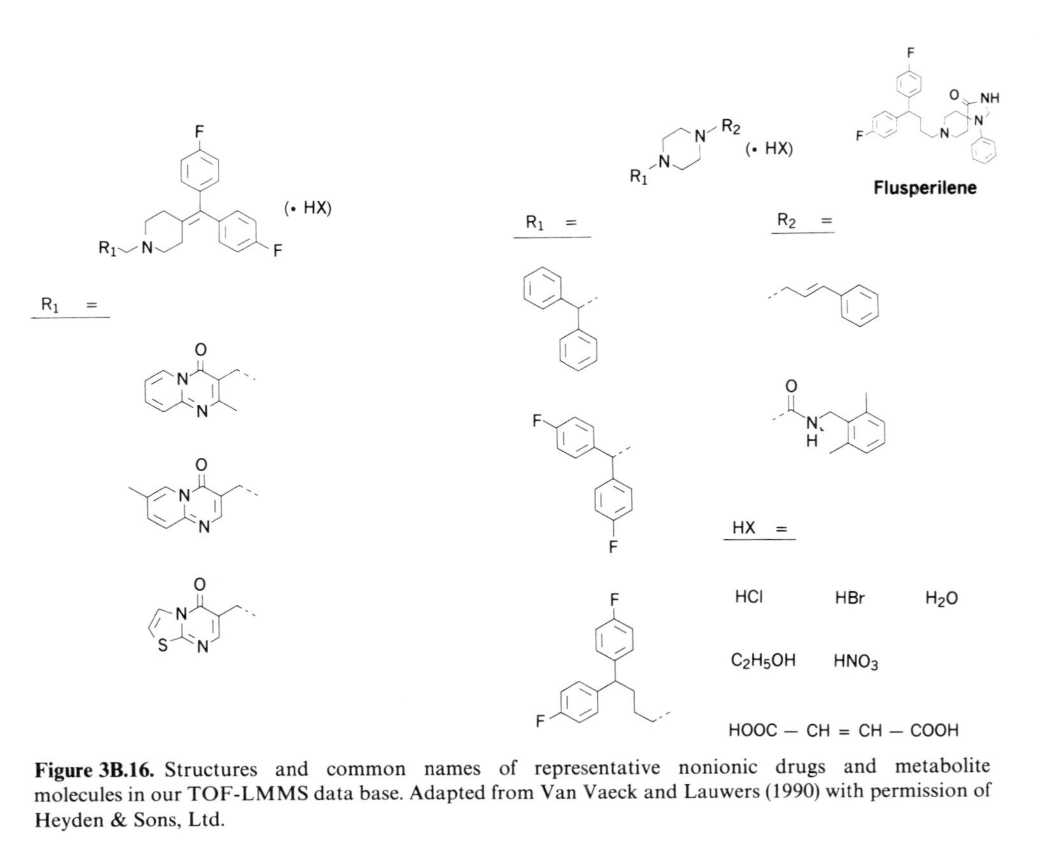

Figure 3B.16. Structures and common names of representative nonionic drugs and metabolite molecules in our TOF-LMMS data base. Adapted from Van Vaeck and Lauwers (1990) with permission of Heyden & Sons, Ltd.

of the structures is given in Figure 3B.16 (Van Vaeck and Lauwers, 1990). A first example concerns a comparison between carnidazole and its *N*-oxide. The mass spectra and structural assignment of the major signals is represented in Figures 3B.17 and 3B.18, respectively (L. Van Vaeck, unpublished results; Van Vaeck et al., 1988b). The results for the nonoxygenated form show several typical features of such multifunctional derivatives. The general appearance of the mass spectra is relatively clean, with a limited number of intense peaks. The MW information is available in the positive and negative ion mass spectra

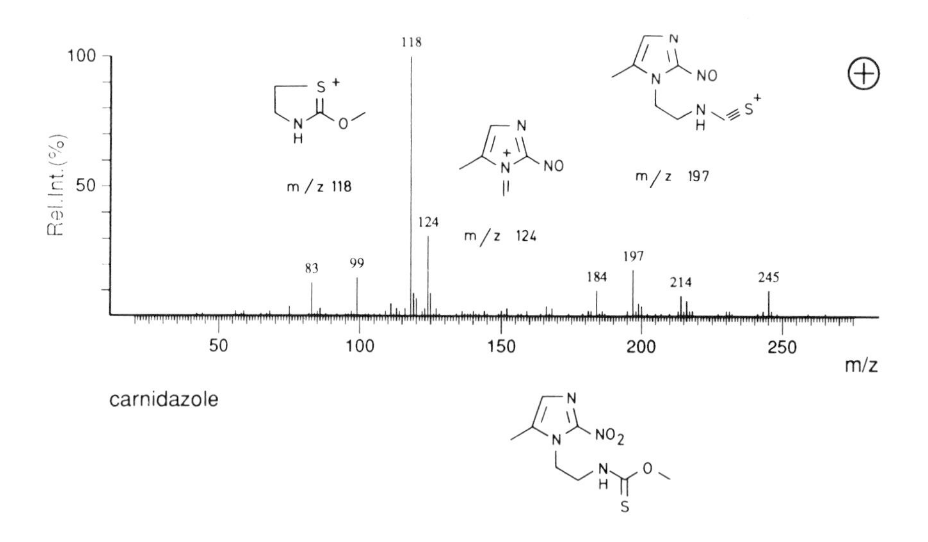

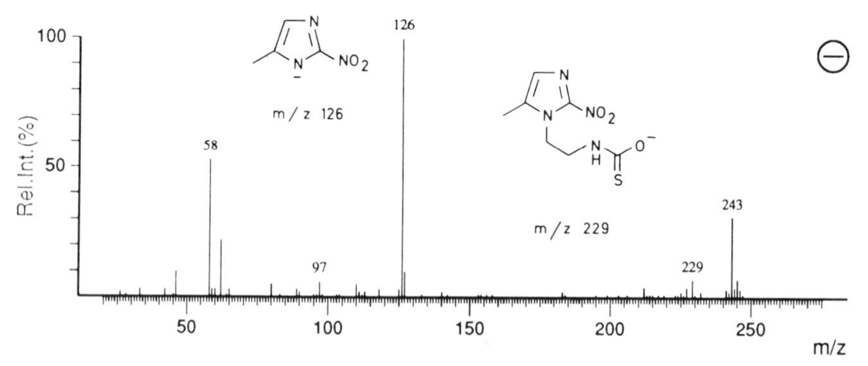

Figure 3B.17. Positive and negative ion mass spectrum of carnidazole (MW 244) recorded by TOF-LMMS (L. Van Vaeck, unpublished results).

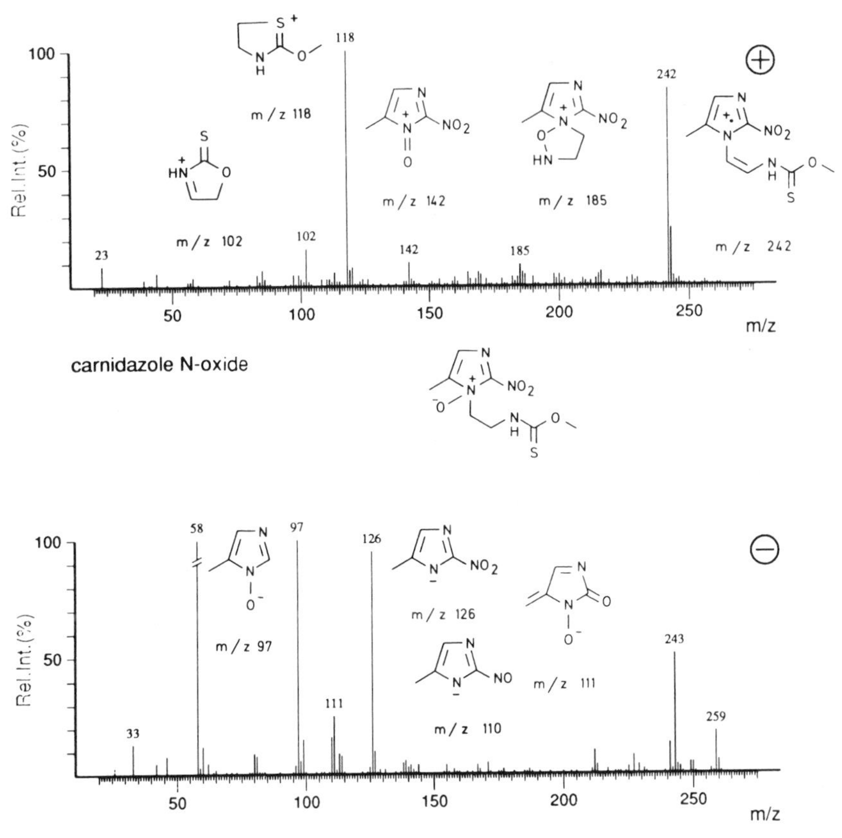

Figure 3B.18. Positive and negative ion mass spectrum of carnidazole *N*-oxide (MW 260) recorded by TOF-LMMS. Reprinted from Van Vaeck et al. (1988b) with permission of John Wiley & Sons.

by means of the $(M + H)^+$ and $(M - H)^-$ signals, respectively. The base peak is due to fragments characterizing the aromatic moiety in the negative ion mass spectrum and the side chain in the positive mode. As a result, complementary information is derived from ions of both polarities. This situation frequently occurs in TOF-LMMS but depends, of course, on the actual nature of the functional groups in the molecule under study. Minor signals in both positive and negative ion mass spectra permit further description of the molecular structure. The formation mechanisms can readily be conceived in terms jibing with common knowledge of molecular behavior in the gas phase. This does not exclude the possibility of direct photofragmentation. Note also the absence of $(M + H - O)^-$ ions, in spite of the nitro functional group of the imidazole system. We have found that formation of the latter ions depends on the applied power density and that the selvedge interactions can be virtually eliminated by very precise application of threshold irradiation. This is important because it enables us to verify the presence of oxygenated impurities in the sample.

The mass spectra for the oxygenated metabolites show characteristic differences with the results from the parent drug. No MW is found in the positive ion mode, but the negative ions clearly indicate the presence of oxygen in the molecule, specifically on the imidazole system. It is interesting to note that the parent ion signal in the positive spectra reveals the loss of water from $M^{+\cdot}$. As a result, these fragments correspond to radical systems. The dehydration permits a complete resonance linkage between the aromatic functional group and the thiocarbamoyl group, and the resulting stabilization explains the high intensity of the corresponding signal. The relevance of this example is twofold. First, application of threshold laser irradiation enhances the informative nature of the data by minimizing the complicating selvedge phenomena such as $(M - H + O)^-$ ion formation or intermolecular group transfer. Secondly, the data allow fine characterization of the molecular structure on the condition that the possible formation and detection of odd-electron ions is accepted. Otherwise, if one tries to interpret the results on the strict basis of even-electron ions, a substantial part of the necessary information remains inaccessible.

Another example concerns a comparison between flunarizine and its *N*-oxide (Van Vaeck et al., 1989d). The mass spectra are compared in Figure 3B.19, and the structural assignment is surveyed in Figure 3B.20 for the parent drug and in Figure 3B.21 for the oxygenated derivative. In the last case, a few relevant fragmentation mechanisms leading from $M^{+\cdot}$ to the fragments are explicitly mentioned to illustrate the generally applicable schemes that allow us to rationalize the detected signals from the *N*-oxides. As a rather unusual feature, virtually all diagnostic information is carried by the positive ions. No daughter ions are observed below m/z 100, as opposed to the results of conventional EI-MS. Note that flunarizine does not show protonated molecules in the parent region of the mass spectra in spite of the fact that the dihydrochloride form has been analyzed. A minor signal in the molecular region of the mass spectrum refers to $(M - H_2)^{+\cdot}$; these fragments strongly suggest the generation of $M^{+\cdot}$ in TOF-LMMS, unless direct photofragmentation or an exception to the even-electron rule is assumed. The absence of the $M^{+\cdot}$ signal can readily be accounted for by the quantitative nature of the unimolecular decomposition.

In fact, this situation characterizes the discussion about radical ions in TOF-LMMS. Fragments pointing to odd-electron species are detected for a substantial number of molecules, not only for selected structures and/or PAH derivatives. The corresponding signals show significant intensity but rarely

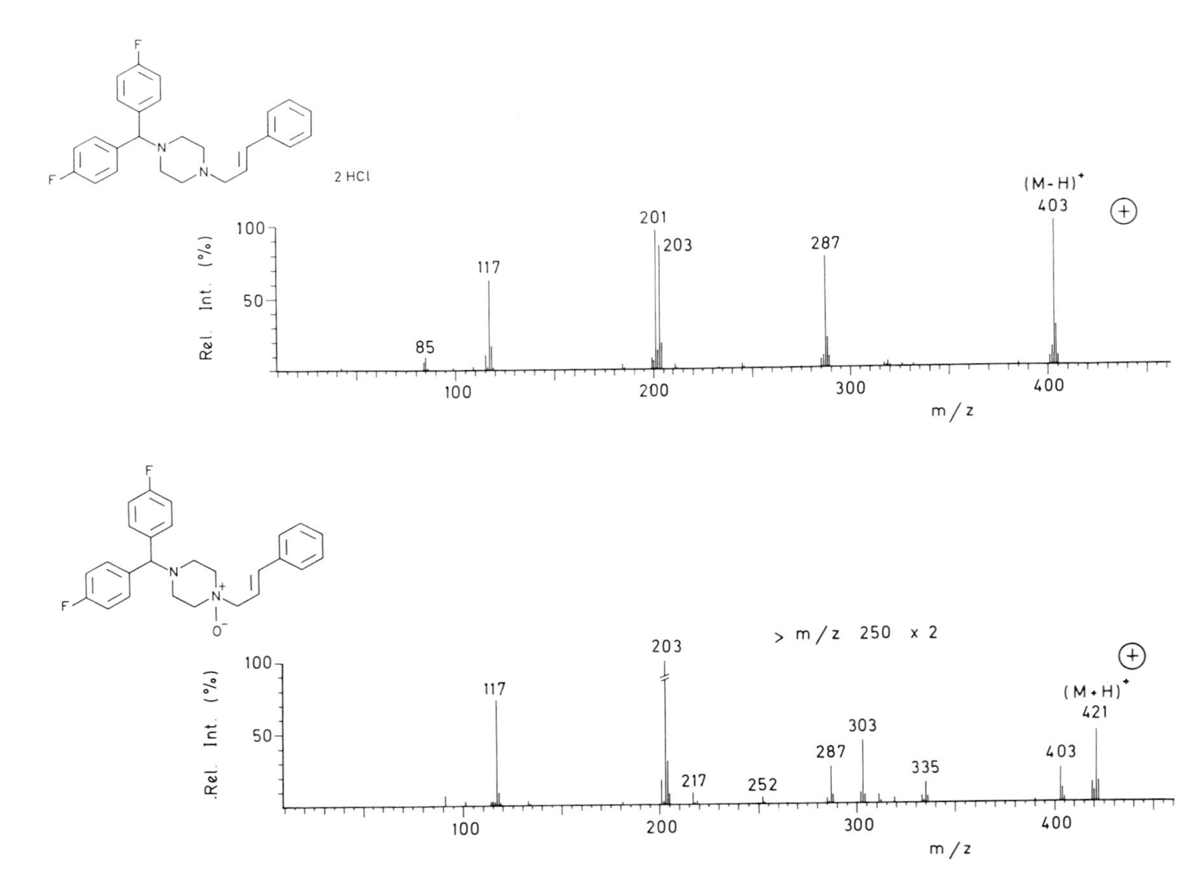

Figure 3B.19. Positive ion mass spectrum of flunarizine (MW 404) and flunarizine *N*-oxide (MW 420) recorded by TOF-LMMS. Reprinted from Van Vaeck et al. (1989d) with permission of John Wiley & Sons, Ltd.

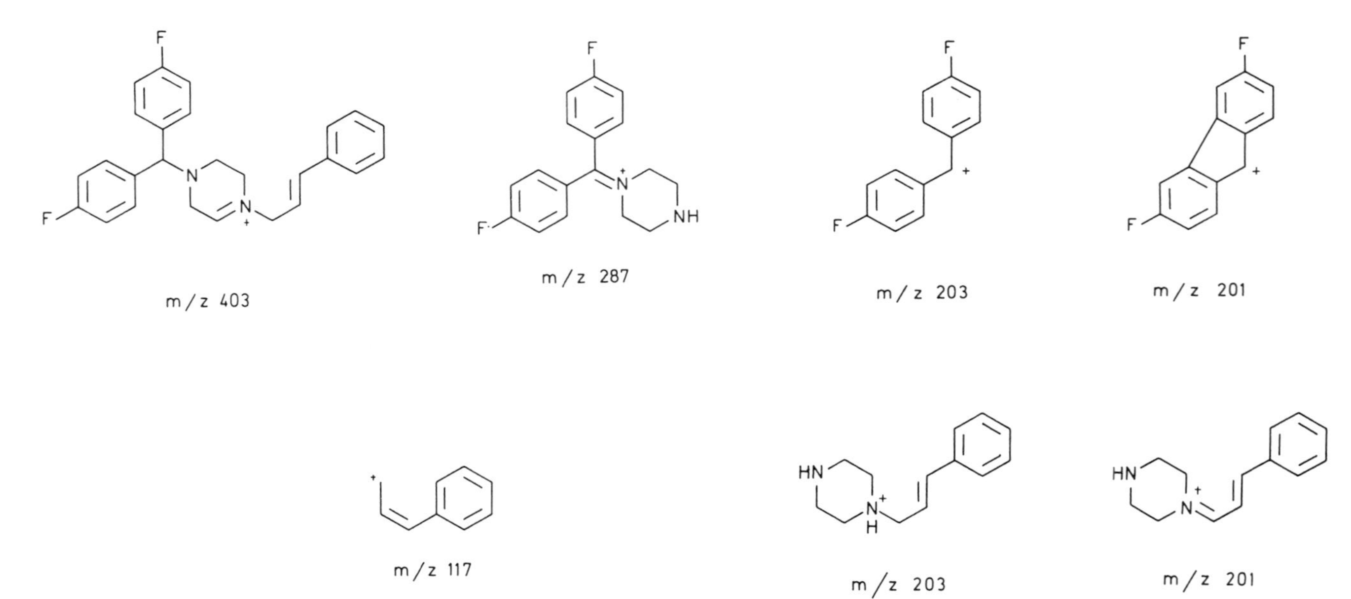

Figure 3B.20. Structural assignment of the major signals in the positive ion mass spectrum of flunarizine. Reprinted from Van Vaeck et al. (1989d) with permission of John Wiley & Sons, Ltd.

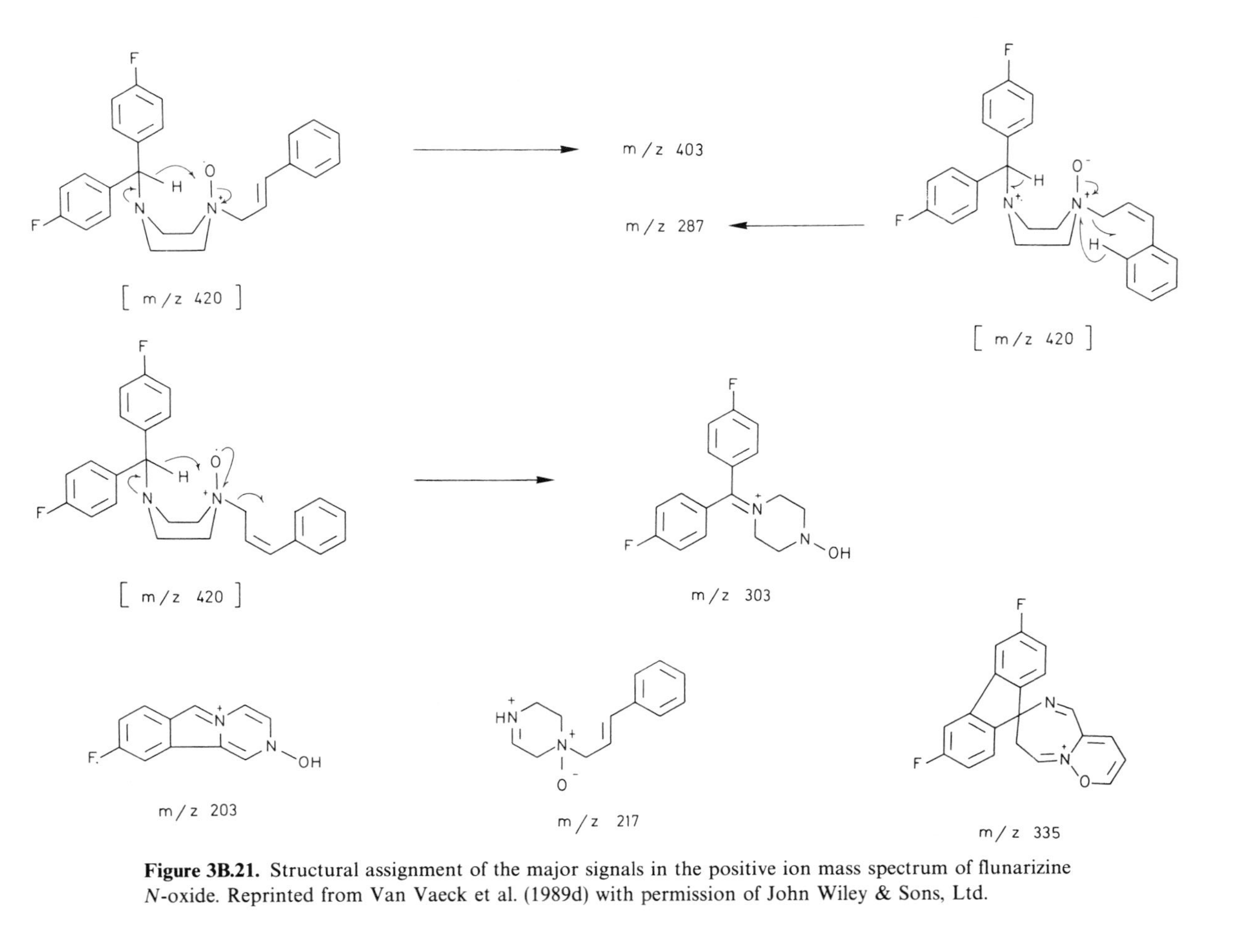

Figure 3B.21. Structural assignment of the major signals in the positive ion mass spectrum of flunarizine *N*-oxide. Reprinted from Van Vaeck et al. (1989d) with permission of John Wiley & Sons, Ltd.

dominate the mass spectrum. The same applies essentially to the majority of EI-MS data as well. However, detection of $M^{+\cdot}$ as a prevalent peak is less common in TOF-LMMS than in EI-MS. It is readily conceived that the internal energy distribution of the $M^{+\cdot}$ can be shifted toward higher values in the case of laser DI on solids as compared to the gas phase situation in EI-MS. In our opinion, the absence of the corresponding signal does not represent a serious argument ruling out the actual formation of these radical ions and their role as actual precursors for the detected fragments.

The generally accepted notion that the parent peak in positive TOF-LMMS simply refers to the adduct ions is not confirmed by the results in our data base. This applies even to closely related structures. A striking illustration is provided by the mass spectra of benperidol, droperidol, and domperidone in Figure 3B.22 (Van Vaeck and Gijbels, 1990b). Whereas the parent peak from benperidol refers to the $(M+H)^+$, the same molecule except for an additional double bond in the piperidine functional group only produces $(M-H)^+$. Domperidone does not provide MW information at all in the positive detection mode. Fortunately, additional and/or supporting information about the MW is frequently available from the negative mode, but the occurrence of proton attachment depends on factors that are not yet well understood. One of the ways that we believe gives some indication as to proton attachment is via careful comparison of the threshold irradiation level. The ranking order of laser threshold conditions is benperidol $<$droperidol$<$ domperidone. This trend parallels that seen for the thermal evaporation by resistive heating. The tentative explanation for the observed presence or absence of $(M+H)^+$ ions is related to the relatively higher density of the selvedge region, which can be achieved in particular for the "'more volatile" analogues.

The formation and structural interpretation of negative ions of medium-sized polyfunctional molecules remain an intriguing aspect of TOF-LMMS analysis. Extensive background information from established MS techniques often facilitates the assignment of fragments in the positive ion detection mode, but there is much less help for the interpretation of negative spectra. Figure 3B.23 illustrates the positive and negative mass spectra of astemizole (Van Vaeck et al., 1989b). The structural assignment of the positive fragments in Figure 3B.24 shows the prominent role of the piperidine and methoxyphenyl functional groups in producing the most intense signals. The relatively modest fraction of the total ion current carried by fragments containing the bicyclic heteroaromatic system may at first glance seem unexpected. However, the explanation is readily found in the structures and pathways of the negatively charged fragments in Figure 3B.25, where the benzimidazole functional group participates in all routes to lower daughters (Van Vaeck et al., 1989b). The role of a given trigger group depends on the tendency to stabilize a charge

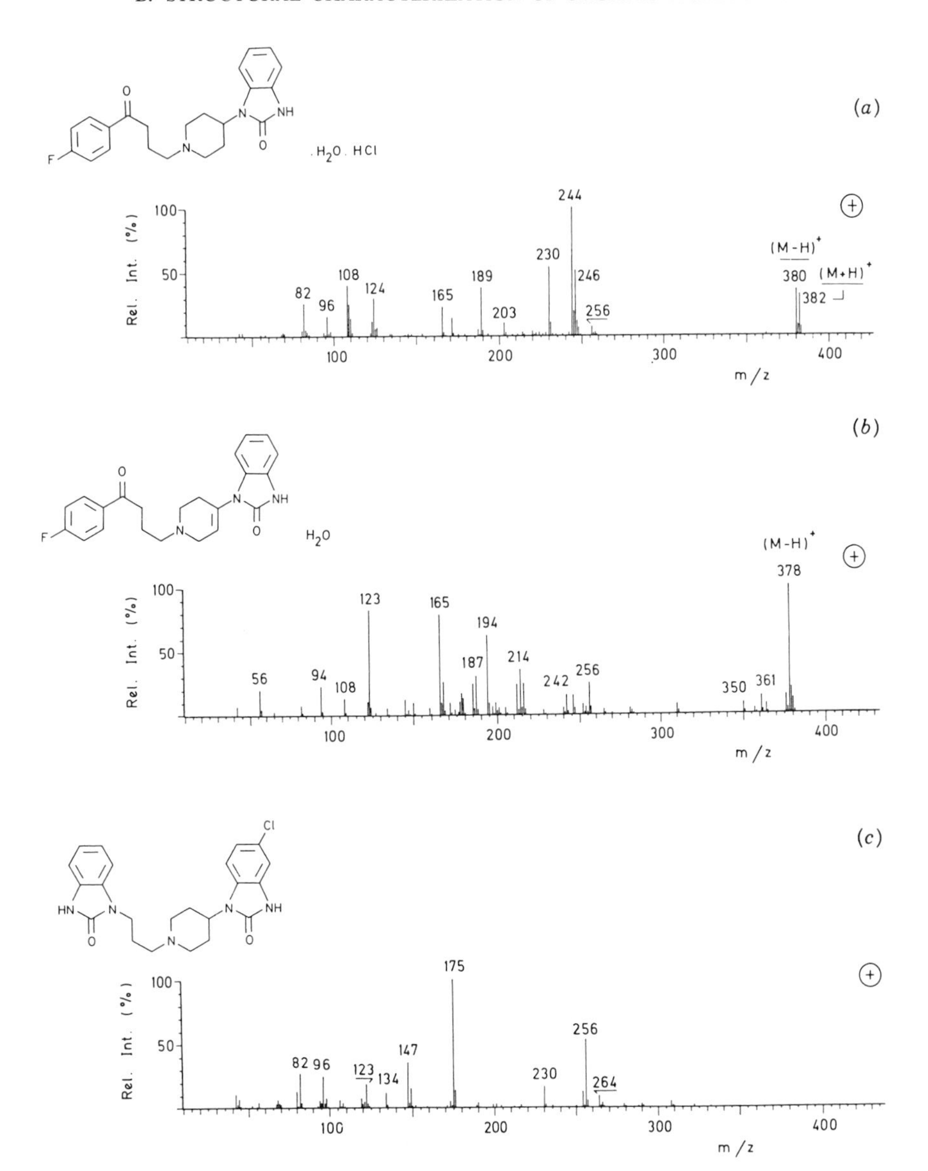

Figure 3B.22. Positive ion mass spectra of benperidol (a, MW of free base 381), droperidol (b, MW of free base 379) and domperidone (c, MW 425) recorded by TOF-LMMS. Reprinted from Van Vaeck and Gijbels (1990b) with permission of Springer-Verlag.

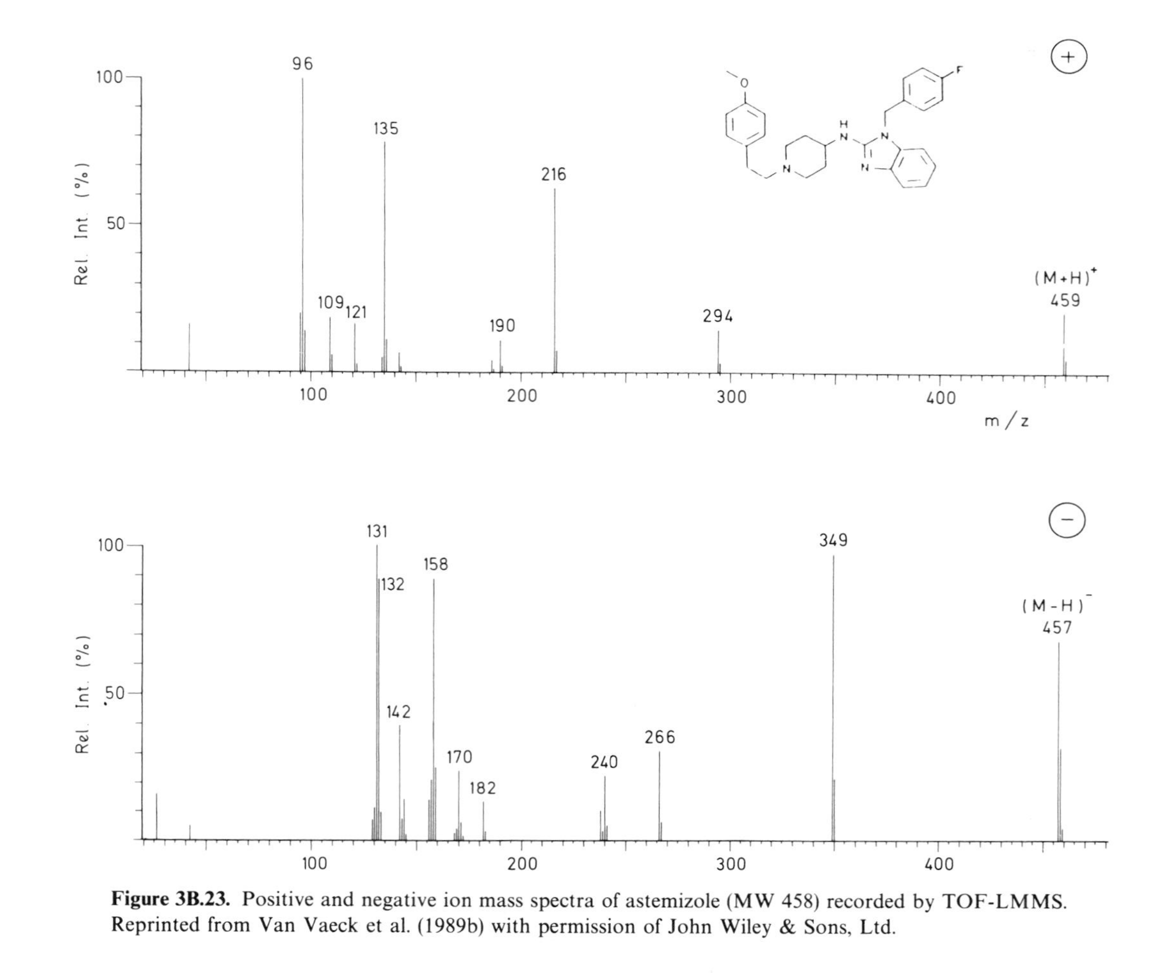

Figure 3B.23. Positive and negative ion mass spectra of astemizole (MW 458) recorded by TOF-LMMS. Reprinted from Van Vaeck et al. (1989b) with permission of John Wiley & Sons, Ltd.

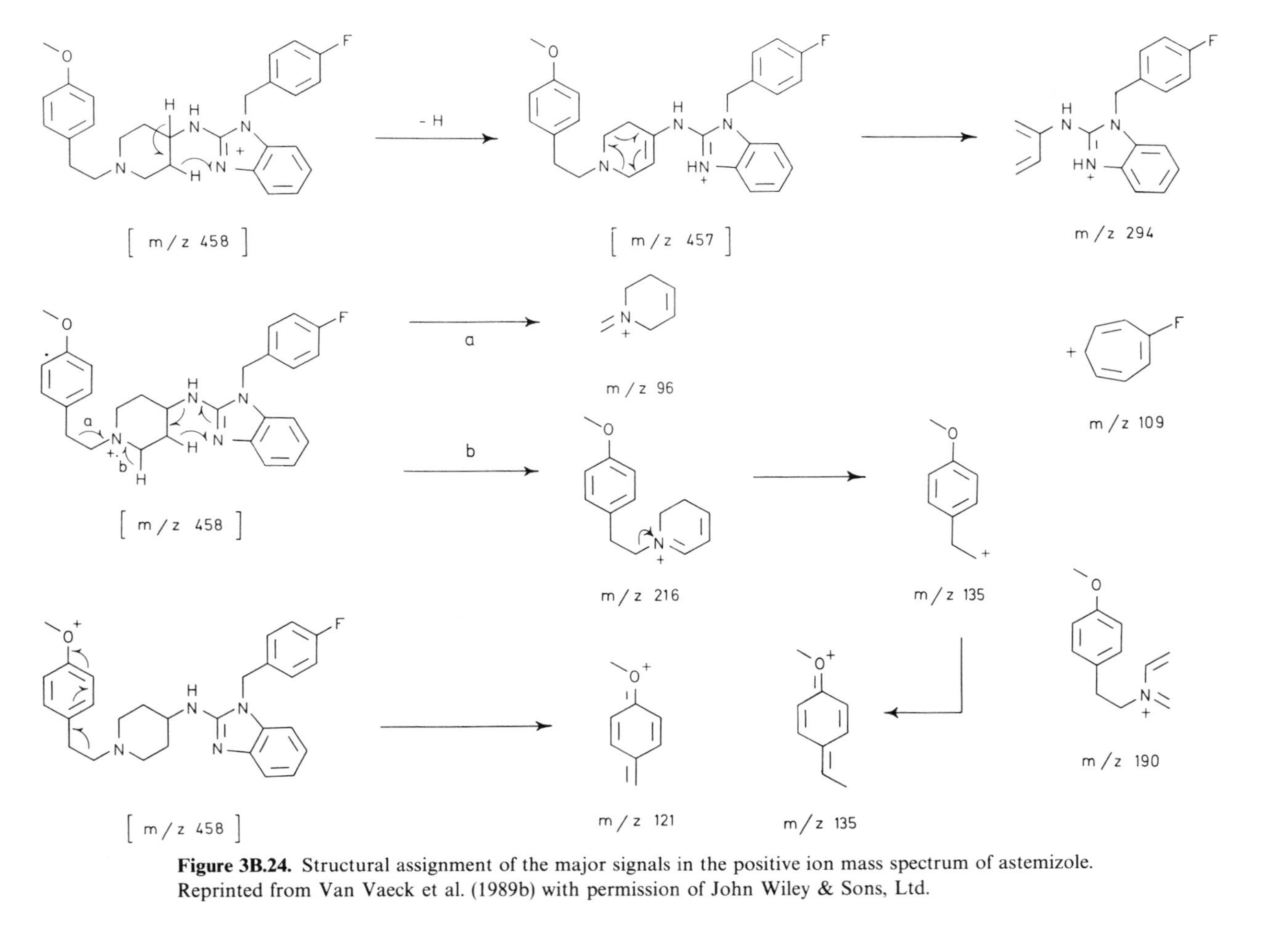

Figure 3B.24. Structural assignment of the major signals in the positive ion mass spectrum of astemizole. Reprinted from Van Vaeck et al. (1989b) with permission of John Wiley & Sons, Ltd.

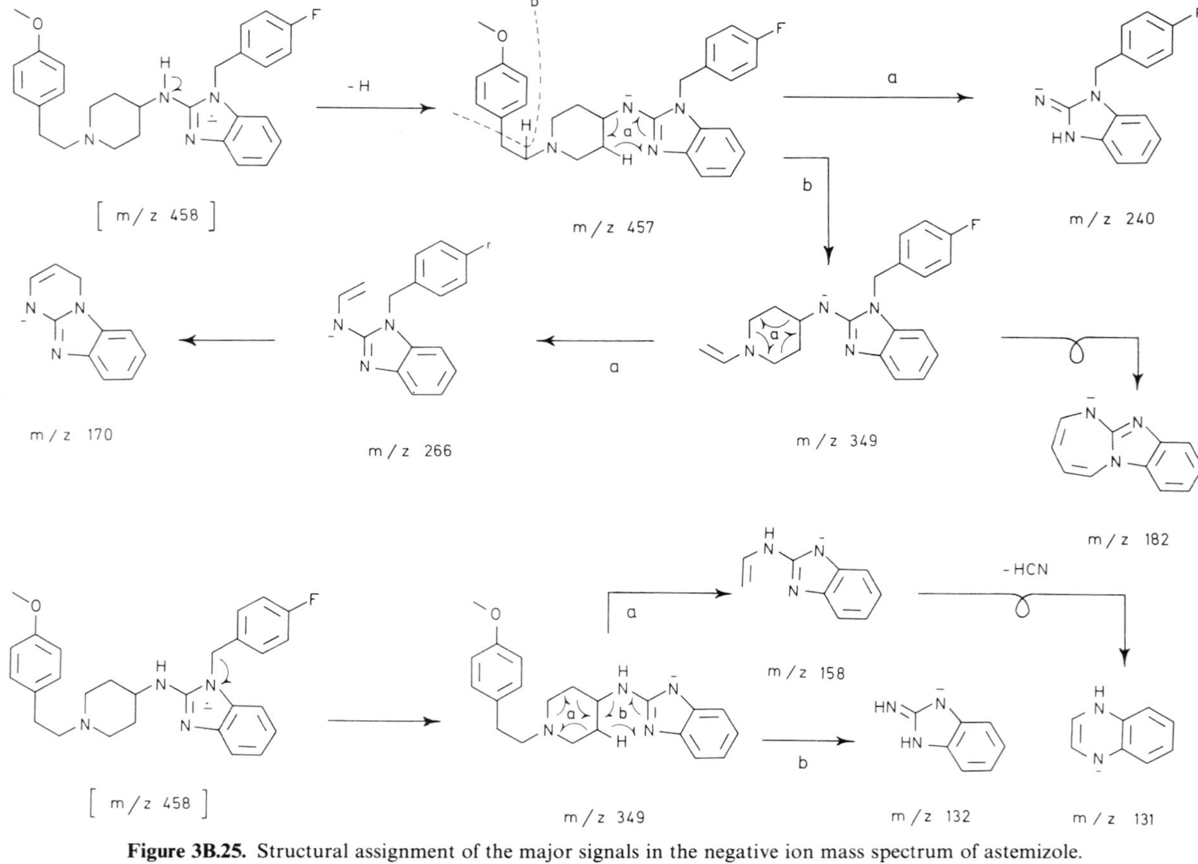

Figure 3B.25. Structural assignment of the major signals in the negative ion mass spectrum of astemizole. Reprinted from Van Vaeck et al. (1989b) with permission of John Wiley & Sons, Ltd.

deficiency or excess in a much more pronounced way than is observed in conventional gas phase methods. This means that information obtainable from positive and negative mass spectra in LMMS often shows a striking complementarity. As we see in Figure 3B.25, the negative ion signals in the spectrum of astemizole can be assigned entirely to even-electron species but it is likely that the radical aminobenzimidazole fragments also contribute to the signal at m/z 131.

However, the occurrence of radical fragments must be considered as well in negative ion mode LMMS data. Convincing evidence is provided by the results for closantel in Figure 3B.26 (Van Vaeck et al., 1989c). The base peak refers to the loss of a hydrogen molecule from $M^{\overline{\cdot}}$. The absence of low mass fragments is remarkable. The molecule indeed contains a large aromatic system, which permits resonance stabilization of the internal energy. Several possibilities for benzylic cleavage are available but not used. This behavior is associated with the efficiency of third-body collisions in the selvedge region to remove virtually all excess energy from $M^{\overline{\cdot}}$. Dissociative ECI can be considered as well, since no $M^{\overline{\cdot}}$ are effectively detected in this mass spectrum but the $M^{\overline{\cdot}}$ signal is found for various polyfunctional derivatives similar to those discussed earlier. Careful consideration of the isotope patterns in the parent region is required to reveal the presence of $M^{\overline{\cdot}}$ ions, since the signal becomes superimposed with the usually abundant ^{13}C satellite of the $(M-H)^-$ ions. It has recently been shown that reliable measurement of isotope ratios in TOF-LMMS is feasible with special precautions so that deviations above 10% from the expected values can be interpreted as significant for superimposed contributions (Van Vaeck and Gijbels, 1990a). Reproducible detection of the $M^{\overline{\cdot}}$ signals requires more careful adjustment of the local energy regime than do most of the even-electron ion signals. Selected compounds provide evidence as to the inverse relationship between parent and fragment ions, but the TOF-LMMS instrument in its standard form does not provide sufficient experimental capabilities for a systematic study (L. Van Vaeck, unpublished results). In practice, sample irradiation at power densities slightly above the threshold often leads to the complete disappearance of these signals from the mass spectra. The same applies to the positive radical ions as well. One of the major problems, however, is that the determination of threshold irradiating conditions also depends on parameters that have nothing to do with ion formation. Specifically, it is the sensitivity of the detection system that determines how many neutrals and ions have to be generated in the selvedge region to obtain a measurable signal.

Several of these low-MW drugs and metabolites can be characterized successfully with conventional MS as well. Apart from the interest in DI mechanisms under focused laser irradiation, possible microanalytical applications motivate these investigations. Figure 3B.27 shows the parent region of the

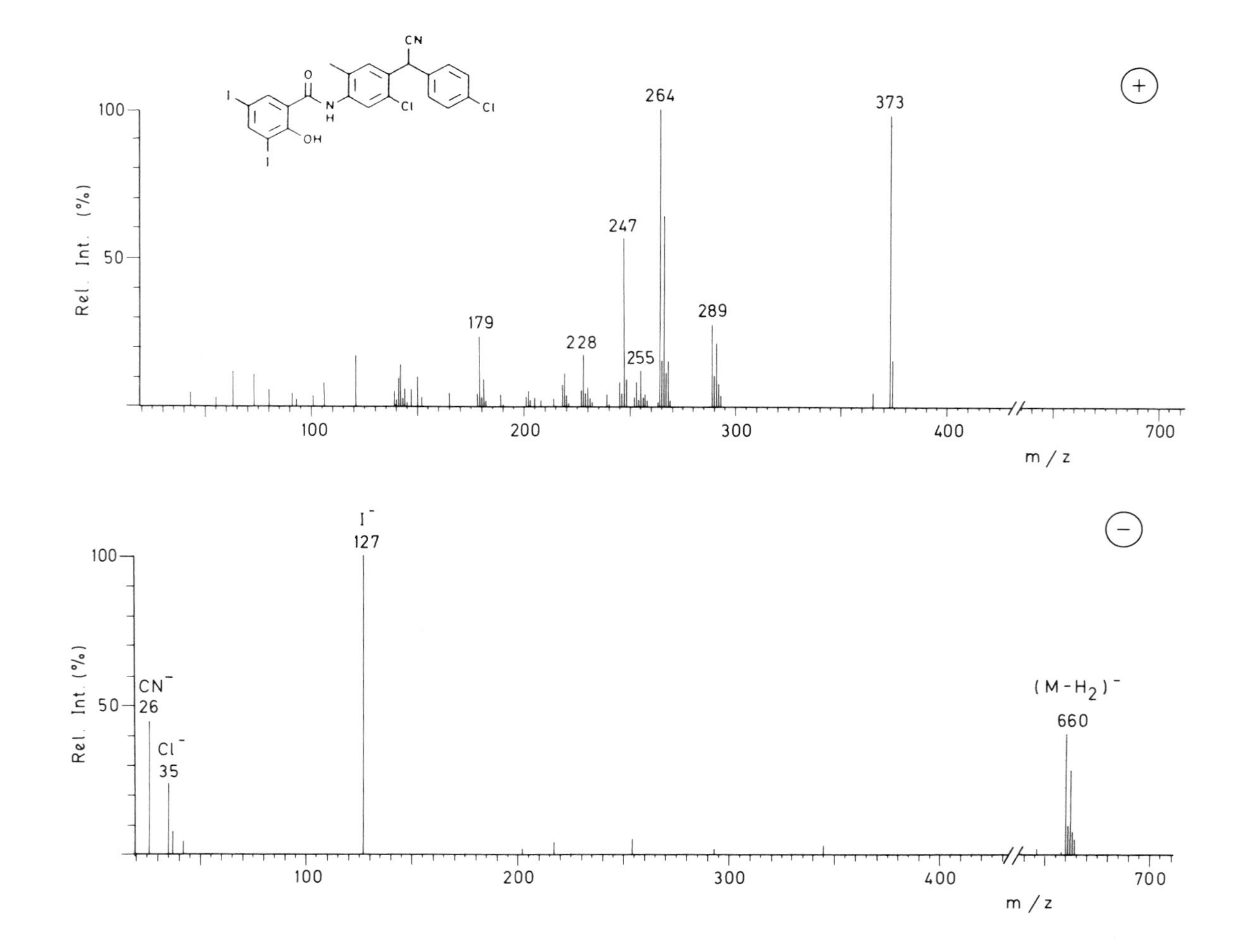

+
100
50
Rel. Int. (%)
CN
O
I
N
H
Cl
Cl
OH
I
264
373
247
179
289
228
255
100
200
300
400
700
m / z
I⁻
127
−
100
50
Rel. Int. (%)
CN⁻
26
Cl⁻
35
(M - H2)⁻
660
100
200
300
400
700
m / z

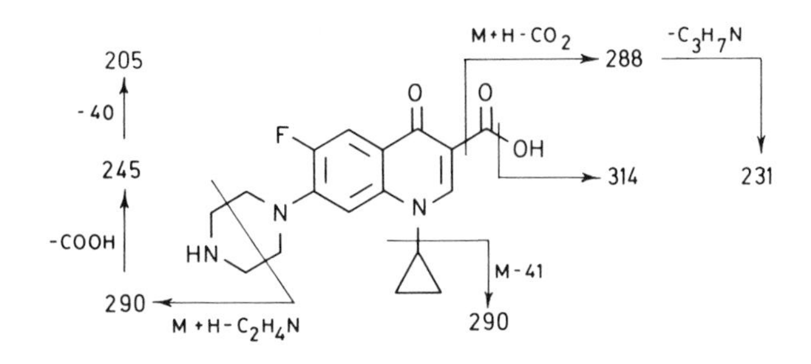

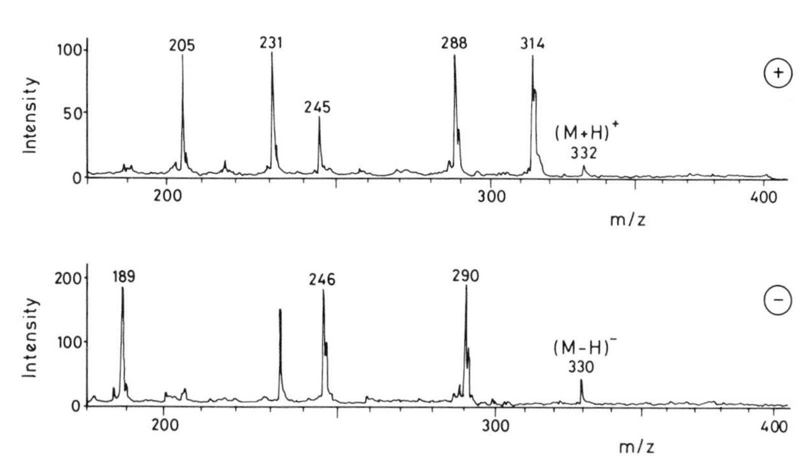

Figure 3B.27. Positive and negative ion mass spectra of ciprofloxacin (MW 331) recorded by TOF-LMMS. Reprinted from Holm and Holtkamp (1989) with permission of San Francisco Press, Inc.

TOF-LMMS mass spectra recorded from ciprofloxacin (Holm and Holtkamp, 1989). A bar graph–type representation of the data is not possible here because of the insufficient mass resolution. However, the main information about the different substituents still remains available from the intense high-mass fragments, while the MW can be derived from smaller signals in both the positive and negative ion detection mode. The compound was indentified as a major constituent of the crystallites in human urine samples.

Figure 3B.26. Positive and negative ion mass spectra of closantel (MW 662), analyzed in the neutral form by TOF-LMMS. Reprinted from Van Vaeck et al. (1989c) with permission of John Wiley & Sons, Ltd.

Table 3B.3. Predicted and Observed Fragments for Five Organophosphorus Pesticides Studied with TOF-LMMS

	Coumaphos		Iodofenphos	
MW: Mass Range Studied:	362 135–750		412 105–850	
	Predicted Relative Intensity	Observed Relative Intensity	Predicted Relative Intensity	Observed Relative Intensity
Positive ions:				
M + H	64 ± 34	50 ± 18	< 2	0.5 ± 0.5
M − R	30 ± 16	25 ± 10	N/A	
M − Cl	10 ± 5	10 ± 4	1 ± 1	1 ± 1
M − RO	20 ± 10	0	20 ± 10	15 ± 8
M − OZ	20 ± 10	100 ± 30	100 ± 0	100 ± 0
M − SZ	10 ± 8	15 ± 6	16 ± 5	12 ± 4
OZ + 2H	100 ± 41	82 ± 28	N/A	
SZ + 2H	4 ± 4	3 ± 1	N/A	
M + H − CH_3Cl	0	10 ± 4	N/A	
M + R	N/A		2 ± 2	7 ± 7
M + H	N/A		N/A	
M − SZ	N/A		N/A	
M + Z	N/A		N/A	
M + H − 2CO	N/A		N/A	
Z	N/A		N/A	
PhCN	N/A		N/A	
Negative ions:				
M − H	5 ± 5	5 ± 5	0.5 ± 0.5	0
M − R	22 ± 13	18 ± 12	50 ± 45	0
M − Z	100 ± 0	5 ± 5	100 ± 41	100 ± 0
OZ	37 ± 18	100 ± 0	37 ± 18	20 ± 7
SZ	6 ± 6	5 ± 5	6 ± 5	2 ± 2
M − H − HCl	N/A		< 10	2 ± 2
M − H − HI	N/A		15 ± 8	12 ± 10
Z − $CHCH_2Cl$	N/A		N/A	
Z − CH_2	N/A		N/A	

Source: Reprinted from Morelli et al. (1987) with permission of Elsevier Science Publishers.

Table 3B.3. (*Continued*)

Ronnel		Dialifor		Phosalone	
330 100–650		393 100–800		367 100–800	
Predicted Relative Intensity	Observed Relative Intensity	Predicted Relative Intensity	Observed Relative Intensity	Predicted Relative Intensity	Observed Relative Intensity
>2	0	N/A		N/A	
N/A		N/A		N/A	
1 ± 1	1 ± 1	N/A		N/A	
20 ± 10	5 ± 5	N/A		N/A	
100 ± 0	50 ± 50	N/A		N/A	
N/A		5 ± 5	5 ± 5	5 ± 5	5 ± 5
N/A		N/A		N/A	
N/A		N/A		N/A	
N/A		N/A		N/A	
N/A		5 ± 5	0	5 ± 5	4 ± 4
2 ± 2	100 ± 30	N/A		N/A	
16 ± 8	3 ± 3	N/A		N/A	
N/A		20 ± 6	13 ± 5	20 ± 6	13 ± 5
N/A		0.5 ± 0.5	0	0.5 ± 0.5	0
N/A		100 ± 0	100 ± 0	100 ± 0	100 ± 0
N/A		1 ± 1	2 ± 2	1 ± 1	1 ± 1
0.5 ± 0.5	0	0.5 ± 0.5	0.5 ± 0.5	0.5 ± 0.5	1 ± 1
50 ± 45	5 ± 5	1 ± 1	1 ± 1	1 ± 1	5 ± 5
100 ± 41	100 ± 0	N/A		N/A	
37 ± 18	40 ± 19	100 ± 0	100 ± 0	100 ± 0	100 ± 0
6 ± 5	12 ± 10	N/A		N/A	
10 ± 5	5 ± 5	N/A		N/A	
N/A		N/A		N/A	
N/A		10 ± 5	0	N/A	
N/A		N/A		10 ± 5	19 ± 7

A number of antidepressant drugs have been investigated with LD atmospheric pressure ionization (API) MS (Kolaitis and Lubman, 1986). Radical molecular ions are produced in the virtual absence of fragments of analogues analyzed as nonionic free bases as well as of those measured in the hydrochloride form. It is clear that these data cannot be compared directly with the other LD-MS and LMMS results.

Miscellaneous Compounds. Interesting research has been conducted on the mass spectra from organophosphorus pesticides and related compounds (Morelli and Hercules, 1984; Morelli et al., 1987). Most pesticides can be described by a general structural formula $(RX)_3PX$, in which X corresponds to O or S. Two of the three R groups correspond normally to small aliphatic groups, whereas the third may vary from a short aliphatic chain to heterocyclic aromatic ring systems and often contains several functional entities such as halogens, amines, or carboxyls. The largest R group will be denoted Z. The low-mass region of the positive and negative mass spectra shows a complex pattern of signals that so far has proved difficult, if not impossible, to interpret. However, if attention is focused on the signals above m/z 180, a set of relatively simple mechanisms can be formulated to describe the data. Variations in fragmentation patterns can be understood in terms of simple organic reactions. The data of 20 compounds have been classified according to four major groups: halogenated and nitrated analogues, derivatives of heteronitrogen ring systems, phosphoramido(othio)ic acids, and the potassium metabolite salts. The relatively simple but highly consistent fragmentation behavior within each subset has permitted prediction of mass spectra of additional analogues not included in the first data base. Table 3B.3 lists the characteristic fragments and predicted ion peak intensities along with the experimentally observed ones (Morelli et al., 1987). Generally, the predictions are quite accurate within the reproducibility of the relative peak intensity measurements, i.e., between 33% and 55%. The relatively small size of the original data set accounts for the detection of fragments that have not yet been predicted, but the score of results is remarkably good in respect of predictability. Such approaches hold a great deal of promise for the direct in situ analysis of organic targets in microprobe studies. For example, pesticides have been identified on the surface of plant leaf sections (Morelli and Hercules, 1986).

Several sets of data have been reported for steroids. TOF-LMMS of pure cholesterol shows only fragment ions, whereas addition of NaCl permits detection of the cationized molecules in the complete absence of fragments (Dutta and Talmi, 1981). However, FTMS studies on several steroids without addition of alkali salts have produced relatively abundant $(M+H)^+$ signals of more than 50% of the base peak. In contrast, $M^{+\cdot}$ yields the most abundant contribution in the molecular ion region of the mass spectrum of cholesterol, but the total intensity remains low at around 8%. For other analogues, the

M^{+} is a low-intensity signal compared with fragment ions, which are comparable with those obtained in conventional EI-MS (Fung and Wilkins, 1988). A two-step LD at 10.6 μm and multiphoton ionization (MUPI) at 266 nm combined with a 30 cm TOF-MS allowed observation of the M^{+} in the quasi-absence of fragments from only 50 fmol of estradiol (Hahn et al., 1987). Results obtained with RETOF (Bruker, Bremen, Germany) for several steroids show effective control over the degree of fragmentation that can be achieved by selecting the proper intensity of the ionizing 270 nm beam (Grotemeyer and Schlag, 1988b). The signal from M^{+} remains quite abundant, while a limited number of fragments provides characterization of the main structural parts. The clean pattern contrasts with the complex set of low-mass ions in the spectra obtained by resistive heating on a direct probe.

Various drugs actually correspond to glycoside-type structures and will be discussed under that heading later on. To conclude this subsection, we mention the comparative study of several desoxinojirmycin and folic acid derivatives with EI, direct chemical ionization (DCI), TOF-LMMS, SIMS, FD, and FAB-MS (Wünsche et al., 1984). The soft ionization methods produce significantly different mass spectra. As to the MW information, FD and FAB are best suited to these particular compounds. However, FD gives the lowest anion yield with a low degree of fragmentation. Also, the extremely delicate sample preparation hinders the routine application of this method. The other soft ionization methods produce a sufficient number of fragment peaks, but the reproducibility is low in comparison with the established MS techniques. We want to emphasize that the practical analyst who applies one of these techniques to solve a particular problem really needs more information about the specificity and the systematics of the involved breakdown or fragmentation patterns.

3B.3.4. Oligosaccharides

Monosaccharides and disaccharides were among the first organic compounds investigated by LD-MS. Data on the simplest analogues such as glucose, galactose, and mannose have been reported (Kubis et al., 1989; Wieser and Wurster, 1983; Krueger and Schueler, 1980; Roczko et al., 1989). As to sucrose, there is an appreciable assemblage of papers dealing with the generation of MW information by MS. There is some controversy in the literature as to whether or not a radical molecular ion of sucrose and other saccharides really exists upon field desorption as well as EI ionization (Veith and Röllgen, 1985; Rogers and Derrick, 1984; Derrick et al., 1985). Application of the two-step LD-MUPI technique in a RETOF instrument produces abundant M^{+} signals, although the analyte does not contain a chromophoric group to absorb the ionizing wavelength of 250 nm. The mass spectrum is shown in Figure 3B.28, and the assignment of the fragments is summarized in

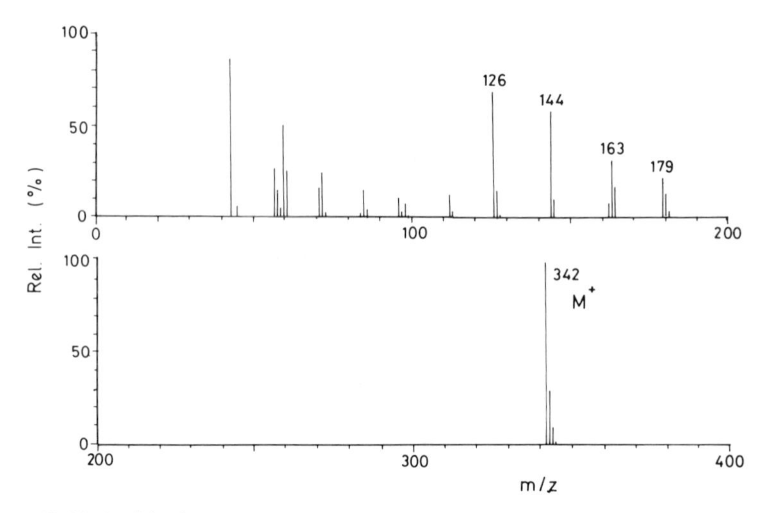

Figure 3B.28. Positive ion mass spectrum obtained from sucrose (MW 342) by partial hard ionization in the two-step LD-MUPI RETOF instrument. Reprinted from Grotemeyer and Schlag (1988b) with permission of John Wiley & Sons, Ltd.

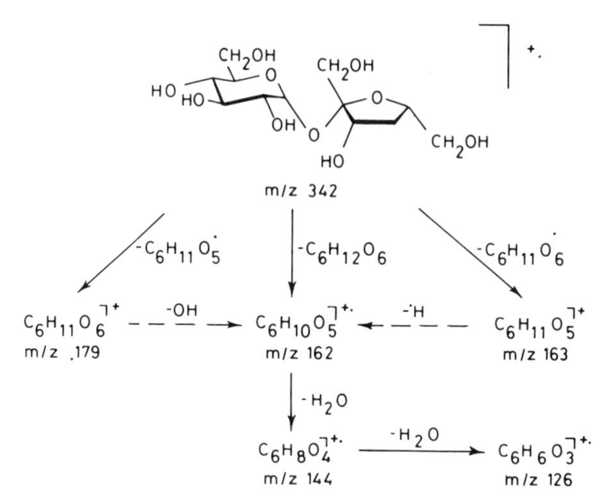

Figure 3B.29. Fragmentation pattern of sucrose observed by multiphoton absorption. Reprinted from Grotemeyer and Schlag (1988b) with permission of John Wiley & Sons, Ltd.

Figure 3B.29 (Grotemeyer and Schlag, 1988b). Although the aforementioned experiments offer evidence of the existence of $M^{+\cdot}$ in the gas phase with sufficient stability to permit MS detection, application of one-step laser DI yields cationized molecules only in the parent region of the positive mass spectrum.

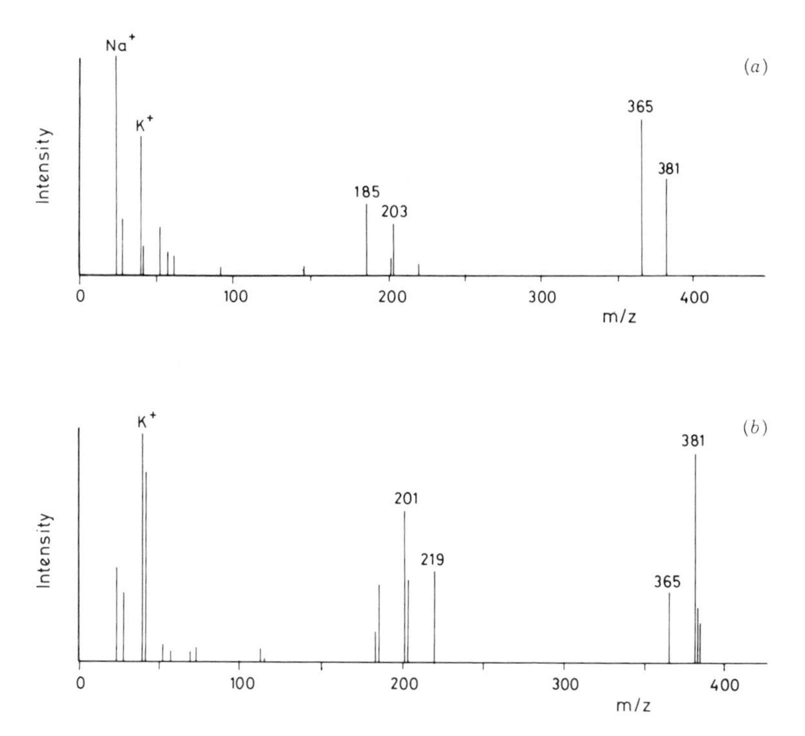

Figure 3B.30. Positive ion mass spectrum of sucrose obtained with a single magnetic sector instrument (a) and a double-focusing analyzer (b). Irradiation is performed with 10.6 μm CO_2 pulses in the former case and a Nd:glass laser emitting 1.06 μm 100 μs pulses in the latter case. Reprinted from Posthumus et al. (1978) with permission of the American Chemical Society.

A lot of work has been performed on both analysis and elucidation of the adduct ionization process. The latter experiments have already been mentioned (see Section 3B.2.3). The mass spectrum in Figure 3B.30 is obtained from LD in conjunction with two different magnetic sector analyzers and irradiation conditions (Posthumus et al., 1978). The base peak is due to the $(M + Na)^+$ ions, while an accompanying signal refers to $(M + K)^+$. These alkali ions are generated from impurities in the sample, and the relative contribution of both types of cationized species varies with the particular sample. Interestingly, the only fragments observed with more than 10% intensity lie at *m/z* 203 and 185 or at *m/z* 201 and 219, respectively, depending on the Na^+ or K^+ contamination of the sample. The signals are obviously associated with the alkali cation attachment to the monomeric unit and the corresponding dehydration product. Note the difference between these daughter ions and those observed in the two-step LD-MUPI data of Figure 3B.28. It was reported later that use of a Nd:glass laser instead of a CO_2 laser yielded additional fragments at *m/z* 163, 127, and 145. The signals correspond to the even-electron equivalents of the radical structures mentioned in Figure 3B.29 and are also observed in, for instance, FD and CI of oligosaccharides (Kistemaker et al., 1979).

The mass spectrum recorded from sucrose by irradiation with 300 ps pulses from a 337 nm nitrogen laser and subsequent TOF mass analysis is essentially identical to that shown in Figure 3B.30, in spite of the fact that largely different time domains are involved in both ion generation and mass analysis (Posthumus et al., 1978; Huang et al., 1988). The short pulses do not cause decomposition, as seen in experiments with peptides (Karas et al., 1985). However, the role of power density in the microenvironment from which the detected ions are generated may influence the data. Because the energy deposition phenomena are not well understood, a careful interpretation is required of laser desorption data obtained on other instruments and under different irradiation conditions.

Cationization of the monomeric units by Ag^+ attachment is observed for sucrose layers deposited on the corresponding metal substrate. The formation of these fragments by unimolecular decay of the silver cationized molecules has been evidenced by the metastable ion kinetic energy spectrum (MIKES), recorded on a double-focusing MS in the reversed geometry (Zakett et al., 1981).

It is interesting to note that the generation of these fragments is related to the thermal nature of the LD process. During experiments with sucrose mixed with NaCl, no fragments are detected under irradiation at 10^9–10^{10} W·cm^{-2}, whereas abundant signals of more than 80% of the base peak arise at 10^5–10^6 W·cm^{-2} (Chiarelli and Gross, 1987). The measurements performed in reflection geometry with a pulsed CO_2 laser at 10.6 μm and FTMS show a more pronounced effect than the data obtained in the transmission mode with 266 nm laser irradiation and TOF-LMMS analysis (Lindner and Seydel, 1985). In contrast, the analysis of 1 and 20 μm thick sample layers, irradiated at 266 nm at power densities of 10^{11} W·cm^{-2}, has produced a signal at *m/z* 203, which is roughly 25% of the intensity observed for the cationized molecules. The intensity of the fragment exceeds that for the parent peak from a 1 μm layer irradiated with 10^8 W·cm^{-2}.

Additional data are available on sucrose with and without addition of alkali salts (Stoll and Röllgen, 1979; Heinen et al., 1981; McCrery and Gross, 1985b; Balasanmugan et al., 1981; Heresch et al., 1980). The influence of the sample micromorphology and laser focus on the production of cationized molecules in TOF-LMMS has been evaluated (Wieser and Wurster, 1983). Negative ion mass spectra as well as photofragmentation data have been recorded on FTMS with IR desorption (Watson et al., 1987).

Non-reducing analogues, e.g. lactose, maltose, or melibiose, are characterized by the stepwise fragmentation of the sugar moiety by loss of CH_2O groups, in contrast to the reducing sugars, such as sucrose, which undergo cleavage of the glycosidic bond (McCrery and Gross, 1985a). The addition of NaCl, KCl, or Ag_2O has been found necessary in order to analyze malto-oligosaccharides (Coates and Wilkins, 1985). Prominent cleavages occur within the D-glucose rings (Coates and Wilkins, 1985). CO_2 laser desorption of 10.6 μm and FTMS have been applied. The structural information content of the LD mass spectra appears to be superior to those obtained by FAB and PD-MS. The regularity of the observed ion series in the mass spectra of polysaccharides suggests predominance of orderly cleavage processes, some of which have been described in detail (Coates and Wilkins, 1987; Lam et al., 1988). The different fragmentation patterns are very helpful in characterizing the structure of underivatized bacterial capsular polysaccharides (Lam et al., 1987). The FTMS mass accuracy of better than 10–20 ppm represents a valuable asset. The capabilities of LD-MS on polysaccharides are particularly promising with regard to the current interest in structural elucidation of such compounds as glycolipids, which will be discussed in a later subsection.

A systematic investigation of fragmentation in underivatized and *N*-acetylated di- up to tetrasaccharides has been performed using LD-TOF-MS (Martin et al., 1989). Cationization does not prevail for all compounds. Maltotretose does not undergo alkali attachment, whereas another tetrasaccharide, stachyose, exhibits abundant $(M + Na)^+$ and $(M + K)^+$ signals. The observed fragmentation depends on generation of cationized or protonated molecules. Glycosidic bond cleavages tend to become prominent for the $(M + H)^+$ ions, whereas two-bond cleavages all produce ions that retain the initially attached Na^+ or K^+. Comparison with FAB and PD-MS results evinced the need to combine complementary information from the different techniques. For instance, the glycosidic cleavages are highly indicative of the structure but are often weak signals in FAB as opposed to LD-MS. Later, a systematic investigation of fragmentation of oligosaccharides under IR irradiation desorption conditions was performed within the same group (Spengler and Cotter, 1990). Fragmentation occurs primarily within the sugar moieties, rather than at the glycosidic bonds. Hence, distinction between isomeric saccharides becomes feasible on the basis of usually intense signals in the mass spectra.

The potential for characterization of larger polysaccharides seems to depend on the experimental conditions. In contrast to the aforementioned ion patterns, observed for γ-cyclodextrin under IR irradiation, only cationized molecules are observed without fragments when TOF-LMMS is applied on 20 μm thick layers irradiated at 10^{11} $W \cdot cm^{-2}$ without perforation (Lindner and Seydel, 1985). The latter conditions are representative of those for nonthermal ionization.

3B.3.5. Nucleosides and Nucleotides

Nucleosides and nucleotides constitute an important class of biological compounds; they are the primary constituents of DNA and RNA and are found in antibiotics as well. Moreover, the corresponding LD-MS data provide a nice tie-in with the previous discussion on saccharides and the following one on the more complex structures in the subsection on glycosides. Again nucleosides in conjunction with nucleotides have been studied in a variety of instrumental configurations, demonstrating the successful application of LD-MS for qualitative and sometimes also quantitative analysis of thermolabile compounds. Additionally, these compounds represent the obvious reference base from which it is possible to attempt characterization of the phosphate derivatives and oligonucleotides. The latter substances undoubtedly hold great interest for the biosciences but pose several problems in respect of their characterization by MS.

LD-MS results on the pyrimidine and purine bases reveal no unexpected features. In the positive ion detection mode, the signal from the protonated molecules predominates and the fragment ions are usually negligible, while the negative mass spectra exhibit a strong signal from the deprotonated molecules (Kubis et al., 1989). Applied as a thin film on metal substrates, adenine shows prominent cation attachment (Schueler et al., 1981). The high sensitivity has been exploited in off-line TLC applications with an initial amount of sample in the range of 2 ng to as low as 50 pg (Kubis et al., 1989; Feigl et al., 1984). Mass spectra obtained by application of LD-API-MS show the formation of protonated molecules for the purines while the dimeric clusters prevail for the pyrimidines (Kolaitis and Lubman, 1986). Molecular ions are detected by two-step LD-MUPI, and controlled fragmentation can be induced to yield more structural information (Hahn et al., 1987; Li and Lubman, 1989a). The data for creatinine contrast with those reported for TOF-LMMS. Laser irradiation under atmospheric conditions has been applied as well by using the external sample mounting technique; no higher organic clusters are detected and fragmentation is significant (Holm et al., 1984).

Comparison of the mass spectra of nucleosides has revealed some apparently conflicting indications as to the predominant ion formation and fragmentation process. An early study on nucleosides compares the results from TOF-LMMS and PD-MS (Schueler and Krueger, 1980). The relative abundance of the cationized vs. protonated molecules in the parent region of the positive ion mass spectra from LMMS exceeds those in PD-MS. No alkali salts have been added, and the average impurity content of the samples is less than 10 ppm. Table 3B.4 summarizes the main signals from positive and negative ion mass spectra recorded under application of the described procedures for threshold analysis in TOF-LMMS (Van Vaeck et al., 1989d, 1990). The neutral form of the purine or pyrimidine molecule is denoted as BH, while

Table 3B.4. Survey of Diagnostic Information from Positive and Negative Mass Spectra of Nucleosides in TOF-LMMS

	m/z Value and Relative Peak Intensity						
	Adenosine	Guanosine	Inosine	Cytidine	Thymidine	Deoxyadenosine	Uridine
$[M + Na]^+$	290 (7)	306 (4)	291 (4)	266 (5)	265 (3)	—	—
$[M + H]^+$	268 (16)	—	—	—	—	252 (5)	—
$[M + H - HCNO]^+$	—	—	—	—	200 (8)	—	202 (5)
$[M + Na - 90]^+$	—	—	201 (1)	—	—	—	—
$[M + H - 90]^+$	178 (2)	194 (1)	—	—	—	162 (3)	155 (6)
$[B + H - N]^+$	—	174 (13)	159 (8)	134 (3)	—	—	133 (4)[a]
$[BH_2]^+$	136 (100)	152 (100)	137 (100)	112 (100)	127 (100)	136 (100)	113 (100)
$[BH_2 - CO]^+$	—	—	—	—	99 (11)	—	—
$[BH_2 - HCNO]^+$	—	—	—	—	84 (36)	—	70 (97)
$[S]^+$	—	—	133 (2)	—	117 (59)	117 (7)	133 (4)[a]
$[M - H]^-$	266 (3)	282 (8)	267 (100)	242 (5)	241 (98)	250 (4)	243 (12)
$[M - H - 90]^-$	—	—	—	152 (8)	151 (4)	160 (2)	153 (7)
$[B]^-$	134 (100)	150 (100)	135 (74)	110 (100)	124 (100)	134 (100)	111 (100)
$[S]^-$	—	133 (10)	—	—	—	117 (5)	—

Source: Reprinted from Van Vaeck et al. (1989d) with permission of John Wiley & Sons, Ltd.

[a]Isobaric fragments.

SH stands for the intact and uncharged sugar structure. The fragment denoted as $(B + H - 90)^+$ arises from the well-known rearrangement within the sugar moiety, formally leading to the expulsion of CH_2O and subsequent breakdown of the ring (Biemann and McMCloskey, 1962). According to Table 3B.4, MW information is given in the positive ion mode primarily by the cationized molecules, except for adenosine and deoxyadenosine, where proton attachment yields substantial signals. Only fragments are detected for uridine in the positive mode, but the MW is now available from the anions. Inosine and thymidine give abundant signals for the deprotonated molecules, while for the remaining analogues the corresponding peak is comparable to the adducts in the positive mass spectra. The base peak in both positive and negative mass spectra refers to the purine or pyrimidine moiety under the protonated or deprotonated form. No cationized fragments are detected. Under strict application of the procedure for threshold analysis in TOF-LMMS, mixing of nucleosides with alkali salts does not improve the sensitivity or change the resulting mass spectra significantly, except for the increased contribution of the $(M + Na)^+$ signal—sometimes up to 30% of the base peak (L. Van Vaeck, unpublished results). Detailed analysis of the peak shape in TOF-LMMS gives some indication as to the time profile of the ion formation. Cationization of most compounds we have studied so far yields a peak broadening, making the mass resolution lower than feasible under the voltage settings used. At the same time, the signals show a clear tendency to shift the mass range upward in respect of the calibration carbon foil. Both effects do not occur within the class of nucleosides. In contrast to our data, detection of cationized fragments by TOF-LMMS has been reported and a study of the metastable decomposition in the first part of the flight tube has been attempted (Hercules et al., 1987).

Within given limits, the time profile of the ion formation process does not matter to the mass analysis in FTMS. The main exception stems from short-lived fast-decaying species. As a result, it is worthwhile examining the data from both methods. The influence of the laser irradiation by 266 and 1064 nm at threshold and increased power densities on the mass spectra of four nucleosides has been investigated systematically in a single-cell LD-FTMS instrument (Chiarelli and Gross, 1989a). Additional experiments using a split probe, permitting desorption of nucleosides and alkali salts from separated wires at a distance of about one millimeter, have been conducted to assess the importance of gas phase cationization. Figure 3B.31 illustrates the positive ion mass spectra for guanosine mixed with NaCl under irradiation with 266 nm at threshold and higher power density conditions (Chiarelli and Gross, 1989a). The so-called doubly sodiated molecules, i.e., $(M - H + 2Na)^+$, yield a prominent signal of which the intensity is at least comparable to the $(M + H)^+$ peak. Unlike TOF-LMMS, mixtures and application of an increased power

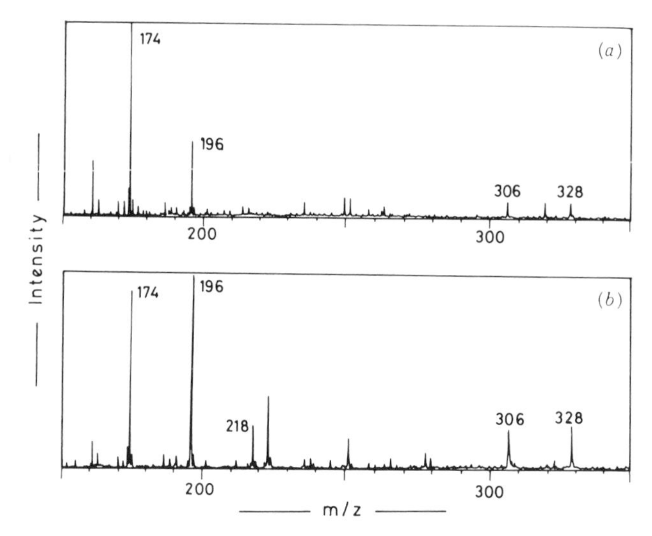

Figure 3B.31. Positive ion mass spectrum of guanosine (MW 283) recorded by LD-FTMS under resonant conditions at near-threshold power (a) and higher power (b). Reprinted from Chiarelli and Gross (1989a) with permission of the American Chemical Society.

density regime can be handled without any problem as regards the MS peak shape, while the calibration remains accurate within the ppm range.

Under laser irradiation with 266 nm near the threshold power density of $10^8\ \mathrm{W\cdot cm^{-2}}$, guanosine forms doubly sodiated molecules but pyrimidine analogues do not. As to fragments, the former product yields a base peak from the cationized purine $(B + Na)^+$ under low-irradiance conditions and the doubly sodiated moiety under high-irradiance conditions, whereas the $(M + Na)^+$ ions predominate for the pyrimidines. Desorption at 1064 nm occurs at lower power densities of $1\text{–}5\ 10^5\ \mathrm{W\cdot cm^{-2}}$. The doubly sodiated ions never exhibit a large abundance. Unlike the LD of sucrose, there is no evidence for formation of di- or even monosodiated molecules in the gas phase for nucleosides at 1064 nm for the power densities used. The formation of doubly sodiated molecules under resonant threshold conditions has been related to the lactim–lactam equilibrium in the ground vs. excited state. Uracil and thymidine, both pyrimidine analogues, preferentially exist as lactam in the ground state and as lactim upon excitation to S_1. Purine derivatives favor the lactam (keto) form in both the ground and excited state. The mechanism of doubly sodiated molecules is then envisioned as follows. Once the pyrimidine is excited, the imide proton vacates its position on the nitrogen and moves to a carbonyl oxygen, making it possible for a sodium ion to associate with the imide nitrogen. Following deexcitation, the nitrogen will associate more strongly with the Na^+, and the return of the hydrogen released from the oxygen is blocked. Adenosine, which does not have imide nitrogens, gives no doubly sodiated ions (Chiarelli and Gross, 1989a). However, the signal is clearly present in the mass spectrum of another publication (Posthumus et al., 1978). The negligible role of gas phase cationization is associated with the preferential stacking of the molecules in the solid state, which leads to self-association. The direct ejection of large aggregates, consisting of an alkali ion and several neutral nucleosides, followed by decomposition into the monocationized molecules, has been proposed (Posthumus et al., 1978). Nevertheless, the problem with LD and subsequent CI of thymidine in an FTMS cell appears to relate to the experimental conditions during laser irradiation. Successful analysis is feasible with a sufficiently high density in the selvedge (McCrery and Gross, 1985b; Amster et al., 1989). The negative mass spectra obtained by LD-FTMS are slightly different from those measured in TOF-LMMS (McCrery and Gross, 1985a).

LD on nucleosides on a modified moving belt LC-MS with subsequent mass analysis by a quadrupole has permitted the detection of protonated dimers and combinations consisting of the cationized molecules with KI, as well as doubly cationized molecules and base fragments (Hardin and Vestal, 1981; Hardin et al., 1984). The time profiles observed for protonated molecules and base fragments point to a prominent contribution arising from the decomposition of the protonated dimer. The formation of protonated bases in the ion source remains almost negligible. In contrast, cationized fragments indicate that generation of these ions directly from the surface or in the selvedge accounts for about 40% of the observed intensity, and decomposition of the cationized molecules for another 50%, the remaining fraction being explained by decay of the $(M + K + KI)^+$ entities. These results contrast with the observed lack of cationized fragments in the mass spectra we obtained from TOF-LMMS. The positive ion TOF-LMMS mass spectrum from thymidine on silver does not show prominent signals from the silver-cationized molecules and from the sugar fragment, whereas minor signals refer to the cationized base and the dehydration of both these fragments as well as of the parent ions (Hillenkamp et al., 1982).

Alkylation of DNA is one form of DNA modification induced by carcinogenic agents, and the O^6-methylation of guanosine is known to be one of the biologically important products generated by alkylating agents. The structural characterization of methyl adducts from guanosine by LD-FTMS has been described in an effort to decide whether the methylated analogues are substituted on the nucleic acids or on the sugar ring (McCrery and Gross, 1985a). The differentiation of isomers with an additional methyl group on the purine system is less obvious and requires application of subsequent collision-induced dissociation (CID). The high-mass resolution and accuracy of the m/z determination in FTMS readily permit elucidation of the fragmentation mechanisms without need for extensive comparative studies. Moreover, limited amounts of sample are sufficient. Whereas the positive daughters primarily yield the loss of NH_3 for a methylated guanosine series, the negative daughter ions yield the necessary structural information to differentiate all the investigated isomers (Hettich, 1989).

Another approach to differentiating methylated isomers consists in the use of two-step LD-MUPI. Production of $M^{+\cdot}$ in the gas phase and control over the actually imparted amount of internal energy allows distinctive fragments to be generated in the so-called hard ionization mode (Li and Lubman, 1989a).

Nucleotides obviously represent another interesting class of compounds that pose some problems in conventional techniques. This section will deal only with these analogues in the neutral form. Ionic analogues will be described in the corresponding section on salts. Adenosine monophosphate (AMP) is one of the compounds that has been used from the very beginning of LD-MS to evince its potential (Posthumus et al., 1978). The positive mass spectrum shows protonated and cationized molecules. Intense fragments

Table 3B.5. Survey of the Diagnostic Information from Positive and Negative Mass Spectra of Nucleotides in TOF-LMMS

	m/z Value and Relative Peak Intensity		
	Adenosine 5′-monophosphate	Adenosine 3′,5′-cyclic monophosphate	Cytidine 2′,3′-cyclic monophosphate
$[M+H]^+$	—	330 (13)	306 (37)
$[M+H-HPO_3]^+$	—	—	226 (15)
$[M+H-H_3PO_4]^+$	250 (13)	232 (3)	—
$[M+H-H_3PO_4-H_2O]^+$	—	214 (5)	—
$[M+H-H_3PO_4-CH_2O]^+$	—	202 (6)	—
$[M+H-BH-H_2O]^+$	178 (4)	—	178 (56)
$[B{=}CH_2]^+$	148 (2)	148 (4)	—
$[BH_2]^+$	136 (100)	136 (100)	112 (100)
PO_3^-	79 (100)	79 (15)	79 (100)
$H_2O{\cdot}PO_3^-$	97 (74)	97 (10)	97 (15)
B^-	134 (47)	134 (100)	110 (21)
$HPO_3{\cdot}PO_3^-$	159 (90)	159 (22)	159 (89)
$H_2O{\cdot}HPO_3{\cdot}PO_3^-$	177 (13)	177 (4)	177 (13)
$HPO_3{\cdot}HPO_3{\cdot}PO_3^-$	239 (16)	239 (2)	239 (4)
$[M-H]^-$	346 (12)	—	—

Source: Reprinted for Van Vaeck et al. (1989d) with permission of John Wiley & Sons, Ltd.

refer to the loss of phosphoric acid and water as well as the cationized and protonated purine moiety.

Table 3B.5 lists the signals of major diagnostic interest in the positive and negative mass spectra recorded by TOF-LMMS and irradiation of the pure compounds with 266 nm under threshold conditions (Van Vaeck et al., 1989d). Comparison of the data with those for nucleosides reveals several characteristic differences. The MW information is now given by the negative or positive ions and is not confirmed by the results in the opposite polarity as for the nucleosides. The parent peak intensity tends to be higher. The mass spectra of AMP and the cyclic dehydration product show unexpected differences, such as the formation of $(M + H)^+$ from AMP and $(M - H)^+$ from the cyclic derivative. As yet, there are not enough comparative data for us to attempt a complete interpretation. The base-related fragments still dominate the positive mass spectra but are certainly less prominent in the negative mode. Apart from the $(M - H)^-$ and the B^- ions, only phosphate-type clusters are detected in the negative mode. The positive fragments indicate the preferential loss of this functionality. Note the differences with the results from the early LD-MS study with CO_2 laser irradiation (Posthumus et al., 1978). No cationization is observed in these mass spectra, and the elimination of H_3PO_4 and H_2 yields only a minor peak.

Pulsed IR laser desorption with subsequent MUPI of the generated neutrals in the RETOF instrument has been applied to a series of small protected nucleosides and oligonucleotides (Lindner et al., 1990). Trimethoxyphenyl, cyanoethyl, and 2-chlorophenyl functional groups have been used as protective reagents. The compounds have been measured as a mixture of Na_3PO_4 and sample, embedded in polyethylene, to increase the molecular ion yield. The fragmentation pathways indicate that photoionization occurs in the aromatic protective groups as well as in the nucleobases. Nevertheless, abundant molecular ions can be generated. The degree of control over the extent of fragmentation by means of the intensity and wavelength of the ionizing beam is not as fine-tuned as is feasible for simple molecules. Practically speaking, the generation of some fragments at the cost of parent peak intensity cannot be avoided, but still a substantial flexibility is available to increase the relative contribution of the $M^{+\cdot}$ signal to the total ion current by a factor of 5–10. As a result, molecular ions are always clearly detectable.

3B.3.6. Glycosides

Digitonin, a steroidal pentaglycoside, and digitoxin, one of the most widely prescribed drugs for treatment of congestive heart failure, have become popular means to prove the benefits of LD-MS. One of the very first publications has demonstrated the predominance of cationization in the parent region of the positive mass spectra (Posthumus et al., 1978). However, most of the ion current is still carried by a series of high-intensity fragments, due to cationized sugar moieties. The previously mentioned compounds were used later as references to test the performance of given LD-MS instruments (Seydel and Lindner, 1983; Heresch et al., 1980; Tabet and Cotter, 1984; McCrery and Gross, 1985a; Shomo et al., 1985, 1986, 1988; Coates and Wilkins, 1986).

CO_2 laser irradiation at 10.6 μm in conjunction with FTMS has been applied on a complete set of dihydro- and sugar-hydrolyzed metabolites from digitoxin and digoxin. The positive mass spectra show the predominance of cationized molecules and relatively small fragments, corresponding to the loss of one to several sugar moieties. A mass accuracy within about 5 ppm is achieved routinely for the mass region between 400 and 800. In combination with the capability to perform MS/MS experiments, LD-FTMS became a highly attractive method for identifying these cardenolides and their metabolites directly in blood and urine samples with a minimum of preseparation steps (Shomo et al., 1988).

LD-FTMS on several glucoconjugates of alkaloids and steroids in the presence of KCl and KBr has shown that the former salt produces anion attachment, the latter not (Coates and Wilkins, 1986). Only the mass spectrum of digitonin essentially agrees with the data discussed above. The other analogue shows practically no fragmentation in the positive ion mode, whereas the negative mass spectra contain relatively abundant signals indicating the loss of one or more sugar functional groups.

Several analogues within the class of antibiotics have also been often used as test cases. Erythromycin is studied by laser desorption on a moving belt LC-MS interface (Hardin et al., 1984). Injection of 1 μg has permitted detection of intense signals from sodium- and potassium-cationized molecules in conjunction with a few intense fragments. The compound has been used to demonstrate that mixing with the UV-absorbing tryptophan improves energy deposition and hence the production of cationized molecules (Karas et al., 1987). Erythromycin and other alkaloids have been studied in the TLC application of TOF-LMMS (Kubis et al., 1989). A comparison of LD-FTMS with FAB and FD magnetic sector MS on a series of glycosidic antibiotics reveals the superior characteristics of the former method in respect to the much lower background and the better yield of molecular or adduct ions with very little fragmentation (Shomo et al., 1985, 1986). Note that the availability of fragments may complicate mixture analysis. Furthermore, FTMS readily permits identification by high-mass resolution and the possibility of subsequent photodissociation or CID. The generation of fragments is not really mandatory here, unlike the needs of simple transport-type analyzers.

LD with CO_2 pulses at 10.6 μm has been used to desorb several compounds such as antibiotics, alkaloids, and other molecules (Nuwaysir and Wilkins, 1989). An excimer laser is then used to photodissociate the trapped ions in the cell. Interestingly, the formation of $M^{+\cdot}$ from the cationized molecules has been

Table 3B.6. Positive Ion Mass Spectra of Amino Acids by TOF-LMMS

a. Positive Ion Mass Spectra

Compound (MW)	$(M+H)^+$	$(M+H-NH_3)^+$/ $(M+H-H_2O)^+$	$(M+H-CO_2H_2)^+$	m/z 30	Other
Glycine (75)	100	—	89 (m/z 30)	—	—
Alanine (89)	100	—	92	—	—
β-Alanine (89)	100	3/7	5	58[a]	—
Valine (117)	100	—	23	36	—
Leucine (131)	100	—	43	12	—
Isoleucine (131)	64	—	100	5	—
Proline (115)	40	—	100	—	—
3-Aminobutyric acid (103)	100	—	11	—	85 (m/z 44)[a]
4-Aminobutyric acid (103)	95	0/13	—	100[b]	11 (m/z 44)[a]
6-Aminobutyric acid (131)	100	13/25	3	63	—
Histidine (156)	100	—	99	19	—
3-Methylhistidine (170)	100	—	80	12	—
Phenylalanine (165)	100	—	100	—	22[c]
Tyrosine (181)	75	8/0	45	20	100[d]
Tryptophan (204)	7	22/0	17	—	100[d]
Tryptophan (204)[i]	21	41/0	9	—	100[d]
DOPA (197)	100	47/0	51	—	93[d]
Phenylalanine methylester (179)[i]	100	—	11[f]	—	11[e]
Proline methylester (129)	100	—	54	—	—
Glycine methylester (103)	100	—	48[g] (m/z 30)	—	—
N,*N*-Dimethylglycine	6	—/11[i]	100	—	4[h]

b. Negative Ion Mass Spectra

Compound	$(M-H)^-$	$(M-H-NH_3)^-$	m/z 74	$(M-H-RH)^-$ (m/z 72)	Other
Glycine	100	—	—	—	—
Alanine	100	—	—	—	—
β-Alanine	100	4 (71)	—	—	—
Valine	100	—	—	—	—
Leucine	100	—	—	—	—
Proline	100	—	—	—	—
3-Aminobutyric acid	100	1	—	—	—
4-Aminobutyric acid	100	—	—	—	7[m] (71)
Histidine	100	3 (138)	—	—	43 (65)[n]
3-Methylhistidine	100	—	—	72	68 (65)[n] 5[a] (141)
Phenylalanine	100	9 (147)	—	35	—
Tyrosine	100	2 (163)	7	24	1 (106),[p] 4 (93),[q] 4 (119),[f] 2 (134)[s]
Tryptophan	100	2 (186)	10	26	12 (116),[q] 8 (142),[r] 2 (157),[s] 5 (159)[t]
DOPA	100	33 (179)	8	40	34 (122),[f] 17 (109),[q] 61 (135),[r] 17 (150)[s]

Source: Reprinted from Parker and Hercules (1985) with permission of the American Chemical Society.

[a]$(M+H-H_3CCO_2H)^+$. [b]$(M+H-H_3CCH_2CO_2H)^+$. [c]α fragment at m/z 74, $H_2N^+{=}CHCOOH$. [d]$(M+H-H_2NCH_2CO_2H)^+$. [e]α fragment at m/z 88, $H_2N{=}CHCOOCH_3$. [f]$(M+H-HCO_2CH_3)$. [g]$(M+H-HCO_2CH_2CH_3)^+$. [h]$(M+H-HCO_2CH_3)^+$. [i]$(M+H-H_3COH)^+$. [j]LAMMA-1000 data. [k]Mass value of ions are given in parentheses. [l]α-fragment $(HN{=}CHCOO)^-$ of α–β fission. [m]$(M-H-H_3CNH_2)$. [n]Decomposition of imidazole ring, C_3N_2H, $^-C{\equiv}CNHC{\equiv}N$. [o]$(M-H-CH_2)^-$. [p]$(M-H-CO_2H_2CNH_2)$. [q]$(M-H-CO_2-C_2H_3NH_2)^-$. [r]$(M-H-CO_2-NH_3)^-$ [s]$(M-H-CO_2H_2)^-$. [t]$(M-H-CO_2)^-$.

Numbers in parentheses for all compounds are formula weights.

achieved. The photon excitation combined with the ion manipulation capabilities of FTMS provides an elegant alternative means to increase the structural information from compounds where the laser solid interaction only produces molecular ions. It should be noted that the photodissociation approach has become particularly valuable for high-mass ions because of the low cyclotron frequency, which makes CID less efficient in FTMS.

3B.3.7. Amino Acids

The MS characterization of amino acids by conventional techniques typically requires derivatization. Practically all soft ionization methods for solids have been applied extensively from the very beginning. Amino acids exist in the solid state as dimers of zwitterions. As a result, the present discussion of these molecules will anticipate the systematic description of ionic compounds and internal salts later on.

An extensive study of mass spectra from aliphatic and aromatic amino acids by irradiation at 266 nm and subsequent analysis by TOF-LMMS have revealed several interesting aspects of the initial ion formation process (Parker and Hercules, 1985). The mass spectra are extremely simple, as evidenced by Table 3B.6, listing the main signals in the positive and negative spectra. The $(M + H)^+$ and $(M - H)^-$ signals not only yield the base peak in the respective detection modes but also exhibit a high absolute intensity. For the cations, fragmentation of the aliphatic analogues is limited to elimination of carboxyls and water, while the formation of the low-mass-range ion CH_2NH_2 at m/z 30 is observed. This contrasts with the general trend of polyfunctional molecules mentioned earlier. However, the amino acids, in addition to the ammonium salts, are practically the only compounds in the MW range up to 500 that do not bear aromatic chromophores and still can be readily analyzed by TOF-LMMS without resorting to matrix-assisted desorption. Aromatic amino acids show the retention of the ring in the major fragments (Parker and Hercules, 1985). Negative ion mass spectra show limited fragmentation, and the applied laser energy has to be increased without the appearence of severe cluster ions or extensive ion–molecule reactions. In particular, this holds true for the aromatic analogues. Deuterium labeling has permitted verification that the formation of $(M + H)^+$ specifically involves proton transfer from the amino group and does not occur through gas phase protonation by a species from the nonspecific disintegration of the organic molecule. Analysis of aliphatic acids with perdeuterated alkyl or aryl functional groups and only H on the amino and acid groups has revealed the elimination of undeuterated acid (Parker and Hercules, 1986). The mass spectra from TOF-LMMS have been compared with those from SIMS in the so-called static mode, PD and EI MS (Hercules et al., 1983).

Additional TOF-LMMS analyses of selected analogues within the class of amino acids have yielded confirmatory data under threshold irradiation (Krueger and Schueler, 1980; Schiller et al., 1981). Application of increased power density favors formation of cationized molecules in the positive mode and attachment of nonspecific clusters such as CN^- and CNO^- in the negative mass spectra (Kupka et al., 1980). However, it is difficult to compare threshold conditions and increased irradiance values for different instruments and sample configurations. When amino acids are adsorbed on metal surfaces, extensive cationization occurs (Schueler et al., 1981). Lack of experimental details on sample preparation makes it difficult to explain why these results are different from those obtained in static TOF-SIMS, where protonated and deprotonated molecules dominate the parent region (Benninghoven, 1985). The latter data have led to the development of the so-called precursor model for ion formation under primary ion bombardment at low density of impinging particles (Benninghoven, 1983). Irradiation of leucine with wavelengths in the visible range yields nonspecific fragments in addition to $(M - H)^-$ ions in the negative mode (Schiller et al., 1981).

The influence of wavelength and power density on the mass spectra of aromatic amino acids has been investigated systematically under identical conditions in respect of instrumentation and sample preparation (Karas et al., 1985). The threshold irradiance at 266 nm significantly exceeds that of tryptophan, which has an absorption band at this wavelength. However, the situation becomes comparable at 355 nm. The mass spectra of tyrosine and phenylalanine confirm the generation of protonated molecules near threshold, whereas cationization and more extensive fragmentation occur at higher irradiances with 266 nm. Excitation with 355 nm reduces the $(M + H)^+$ signal to a minor peak while cationization is favored and the fragments dominate the mass spectrum. Aliphatic amino acids show abundant cationization at threshold analysis, whereas increase of the power density removes the organic information from the mass spectrum, which becomes dominated by alkali-cluster ions and nonspecific fragments. Application of shorter irradiating pulses favors generation of nonspecific fragment ions. A later study on several amino acids has used sample layers deposited by a molecular beam technique on quartz and silver substrates (Hillenkamp et al., 1986). Lack of absorption by tryptophan at 355 nm and alanine at 266 nm causes higher threshold irradiances, and at the same time cationization by Ag^+ is observed for the latter case. Absorption within even very thin layers of strongly absorbing compounds prevents the active contribution of the Ag^+ to formation of parent ions from tryptophan.

Laser irradiation of amino acids on the moving belt LC interface produces a prevalence of cationized molecules in the monomeric and dimeric entities (Hardin and Vestal, 1981). Application of LD-API yields $(M + H)^+$ (Kolaitis and Lubman, 1986). The analysis of amino acids by the two-step LD-MUPI takes advantage of the generation of radical molecular ions from the desorbed neutrals and the controlled degree of fragmentation (Grotemeyer and Schlag, 1988c). The requirement of a chromophoric group limits the applicability to the aromatic analogues (Grotemeyer et al., 1989; Engelke et al., 1987). Derivatization by phenylhydantoin or carbobenzyloxy groups allows study of the aliphatic analogues as well (Grotemeyer et al., 1987; Li and Lubman, 1988). Analysis of phenylhydantoin derivatives permits quantitative determinations with a linear range over several orders of magnitude and with a detection limit in the picomole range or less (Engelke et al., 1987), as well as distinction between isomeric acids, e.g., leucine and isoleucine (Grotemeyer et al., 1987). Sample matrix effects obviously exist during the laser evaporation prior to the MUPI step. Addition of sugars seems to suppress the decomposition of small peptides while IR absorbing matrices tend to increase the contribution from pyrolysis (Beavis et al., 1988b,c). Characterization of a 15-component mixture of amino acids combined with an equal weight of NaCl has been achieved using 1.06 μm laser irradiation at $10^8\ W \cdot cm^{-2}$ and FTMS (Chiarelli and Gross, 1989b). The individual constituents only yield the respective $(M - H + 2Na)^+$ signal in the positive ion detection mode. The relative peak intensities correlate with the sublimation enthalpy of the pure constituent, which is in turn rationalized by desorption of the analyte in the neutral form, i.e., with Na attached to the carboxylate functional group and no charge on the amine group. Subsequently, gas phase cationization occurs. The reaction heat for the H–Na exchange reaction is nearly constant for all amino acids, and hence the sublimation enthalpy becomes the major factor determining the ion yield. Protonated and simply cationized molecules overtake the signal from the doubly sodiated and analyte under low-power conditions.

Surprisingly, a recent study has evidenced the role of internal lactam formation for terminal amino acids with C_6 or C_7 skeletons. Samples have been analyzed as thin layers on Ag and quartz under irradiation with 266 nm at power densities between 10^6 and $10^8\ W \cdot cm^{-2}$. No cationization with Ag^+ is observed in contrast to the case with α-analogues. The loss of water from the amino acids indicates thermal decomposition into a protonated lactam, and the signal peak intensity correlates well with thermodynamic data on the heat of formation (Rosnack et al., 1989).

3B.3.8. Oligopeptides

Using the above encouraging results for amino acids, several investigators focused their research primarily on oligopeptides. Later, the matrix-assisted desorption technique permitted a dramatic extension of the working range to large biomolecules, but this topic is within the preceding review by Vertes and Gijbels (Part A of this chapter). This section will deal first with small molecules from dipeptides up to decapeptides, including some highly popular test compounds such as gramicidin and angiotensin. These data permit comparison of the different approaches in the field of LD-MS.

The previously mentioned research of Hillenkamp's group on the influence of wavelength and power density in TOF-LMMS has been extended to a series of di- and tripeptides (see Spengler et al., 1987). The general trend largely confirms the findings for amino acids. Use of shorter wavelengths under threshold conditions, where the aromatic amino acids exhibit increasing absorption, yields a mass spectrum with only one fragment, due to the loss of the carboxyl functional group. Protonated molecules are detected at increased irradiance, whereas at still higher irradiance levels this signal disappears and the doubly sodiated molecules start dominating the parent region. Increasing wavelength, e.g., 248 and 266 nm, gives mass spectra with cationized, protonated, and decarboxylated molecules $(M + Na)^+$ even near the threshold. Higher irradiance promotes the formation of $(M - H + 2Na)^+$, but the working range becomes much smaller, namely, 4 times the threshold value as opposed to 52 times in the case of 193 nm (Spengler et al., 1987; Karas et al., 1985; Hillenkamp et al., 1987).

To our knowledge, there is virtually no information about negative mass spectra from small peptides in TOF-LMMS, apart from an early publication that describes the production of an intense $(M - H)^-$ signal from a dipeptide as well as a series of fragments, essentially referring to the amino acids (Schiller et al., 1981).

The recently emerging interest in LD-FTMS has, of course, yielded noteworthy studies in this field as well. The great flexibility of the instrument which enables us to combine several approaches, is nicely illustrated by LD-CI experiments on tripeptides, exploiting the intact desorption of the analyte by the extremely rapid heating and the benefits of the well-established CI method. In fact, this approach was pioneered in the early experiments on a magnetic sector instrument for steroid glucuronides (Cotter, 1980). Recent FTMS studies have exploited the possibility of influencing the chemical form and MS behavior of the products of decomposition of the ion–molecule complex by the choice of reagent gas (Speir et al., 1991). Selectivity can be influenced by properly matching the acid–base characteristics of the analyte and those of the reagent ions. The extent of fragmentation depends also on the exothermicity of the proton transfer within the initial ion–molecule complex. Use of different reagent species, e.g., NH_4^+, $C_2H_5^+$, or N_2OH^+, provides the flexibility to affect the extent of fragmentation of the $(M + H)^+$ from 0 to 100% by changing the proton affinity of the reagent gas over a range of only 3.3 eV. It has been found that direct ionization without CI yields only fragments from the cleavage between the amino acids. Higher irradiance at 193 nm and power density of 10^5–10^6 W·cm^{-2} give rise to the protonated molecules. The "soft" and "hard" ionization capabilities obtained by different reagent species parallel those in the two-step LD-MUPI method but are not limited to the positive ions (Speir et al., 1991). Direct ion formation by laser irradiation has been performed and yields no fragmentation (McCrery et al., 1982).

The efficient postionization in the gas phase under resonance-enhanced multiphoton ionization (REMPI) conditions up to saturation and the generation of many more neutrals than ions from laser irradiation of solids give the two step-approach favorable prospects with regard to sensitivity. Fine tuning of the internal energy deposited in the molecular ions allows generation of a sufficient number of structural fragments while the ion current still remains concentrated in a few intense peaks. Unless the molecules already contain an acid with an aromatic group, e.g., tyrosine or phenylalanine, derivatization is required by, for instance, benzyloxycarbonyl (CBZ), as well as other groups (Tembreull and Lubman, 1987; Grotemeyer et al., 1986b; Segar and Johnston, 1989). Segar and Johnston (1989) have studied the fragmentation of CBZ derivatives and the use of metastable ions to distinguish isomers. Analysis of mixtures points to the independent desorption of neutrals, and the mass spectra do not suffer from hydrophilicity problems as they do in FAB for mixtures of highly and less polar dipeptides (Grotemeyer and Schlag, 1988a). Analysis of the more volatile and/or protected analogues by RETOF instead of conventional MS is motivated by the superior capabilities for diagnostic characterization. This applies not only to the intensity of the molecular ions but also to the structure and intensity of the fragments. Nevertheless, the formation of thermal degradation products occurs during IR laser irradiation with 10.6 μm at modest power density and may complicate the interpretation of the results. For instance, cyclization into a 2,5-diketopiperazine occurs for amino acid moieties in the positions 1 and 2 from linear oligopeptides (Li and Lubman, 1989b).

The pioneering experiments on LD-MS of penta- to octapeptides evinced the potential to obtain MW information from $(M + H)^+$ and $(M + K)^+$ while fragments are virtually absent (Posthumus et al., 1978). In contrast, LD-FTMS analysis yields a distinct pattern of intense fragment signals, permitting significant sequence information to be derived (Yang and Wilkins, 1989). Nevertheless, it is often necessary to apply the multistage capabilities of FTMS. Typically 1 μg of sample is sufficient to perform a CID experiment and to obtain the daughter ion mass spectrum. Linear and cyclic oligopeptides have been distinguished. Correct sequencing of the 15 amino acids in gramicidin D has been achieved (Yang and Wilkins, 1989).

The latter experiment brings us to the survey data on selected molecules such as bradykinin, angiotensin, insulin, and the aforementioned gramicidin. These compounds are of significant importance in the biosciences but they also encounter much interest in the field of LD-MS. Nearly all approaches using single or two-step laser DI, with or without postionization and using different types of mass analyzers, have been employed more or less successfully for the analysis of these molecules. The same applies to other forms of direct excitation and ion formation from organic solids, e.g., SIMS, FAB, and PD-MS. For the sake of brevity, this section will not attempt detailed comparison of the individual results.

Mass analysis of the directly generated ions in the presence of a matrix to facilitate desorption essentially yields the MW information in TOF-LMMS instruments for polypeptides above 100 kDa (Tanaka et al., 1988; Karas et al., 1989). The excess of a strongly absorbing compound, e.g., nicotinic acid at 266 nm, and the decomposition of the peptide poison the low-mass range by a bunch of high-intensity signals that are without any use for further structural characterization (Vertes et al., 1990; Doktycz et al., 1991). The strength of the technique lies in the possible detection of high-mass ions, not in the resolution of these signals or the production of sequence information. Calibration of the flight time in terms of a relatively accurate m/z value is not facilitated by the time domain in which ion formation occurs (Van Vaeck et al., 1986a). More information on high-mass analysis will be found in the preceding review by Vertes and Gijbels (Part A of this chapter). Interesting data as to the ion formation kinetics have been obtained from several peptides in the more modest MW range by the application of drawout pulses after different delay times vs. the laser pulse. The predominance of fragments is generated initially, whereas clean mass spectra with only protonated and cationized molecules are recorded after several tens of microseconds (Tabet and Cotter, 1984). Additionally, the peak width decreases significantly, reflecting a narrower energy distribution. As a result, a much better resolution up to m/z 1000 can be achieved (Cotter and Tabet, 1983).

The other laser techniques are of course confined to the lower mass range, and the remainder of this contribution again refers primarily to the standard set of frequently used test compounds. RETOF data on bovine insulin show extremely clean mass spectra, dominated by the radical molecular ions and only two fragments, due to the cleavage of the A and B chain. Two full-scale peaks at m/z 5729 are separated with a 50% valley (Grotemeyer and Schlag, 1987). Soft ionization yielding only $M^{+\cdot}$ in the mass spectrum has been shown for angiotensins, whereas increasing the ionizing beam intensity gives full sequence information (Grotemeyer and Schlag, 1988a; Grotemeyer et. al., 1986a). Gramicidin has been tested to assess the potential of LD on the moving belt LC-MS interface (Hardin et al., 1984).

FTMS data on several bradykinins permit detection of abundant cationized molecules and high-mass fragments with an accuracy of 2–7 ppm on the m/z value (Wilkins and Tang, 1986). Matrix assistance can be very useful especially in this technique, where the precise time of ion formation and the rapid decay of the initially ejected large clusters of analyte and matrix does not affect the measurement at all. Essential conditions to be taken into account concern the stability of the species to be detected and the upper limit for the kinetic energy distribution component, which must be under the trapping potential and which motivates current research on prior jet cooling in FTMS analysis as well (Hettich and Buchanan, 1990). The work on biomolecules generated by primary ion bombardment in FTMS clearly demonstrates the capabilities of this analyzer to perform sensitive and extensive characterization of high-mass molecules up to m/z 5000 routinely (Shabanowitz and Hunt, 1990). As a result, it is evident that ongoing research on the optimization of ion formation with regard to the requirements of FTMS make it likely that this instrument will become the method of choice for MS of polypeptides up to 10 kDa (Hanson et al., 1987). Accuracy of mass determination, high resolution, and the capability to perform multistage experiments, either by CID or photodissociation, are undoubtedly valuable assets that are at the given sensitivity level not provided by other instruments.

3B.3.9. Lipopolysaccharides

A particularly promising area of research with LD is the analysis of lipid A from the liposaccharides (LPS) obtained from gram-negative bacteria. LPS are composed of three structural regions: the O-specific polysaccharide, the common core, and a component, called lipid A, which is a phosphorylated glucosamine disaccharide containing both normal and hydroxy fatty acids. The "free" lipid A, which can be obtained from LPS by mild hydrolysis, is of great pathophysiological interest because of its toxicity. The fatty acid distribution can be obtained by hydrolytic cleavage from the disaccharide backbone. There is a lot of interest in obtaining information on the positions of the fatty acids from MS analysis of intact lipid A.

Figure 3B.32 illustrates the positive mass spectrum above m/z 1150 of methylated diphosphoryl lipid A from *E. coli* (Cotter et al., 1987). The data are recorded on a linear TOF instrument with drawout pulse extraction and 10.6 μm laser irradiation. The parent peak refers to $(M+K)^+$. The first set of fragments points to the easy cleavage of the reducing-end phosphate functional group. The structural interpretation of these signals is explained in Figure 3B.33. The next group of fragments shows a mass difference of m/z 200, 224, and 244, characterizing the presence of myristoxy-, lauroxy-, and hydroxymyristic acid chains, respectively.

Similar results have been reported for the liposaccharides from different bacterial strains (Qureshi et al., 1988). Analysis of these compounds as ammonium salts yields the protonated molecules (Tabet and Cotter, 1984). An early report on the application of TOF-LMMS on a synthetic lipid A–like compound mentions the detection of similar adduct and fragment ions for

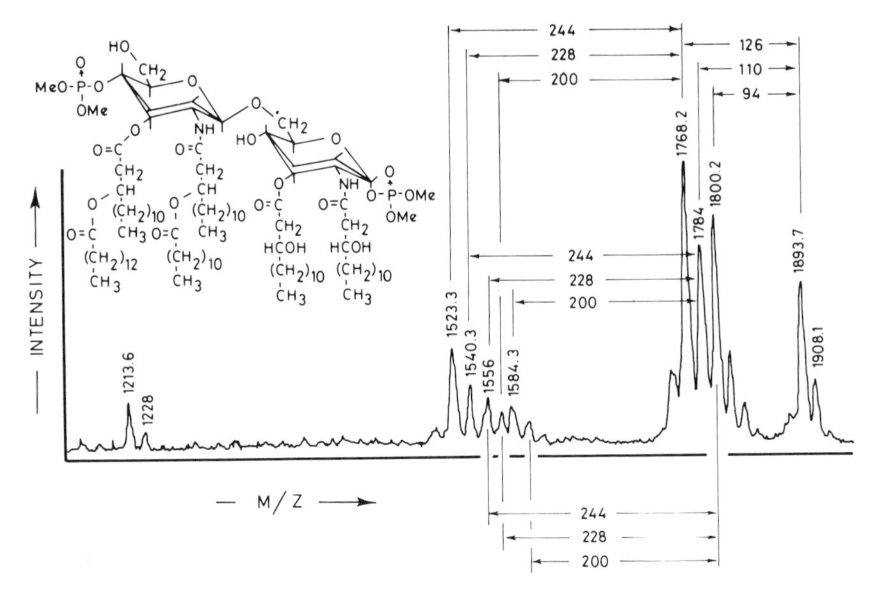

Figure 3B.32 Positive ion mass spectrum above *m/z* 1150 of methylated diphosphoryl lipid A from *Escherichia coli*, recorded by linear TOF-MS. Reprinted from Cotter et al. (1987) with permission of John Wiley & Sons, Ltd.

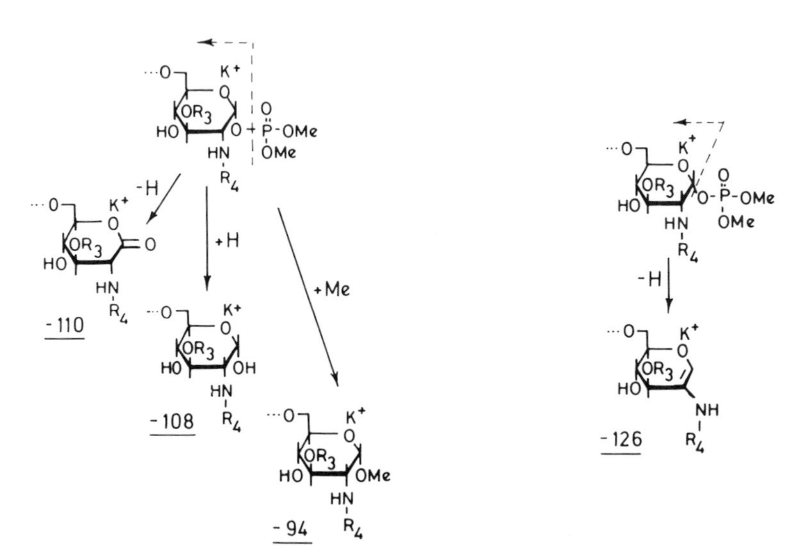

Figure 3B.33. Structural assignment of the major fragments in the positive ion mass spectrum of methylated diphosphoryl lipid A from *E. coli*. Reprinted from Cotter et al. (1987) with permission of John Wiley & Sons, Ltd.

the non phosphorylated analogues (Seydel and Lindner, 1983). Presence of the phosphate functional group yielded numerous low-intensity fragments, but no interpretable spectra were obtained. However, some successful applications were reported later (Krauss et al., 1989; Seydel et al., 1984; Weintraub et al., 1989; Wollenweber et al., 1984; Lindner et al., 1988; Helander et al., 1988). The steadily increasing number of papers in the literature reflects the high interest from a biochemical point of view, but it is not possible to describe the individual results in detail within the framework of this subsection.

3B.3.10. Polymers

Structural characterization of polymer systems is by definition largely different from that of smaller molecules. MS inevitably involves the generation of ions in the vapor phase. This has been applied to earlier experiments with the different methods of pyrolysis MS but still remains valid for the newer methods with DI directly on solids. It should be realized that even the range of giant ions above 100 kDa, accessible today for some bio-molecules with the aid of matrix-assisted desorption, actually corresponds to relatively small fragments from currently used "polymer molecules." It is common practice to use a comparative approach, based on the so-called fingerprint mass spectra of representative reference samples. The analytical potential of this method depends, then, on the reproducibility with which these small ions can be generated. In this respect, the conventional pyrolysis of bulk materials still remains competitive with more recent DI techniques on solid samples.

The possible role of LD-MS in the field of polymer analysis has certainly not yet become clear. Data obtained in TOF-LMMS appear to be fundamentally different from those recorded by LD-FTMS and, even though laser irradiation is not performed under exactly similar conditions of wavelength and power density, it seems unlikely that the laser–solid interaction process itself accounts for the observed difference. To evaluate the application of LD-MS on polymers properly, we might well profitably mention the characteristic features of results from conventional pyrolysis and static SIMS to illustrate the range of structural information obtained from current techniques.

Conventional pyrolysis MS often characterizes the material by means of small decomposition molecules, as illustrated in Table 3B.7 (Wuepper, 1979). The application of modern DI methods is primarily intended to detect more or less intact combinations of the repeat units up to m/z 10,000. Also, local and surface analysis can be attempted. FD-MS was the first method to successfully explore the field of polymer analysis (Schulten and Lattimer, 1984; Prokai, 1990). However, today static SIMS is considered to be one of the most efficacious methods for these applications. The mass spectra indeed contain numerous signals, most of which are really indicative of important structural features. The positive mass spectra from poly(styrene), for instance, essentially

Table 3B.7. Diagnostic Peaks in Positive Ion Mass Spectra of Rubber Samples: Analysis by Pyrolysis MS

Material	Diagnostic Peaks in Pyrogram (m/z Ratio of Molecular Ion)
SBR[a]	Butadiene (54), benzene (78), toluene (92), styrene (104), and azulene (128). Ratio of butadiene to styrene is approximately 2–4:1.
Poly(chloroprene)	2-Chloro-1,3-butadiene (88). Ratio of chloroprene to any other peak is about 10:1.
Nitrile	Acrylonitrile (53), ratio of C_4^{2-} to acrylonitrile is 2:1.
EPDM[b]	Propylene (42), butene (56), benzene, and toluene can be similar to SBR, but styrene intensity is negligible.
Butyl	Isobutylene (56) dominates pyrogram.
Synthetic poly(isoprene)	C_5 diene (68) dominates gas chromatogram. Absence of diagnostic peaks such as styrene, acrylonitrile, and chloroprene. Aromatics are minor.

Source: Reprinted from Wuepper (1979) with permission of the American Chemical Society.
[a]SBR = rubber trade name.
[b]EPDM = rubber trade name.

comprise a superposition of three sets of ion signals (Van Leyen et al., 1989). Intact oligomers are observed in the singly or doubly cationized form with one or two charges in the m/z range between 600 and 10,000. These signals form a regular pattern, allowing determination of the MW distribution for the oligomers. Secondly, chain scission of the polymer backbone induced by the primary ion beam in the solid produces fragments that are detected under the cationized form in a regular pattern between m/z 600 and 6000 (Lub et al., 1989). Finally, the low-mass range up to m/z 200 contains numerous signals from stable aromatic ions. Depending on the material under study, detailed information about the functional groups at the surface can be derived from these low-mass signals (Lub et al., 1988). The presence of high-mass ions depends on the average MW (M_n) of the polymer, but emission of the mid-range fragments seems to be unaffected by the chain length for polymers with M_n above 10,000 (Van Leyen et al., 1989).

Initial results of TOF-LMMS on polymers have shown that focused laser irradiation produces primarily low-mass ions in the range up to m/z 300 (Gardella and Hercules, 1980, 1981). This applies to the positive as well as the negative detection mode. Simple carbon–carbon backbone polymers such as poly(propylene) produce ions of the carbon cluster type. Poly(vinyl chloride) gives essentially the same ions but with a different intensity pattern. Only very few signals actually point to the presence of chlorine. Teflon and less fluorinated poly(ethylenes) exhibit a series of alkyl fragments with F or H depending on the analyzed material (Mattern and Hercules, 1985). Aromatic and carbonyl groups as incorporated in poly(styrene), poly(amide)s and poly(acrylate)s give rise to fragments related to the corresponding functional groups. Polymers with an amide backbone such as poly(caprolactam) show detection of the repeat unit under protonated form in the positive mode while the negative mass spectra contain several clusters, containing carbon, nitrogen, and oxygen. The structural assignment of the peaks also indicates that the structural integrity of the polymeric repeat units is lost (Gardella and Hercules, 1980). Poly(4-vinylpyridine) and poly(2-vinylpyridine) polymers have been distinguished using fingerprints showing the protonated monomer and $(2M-H)^+$, with M referring to the repeat unit in the positive ion mass spectra (Fletcher and Fatiadi, 1986).

This points to the weakness of the comparative approach in LMMS. The method permits detection of elemental and organic ions in a variety of analytical problems with minimal sample preparation. This flexibility makes detailed control of possible artifacts difficult. The low specificity of the low m/z ions, in general, and the observed tendency of these ions to loose their structural integrity when generated from polymer, in particular, aggravate the situation. However, comparison of the mass spectral pattern can be useful in selected applications. The characterization of a poly(methacrylate) inclusion in poly(carbonate) has been reported (Holtkamp et al., 1991). Application of principal component analysis to TOF-LMMS data on a series of polymers has been reported to yield a viable method for selected applications (Odom et al., 1989). The abundant generation of nonspecific carbon clusters does not hinder the detection of low-MW compounds in the polymeric system, as evidenced by the identification of styrene sulfonic acid in ion exchange resins (Scanlan et al., 1989).

Sputtering of gold as a thin film on the polymer dramatically improves the detection of structural fragments in the low-mass range. This is tentatively explained by the formation of excited states and/or radicals in the upper layer during the first part of the laser pulse, which then causes a drastic increase of the absorbance. These species then achieve very efficient transfer of the laser energy into a very shallow sample depth during the rest of the irradiation period. The extremely high heating rate finally causes the ablation and ion formation processes (Holtkamp et al., 1991).

An early report mentioning the detection of $M^{+\cdot}$ from 3% carbazole in poly(ethylene) has suggested that the measurement of oligomers from polymer samples by TOF-LMMS is feasible (Heinen, 1981). Analysis of a poly(amide) sample that effectively contains substantial amounts of oligomers with up to 7 repeat units has permitted detection of the corresponding molecules under the cationized form (Holm et al., 1987). The peak intensities approximately

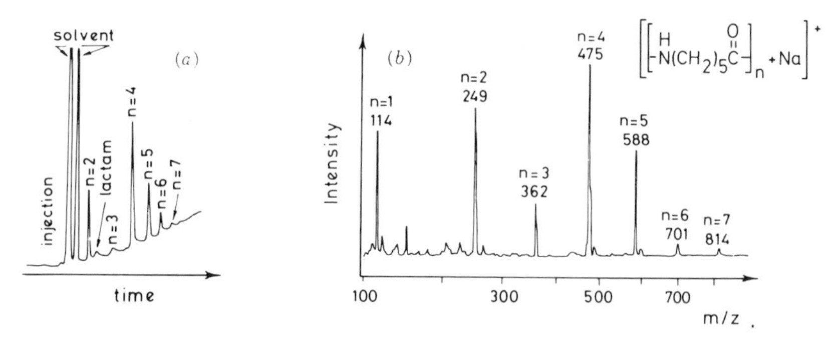

Figure 3B.34. Molecular weight distribution of an oligomer extract from a poly(amide)-6 sample, determined from HPLC analysis (a) and positive ion mass spectra recorded by TOF-LMMS (b). Reprinted from Holm et al. (1987) with permission of the American Chemical Society.

reflect the relative concentration of the individual monomers. Figure 3B.34 illustrates the positive mass spectrum and the corresponding high-performance liquid chromatography (HPLC) profile (Holm et al., 1987). Additionally, oligomer patterns have been measured from poly(dimethylsiloxane) and poly(styrene) samples with an average MW around 1500. In the former case, the distribution peaks for oligomers with 14 units and the ions with up to 40 monomeric entities still exceed the detection limit. As a result, the method parallels the high- and low-mass-range ions, which characteristically occur in static SIMS. However, the third series of midrange fragments is not observed in TOF-LMMS (Holm et al., 1987).

Study of several poly(ethylene glycol) and poly(propylene glycol) (PEG and PPG) samples has allowed comparison of the analytical capabilities of TOF-LMMS with several other MS techniques using soft ionization, namely, electrohydrodynamic ionization (EH) MS, FD-MS, and FAB-MS (Mattern and Hercules, 1985). The polymers are doped with alkali salts and smeared as a thin film on a zinc foil. The positive mass spectra recorded by TOF-LMMS primarily show the cationized oligomer distribution under low-power-density irradiation. Occasionally, i.e., in about 20% of the mass spectra, low-mass fragments under 150 result from pyrolysis and have practically no diagnostic value, whereas those in the mid-mass range result from bond cleavage along the poly(glycol) chain. The presence of these signals is associated with the application of higher-power-density irradiation. These fragments correspond to the ones observed in FAB-MS and are related to the initial higher-energy portion of the fast-atom-induced collisional cascade. Hence, direct ionization in the high-energy laser–solid interaction zone is proposed to be responsible for these signals in TOF-LMMS. In contrast to EH-MS, no multiply charged ions formed by attachment of several alkali ions are observed. It is not clear whether this is an effect of the laser ionization or of the geometry of TOF-MS used here. The distribution of cationized oligomers permits determination of the M_n. The values compare favorably with those obtained from end-group titration, EH-MS, FD-MS, and FAB-MS, on the condition that additional postacceleration is applied. Otherwise, the high-mass discrimination on detection tends to shift the centroid of the oligomer distribution toward lower numbers. Unlike bulk methods, the chain length dependence of the selectivity for cation attachment can be studied from the peak intensity of individual oligomers. MS provides access to the chain length–dependent selectivity of oligomers for cation attachment. Comparison of the data for Na, Li, K, and Cs shows significant differences, which can be explained by structural features such as the helix form of PEG and the non-planar zigzag conformation of PPG (Mattern and Hercules, 1985).

Figure 3B.35 shows the oligomer distributions for a mixture of cyclic and linear siloxanes sprayed on a layer of NaCl microcrystals on a silver foil (Holm and Holtkamp, 1989). Comparison of the data for the linear component shows that the ions with a cyclic structure are really indicative of the sample composition and do not result from fragmentation of the linear oligomer ions. In this respect, the TOF-LMMS method seems to have capabilities very similar to those of static SIMS, but sample preparation still remains critical.

PEG and PPG samples have been analyzed in the presence of an ultrafine metal cobalt powder and with irradiation at 355 nm instead of 266 nm as in the aforementioned experiments (Tanaka et al., 1988). The mass spectra from samples with relatively low M_n show only signals from the cationized oligomers,

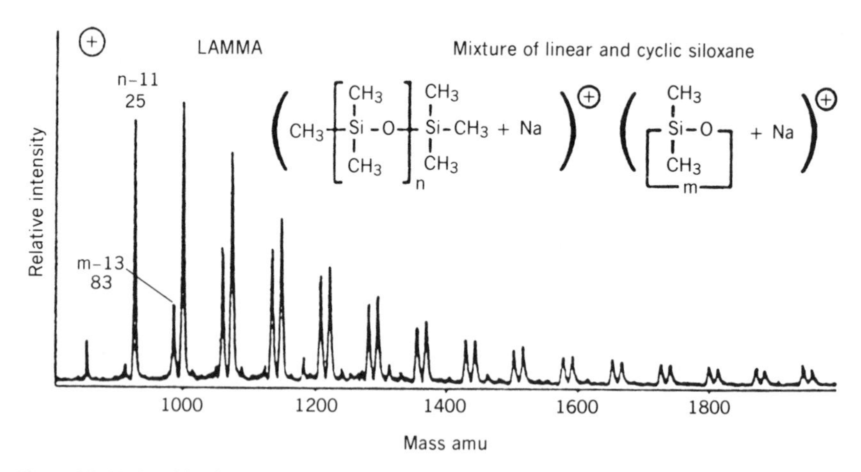

Figure 3B.35. Positive ion mass spectrum of the siloxane oligomers, obtained by TOF-LMMS from a solution residue on a NaCl-coated Ag substrate. Reprinted from Holm and Holtkamp (1989) with permission of San Francisco Press, Inc.

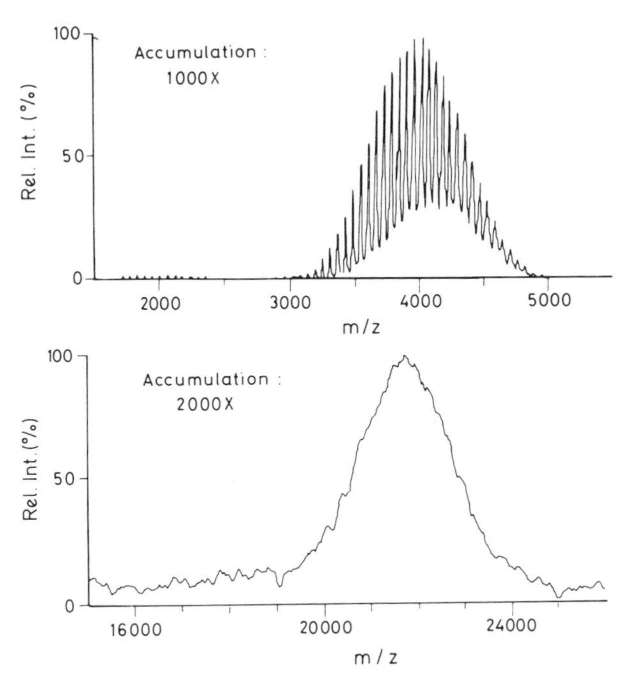

Figure 3B.36. Positive ion mass spectrum of poly(propylene glycol) with average MW of 4 kDa (top) and of poly(ethylene glycol) with average MW of 20 kDa (bottom), recorded by TOF-MS and laser ionization. Reprinted from Tanaka et al. (1988) with permission of Heyden & Son, Ltd.

whereas low- and mid-mass-range fragments ions are negligible. Figure 3B.36 illustrates the results for a PPG sample with an M_n of 4 kDa and PEG with an M_n of 20 kDa (Tanaka et al., 1988). The data point to the ultimate dilemma we face at the present state of the art. More or less adequately resolved signals can be obtained up to *m/z* 5000 whereas laser ionization is capable of generating structurally meaningful ions in the much higher mass range. However, detection is limited to a broad but still quite symmetric signal, which is believed to represent the distribution envelope of the unresolved cationized oligomer peaks. The centroid of this signal should then still provide an indication as to the M_n. To our knowledge, TOF-SIMS data have not yet shown such high-mass ions. Moreover, it is not clear just what the practical applicability of LD-MS really is on the high-mass ions from polymers in view of the actual instrumental performances. Dramatic improvements of TOF-MS analyzers must be achieved before this type of instrument can adequately cope, in respect of resolution and detection, with the wealth of informative ions that the laser–solid interaction makes available from polymer samples.

The obvious question now arising concerns the possible progress to be expected from recent developments in the field of polymer analysis by LD-FTMS. So far, most of the results have been obtained using CO_2 laser irradiation at 10.6 μm; irradiation modes will be explicitly mentioned below. An excellent survey of this topic is available (Cody et al., 1990). This discussion will focus first on the high-mass ions. Figure 3B.37 illustrates the high-mass region of the positive ion mass spectrum from a PEG-8000 sample, showing the cationized oligomers up to *m/z* 10,000 (Ijames and Wilkins, 1988). The data were obtained by co-addition of single scan spectra from 50 laser shots. Ejection of the unnecessary ions and analysis in the heterodyne mode permits attainment of a resolution of 160,000 at *m/z* 3200 and 60,000 at *m/z* 5922, whereas the accuracy of the *m/z* determination is better than 10 ppm between *m/z* 5500 and 6000. This performance is superior to what is presently feasible in TOF-MS. Similarly spectacular patterns are recorded in the negative ion detection mode from KRYTOX® 16140, a fluorinated polyether, where the distribution goes up to *m/z* 7000 (Wilkins et al., 1985). The results on Teflon samples give a series of fluorinated alkyl fragments up to *m/z* 1500 (Cody et al., 1990) and hence provide a clear contrast with the data from TOF-LMMS mentioned earlier. Mass spectra on poly(dimethylsiloxane)s (Cody et al., 1990) show oligomer distributions up to *m/z* 8000, which points to an at least equivalent potential of LD-FTMS for high-mass ion generation and detection in comparison with static SIMS (Bletsos et al., 1988). Determination of the M_n values from oligomer ion distributions in FTMS has been attempted for alkoxylated pyrazole and hydrazine polymers, used as corrosion inhibitors in acids, antifreezes, and hydraulic fluids (Nuwaysir and Wilkins, 1988). Com-

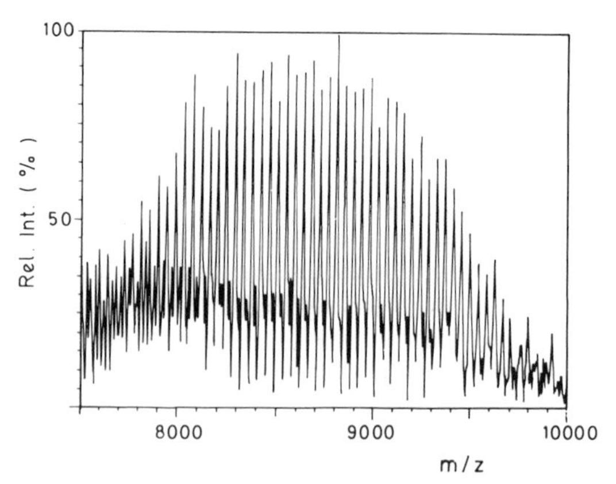

Figure 3B.37. High *m/z* region of the positive LD-FT mass spectrum from poly(ethylene glycol)-8000. Reprinted from Ijames and Wilkins (1988) with permission of the American Chemical Society.

parison with reported FAB-MS and EI-MS data again reveals the trend to obtaining higher M_n values by FTMS due to the absence of detector discrimination. The authors suggest an additional application for these compounds: the ease by which the mass spectra are obtained makes them suitable candidates for calibration use in LD-FTMS. Siloxanes have been identified by means of accurate mass assignments in the spectra from a surface-contaminated magnetic disc by application of IR laser irradiation and subsequent EI of the neutrals (Land et al., 1987).

Despite all this, several polymers are also intractable to analysis by LD-FTMS and only produce high-mass carbon cluster ions up to several thousands of mass. It was mentioned that the generation of these nonspecific ions may occur for given polyester samples and not for other products with similar chemical composition (Cody et al., 1990). The morphology of the specimen may be involved, for instance, thick layers or larger particles appear to favor carbon cluster production more than thin films do. UV laser irradiation at 266 nm of polyimides in an FT-LMMS setup, permitting analysis of 5–8 μm areas, has also generated carbon cluster–dominated mass spectra going up to m/z 3000 with C_{250}^+ as the largest entity observed (Creasy and Brenna, 1988).

Although UV irradiation at the same wavelength is used, significant differences exist between TOF-LMMS and FT-LMMS data on polymers (Brenna, 1989). In general, the latter method shows ions extending to higher mass range. A pronounced tendency toward extensive rearrangements is observed as well, and this significantly complicates the structural interpretation. Possible explanations refer to the detector discrimination effects in TOF-LMMS for high-mass ions. Also, the instrument employed for FT-LMMS generates the ions inside the magnet. The charged particles moving along small radius orbits in the beginning remain exposed to second-order interactions with ions within the cloud of laser-generated species. In contrast, the high-extraction field in the TOF-LMMS limits the residence time of the ions in the selvedge region to a few tens of nanoseconds as opposed to tenths of microseconds in FTMS. The effects related to strong gas phase absorption within the laser-generated microplasma may additionally affect the ablation of material from the underlying sample, as proposed from comparison of mass spectra for poly(styrene) and poly(terephthalate) at different wavelengths by Krier et al. (1989).

A series of papers describe the application of LD in combination with FTMS for characterization of various conducting polymers and related compounds in a variety of synthesis mixtures (Nuwaysir and Wilkins, 1988; Brown et al., 1985, 1986a, b, c, 1988a, b; Miller et al., 1986). For the sake of brevity we shall not go into detail here, but it should be noted that LD-FTMS has proved able to yield a detailed characterization of samples that cannot be obtained by other currently available techniques. Specifically, there are no comparable data from TOF-LMMS on the detection of intact oligomers in the mass range up to *m/z* 2000 for analogous samples. Moreover, mass resolution is important; accurate *m/z* determination and correct isotope pattern measurements are essential, even though the ultrahigh resolving power and extreme ppm accuracy of the *m/z* values are not strictly required.

It now seems worthwhile to consider the possible reasons for the apparently better potential of FTMS vs. TOF-LMMS, although detailed experiments are still required to fully resolve this issue. First of all, the laser irradiation wavelength is different from that of the commercial TOF-MS and no similar data are known from the in-house-built setups on these applications. Secondly, the ion generation within the FTMS cell favors second-order interactions between the laser-generated ions and neutrals. Similar experiments in an instrument with an external source would be very informative on this point. Finally, the time domain is an important factor as well. FTMS often uses ion accumulation prior to excitation over a time period exceeding the entire flight time of ions in TOF-MS. In this way, detection in the latter method may occur even before the measurement in FTMS begins. If it is assumed that the base peak in a TOF spectrum is due to a short-lived ion, then it is easily conceivable that the corresponding ions are not detected by FTMS. The latter method will then evince the minor signals, part of which are under the detection limit of TOF-MS. The systematic differences between the mass spectra in TOF-MS and FTMS from polymeric samples are the most obvious ones in comparison with the other classes discussed so far. It would no doubt be profitable to focus future research on the actual reasons, which would permit more precise evaluation of the advantages and limitations of both methods.

3B.3.11. Ionic Compounds and Salts

This section describes compounds with so-called preformed organic ions and will include a survey of complexes and hydrochlorides. Some examples of the latter class have already been discussed because, as far as TOF-LMMS is concerned, hydrochlorides and free bases yield essentially similar results. A key element in the discussion is the possible occurrence of thermal degradation, which may indicate the prevalent thermal or nonthermal nature of the initial laser–solid irradiation. Also the generally accepted notion deserves attention that ionic compounds are better suited to be analyzed by DI. Hence, comparison of the ionic derivatives with the corresponding neutral analogue becomes highly relevant.

Note that in the following subsections *C*, printed in italic, will be used to indicate the intact preformed cation, whereas the usual (roman) C will refer to carbon.

Ammonium Salts. Considerable attention has been paid to the MS analysis of quaternary ammonium compounds, as many of these salts are of biological and pharmaceutical interest. Resistive heating causes several thermal decomposition reactions. The most commonly observed reaction leads to the formation of tertiary amines with the preferential loss of the largest substituent, which becomes combined with the counterion. The typical Hoffman-type degradation involves the formation of the corresponding alkene while the attack of the counterion often induces its incorporation in conjunction with the cleavage of the C—N linkage (Budzikiewicz et al., 1967). In the past, FD has proven to be the first method capable of detecting the intact preformed cations from the salts. The absence of fragments necessitates subsequent CID to obtain adequate structural information (Gierlich et al., 1977). There has been considerable debate in the literature concerning the need for a field to produce the C^+ ions from an emitter. It has been shown that the lower aliphatic ammonium salts up to tetrabutylammonium iodide emit the preformed ions intact when coated on an electrically heated wire in the absence of a high ionizing field (Stoll and Röllgen, 1980; Cotter and Yergey, 1981a,b). The desorption temperature lies below that for Na^+ or K^+. No postionization is required to observe several fragments, which are believed to be generated directly in charged form from the solid. These surface reactions also to some extent involve the alkyl transfer between different constituents in a mixture (Cotter and Yergey, 1981a). PD-MS produces intense signals from the C^+ ions as well as several significant fragments arising from the heterolytic cleavage of the N—C bond together with α-cleavage, leading to iminium ions. It is interesting to note that dimeric forms have yielded peaks with substantial intensity above 20% in both PD-MS and FD-MS data for tetrabutylammonium iodide (Schueler and Krueger, 1979).

A series of quaternary ammonium salts with aliphatic and aromatic substituents have been investigated by TOF-LMMS in our laboratory to probe the mechanistic aspects of the DI process under the specific conditions of focused laser irradiation at 266 nm (Van Vaeck et al., 1984). Figure 3B.38 compares the results from the laser microprobe with those from EI-MS with conventional direct probe introduction for tetraethylammonium chloride and tetrapropylammonium iodide. The EI-MS data reveal $M^{+\cdot}$ and subsequent fragments of the triethylamine, which corresponds to the thermal dealkylation decomposition product. In contrast, the tetrapropylammonium iodide shows a superposition of the mass spectra for tripropyl amine and propyl iodide. The latter degradation product is responsible for the ions at m/z 170 and 43. The majority of fragments in the TOF-LMMS mass spectra are indeed reminiscent of those in EI- MS, but the relative peak intensities do not agree as well as in the case of the UV stabilizers mentioned earlier. Note that the spectrum of tetraethylammonium chloride shows a signal at m/z 101, which clearly exceeds the ^{13}C contribution from the ions at m/z 100. As a result, the likely detection of $M^{+\cdot}$ from triethylamine and propyliodide is involved.

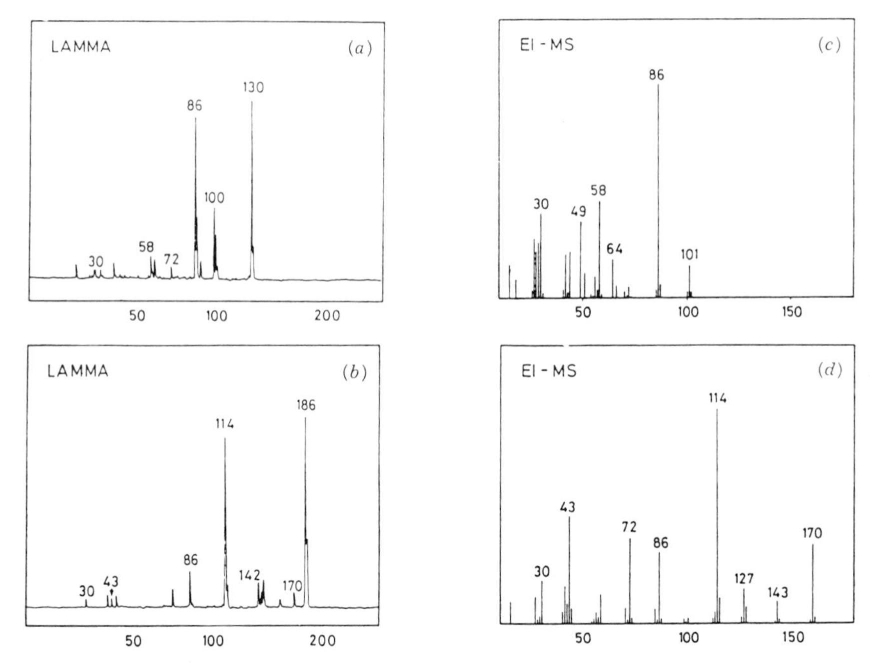

Figure 3B.38. Comparison of positive ion mass spectra recorded by TOF-LMMS and EI-MS from tetraethylammonium chloride (a&c, MW of C^+ 130) and tetrapropylammonium iodide (b&d, MW of C^+ 186). Reprinted from Van Vaeck et al. (1984) with permission of Springer-Verlag.

The generation of odd-electron ions from the intact cations, which are even-electron systems, points to the importance of thermal mechanisms in TOF-LMMS. Indeed, the structure of these analogues makes violation of the even-electron rule unlikely. As a result, the re-ionization of the initially generated neutrals from the organic salts, as proposed in the empiric DI model, provides a convenient and simple way to rationalize the presence of these signals in TOF-LMMS data. It is interesting to note that the corresponding signals from butyl iodide are also obtained by CO_2 laser irradiation and quadrupole mass analysis (Stoll and Röllgen, 1979). The signals are vaguely assigned to decomposition products of the salt, and the authors explicitly mention that generation of the butyl ion at m/z 57 is difficult to understand. The obvious explanation of thermal decomposition and electron ionization is not considered because of the detection of intact cations as a prominent or base peak and the prevalence of even-electron-type fragments. This observation has led to the still-persisting idea that radical ions are not generated in LD except for PAHs and related molecules.

The quaternary ammonium salts are quite sensitive in TOF-LMMS analysis. Detection limits on the order of 10^{-15}–10^{-16} g, corresponding to 10^6–10^7 molecules, have been reported (Feigl et al., 1984). Comparison between TOF-LMMS and FAB-MS yields a detection limit of 2×10^{-15} mol in the microprobe technique as opposed to 10^{-9} mol in the latter technique, which however provides a much better mass resolution of 10,000 against 1000 in LMMS (Kelland et al., 1987). The quantitative analysis of benzylalkonium salts in mixtures reveals that the relative peak intensities of the intact cation signals reflect the composition ratio within 10% (Balasanmugan and Hercules, 1983a). Additionally, the mass spectra seem less affected by subtle differences in the locally applied energy density regime. These compounds have been used to assess the surface adsorption characteristics of asbestos samples (De Waele et al., 1983c).

The generation of intact cations from solids was exploited quite early for the structural characterization of, for instance, naturally occurring quaternary compounds and alkaloids using high-resolution MS and tandem instruments. Multistage experiments have been performed on selected substances in cactus extracts at the 1 ppt level, and the use of primary ion bombardment has been compared to laser ionization for a range of salts of interest (Davis et al., 1983; Busch et al., 1982). A comparison with SIMS data reveals a better structure specificity of the signals in LD-MS (Hand et al., 1989).

Figure 3B.39 illustrates the positive and negative ion mass spectra recorded by TOF-LMMS from 1-(3,5-di-*O*-benzoyl-β-D-ribofuranosyl)-3,5-dimethylpyridinium chloride (Van Vaeck et al., 1989d). The structural assignment of the major ions is summarized in Figure 3B.40. The parent peak in the positive ion detection mode refers to the intact cations. The corresponding intensity is modest but still significant in spite of the fact that the polarized C—N bond and the benzoyl functional group provide adequate leaving entities. The high-mass fragments are again not very intense peaks but still permit establishment of a reasonable linkage with the structure, except for the signals at m/z 213 and 184. The former signal can be formally explained by attachment of neutral dimethylpyridinium to an azatropylium moiety, but occurrence of dimeric forms of fragments is not very often seen in our data base. Also the tentatively proposed structure for m/z 184 is not entirely convincing. (The need for subsequent ion studies to have high-mass resolution and m/z determination with ppm accuracy, as feasible in FTMS, is often expressed by those who are seeking to exploit the potential of LD for analysis of conventionally unamenable compounds.) The information from the negative ion mode mass spectra is confined to the benzoate functional group.

Additional results are available for nicotinamide dinucleotide (NAD), irradiated at 266 nm and 347 nm (Schueler and Krueger, 1980). These wavelengths correspond to the absorption bands in aqueous solution.

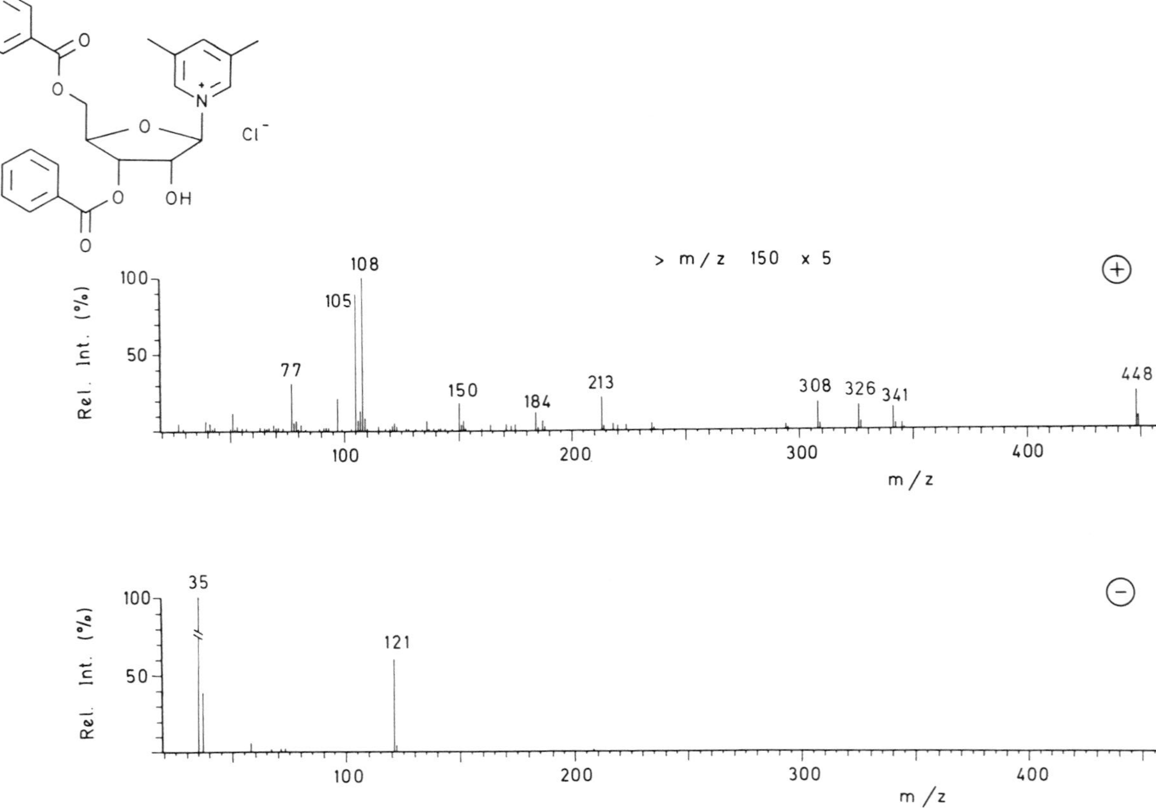

Figure 3B.39. Positive and negative ion mass spectra of 1-(3,5-di-*O*-benzoyl-*β*-D-ribofuranosyl)-3,5-dimethylpyridinium chloride recorded by TOF-LMMS. Reprinted from Van Vaeck et al. (1989d) with permission of John Wiley & Sons, Ltd.

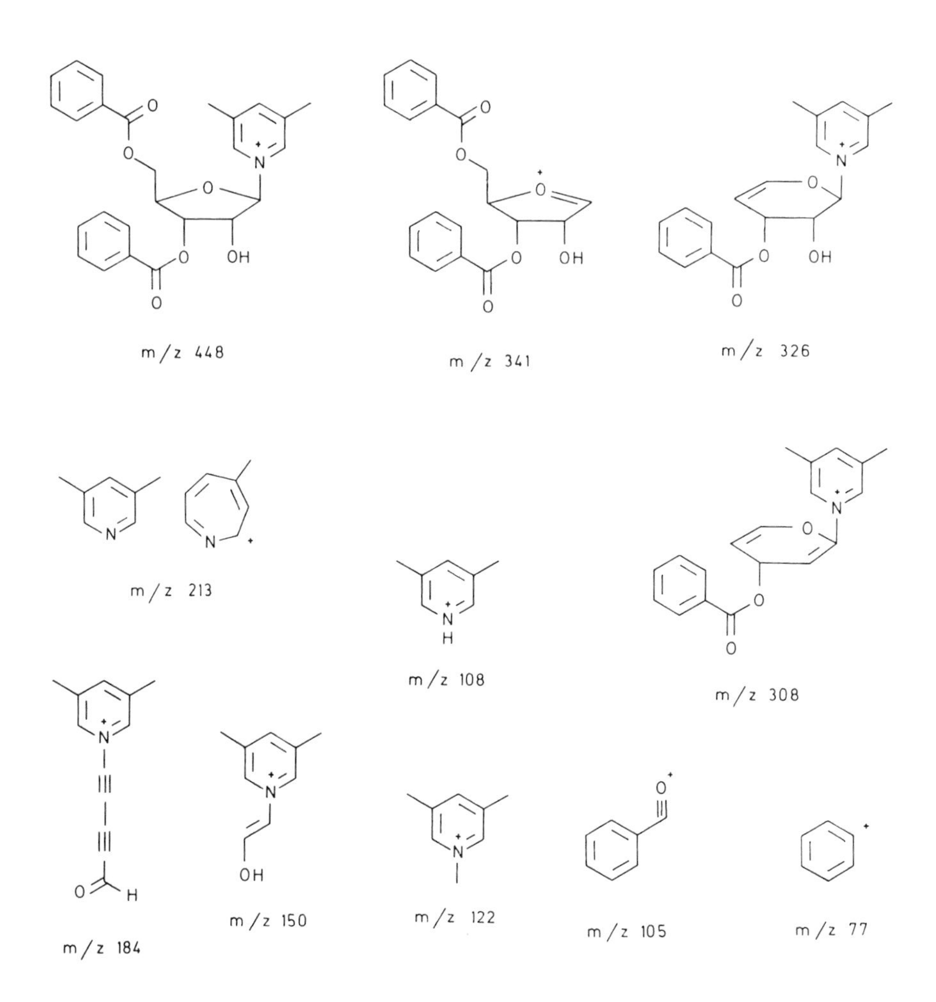

m / z 121

Figure 3B.40. Structural assignment of the major signals in mass spectra of 1-(3,5-di-*O*-benzoyl-β-D-ribofuranosyl)-3,5-dimethylpyridinium chloride. Reprinted from Van Vaeck et al. (1989d) with permission of John Wiley & Sons, Ltd.

Significant differences in the relative intensities occur. The presence of some fragments in the positive and negative ion detection mode also depends on the wavelength. No information is obtained in the MW region of the mass spectra, containing only fragments up to *m/z* 303.

The characterization of diquaternary ammonium salts by laser DI has been attempted as well. The mass spectrum of trimethylenebis(pyridinium bromide) in Figure 3B.41 shows no doubly charged intact cations (Cotter and Tabet, 1983). The data have been recorded on a linear TOF system. According to the structural assignment in Figure 3B.42, the main fragments refer to the

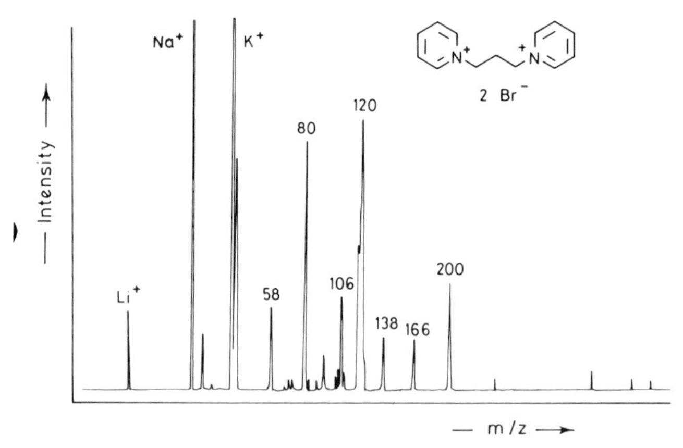

Figure 3B.41. Positive ion mass spectrum of trimethylenebis(pyridinium bromide) (MW of C^{2+} 200) recorded by TOF-MS. Reprinted from Cotter and Tabet (1983) with permission of Elsevier Science Publishers.

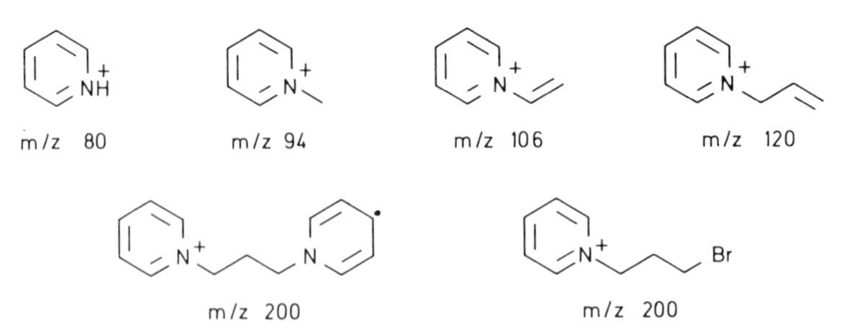

Figure 3B.42. Structural assignment of the positive fragment ions from trimethylenebis(pyridinium bromide). Reprinted from Cotter and Tabet (1983) with permission of Elsevier Science Publishers.

Table 3B.8. Summary of Data on Diquaternary Ammonium Salts, Analyzed by TOF-LMMS

	Intercharge Distance (Å)	$(M-H)^+$ LMMS	$(M-R)^+$ LMMS	Halide Substitution LMMS
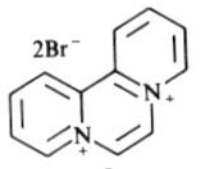 I ($2Br^-$)	2.8	yes	no	no
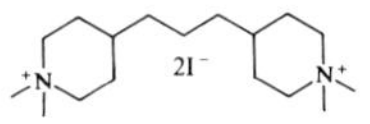 II ($2I^-$)	10.9	yes	yes	no
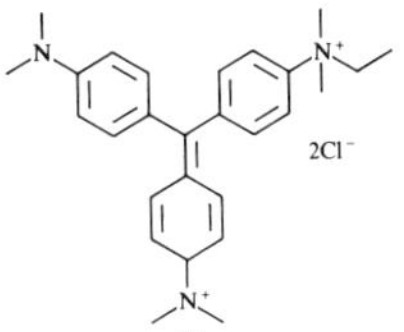 III ($2Cl^-$)	13.4	yes	yes	no
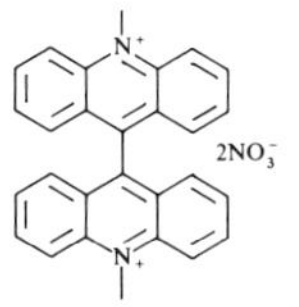 IV ($2NO_3^-$)	7.3	yes	yes	no
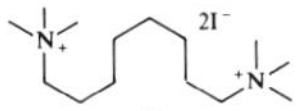 V ($2I^-$)	10.0	no	yes	yes
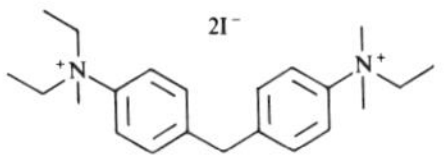 VI ($2I^-$)	9.5	no	yes	no
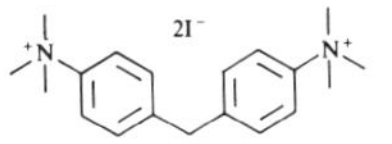 VII ($2I^-$)	9.5	no	yes	no
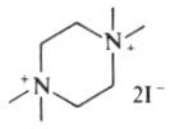 VIII ($2I^-$)	3.0	no	yes	no

Table 3B.8. (*Continued*)

	Intercharge Distance (Å)	$(M-H)^+$ LMMS	$(M-R)^+$ LMMS	Halide Substitution LMMS
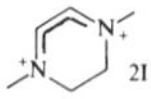 IX ($2I^-$)	3.0	no	yes	no
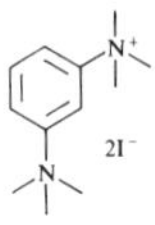 X ($2I^-$)	5.1	no	yes	no
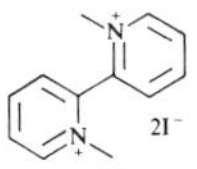 XI ($2I^-$)	5.2	no	yes	no
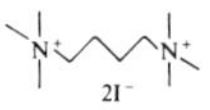 XII ($2I^-$)	6.3	no	no	yes
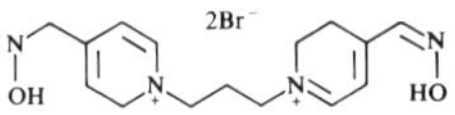 XIII ($2Br^-$)	4.9	no	no	yes
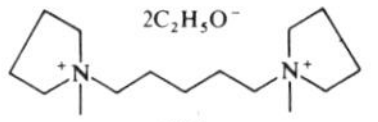 XIV ($2C_2H_5O^-$)	7.0	no	no	no

Source: Reprinted from Dang et al. (1984a) with permission of Elsevier Science Publishers.

expected decomposition reactions that remove one of the charge centers, with and without halide substitution. Additionally, an unresolved signal at m/z 200 is assigned in part to the singly charged radical ion arising from the uptake of an electron. The proposed interpretation is supported by comparative data from FAB-MS. A comprehensive series of diquaternary ammonium salts has been studied by TOF-LMMS (Dang et al., 1984a). Table 3B.8 summarizes the structures, the intercharge distance, the occurrence of dealkylation, deprotonation, halogen substitution, and the charge separation process. The last-named process involves decomposition of the dication by cleavage of a C—N bond and transfer of a hydrogen in a four-center mechanism. As a result, two monocation species are generated. The intercharge distance is a measure for the Coulomb repulsion. The increased chain length between the two quaternary nitrogens favors the charge separation process, whereas connection by a cyclic or bicyclic structure promotes expulsion of a group directly from the nitrogen. The preferential loss of an R or an H group can be readily conceived in terms of the corresponding structure. However, the apparent inability to detect the doubly charged dication in LD-MS complicates the diagnostic analysis, particularly since a different parent ion formation mechanism is involved.

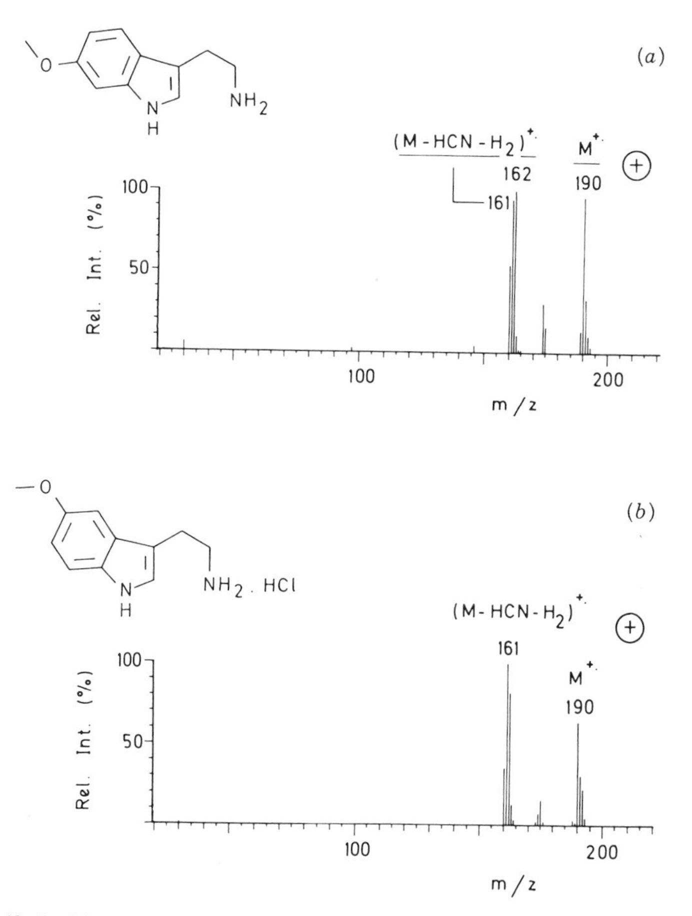

Figure 3B.43. Positive ion detection mode results recorded by TOF-LMMS from methoxytryptamine in the neutral form (a, MW 190) and as a hydrochloride (b). Reprinted from Van Vaeck et al. (1990) with permission of Springer-Verlag.

This not totally unexpected result parallels the formation of $(M+H)^+$ and $(M-H)^+$ ions from polyfunctional molecules.

The monitoring of *N,N*-dimethylaniline and its catalytic conversion into methyl violet–related dyes have been exploited. These compounds have been used as probes to assess the surface properties of asbestos fibers. Apart from the detailed structural information available for organic molecules, lateral resolution on the order of micrometers, good visualization, and excellent surface sensitivity are mandatory for these studies (De Waele et al., 1983b).

We shall conclude this subsection by turning to a comparison between a hydrochloride and the corresponding free base. Figure 3B.43 illustrates the positive ion mass spectra recorded by TOF-LMMS from methoxytryptamine in the neutral form and as the hydrochloride (Van Vaeck et al., 1990). The results for the free base show prominent signals from the $M^{+\cdot}$ and $(M-CH_2NH_2)^{+\cdot}$ (Van Vaeck et al., 1990). It is readily apparent that the majority of the ion current is carried by radical species. Surprisingly, the same essentially holds true for the corresponding hydrochloride. The major differences lie in the relative abundance of the molecular vs. the fragment ions, which can be partly related to the slightly different threshold irradiation values. Assuming that the radical fragments initially issue from the $M^{+\cdot}$ and not from the intactly desorbed cations or protonated free-base molecules at *m/z* 191, it can be calculated that the total contribution from even-electron parents to the total ion current is 13% for the hydrochloride and 10% for the free base. The adduct ionization of the free base, desorbed as neutral from the hydrochloride, is of course indistinguishable from the intact release of the preformed ion itself. However, these figures support the idea that the role of even-electron parents in TOF-LMMS should not be exaggerated and that the formation of odd-electron ions is certainly to be considered. Several compounds within our data base confirm the major trend. These peaks remain highly relevant for the mechanistic aspects of laser DI, in particular the role of thermal processes. However, the contrast with the data recorded on so-called integrating mass analyzers, such as a magnetic sector, quadrupole, and FTMS, suggests that radical ion production is relatively more important for the promptly formed species and hence more prominently observable in TOF-LMMS.

Phosphonium Salts. These analogues are important intermediates in organic synthesis, the Wittig reaction being the best known example. The MS characterization of the ionic intermediates is a twofold challenge. It is essential that, in addition to the detection of intact cations, the mass spectra contain fragments to permit monitoring of the numerous rearrangements, side reactions, and sometimes completely unexpected products. Soft ionization techniques such as FD-MS require the additional use of CID experiments to yield sufficient information (Veith, 1978; Sammons et al, 1975; Weber et al., 1978). In contrast, LD-MS as well as FAB-MS yield adequate molecular and structural data by means of the directly emitted ions, but multistage experiments may well remain useful or necessary to elucidate specific details in the mass spectrum.

A series of quaternary triphenylphosphonium salts have been investigated during our study of organic polyfunctionals by TOF-LMMS (Van Vaeck et al., 1986a, b; Claereboudt et al., 1987). Apart from the straightforward

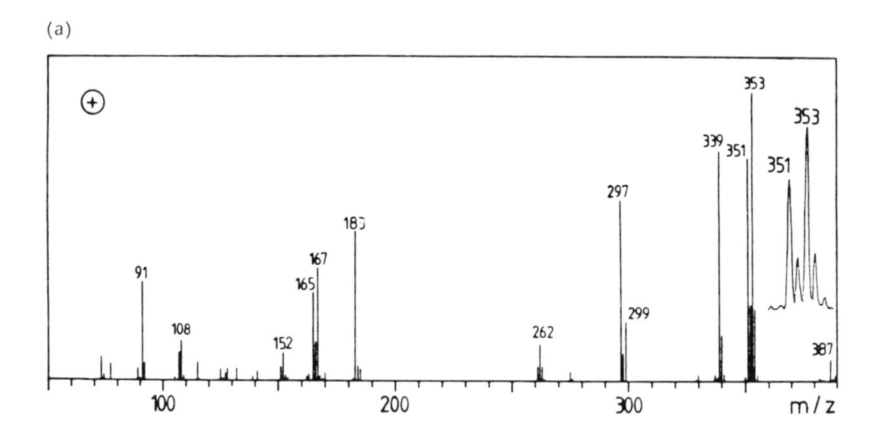

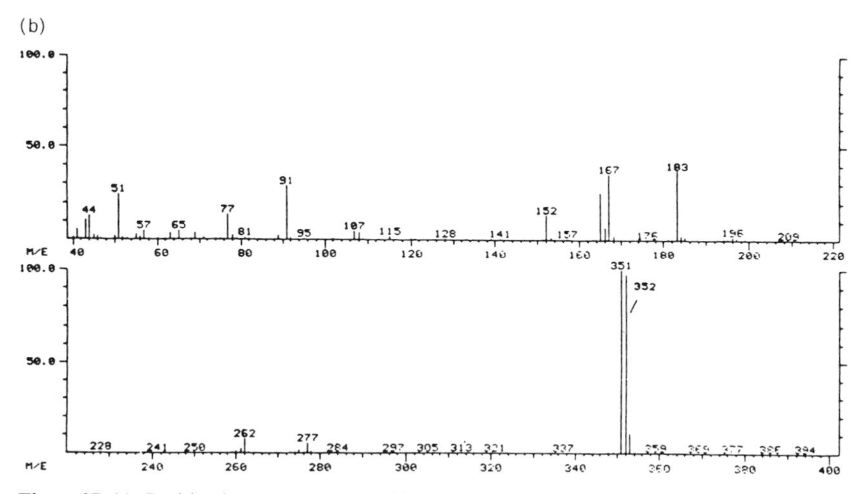

Figure 3B.44. Positive ion mass spectra of benzyltriphenylphosphonium chloride recorded by TOF-LMMS (a) and EI-MS (b). Reprinted from Claereboudt et al. (1987) with permission of Elsevier Science Publishers.

interest in the characterization of the ionic intermediates during the synthesis of conducting polymers, the mass spectra from these analogues yield interesting information about the laser-induced DI process. Figure 3B.44 compares the positive ion mass spectra of benzyltriphenylphosphonium chloride, recorded by TOF-LMMS and EI-MS with resistive heating on a direct probe (Claereboudt et al., 1987). The parent region of the latter mass spectrum contains abundant signals, due to $(C-H)^{+\cdot}$ and $(C-H-H_2)^{+\cdot}$, whereas the laser microprobe yields a base peak for the intact cation. The lower mass fragments below m/z 275 are essentially similar, and the corresponding structural assignment in Figure 3B.45 is readily rationalized (Claereboudt et al., 1987). Significant differences exist between both methods in respect of the mid-mass-range ions, where the conventional technique is not very informative. The laser microprobe data contain abundant signals associated with the tetraphenylphosphonium ions and the attachment of Cl to the triphenylphosphonium moiety. Both types of ions are typically affected by the locally applied energy density regime and are negligible for other analogues (Van Vaeck et al., 1986a,b). Threshold values depend on the structure and lattice properties of the analyte. Depending on the analogue actually studied, more or less energy can be required. As a result, according to the empirical model for DI, these ions have been associated with generation of a more densely populated

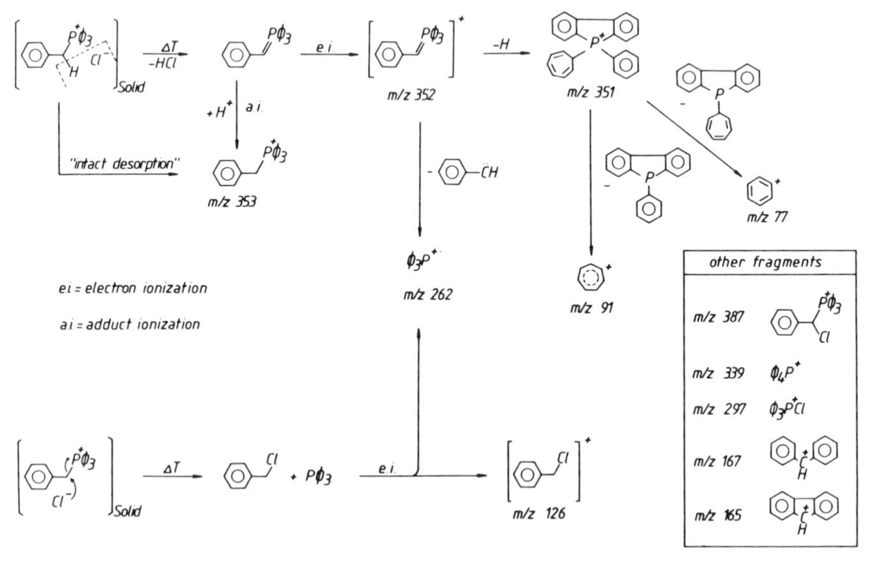

Figure 3B.45. Structural assignment and possible formation of the major ions in positive ion laser microprobe mass spectra of benzyltriphenylphosphonium chloride. Reprinted from Claereboudt et al. (1987) with permission of Elsevier Science Publishers.

selvedge, where additional interactions occur. In spite of the fact that even-electron ions prevail in the laser microprobe mass spectrum, minor signals still refer to radical systems, in particular at *m/z* 262 and 152. Questions as to possible violation of the even-electron rule, direct photofragmentation, and/or unimolecular decomposition of $M^{+\cdot}$ generated from the neutral decomposition product ($C-H$) cannot be answered unambiguously using the TOF-LMMS data, but additional information can be obtained from analysis of analogues with polyaromatic systems.

Figure 3B.46 illustrates the positive ion mass spectrum of 9-anthracenylmethyltriphenylphosphonium chloride (Van Vaeck et al., 1990). Clearly the relative contribution of $(C-H)^{+\cdot}$ to the total ion current in the parent region of the mass spectra greatly exceeds the possible fraction carried by the intactly emitted preformed ions. It should be remembered that the peak at *m/z* 453 actually comprises two and most likely three contributions. First of all, the ^{13}C contribution from the signal at *m/z* 452 accounts for about 50% of the observed peak intensity. The remaining fraction probably includes a substantial number of adduct ions from the protonation of ($C-H$) in the selvedge. Again, the data parallel the conclusion from the previous comparison of tryptamine derivatives in the hydrochloride and free-base form. However, use

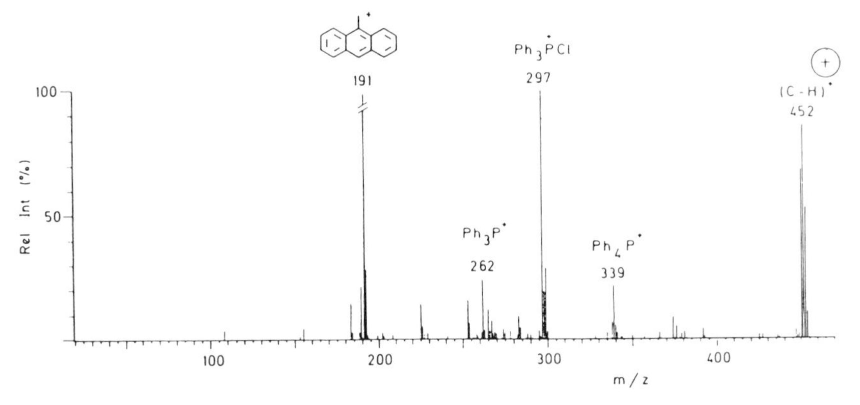

Figure 3B.46. Positive ion mass spectrum of anthracenylmethyltriphenylphosphonium chloride recorded by TOF-LMMS. Reprinted from Van Vaeck et al. (1990) with permission of Springer-Verlag.

of the intact cations vs. the radical molecular ions from the ($C-H$) neutral decomposition product as actual precursors for the lower mass fragments may be questioned. The key argument concerns the detection of such radical structures, unless the even-electron rule does not apply in these particular cases.

Relevant data in this respect have been obtained from investigation of several tetraalkylphosphonium salts by LD-FTMS and FAB ionization on a four sector instrument (McCrery et al., 1985). Additional CID experiments have been performed on both setups. As far as the direct ion formation under laser irradiation is concerned, the LD mass spectra show features essentially similar to those seen with TOF-LMMS, even though a 10.6 μm instead of the 266 nm laser wavelength is used. The base peak issues from the intact cations, the major signals refer to the triphenylphosphine moiety and the corresponding phosphofluorene fragment, while the stepwise breakdown of the aliphatic chain yields additional daughters in the high-mass range. Low-energy CID in the FTMS cell of course favors the remote-to-charge-site fragmentations, as opposed to the high-energy collision experiments on the FAB-generated ions. Figure 3B.47 presents the daughter ion mass spectrum after CID of the parents from *n*-pentyltriphenylphosphonium salts. The small but still significant signal at m/z 262 refers to the detection of radical ions with triphenylphosphine structure. The major signal at adjacent mass obviously points to the same

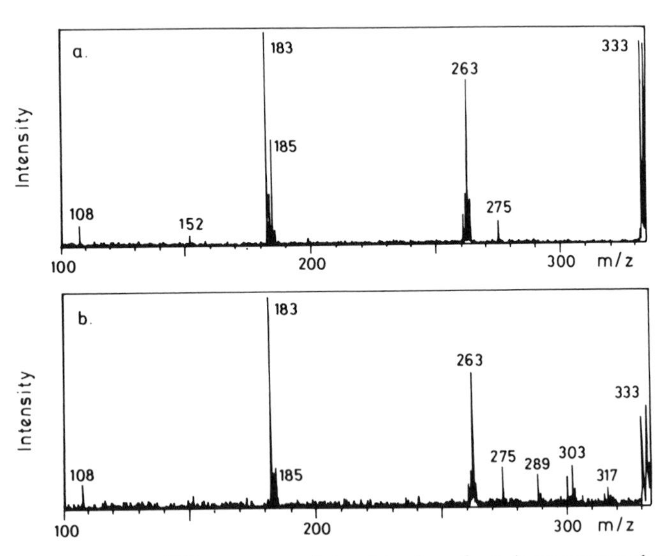

Figure 3B.47. Low-energy CID spectrum of daughter ions from laser-generated *n*-pentyltriphenylphosphonium cations. Spectra are obtained using helium (a) and argon (b) as collision gas. Reprinted from McCrery et al. (1985) with permission of the American Chemical Society.

Table 3B.9. FAB and LD Positive Mass Spectra of (Butyl)triphenylphosphonium Bromides

Compound	Relative Abundance at m/z									
$[C_4H_9P(Ph_3)]Br$	319	317	303	289	275	263	262	199	185	183
n-Butyl (FAB)	100	1.9	—	2.0	1.8	0.8	1.8	3.1	2.9	6.4
n-Butyl (LD)	100	7.6	—	16.1	7.0	12.5	8.5	—	12.5	43.8
Isobutyl (FAB)	100	2.3	1.7	—	1.8	1.3	1.2	2.0	2.4	6.6
Isobutyl (LD)	100	8.2	19.6	—	15.0	62.2	8.7	—	20.1	45.7
sec-Butyl (FAB)	100	1.7	1.2	12	—	7.4	2.5	—	3.9	11.0
sec-Butyl (LD)	52.6	6.3	—	—	—	100	4.5	—	24.7	—

Source: Reprinted from McCrery et al. (1985) with permission of the American Chemical Society.

moiety in protonated form. Generation of radical daughter ions from the even-electron precursor indeed occurs under these conditions. It is not clear, however, to what extent the fragmentation of a collisionally excited parent can be compared directly to the behavior of the directly emitted cations.

Comparison of the directly formed ions at *m/z* 262 and 263 in LD and FAB reveals that the protonated form prevails over the radical form in laser irradiation as opposed to FAB of the solid. At the same time, fragment ion intensity is lower, as shown in Table 3B.9 (McCrery et al., 1985). It is concluded that the internal energy deposited in the FAB-generated ions is centered around lower values but still exhibits a tail at high energy, responsible for the observed fragments. The distribution of laser-desorbed ions includes a relatively larger fraction of ions above the fragmentation threshold and specifically promotes rearrangement into *m/z* 263 even-electron systems. The situation is schematically illustrated in Figure 3B.48 (McCrery et al., 1985). Additionally,

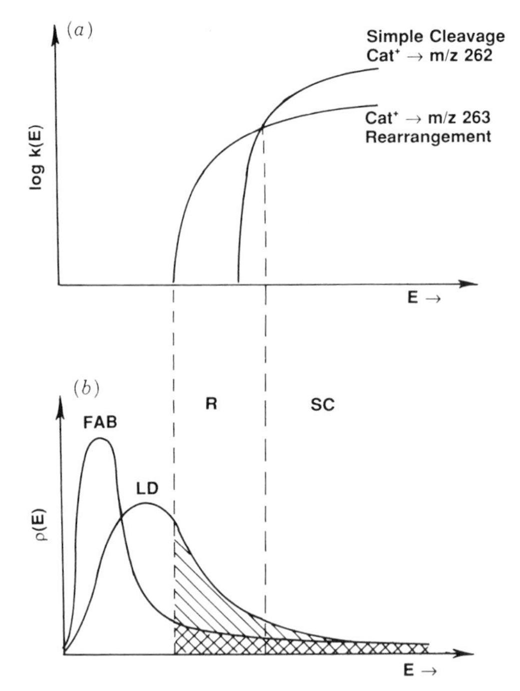

Figure 3B.48. (a) Relationship of the decomposition rate, $k(E)$, for the rearrangement and cleavage processes to the internal energy E of a decomposing cation. (b) Proposed schematic of the probability function $\rho(e)$, describing the distribution of the internal energies E of the phosphonium ions made in LD- and FAB-MS. The regions labeled R and SC represent internal energies at which the rearrangement or simple cleavage is dominant, respectively. Reprinted from McCrery et al. (1985) with permission of the American Chemical Society.

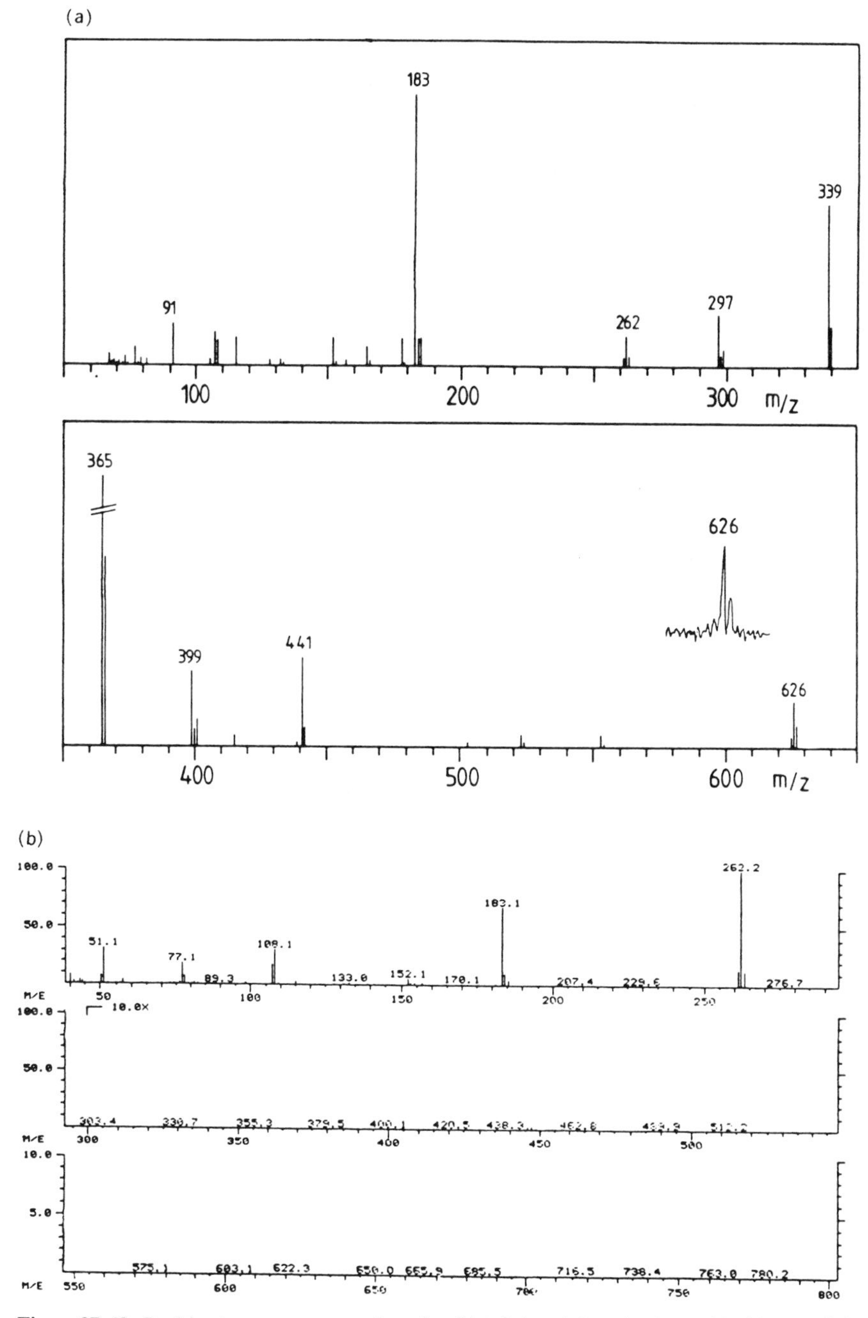

Figure 3B.49. Positive ion mass spectra of *p*-xylenebis(triphenylphosphonium) chloride recorded by TOF-LMMS (a) and EI-MS (b). Reprinted from Claereboudt et al. (1987) with permission of Elsevier Science Publishers.

a different extent of fragmentation is shown by comparison of LD-MS results from a phosphonium salt deposited on smooth and rough probe surfaces by pipetting or electrospraying a solution. The data have been explained by the existence of hot spots, particularly when rough substrates are used. Relatively more excited cations, which undergo fragmentation in the selvedge region, are then formed (McCrery and Gross, 1985b). Conceptually, the interpretation comes very close to the approach followed in the empirical model for DI.

Note again that the intact cations without further fragment peaks have been detected from thermal evaporation of such analogues as tetramethylphosphonium and ethyltriphenylphosphonium salts, deposited on an electrically heated wire (Stoll and Röllgen, 1980). No thermal evaporation of negative ions has been observed. A comparison of LD-MS with SIMS reveals that the former method produces less but more intense and structurally relevant fragments, whereas the primary ion bombardment leads to a stepwise breakdown of the aliphatic substituents (Hand et al., 1989).

TOF-LMMS analysis of diphosphonium salts has been attempted within our group (Claereboudt et al., 1987). Figure 3B.49 compares the results with those from conventional EI-MS with direct probe introduction. Surprisingly,

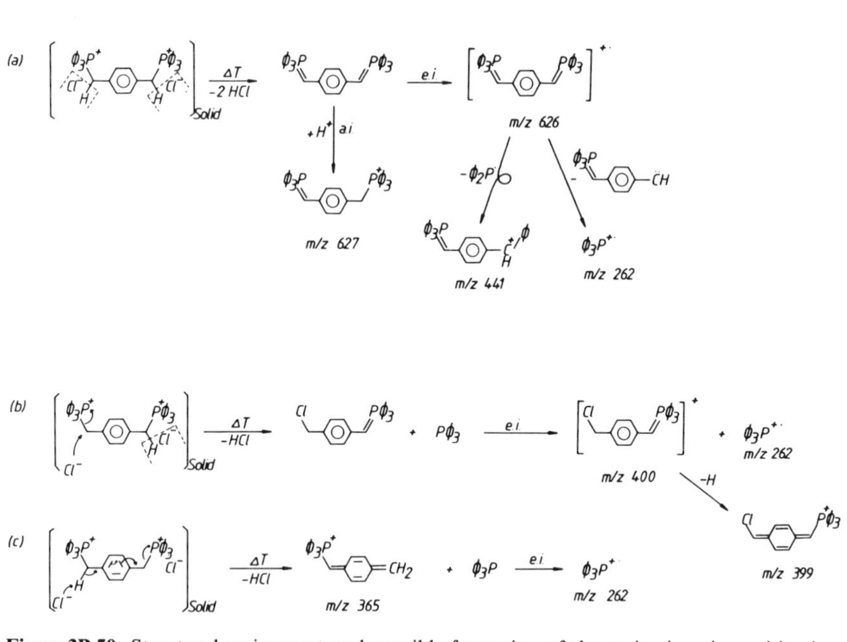

Figure 3B.50. Structural assignment and possible formation of the major ions in positive ion laser microprobe mass spectra of *p*-xylenebis(triphenylphosphonium) chloride. Parts (a), (b), and (c) demonstrate the three possible ways of decomposition. Reprinted from Claereboudt et al. (1987) with permission of Elsevier Science Publishers.

the major peak in the parent region refers to an odd-electron system corresponding to $(C-2H)^{+\cdot}$. The formation and structural assignment of the remaining signals is summarized in Figure 3B.50 (Claereboudt et al., 1987). The information gained from conventional MS is rather disappointing, and that from TOF-LMMS certainly seems more promising. However, the compound illustrated actually marks the limit of analysis under threshold conditions. No higher analogues can be dealt with, as resolution and mass determination suffer from the effects of delayed ion formation. Although we have had only limited experience with such higher analogues, we tend to think that the use of so-called integrating mass analyzers would profitably extend the working range.

Salts of Carboxylic Acids and Phenols. Most reports in the literature simply describe the results of studies with selected salts. However, a comparison of mass spectra of carboxylic acids and phenols in salt form with those from corresponding neutral analogues might well be a highly interesting topic in view of the generally accepted notion that ionic compounds are better suited to direct DI from solids than the corresponding neutral molecules. To our knowledge, no really systematic study has been published on this issue. A few experiments have at least given some useful clues. We shall first describe two test cases from our own TOF-LMMS research.

Figure 3B.51 presents the positive and negative ion mass spectra from closantel in the salt form (Van Vaeck et al., 1989c). Data for the corresponding neutral analogues have been discussed previously in connection with Figure 3B.26. Whereas the neutral molecule permits fine structural analysis from the fragments in the positive ion detection mode, the corresponding salt yields primarily NaI- and NaCl-related clusters. There is no organic information from the cations, whereas the negative mass spectra only contain a signal from the preformed ions, with no apparent fragments. The observed tendency to produce primarily "inorganic" cluster ions is typical for ionic derivatives, in particular the polyfunctional analogues in MW range above 400. Additionally, threshold irradiance values tend to become higher for the ionic derivatives. Similar observations apply to the TOF-LMMS analysis of amidotrizoate (Heinen, 1981).

A quite interesting observation stems from the TOF-LMMS analysis of 1-*N*-benzyl-3-carbomethoxy-4-piperidone in the neutral and ionic form (L. Van Vaeck, unpublished results). No particular problems arise with positive or negative ion mass spectra from the neutral form. The protonated and deprotonated molecules are detected, and the extensive fragmentation follows the pattern we would expect from other results within our data base. In contrast, the parent peak of the positive ion mass spectrum from the salt cannot be calibrated adequately against carbon foil, which is used as a

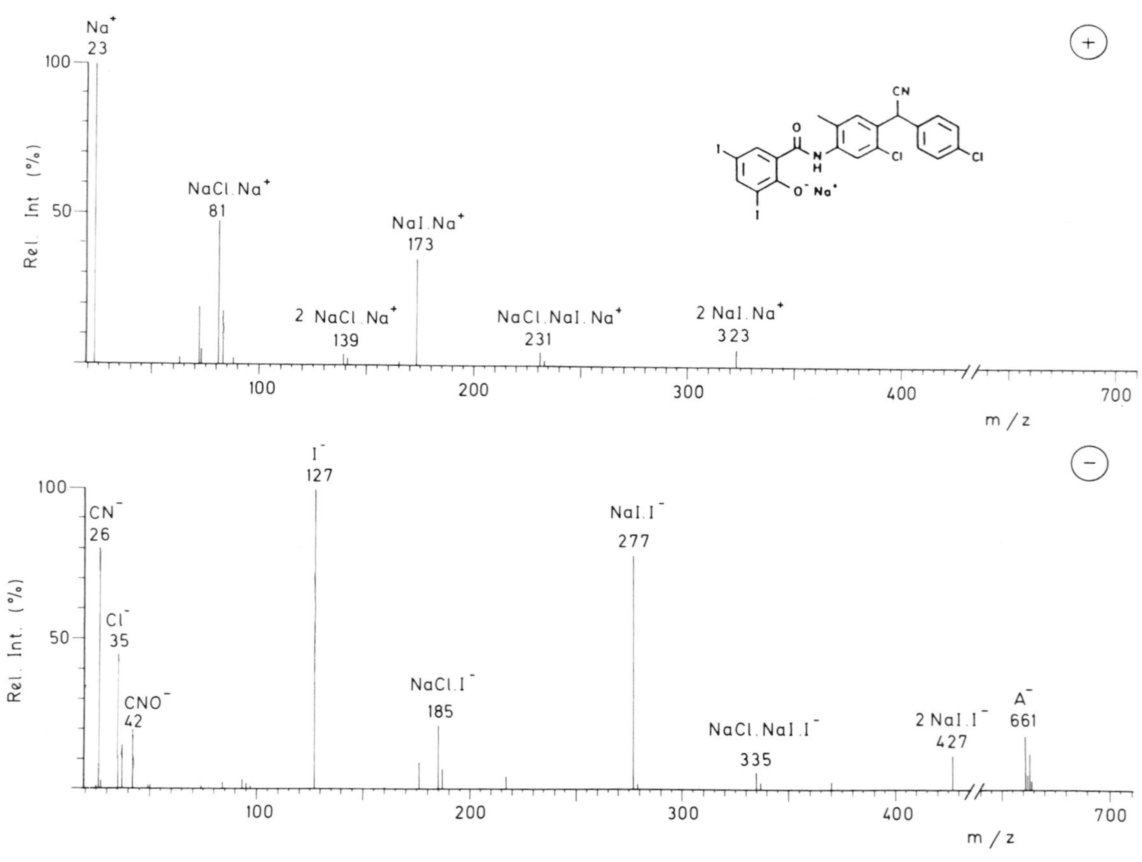

Figure 3B.51. Positive and negative ion mass spectra of closantel, analyzed in the ionic form by TOF-LMMS (MW of intact anions 661). Reprinted from Van Vaeck et al. (1989c) with permission of John Wiley & Sons, Ltd.

reference throughout our entire set of results. The signal lies about 1 m/z unit higher than the nominal value for $(A + 2Na)^+$ when the commonly used accelerating voltage of 1.4 kV is applied. Increasing the extraction field gives an additional shift upward. A correction of the m/z scale calibration using the Na^+ signal within the same individual spectrum permits us to consistently obtain a mass value of the parent peak that comes close to the nominal value of the doubly sodiated anions. Surprisingly, fragments calibrate accurately on carbon foil without correction.

The situation is schematically shown in Figure 3B.52 (L. Van Vaeck, unpublished results). In our opinion, the practical interpretation is fairly straightforward. The calibration with carbon foil fixes a given time window of the ion formation during, or shortly after, the laser pulse irradiation. Apparently, the positive fragments are compatible with this reference situation, whereas the calibration problem with the positive parents reflects a delayed formation mechanism, which still remains in phase with the Na^+ emission. As a result, it seems reasonable to associate the formation of $(A + 2Na)^+$ with gas phase or selvedge recombination of the neutral analyte. The problem is aggravated whenever salts require application of higher irradiances, which is often the case in comparison with the neutral analogues. In fact, this problem is representative of those arising with the majority of salts we have analyzed so far. The same applies to nucleotides in the acidic and salt form (Van Vaeck et al., 1989d).

In conclusion, it should be noted that LD permits generation of organic ions from salts but that ion formation is not always adequately timed to suit TOF-MS unless drawout pulses are applied. The latter statement depends on the structure. It is certainly not true that, under the specific conditions and limitations imposed by TOF-LMMS, ionic salts generally represent an asset for analysis. The information content seems lower, and the analysis at low irradiance is harder to perform.

However, laser irradiation of organic salts in conjunction with the so-called integrating type of analyzers, for instance, magnetic sector, quadrupole, or FTMS, still shows similar trends. The positive ion mass spectra of squaric acid as the sodium salt primarily contain $(M + H)^+$ and $(M + Na)^+$ signals but no structural fragments as opposed to those from the neutral analogue except when ammonium salts are measured. The negative ion mode mass spectra of the neutral form and the salt contain equivalent information in the parent region, but in both cases there are no fragments apart from C_2O^- and C_2HO^- (Byrd et al., 1986).

Several reports mention the LD-MS and LMMS mass spectra from selected salts belonging to miscellaneous classes (Rosmarinowsky et al., 1985; Heinen et al., 1980; Heinen, 1981; Posthumus et al., 1978). Identification of organic salts in tissues on a micrometer scale has been reported (Verbueken et al., 1987).

·LMMS : TIME = MASS !

Calibration of m / z - scale :

external = constant, from e.g. carbon foil

internal = variable, correction by e.g. Na^+

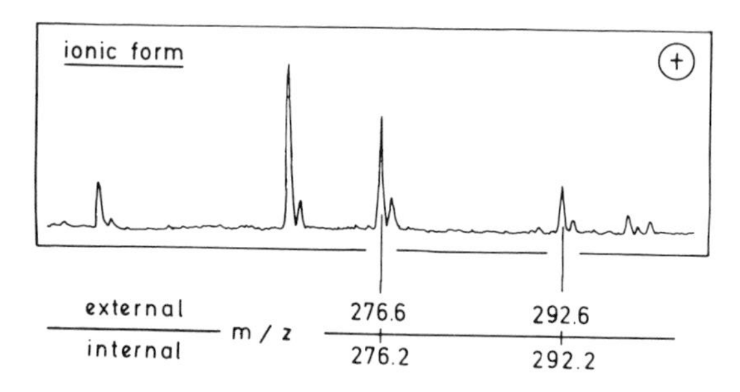

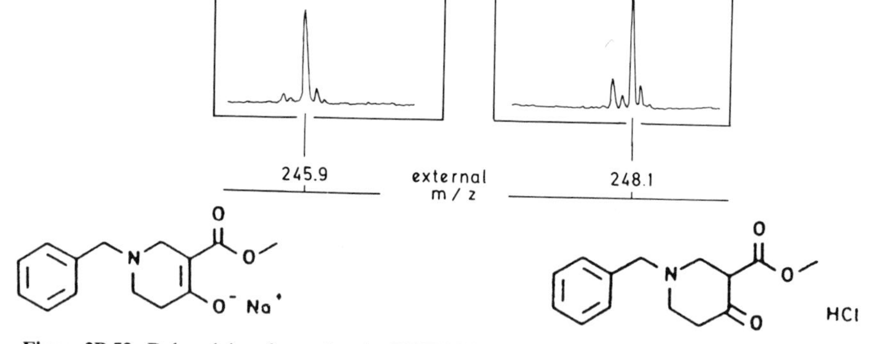

Figure 3B.52. Delayed ion formation in TOF-LMMS, evidenced by the parent ions in the negative mode from 1-*N*-benzyl-3-carbomethoxy-4-piperidone salt, contrasted with the ions from the neutral analogue and the negative fragments from the salt (L. Van Vaeck, unpublished results).

Sulfonate and Phosphate Salts. Study of nucleotides in the ionic form obviously represents a field of significant interest that is as yet relatively poorly documented. The initial report (Posthumus et al., 1978) on LD-MS mentions the generation of molecular and fragment information from the AMP disodium salt. Within the set of phosphate cluster ions, signals refer to the cationized nucleoside and purine moieties. As to TOF-LMMS applicability to these compounds, we found that the salts are difficult or even impossible to analyze correctly whereas the corresponding analogues yield satisfactory information. A series of eight oligonucleotides containing 3–8 units have been investigated by the matrix-assisted technique (Spengler et al., 1990). Nicotinic acid is added to the analyte in a molar ratio of 300:1 up to 1250:1 depending on the compound studied. The MWs of the analogues studied range from 900 to 1800. Intense signals from protonated molecules have been detected by TOF-MS analysis. Matrix signals are used for calibration of the m/z scale. Measured masses lie within 1 amu of the calculated values of 1000–1200.

In contrast, sulfonates seem much better suited to LD-MS characterization. But then, the number of literature reports in this class is rather limited, apart from the initial studies on sodium hexylsulfonates (Vastola et al., 1970). Recent LD-FTMS data on benzene sulfonates have been reported in a study of the quantitative detection capabilities (Chiarelli et al., 1990). The surfactant mixtures have been analyzed directly and adsorbed to cloth samples in a concentration of $100\,\mu g.g^{-1}$, which represents a gain by a factor of 10 in comparison to the conventional analytical procedure using GC with flame ionization detection on extracts. In our systematic TOF-LMMS study of polyfunctional analogues, interesting data were obtained from a series of dyes and food additives.

Figure 3B.53 illustrates the positive and negative ion mass spectra from methyl orange. The structural assignment of the major signals is given in Figure 3B.54 (Van Vaeck et al., 1989a). The product is desorbed at very low irradiance, and the m/z determination of the parent ions in the positive mass spectra is performed without problems using external calibration on carbon foil. The fragmentation behavior of diazo compounds is not very clear, especially in the negative ion detection mode. Expulsion of the central nitrogen has been proposed, in accord with data for methyl red and additional EI-MS results (Van Vaeck et al., 1989b; Bowie et al., 1987). So far, no adequate fragmentation pathway has been elaborated for the proposed diazo cleavage, yielding fragments similar to those observed in FAB data for related analogues (Monaghan et al., 1982). The high-mass fragment at m/z 290 formally corresponds to the intact anions except for a methylene group. It is difficult to imagine a unimolecular decomposition path leading to these fragments. As a result, the formal addition of a H^+ allowing desorption of the neutral form and the subsequent ECI in the selvedge, followed by cleavage of a methyl radical, becomes,

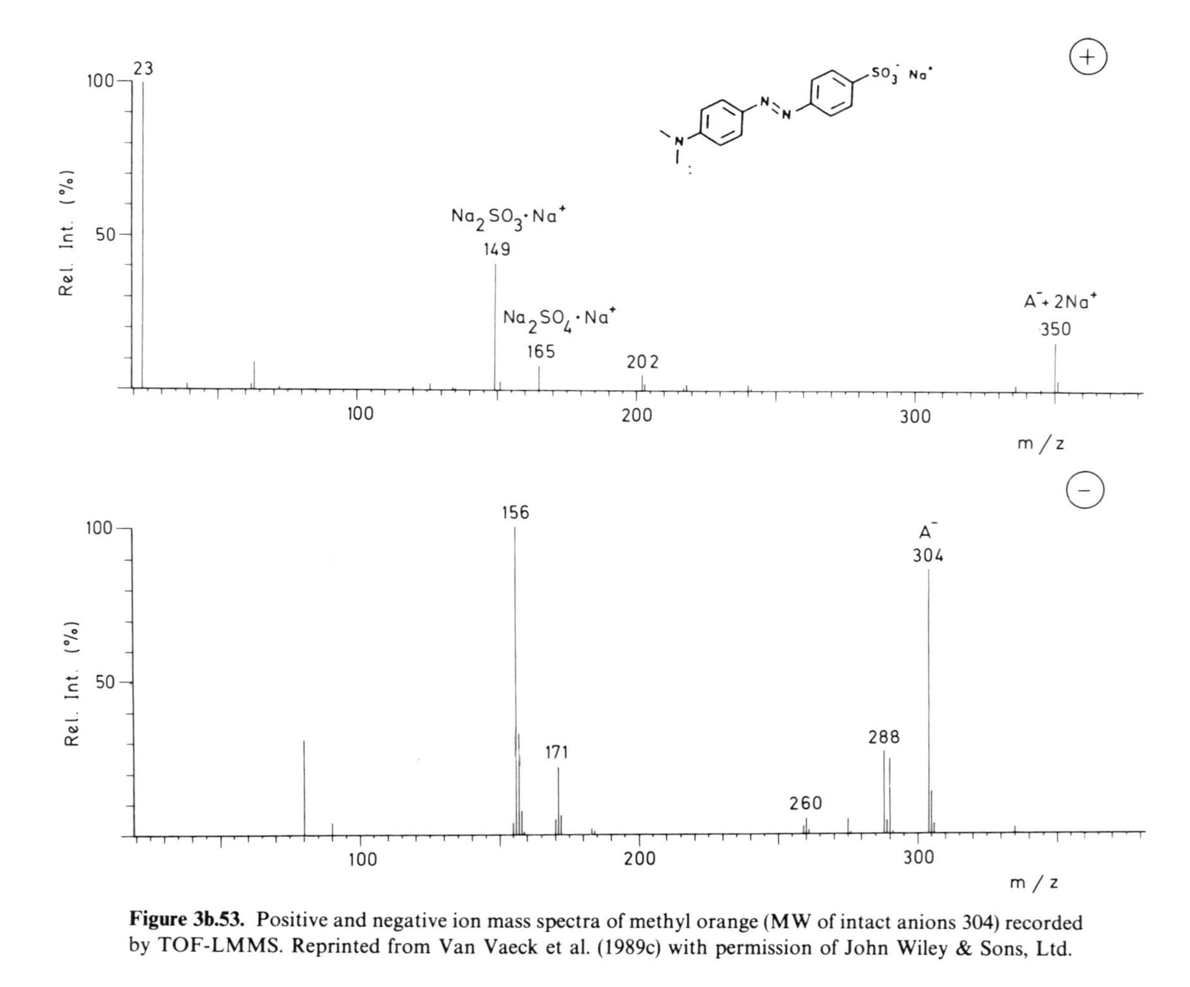

Figure 3b.53. Positive and negative ion mass spectra of methyl orange (MW of intact anions 304) recorded by TOF-LMMS. Reprinted from Van Vaeck et al. (1989c) with permission of John Wiley & Sons, Ltd.

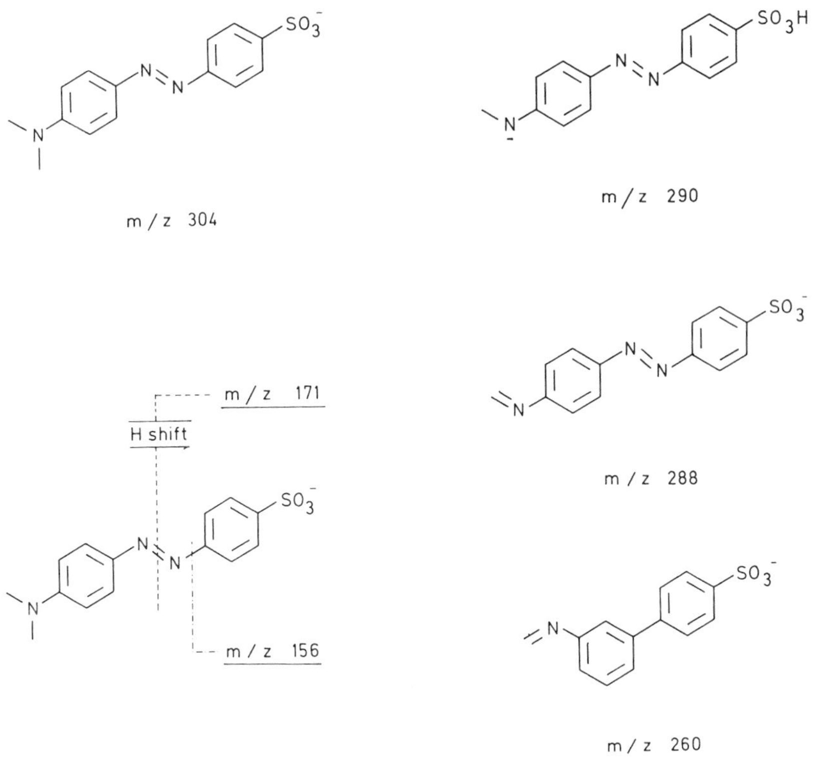

m / z 304

m / z 290

m / z 171

H shift

m / z 288

m / z 156

m / z 260

Figure 3B.54. Structural assignment of the negative fragment ions from methyl range. (Reprinted from Van Vaeck et al. (1989c) with permission of John Wiley & Sons, Ltd.

in our opinion, a more plausible alternative. The occurrence of third-body collisions in the selvedge region may contribute to stabilization of the radical anions. Dissociative ECI can be involved as well. It should be remembered that the empirical model does in practice not distinguish ECI and subsequent fragmentation, on the one hand, and the dissociative process, on the other.

It is interesting to observe the formation of radical cations from salts with organic anions. The corresponding signals from the odd-electron fragments show particularly substantial intensity in the TOF-LMMS mass spectra of phenol red. The highly aromatic nature of the molecule favors the formation of radical systems (Van Vaeck et al., 1989c). Again, this observation can be explained straightforwardly by the assumption of neutral intermediates in the selvedge region where electron ionization, by electron abstraction or attachment, may occur, but the empirical approach does not exclude the possibility of a contribution from direct photofragmentation.

The presence of several anionic functional groups prevents the detection of

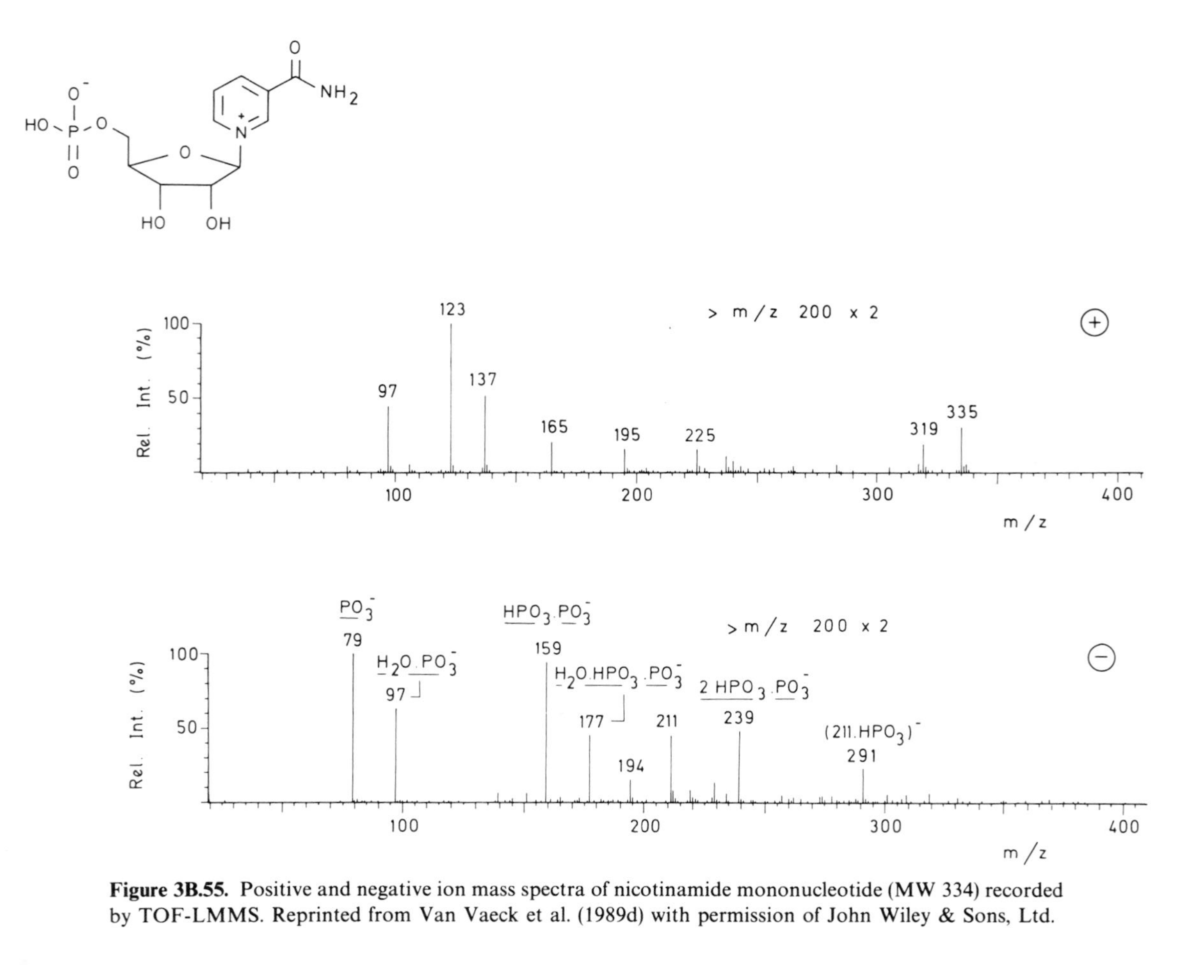

Figure 3B.55. Positive and negative ion mass spectra of nicotinamide mononucleotide (MW 334) recorded by TOF-LMMS. Reprinted from Van Vaeck et al. (1989d) with permission of John Wiley & Sons, Ltd.

intact molecules. Compounds such as tartrazine undergo cleavage and yield only a very limited number of structure specific fragments (Van Vaeck et al., 1990). In fact, the situation becomes identical to that observed for the diquaternary ammonium salts (Cotter and Tabet, 1983; Dang et al., 1984b).

Internal Salts. Strictly speaking, the amino acids can be considered as internal salts—but these compounds have been discussed separately in an earlier subsection. The quaternary ammoniocarboxylic acids will be included here together with other analogues containing different functional groups to stress the potential and characteristics of the LD-MS for diagnostic analysis of internal salts.

Figure 3B.55 contains the positive and negative ion mass spectra recorded by TOF-LMMS from nicotinamide mononucleotide (Van Vaeck et al., 1989d). Proton attachment yields a parent signal with appreciable intensity in the positive ion detection mode. In addition to the MW information, the structural moieties and functional groups are clearly recognizable from the main fragments in the postive ion mass spectra. Figure 3B.56 summarizes the tentatively proposed assignment for the major signals of diagnostic interest. For the sake of brevity no explicit fragmentation pathways are shown, but reasonable decomposition routes can readily be deduced using commonly applicable concepts. A stepwise breakdown of the molecule, starting with the elimination of the phosphate functionality, makes it easy to work back from the positive mass spectrum to the original structure. The negative ion mode mass spectra primarily contain phosphate cluster–type signals. Only two peaks with modest intensity refer to specifically organic fragments. Nevertheless, the mass spectra remain extremely useful, indicating the presence of a phosphorylated sugar, directly evident from the data in the positive mode apart from the high-mass signal. As a result, the predominance of the inorganic clusters in the negative ion detection mode, which characteristically occurs in several ionic derivatives, may still represent a great deal of complementary information (Van Vaeck et al., 1989d).

Another illustration of internal salt analysis by TOF-LMMS is given by the positive and negative ion mass spectra from miconazole methacrylate in Figure 3B.57 (Van Vaeck et al., 1989a). Note that the number above the signals denotes the highest signal within the group whereas the *m/z* value of the fragments given under the structures corresponds to those with only ^{35}Cl. The presence of several chlorine atoms in the molecule and the tendency for ring formation during the fragmentation of this particular molecule necessitate detailed analysis of the isotope patterns. The positive mass spectrum only contains five groups of more or less intense signals but still yields a wealth of information about the original molecule. The MW of the compound is available from the signal for the protonated structure, but the combination of

m/z 335 m/z 225 m/z 195

m/z 165 m/z 137 m/z 123

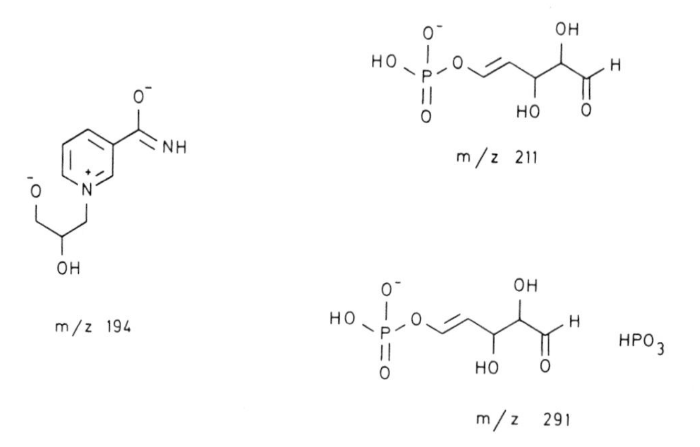

Figure 3B.56. Structural assignment of the major fragments in mass spectra of nicotinamide mononucleotide. Reprinted from Van Vaeck et al. (1989d) with permission of John Wiley & Sons, Ltd.

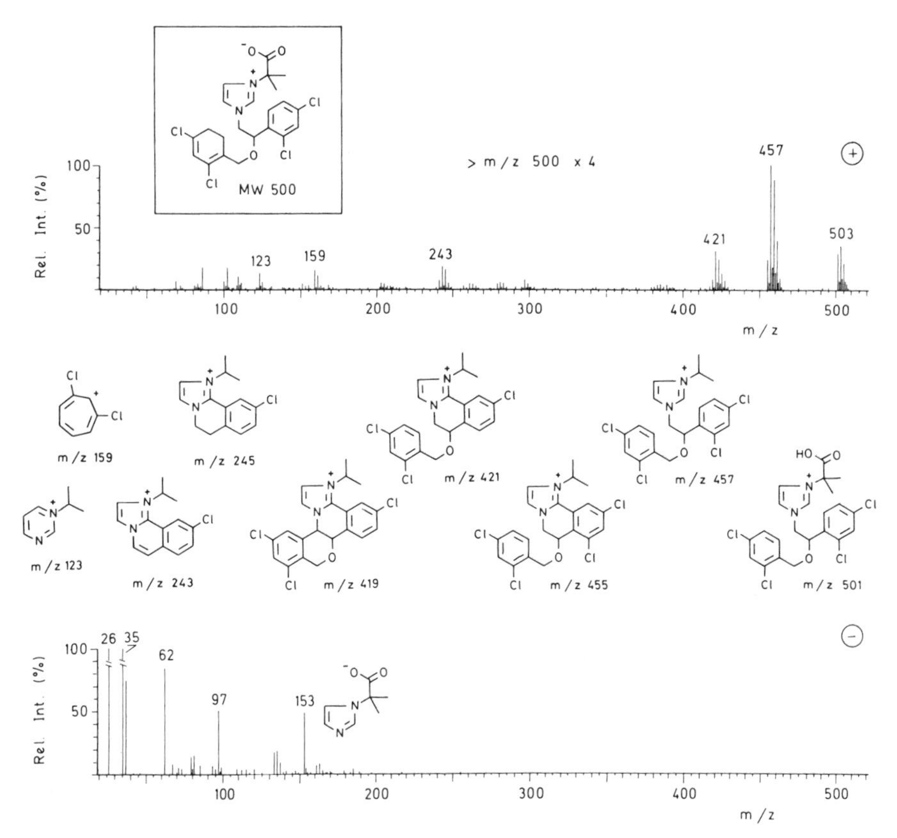

Figure 3B.57. Positive and negative ion mass spectra of miconazole methacrylate (MW 501) recorded by TOF-LMMS. Reprinted from Van Vaeck et al. (1989a) with permission of John Wiley & Sons, Ltd.

fragments from the positive ion mass spectra does not permit us to specify the location of the methacrylate group in the molecule. In principle, this functional group might be present on the imidazole system or somewhere else on the ether chain. The negative ion mode mass spectra provide the "missing" information, as well as an additional illustration of the striking complementarity that often exists in the distribution of structural data between positive and negative ions in TOF-LMMS.

TOF-LMMS studies have been performed on quaternary ammoniohexanoates, hydroxyammonioalkanoates, and sultaines with various aliphatic substituents between methyl and dodecyl in addition to two methyl groups on the nitrogen (Balasanmugan and Hercules, 1983b, 1984). The molecular structure of sultaines corresponds to 3-dimethylalkylammoniumpropanesulfonate. Intermolecular alkyl transfer is observed, yielding relatively strong $(M+CH_3)^+$ and $(M-CH_3)^-$ signals. It is suggested that the process involves a concerted mechanism within a bimolecular complex. However, the extent of alkyl group transfer clearly depends on the application of relatively high laser power irradiation conditions. The presence of a proton donor, such as for small amounts of *p*-toluenesulfonic acid added to the sample, almost completely suppresses the phenomenon. Cleavage of the long chain from the quaternized nitrogen yields the major fragments.

An interesting comparison has been carried out using SIMS, LD-, PD-, FAB-, and FD-MS on a series of sultaines (Busch et al., 1982). All techniques show parent ions corresponding to $(M+H)^+$ and fragment ions corresponding to $(M+H-SO_3)^+$, $(M+H-H_2SO_3)^+$, and $(M+H-C_3H_7SO_3H)^+$. The negative fragments below $(M-CH_3)^-$ are similar in all techniques. However, alkyl transfer occurs to a minor extent only in SIMS and PD-MS as opposed to TOF-LMMS, whereas the process is not observed in FAB- or FD-MS. Cationization only occurs in SIMS and PD-MS; proton attachment prevails in the other methods. The formation of higher organic clusters containing up to 14 sultaine molecules is observed in the positive ion mode results from FAB-MS. Cationized combinations of several molecules, up to three, occur in PD-MS and SIMS. The degree of fragmentation differs from technique to technique. FAB-MS produces abundant signals in the low-mass range, but the specificity of the fragments is poor. PD-MS, SIMS, and LD-MS yield the most useful mass spectra for detailed diagnostic analysis within this series of compounds.

Organic Complexes. One of the presumed advantages of laser irradiation of solids is the potential to obtain either organic or inorganic information, or both, depending on the local energy regime actually applied. In fact, the TOF-LMMS technique has raised high expectations as regards the possible achievement of a total analysis at the micrometer level. It is not surprising that the laser microprobe and related bulk techniques have emerged as possible methods of choice for the MS characterization of organometallic compounds, constituting the interface between organic and inorganic chemistry.

A lot of the systematic work on TOF-LMMS of organometallic compounds was achieved by Muller's group at the University of Metz, France, where the development of an FTMS-based laser microprobe was also pioneered later. Various organometallics have been studied, including phenylbutanedionate, tetrathiacyclotetradecane and undecane, bipyridyl, coronate, and bicyclic cryptand as ligands associated with transition metals (Muller et al., 1981, 1984). This selection covers a rather broad range of structures, which prevents us from giving a detailed description of the characteristic features observed in the corresponding mass spectra. However, in general, significant signals in the positive ion mass spectra refer to an intact entity of cation and ligand, complemented if applicable with counterions to obtain the most straightforward singly charged combination. Lower mass range signals include the fragments issuing from the subsequent loss of the ligands and/or counterions included in the parent species. Also the protonated and cationized ligand molecules yield major signals. As a result, the set of peaks in the mass spectrum displays the major building blocks of the complex in a stepwise manner. This behavior is of course related directly to the nature of the analyte, while the advantage of laser irradiation resides in the potential to generate intact metal–ligand ions from the solids. The positive ion mass spectra for copper bipyridyl complexes in Figure 3B.58 illustrate the capability to distinguish the linkage of the cation to one or two ligands (Muller et al., 1981). Moreover, intense peaks give the relevant information in a very simple and direct way.

However, the actual energy regime imposed on the sample during laser irradiation is critical for the information content of the recorded mass spectra. At threshold irradiance, large clusters containing several metal–ligand combinations can be observed. Increase of the power density on the sample reduces the information to the monomeric ions. Still higher irradiances reduce the organic ion intensity and favor metal ion detection. One of the problems is that the macroscopically measured energy per pulse is not fully adequate to

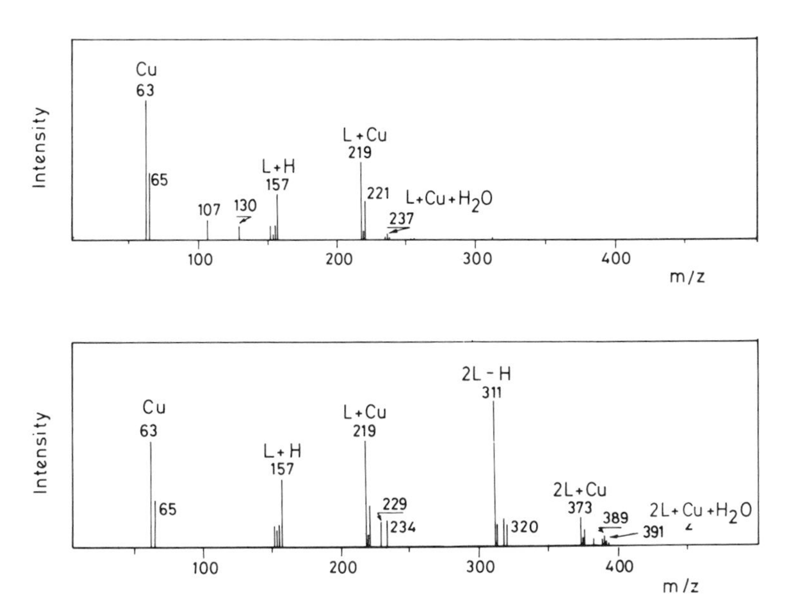

Figure 3B.58. Positive ion mass spectra of copper bipyridyl (bipyr) complexes recorded by TOF-LMMS. *Top*: [bipyr $Cu(H_2O)_2$]SO_4; *bottom*: [$(bipyr)_2Cu$][ClO_4]$_2$. Reprinted from Muller et al. (1981) with permission of Springer-Verlag.

describe the local energy regime imposed on the sample within the irradiated spot. Hence, data recorded at the lowest possible threshold irradiance in different laboratories show significant differences (Heinen, 1981; Muller et al., 1981).

Additional experimental flexibility arises when the 266 nm irradiation is substituted by a tunable dye laser system. Although the achievement of real resonance conditions in the solid state is not entirely obvious, the relative sensitivity of the elemental ions generated from Cd acetonylacetonate (acac) increases by at least a factor of 5 in comparison to the signal from the ligand–metal combination (Rohly et al., 1984). A series of additional bipyridyl and *o*-phenanthroline complexes with Fe, Ag, Ni, Cu, Pb, and Ba has been investigated by TOF-LMMS and essentially confirms previous results (Balasanmugan et al., 1985). Negative ion mass spectra show signals referring to ion–molecule reactions and provide additional information on the oxidation state of the central metal ion of the complex. Transition metal complexes with acac and the methylated ligand as well as with the corresponding Schiff base show that TOF-LMMS data in the positive ion detection mode largely correspond to those obtained in EI-MS for these complexes (Rohly et al., 1984). Negative ion mass spectra provide additionally sustaining information but do not really add diagnostic features. Parent anions in the MW region have been observed in the case of $VO(acac)_2$. The study of transition metal complexes with acac ligands by LD-MS/MS on a reversed geometry double-focusing magnetic instrument has yielded evidence that the parent ions actually are strongly linked entities and that the metal cations are certainly not loosely bonded to the ligand (Pierce et al., 1982).

The two-step laser desorption and subsequent resonance-enhanced MUPI of the neutrals, cooled in a supersonic jet expansion, has been successfully applied to ruthenium bipyridyl acetate (Beavis et al., 1988a). The parent peak in the positive ion mass spectra refers to the metal cation attached to the ligands. The fragments exhibit a much simpler pattern as opposed to the previous data from the one-step techniques. However, further research on the spectroscopy of the cooled species is necessary to exploit the ultimate capabilities of the approach. Recent results on the application of LD-FTMS in the field of complexes has also evinced interesting possibilities for the structural analysis of the true organometallic compounds, where the metal is attached by a sigma bond to the carbon skeleton. However, these compounds are beyond the scope of this subsection (Bjarnason, 1989).

Numerous experiments have been performed on the MS characterization of porphyrin systems without protective groups by means of virtually all soft ionization techniques. An outstanding example is the investigation of corrins, cobyric acid, and vitamin B_{12}, three closely related structures, for which a recent reviewer needed more than 150 pages to cover the experiments (Schiebel and Schulten, 1986). Obviously, the remainder of this subsection must be less extensive.

A series of cobalamins have been studied by means of TOF-LMMS (Graham et al., 1982). The mass spectra characterize the presence of the cation by means of a base peak from the elemental ions in the positive mode, while the ligand is characterized by means of $(M+H)^+$ and $(M-H)^-$ signals in the corresponding detection mode. Mass resolution is of course less than one, and uncertainty of the m/z determination is claimed to be around ± 1 amu in the parent region. As a result, distinction still can be made between the cyano-, hydroxy-, and methylcobalamins. Structure-specific fragmentation is seen in the high-mass range of the positive mass spectra, but the intensity of the corresponding signals is modest. A complex pattern of fragments is found between m/z 400 and 500, but the reproducibility is poor. The negative mass spectra contain fewer fragments, all of which are below m/z 150.

LD-TOF-MS of chlorophyll a and derivatives has been performed on a linear analyzer with the use of a drawout pulse extraction of the ions (Tabet et al., 1985). In contrast to the previous results with TOF-LMMS, irradiation is achieved by a 10.6 μm CO_2 laser. CsI is added, but cationization of the ligand by Na and K yields the major signals in the parent region. Depending on the delay between the laser pulse and introduction of the ion bunch from the source to the MS, different fragment peak patterns are observed in the mass spectra while the peak width of the parent ions becomes smaller. The ions corresponding to the loss of the phytyl chain and small neutral species decrease steadily with increasing delay times, whereas the more extensive fragmentation into relatively low- mass ions reaches a maximum after several microseconds (Tabet et al., 1985).

Two-step LD and MUPI in the gas phase with intermediate cooling of the neutrals by supersonic jet expansion has been applied to native chlorophylls (Grotemeyer et al., 1986c). The excellent mass resolution allows detailed verification of the peak pattern in the high-mass region for the possible occurrence of H transfer or abstraction during the evaporation step. Also, hard and soft ionization can be achieved on these molecules, resulting in an extremely clean and simple spectrum with only three fragment peaks in the latter mode, where the parent refers to $M^{+\cdot}$. Extensive fragmentation goes on at the cost of the molecular ion peak intensity (Grotemeyer and Schlag, 1988c). Two-step LD at 10.6 μm and MUPI at 266 nm combined with a 30 cm TOF-MS has allowed observation of $M^{+\cdot}$ in the virtual absence of fragments from only 50 fmol of porphyrins and adenine (Hahn et al., 1987).

The characterization of chlorophyll a and b by LD-FTMS of course provides a still-unsurpassed mass resolution and accuracy of the m/z determination, on the condition that the ions are long-lived. Previous experiments on the metastable decomposition, performed by TOF-MS, have raised doubt that the intactly desorbed molecules might not survive the long measurement time in FTMS. Nevertheless, the cationized or protonated ligand molecules are detected with very good signal-to-noise ratio from chlorophylls with and without KBr as dopant (R. S. Brown and Wilkins, 1986b). Fragmentation yields a relatively low number of intense signals, above *m/z* 450. The lower mass cutoff is required to improve resolution. It should be noted that the difference in time domain between TOF and FTMS is large enough to make the evidence about short-lived species, obtained from the former technique, no longer applicable. Indeed, FTMS can deal with the results of CI processes, for instance, of which the slow rate makes these phenomena undetectable by the TOF approach. There are several orders of magnitude between the duty cycle of a typical TOF instrument as compared to that of FTMS.

Several synthetic porphyrins and capped analogues have been studied by LD-FTMS (R. S. Brown and Wilkins, 1986a; Marchon et al., 1989). Surprisingly enough in view of the expectations discussed earlier, the production of intact ions from metalloporphyrin systems seems not to cause a major problem, but the generation of sufficient fragments often requires the application of CID or, more elegantly, photodissociation (Nuwaysir and Wilkins, 1989). The latter approach is especially effective in circumventing problems with low-energy collisional activation arising from the relatively low ion frequency at higher *m/z*. Alternatively, the use of surface-induced dissociation has been successfully applied (Ijames and Wilkins, 1990). Basically, the ion trap plate serves as a collision target.

3B.4. LOCAL ANALYSIS BY MEANS OF STRUCTURAL IONS

The use of lasers has introduced the perspective of real microprobe applications in the field of organic MS analysis. It should be recognized that the location of an organic target molecule within the complex environment of a heterogeneous matrix at the micrometer level still remains the ultimate goal, which will require a lot of additional research on both the fundamentals of ion formation and practical methodology. The fundamental aspect refers to possible interaction between the organic target and the accompanying constituents within the irradiated and evaporated microvolume. Apart from the obvious instrumental optimization, we have found that the analytical applicability of the method to practical problem solving largely depends on the surface availability of the targets and on the possibility of deductively interpretating the mass spectral information. The purpose of this section is to highlight some examples of work evidencing the ultimate feasibility of the old dream of achieving local organic analysis with the versatility to characterize a wide variety of molecules that an MS detector inherently offers.

The chance of detecting a target from an evaporated volume on the order of 1 μm^3 not only depends on the instrument sensitivity but also on the efficiency of ion generation from the entire amount of consumed material. Otherwise stated, when the information depth for organic structurally specific ions is less than the crater depth, surface concentration of the target must increase. It has been shown that most elemental ions that reach the detector in TOF-LMMS primarily issue from the surface (Bruynseels and Van Grieken, 1986). To define the thickness of the surface layer from which structural ions are generated, sandwich samples have been prepared consisting of microcrystalline targets between two sections of matrix-simulating material. The total thickness of the sample is kept at 1 μm. Nonporous films of polymeric material or carbon foil whose thickness is as low as 5 nm are used as the covering layer. The thickness of the substrate layer is adapted so that the target in different samples is located at different distances from the surface. The samples are perforated with up to 10 times more irradiance than is usually applied on isolated 1 μm particles in the threshold mode. The approach is illustrated in Figure 3B.59 (Van Vaeck et al., 1990).

The mass spectra in Figure 3B.60 show that the presence of even a very thin covering layer leads to a drastic decrease of sensitivity, while the mass resolution, peak shape, and position on the m/z scale are severely degraded (Van Vaeck et al., 1988a). It is clear that targets present in the lower part of the sample do not contribute to the organic ion current insofar as the latter is measured as a normal signal, i.e., an adequately resolved, symmetric, and properly calibrated peak. In practical terms, use of targets only in the uppermost layers necessitates that the target concentration in the near-surface region be about 10% (Van Vaeck et al., 1988a).

An indirect confirmation of the foregoing results is available from the detection of benzotriazole (BT) and tolyltriazole (TT), two corrosion inhibitors, on copper (Holm and Holtkamp, 1989). Figure 3B.61 shows the positive mass

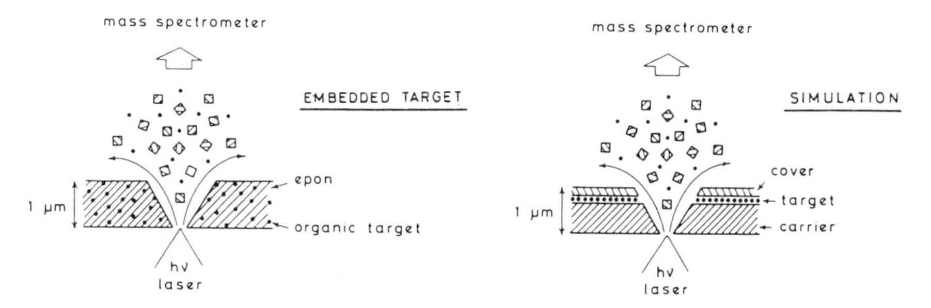

Figure 3B.59. Model sample simulation for the analysis of embedded targets by TOF-LMMS and definition of the carrier and barrier layers. Reprinted from Van Vaeck et al. (1990) with permission of Springer-Verlag.

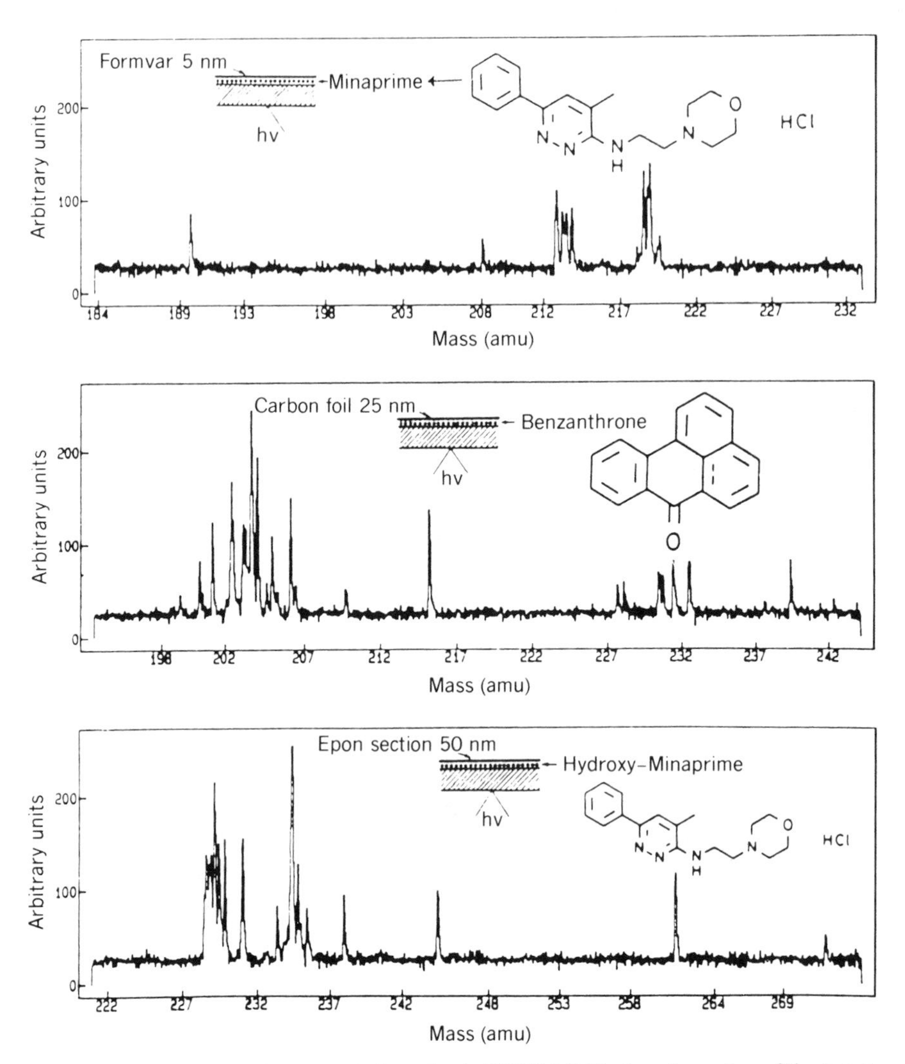

Figure 3B.60. Examples of mass peak distortion in TOF-LMMS when desorption of the target is hindered by presence of a covering layer. Reprinted from Van Vaeck et al. (1988a) with permission of John Wiley & Sons, Ltd.

spectra recorded by TOF-LMMS in reflection. The copper substrate has been immersed for only 5 minutes in tap water containing BT and TT. Auger electron spectroscopy has permitted verification that the surface is covered by maximally five molecular monolayers. Still, abundant signals are generated from a defocused 10 μm spot, containing 4×10^9 molecules. Note the avail-

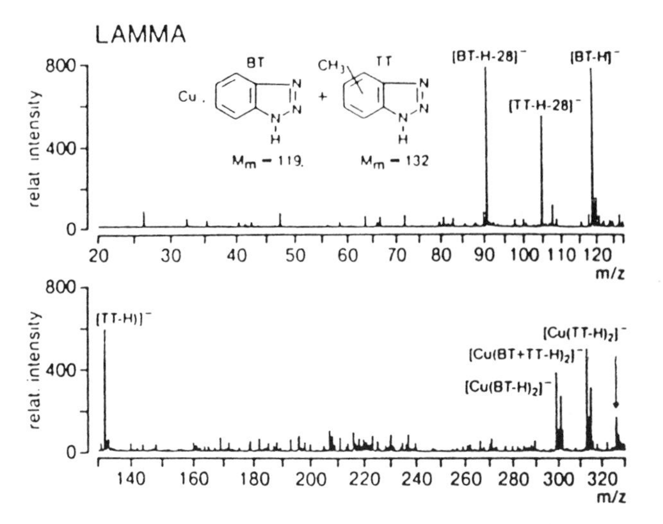

Figure 3B.61. Negative ion mass spectra of a copper substrate immersed in tap water containing 5 mg/L benzotriazole (BT) and tolyltriazole (TT), recorded by TOF-LMMS. Reprinted from Holm and Holtkamp (1989) with permission of San Francisco Press, Inc.

ability of molecular and fragment information exhibiting the characteristic loss of nitrogen from the benzotriazine moiety.

The limited information depth of TOF-LMMS renders the method impractical for analysis of more or less homogeneously distributed targets within a compact material. Local inclusions remain the most suitable applications at the present state of the art. Also cryo-sections as opposed to embedded specimens facilitate the detection. Although these conditions impose severe limitations on the range of possible applications, there still remains a variety of practical situations where organic target characterization at the micrometer level in complex matrices can be successfully realized.

The detection of herbicides on leaf samples and glyceollin at the cellular level in infected soybean samples has demonstrated the feasibility of organic target location within the complex environment of biological materials (Morelli and Hercules, 1986; Kelland et al., 1987; Moesta et al., 1982). Following identification of naturally occurring pigments in microlichen, we attempted the study of phytoalexin accumulations in the vessels of infected carnation plants (Mathey et al., 1987, 1989). Semipreparative isolation of the targets and structure determination was performed prior to TOF-LMMS analysis. Longitudinal and transverse sections of the vessels were analyzed for the presence of dianthalexine (DX) and methoxydianthramide S (MDS). Figure 3B.62 shows the structure of the targets, the fluorescence micrograph of a section, and the corresponding mass spectra (Mathey et al., 1989). Apart from

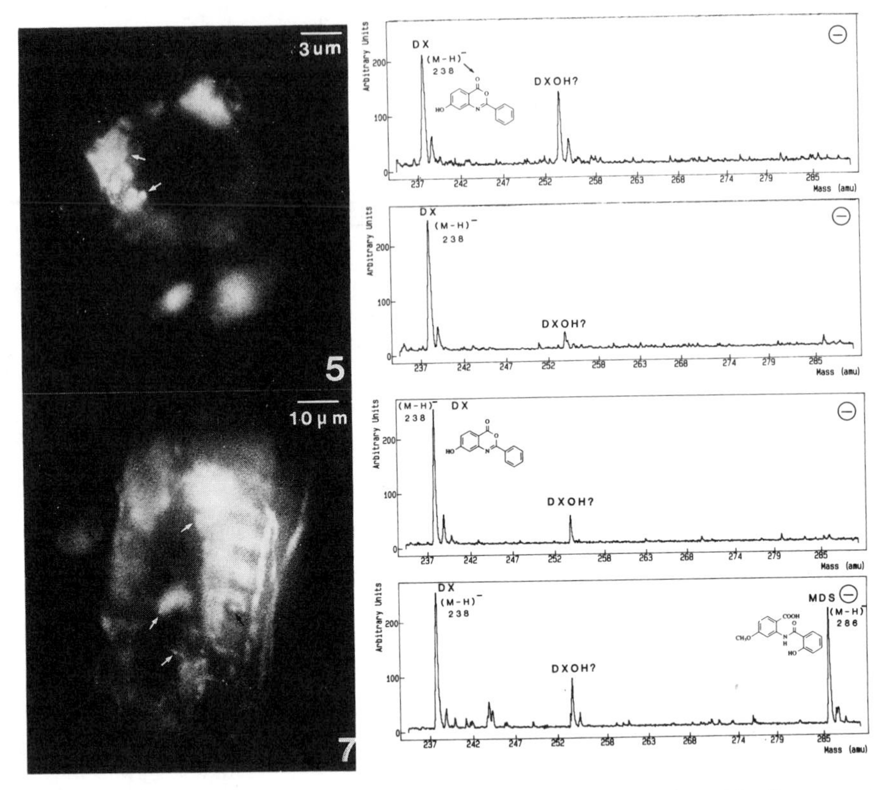

Figure 3B.62. In situ analysis of phytoalexines in carnation cryo-sections. The fluorescence micrographs of a transverse section (top left) and a longitudinal section (bottom left) indicate the sites where the negative ion mass spectra are recorded by TOF-LMMS. Reprinted from Mathey et al. (1989) with permission of San Francisco Press, Inc.

the detection of DX and MDS within the deposits on the vessel walls, the presence of a third compound was traced that had escaped prior chemical isolation. Using the background information from our data base, we tentatively assigned the molecular and fragment ion peaks to the salt of MDS as a more likely alternative to the hydroxylated DX. Such an interpretation needs subsequent verification. However, it illustrates the typical situation in practical TOF-LMMS applications. Knowledge of the target structure as derived from macroscopic methods still lets the microprobe user be confronted with slightly different modifications of the original structure, sometimes as a result of the local pH, or sometimes because the chemical separation has not been adequate. As a result, had the interpretation of LMMS results been based primarily on fingerprint comparison instead of deductive reasoning, most of

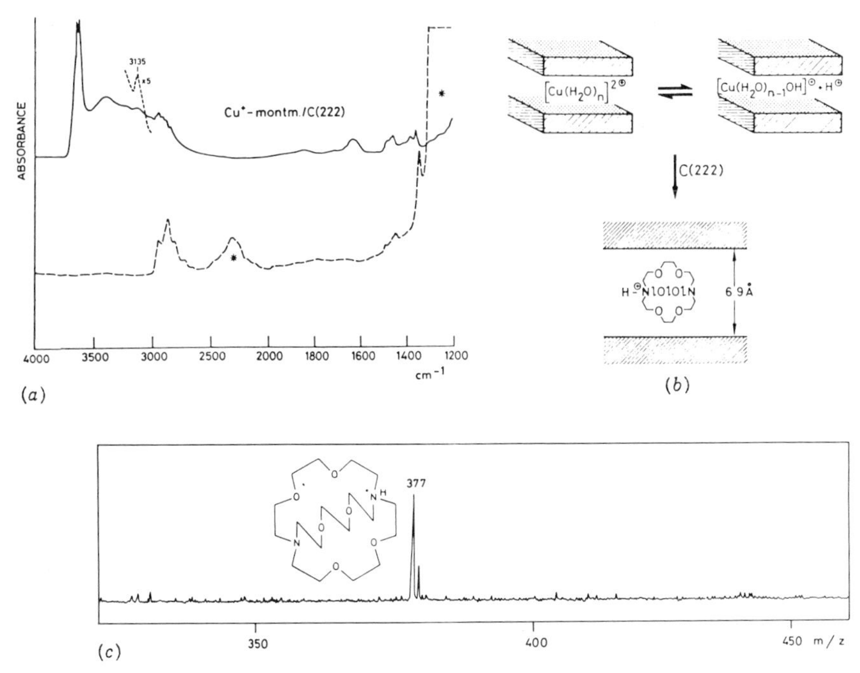

Figure 3B.63. Study of intralamellar Cu^{2+} cryptand complexes in montmorillonite: (a) IR spectroscopic data for the intercalated cryptands in fluorolube oil (dashed line, with asterisks denoting the fluorolube absorption); (b) coordination of cryptands by intracrystalline proton exchange of acidic Cu^{2+} hydrated cations; (c) positive ion mass spectrum recorded by TOF-LMMS from the intercalated ligand. Reprinted from Casal et al. (1988) with permission of Kluwer Academic Publishers.

the successful applications we have achieved would have ended without significant results.

The limited depth of information does not imply that the organic compounds really have to be present at the surface in terms of molecular layers. It has been shown that TOF-LMMS permits analysis of crown ether complexes, encapsulated in layered silicates (Casal et al., 1988). Macrocyclic ligands such as crown ethers and cryptands can be intercalated in homoionic 2:1 layer silicates, yielding very stable intracrystalline polydentate coordination complexes. Figure 3B.63 summarizes the main information concerning the detection of Cu^{2+} cryptand complexes in montmorillonite. IR spectroscopy reveals that the characteristic absorption of the macrocyclic ligand has changed upon intercalation into the layered silicate. The band at 3100 cm^{-1} is tentatively associated with the protonation of the ligand induced by the acidic nature of the hydrated transition metal cation. The positive ion mass spectra show detection of the protonated crown ether. Note that the peak shape and mass resolution of the parent ions are adequate. The situation strongly contrasts with the distorted signals in Figure 3B.60. It is readily conceived that the targets are encapsulated layer by layer, which represents a completely different situation than the presence of a compact covering layer of nonporous polymer material above the targets. Signals referring to the combination of the ligand and copper are detected from other samples but not in the case of the studied mineral. Nevertheless, the dependence of the mass spectra on the locally applied energy regime and the possible decomposition and subsequent reionization of the cationized ligands during DI or in the selvedge must be considered. Interpretation of TOF-LMMS data alone is certainly not conclusive for this particular problem. However, the information remains highly relevant to complement or support the indications from other techniques. In fact, one of the strengths of the laser microprobe actually resides in the possibility of obtaining valuable indications in a relatively short time and with a minimal sample preparation. To date, the instrument has not been provided at enough experimental facilities; moreover, lack of knowledge about DI of organic and inorganic constituents and lack of totally adequate control over the local energy regime, limit the ability of TOF-LMMS to generate definitive proofs. This conclusion does not, however, diminish the usefulness of the approach; it does reinforce the need to be careful with interpretation of the results.

In spite of the foregoing limitations, TOF-LMMS has been found extremely useful in various industrial problem-solving cases. Characterization of local inclusions, an often encountered task in polymer studies, is the research area of choice for the method at the moment. Although the average concentration of the target can be extremely low when the bulk material is considered, the compound may occur as a major component within the localized heterogeneity on a micrometer scale. It should be realized that an inclusion with dimensions of at least the irradiated spot represents the same analytical situation as a pure product. As a result, the challenge often lies in the interpretation of the mass spectra more than in the actual measurement itself. Figure 3B.64 presents a micrograph of a rubber sample containing a poorly dispersed ingredient, as well as the negative ion mass spectra recorded from the heterogeneities and a reference compound (Holm and Holtkamp, 1989). Energy dispersive X-ray microanalysis provides even better lateral resolution, but the information is essentially limited to the detection of S in the local heterogeneities, leaving the possibility of free sulfur or a sulfur-containing molecule. Surface layers of bulk samples are measured directly with TOF-LMMS in the reflection geometry. Within typically less than 30 minutes a lot of information becomes available. Even without comparison to a reference spectrum, the data already point clearly to an organic ingredient with most likely the following features: a MW of 332; a sulfide or thiol function group;

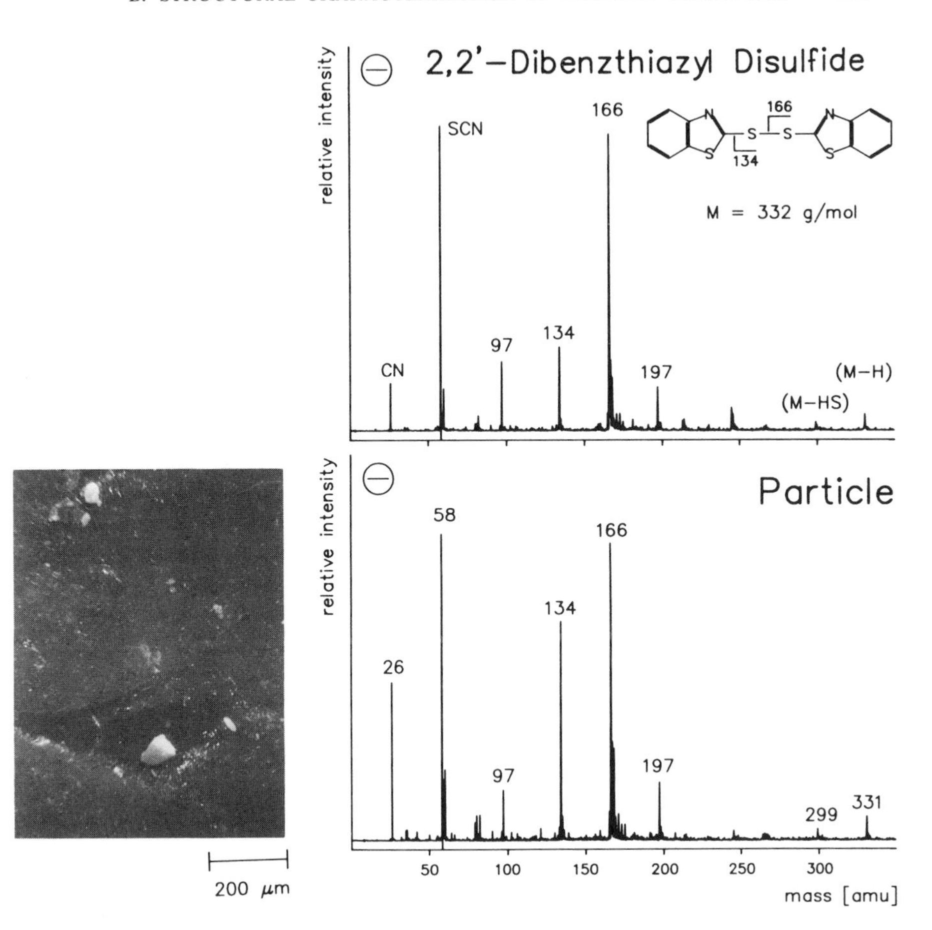

Figure 3B.64. Micrograph of the agglomerates in a rubber sample, and negative ion mass spectra recorded from the inclusion and from pure dibenzothiazyl disulfide. Reprinted from Holm and Holtkamp (1989) with permission of San Francisco Press, Inc.

the additional presence of an SCN group or a heteroaromatic system with nitrogen and sulfur; and a highly symmetric molecule with a central linkage that favors direct cleavage. With this information it is quite easy to track the possible candidates from the reagents in the manufacturing process and to verify the structure of the poorly dispersed ingredient in an extremely short time.

One of the major strengths of TOF-LMMS stems from the experimental flexibility that laser irradiation offers to deal with a variety of samples. Local inclusions embedded deeper in the material can still be analyzed without additional sample preparation. Successive laser irradiation provides an

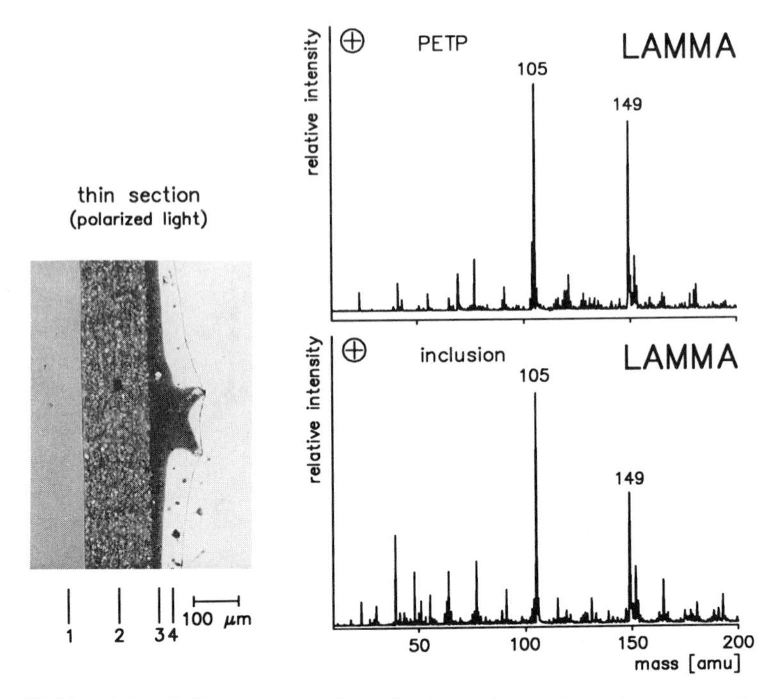

Figure 3B.65. A bright-field micrograph of a defect in a paint coating, taken under polarized light, and positive ion mass spectra recorded by TOF-LMMS from the inclusion and poly(ethylene terephthalate). Reprinted from Holm and Holtkamp (1989) with permission of San Francisco Press, Inc.

elegant means for quick erosion of the covering material. Figure 3B.65 shows a thin section made in the vicinity of a defect in a surface coating (Holm and Holtkamp, 1989). The problem is caused by a structure of about 10 μm in diameter that is, however, not at the surface. Following laser ablation of the uppermost layers, the recorded positive ion mass spectrum corresponds very well to the reference data from poly(ethylene terephthalate) and the inclusion is characterized as a polyester fiber (Holm and Holtkamp, 1989). Note, however, that the resolution of the ablation process is rather rough in comparison to the commonly achieved depth profiling characteristics in SIMS, for example (Hand et al., 1989).

The reported application of TOF-LMMS to the discrimination of indigo-dyed cotton fibers in forensic analysis represents a (literally) capital test case by which to judge the merits of TOF-LMMS (Schmidt and Brinkmann, 1989). Denim fibers from the clothing (jeans) of the defendant were compared to those found on the body of the victim. The indigo dyes in both samples were identical, but differentiation was possible by means of the characteristic pattern of molecular and fragment ions issuing from the respective fabric softeners.

Note that the need for surface availability of the organic targets can very well arise from the application of the TOF analyzer with continuous extraction of the ions from the source region into the drift tube. As mentioned earlier, the local target concentration in the uppermost surface layers must be around 10% to reach the detection limit. In contrast, when LD-FTMS is applied, detection concerns the entire production of low-energy ions over more than 100 μs. Additionally, postionization by EI can be easily implemented to make effective use of the desorbed neutrals. LD-FTMS of colored polymethylmethacrylate samples of the type used in automobile taillight lenses has permitted detection of the corresponding dyes down to a 0.1% weight/weight concentration level with a very adequate signal-to-noise ratio, even without application of postionization (Hsu and Marshall, 1988). The mass measurement is accurate within 0.5 to 20 ppm, facilitating the correct identification of the targets in mixtures and the occurrence of unexpected ions. These data have been recorded with a relatively low lateral resolution of 50 μm, as opposed to typically 1–5 μm in TOF-LMMS, and with a CO_2 10.6 μm laser instead of an Nd:YAG at 266 nm. It is clear, however, that use of a different time domain may offer significant potential to extend the working range of laser microprobe technology for organic applications. Within the next few years we anticipate that a substantial breakthrough in this field may well occur through use of FT-LMMS instruments, which are now becoming commercially available.

ACKNOWLEDGMENTS

Luc Van Vaeck and Wim Van Roy are indebted to the National Science Foundation, Belgium (N. F. W. O.), as research associate and research assistant, respectively.

REFERENCES

Amster, I. J., Land, D. P., Hemmington, J. C., and McIver, R. T., Jr. (1989). *Anal. Chem.* **61**, 184–186.

Anderson, F. W., Heinen, H. J., and Ramsey, J. N. (1985). In *Microbeam Analysis—1985* (J. T. Armstrong, ed.), pp. 325–330. San Francisco Press, San Francisco.

Asamoto, B., Young, J. R., and Citerin, R. J. (1990). *Anal. Chem.* **62**, 61–70.

Balasanmugan, K., and Hercules, D. M. (1983a). *Anal. Chem.* **55**, 145–146.

Balasanmugan, K., and Hercules, D. M. (1983b). *Spectrosc. Lett.* **16**, 1–9.

Balasanmugan, K., and Hercules, D. M. (1984). *Anal. Chim. Acta* **166**, 1–26.

Balasanmugan, K., Dang, T. A., Day, R. J., and Hercules, D. M. (1981). *Anal. Chem.* **53**, 2296–2298.

Balasanmugan, K., Viswanadham, S. K., and Hercules, D. M. (1983). *Anal. Chem.* **55**, 2426–2427.

Balasanmugan, K., Day, R. J., and Hercules, D. M. (1985). *Inorg. Chem.* **24**, 4477–4483.

Balasanmugan, K., Viswanadham, S. K., and Hercules, D. M. (1986). *Anal. Chem.* **58**, 1102–1108.

Balasanmugan, K., Viswanadham, S. K., and Hercules, D. M. (1989). *Talanta* **36**, 117–124.

Beavis, R., Lindner, J., Grotemeyer, J., Atkinson, I. A., Keene, F. R., and Knight, A. E. W. (1988a). *J. Am. Chem. Soc.* **110**, 7534–7535.

Beavis, R. C., Lindner, J., Grotemeyer, J., and Schlag, E. W. (1988b). *Chem. Phys. Lett.* **146**, 310–314.

Beavis, R. C., Lindner, J., Grotemeyer, J., and Schlag, E. W. (1988c). *Z. Naturforsch.* **43A**, 1083–1090.

Benninghoven, A. (1983). In *Ion Formation from Organic Solids* (A. Benninghoven, ed.), Chem. Phys. No. 25, pp. 64–89. Springer-Verlag, Berlin.

Benninghoven, A. (1985). *J. Vac. Sci. Technol.* **A3**, 451–460.

Beuhler, R. J., Flanigan, E., Greene, L. J., and Friedman, L. (1974). *J. Am. Chem. Soc.* **96**, 3990–3999.

Biemann, K., and McCloskey, J. A. (1962). *J. Am. Chem. Soc.* **84**, 2005.

Bjarnason, A. (1989). *Rapid Commun. Mass Spectrom.* **3**, 373–376.

Bletsos, I. V., Hercules, D. M., Magill, J. H., van Leyen, D., Niehaus, E., and Benninghoven, A. (1988). *Anal. Chem.* **60**, 938–944.

Bowie, J. H., Lewis, G. E., and Cooks, R. G. (1987). *J. Chem. Soc. B*, pp. 621–628.

Brenna, J. T. (1989). In *Microbeam Analysis—1989* (P. E. Russell, ed.), pp. 306–310. San Francisco Press, San Francisco.

Brown, C. E., Kovacič, P., Wilkie, C. A., Cody, R. B., Jr., Hein, R. E., and Kinsinger, J. A. (1985). *J. Polym. Sci., Part C: Polym. Lett.* **23**, 453–463.

Brown, C. E., Kovacič, P., Wilkie, C. A., Kinsinger, J. A., Hein, R. E., Yaniger, S. I., and Cody, R. B., Jr. (1986a). *J. Polym. Sci., Part A: Polym. Chem.* **24**, 255–267.

Brown, C. E., Kovacič, P., Wilkie, C. A., Cody, R. B., Jr., Hein, R. E., and Kinsinger, J. A. (1986b). *Synth. Met.* **15**, 265–279.

Brown, C. E., Kovacič, P., Cody, R. B., Jr., Hein, R. E., and Kinsinger, J. A. (1986c). *J. Polym. Sci., Part C: Polym. Lett.* **24**, 519–528.

Brown, C. E., Wilkie, C. A., Smukalla, J., Cody, R. B., Jr., and Kinsinger, J. A. (1988a). *J. Polym. Sci., Part A: Polym. Chem.* **24**, 1297–1311.

Brown, C. E., Kovacič, P., Welch, K. J., Cody, R. B., Jr., Hein, R. E., and Kinsinger, J. A. (1988b). *J. Polym. Sci., Part A: Polym. Chem.* **26**, 131–148.

Brown, R. S., and Wilkins, C. L. (1986a). *Anal. Chem.* **58**, 3196–3199.

Brown, R. S., and Wilkins, C. L. (1986b). *J. Am. Chem. Soc.* **108**, 2447–2448.

Bruynseels, F., and Van Grieken, R. E. (1986). *Int. J. Mass Spectrom. Ion Processes* **74**, 161–177.

Budzikiewicz, H., Djerassi, C., and Williams, D. H. (1967). *Mass Spectrometry of Organic Compounds* Holden Day, San Francisco.

Busch, K. L., Unger, S. E., Vincze, A., Cooks, R. G., and Keough, T. (1982). *J. Am. Chem. Soc.* **104**, 1507–1511.

Byrd, G. D., Fatiadi, A. J., Simons, D. S., and White, E. V. (1986). *Org. Mass Spectrom.* **21**, 63–68.

Campana, J. E. (1987). *Mass Spectrom. Rev.* **6**, 395–442.

Casal, B., Ruiz-Hitzky, E., Van Vaeck, L., and Adams, F. C. (1988). *J. Inclusion Phenom.* **6**, 107–118.

Chiarelli, M. P., and Gross, M. L. (1987). *Int. J. Mass Spectrom. Ion Processes* **78**, 37–52.

Chiarelli, M. P., and Gross, M. L., (1989a). *J. Phys. Chem.* **93**, 3595–3599.

Chiarelli, M. P., and Gross, M. L. (1989b). *Anal. Chem.* **61**, 1895–1900.

Chiarelli, M. P., Gross, M. L., and Peake, D. A. (1990). *Anal. Chim. Acta* **228**, 169–176.

Claereboudt, J., Van Vaeck, L., Baeten, W., Geise, H., Gijbels, R., and Claeys, M. (1987). *Anal. Chim. Acta* **195**, 343–350.

Coad, P., Coad, R. A., Yang, L. C., and Wilkins, C. L. (1987). *Org. Mass Spectrom.* **22**, 75–79.

Coates, M. L., and Wilkins, C. L. (1985). *Biomed. Mass Spectrom.* **12**(8), 424–428.

Coates, M. L., and Wilkins, C. L. (1986). *Biomed. Environ. Mass Spectrom.* **13**, 199–204.

Coates, M. L., and Wilkins, C. L. (1987). *Anal. Chem.* **59**, 197–200.

Cody, R. E., Bjarnason, A., and Weil, D. A. (1990). In *Lasers and Mass Spectrometry* (D. M. Lubman, ed.), pp. 316–339. Oxford University Press, New York.

Conzemius, R. J., and Capellen, J. M. (1980). *Int. J. Mass Spectrom. Ion Phys.* **34**, 197–271.

Cooks, R. G., and Busch, K. L. (1983). *Int. J. Mass Spectrom. Ion Phys.* **53**, 111–124.

Cotter, R. J. (1980). *Anal. Chem.* **52**, 1767–1770.

Cotter, R. J. (1984). *Anal. Chem.* **546**, 485A–504A.

Cotter, R. J., and Tabet, J. C. (1983). *Int. J. Mass Spectrom. Ion Phys.* **53**, 151–166.

Cotter, R. J., and Yergey, A. L. (1981a). *J. Am. Chem. Soc.* **103**, 1596–1598.

Cotter, R. J., and Yergey, A. L. (1981b). *Anal. Chem.* **53**, 1306–1307.

Cotter, R. J., Honovich, J., Qureshi, N., and Takayama, K. (1987). *Biomed. Environ. Mass Spectrom.* **14**, 591–598.

Creasy, W. R., and Brenna, J. T. (1988). *Chem. Phys.* **126**, 453–468.

Dang, T. A., Day, R. J., and Hercules, D. M. (1984a). *Anal. Chim. Acta* **158**, 235–246.

Dang, T. A., Day, R. J., and Hercules, D. M. (1984b). *Anal. Chem.* **58**, 866–871.

Daves, G. D. (1979). *Acc. Chem. Res.* **12**, 359–365.

Daves, G. D., Lee, T. D., Anderson, W. R., Barofsky, D. F., Maissey, G. A., Johnson, J. C., and Pincus, P. A. (1980). *Adv. Mass Spectrom.* **8**, 1012–1018.

Davis, D. V., Cooks, R. G., Meyer, B. N., and McLaughlin, J. L. (1983). *Anal. Chem.* **55**, 1302–1305.

Denoyer, E., Mauney, T., Natusch, D. F. S., and Adams, F. (1982). In *Microbeam Analysis—1982* (K. F. J. Heinrich, ed.), pp. 191–196. San Francisco Press, San Francisco.

Derrick, P. J., Tang-Trong, N., and Rogers, D. E. C. (1985). *Org. Mass Spectrom.* **20**, 690.

De Waele, J. K., Gybels, J. J., Vansant, E. F., and Adams, F. C. (1983a). *Anal. Chem.* **55**, 2255–2260.

De Waele, J. K., Vansant, E. F., and Adams, F. C. (1983b). *Mikrochim. Acta* **3**, 367–384.

De Waele, J. K., Verhaert, I., Vansant, E., and Adams, F. C. (1983c). *Surf. Interface Anal.* **5**, 186–192.

Doktycz, S. J., Savickas, P. J., and Krueger, D. A. (1991). *Rapid Commun. Mass Spectrom.* **5**, 145–148.

Dutta, P. K., and Talmi, Y. (1981). *Anal. Chim. Acta*, **132**, 111–118.

Dutta, P. K., Rigaus, D. C., Hofstader, R. A., Denoyer, E., Natusch, D. F. S., and Adams, F. C. (1984). *Anal. Chem.* **56**, 302–304.

Emary, W. B., Hand, O. W., and Cooks, R. G. (1990). In *Lasers and Mass Spectrometry* (D. M. Lubman, ed.), pp. 223–248. Oxford University Press, New York.

Engelke, F., Hahn, J. H., Henke, W., and Zare, R. N. (1987). *Anal. Chem.* **59**, 909–912.

Feigl, P. K. D., Krueger, F. R., and Schueler, B. (1984). *Mikrochim. Acta* **2**, 85–96.

Fletcher, R. A., and Currie, L. A. (1988). In *Microbeam Analysis—1988* (D. E. Newbury, ed.), pp. 367–370. San Francisco Press, San Francisco.

Fletcher, R. A., and Fatiadi, A. J. (1986). *J. Trace Microprobe Tech.* **4**, 216–226.

Forest, E., Marchon, J.-C., Wilkins, C. L., and Yang, L. C. (1989). *Org. Mass Spectrom.* **24**, 197–200.

Fung, E. T., and Wilkins, C. L. (1988). *Biomed. Environ. Mass Spectrom.* **15**, 609–613.

Furstenau, N., and Hillenkamp, F. (1981). *Int. J. Mass Spectrom. Ion Phys.* **37**, 135–151.

Gardella, J. A., Jr., and Hercules, D. M. (1980). *Spectrosc. Lett.* **13**, 347–360.

Gardella, J. A., Jr., and Hercules, D. M. (1981). *Fresenius' Z. Anal. Chem.* **308**, 297–303.

Gierlich, H. H., Röllgen, F. W., Borchers, F., and Levsen, K. (1977). *Org. Mass Spectrom.* **12**, 387–390.

Graham, S. W., Dowd, P., and Hercules, D. M. (1982). *Anal. Chem.* **54**, 649–654.

Greenwood, P. F., Strachan, M. G., Willett, G. D., and Wilson, M. A. (1990). *Org. Mass Spectrom.* **25**, 353–362.

Grotemeyer, J., and Schlag, E. W. (1987). *Org. Mass Spectrom.* **22**, 758–760.

Grotemeyer, J., and Schlag, E. W. (1988a). *Org. Mass Spectrom.* **23**, 388–396.

Grotemeyer, J., and Schlag, E. W. (1988b). *Biomed. Environ. Mass Spectrom.* **16**, 143–149.

Grotemeyer, J., and Schlag, E. W. (1988c). *Angew. Chem. Int. Ed. Engl.* **27**, 447–458.

Grotemeyer, J., Boesl, U., Walter, K., and Schlag, E. W. (1986a). *Org. Mass Spectrom.* **21**, 595–597.

Grotemeyer, J., Boesl, U., Walter, K., and Schlag, E. W. (1986b). *Org. Mass Spectrom.* **21**, 645–653.

Grotemeyer, J., Boesl, U., Walter, K., and Schlag, E. W. (1986c). *J. Am. Chem. Soc.* **108**, 4233–4234.

Grotemeyer, J., Walter, K., Boesl, U., and Schlag, E. W. (1987). *Int. J. Mass Spectrom. Ion Processes* **78**, 69–83.

Grotemeyer, J., Boesl, U., Walter, K., Lindner, J., Beavis, R. C., and Schlag, E. W. (1989). *Adv. Mass Spectrom.* **11A**, 314–315.

Hahn, J. H., Zenobi, R., and Zare, R. N. (1987). *J. Am. Chem. Soc.* **109**, 2842–2843.

Hand, O. W., Emary, W. B., Winger, B. E., and Cooks, R. G. (1989). *Int. J. Mass Spectrom. Ion Processes* **90**, 97–118.

Hanson, C. D., Castro, M. E., Russell, D. H., Hunt, D. F., and Shabanowitz, J. (1987). *ACS Symp. Ser.* **359**, 100–115.

Hardin, E. D., and Vestal, M. L. (1981). *Anal. Chem.* **53**, 1492–1497.

Hardin, E. D., Fan, T. P., Blakey, C. R., and Vestal, M. L. (1984). *Anal. Chem.* **56**, 2–7.

Heinen, H. J. (1981). *Int. J. Mass Spectrom. Ion Phys.* **38**, 309–322.

Heinen, H. J., and Holm, R. (1984). *Scanning Electron Microsc.* **3**, 1129–1138.

Heinen, H. J., Meier, S., Vogt, H., and Wechsung, R. (1980). *Adv. Mass Spectrom.* **8A**, 942–953.

Heinen, H. J., Meier, S., Vogt, H., and Wechsung, R. (1981). *Fresenius' Z. Anal. Chem.* **308**, 290–296.

Helander, I. M., Lindner, B., Bradew, H., Altmann, K., Lindberg, A. A., Rietschel, E. T., and Zahringer, U. (1988). *Eur. J. Biochem.* **177**, 483–492.

Hercules, D. M., Day, R. J., Balasanmugan, K., Dang, T. A., and Li, C. P. (1982). *Anal. Chem.* **54**, 280A–305A.

Hercules, D. M., Parker, C. D., Balasanmugan, K., and Viswanadham, S. K. (1983). In *Ion Formation from Organic Solids* (A. Benninghoven, ed.), Chem. Phys. No. 25, pp. 222–228. Springer-Verlag, Berlin.

Hercules, D. M., Novak, F. P., Viswanadham, S. K., and Wilk, Z. A. (1987). *Anal. Chim. Acta* **195**, 61–71.

Heresch, F., Schmidt, E. R., and Huber, J. F. K. (1980). *Anal. Chem.* **52**, 1803–1807.

Hettich, R. L. (1989). *Biomed. Environ. Mass Spectrom.* **18**, 265–277.

Hettich, R. L., and Buchanan, M. V. (1990). *J. Am. Chem Soc.* **2**, 22–28.

Hillenkamp, F. (1983). In *Ion Formation from Organic Solids* (A. Benninghoven, ed.), Chem. Phys. No. 25, pp. 190–210. Springer-Verlag, Berlin.

Hillenkamp, F., Feigl, P., and Schueler, B. (1982). In *Microbeam Analysis—1982* (K. F. J. Heinrich, ed.), pp. 359–364. San Francisco Press, San Francisco.

Hillenkamp, F., Karas, M., Holtkamp, D., and Klüsener, P. (1986). *Int. J. Mass Spectrom. Ion Processes* **69**, 265–276.

Hillenkamp, F., Bahr, U., Karas, M., and Spengler, B. (1987). *Scanning Microsc. Suppl.* 1, pp. 33–39.

Hillenkamp, F., Karas, M., Bahr, U., and Ingendoh, A. (1989). In *Ion Formation from Organic Solids (IFOS V)* (A. Hedin, B. U. R. Sundqvist, and A. Benninghoven, eds.), pp. 111–118. Wiley, Chichester.

Holm, R., and Holtkamp, D. (1989). In *Microbeam Analysis—1989* (P. E. Russell, ed.), pp. 325–329. San Francisco Press, San Francisco.

Holm, R., Kampf, G., Kirchner, D., Heinen, H. J., and Meier, S. (1984). *Anal. Chem.* 56, 690–692.

Holm, R., Karas, M., and Vogt, H. (1987). *Anal. Chem.* **59**, 371–373.

Holtkamp, D., Bayer, G., and Holm, R. (1991). *Mikrochim. Acta* **1**, 245–260.

Hsu, A. T., and Marshall, A. G. (1988). *Anal. Chem.* **60**, 932–937.

Huang, L. Q., Conzemius, R. J., Junk, G. A., and Houk, R. S. (1988). *Anal. Chem.* **60**, 1490–1494.

Ijames, C. F., and Wilkins, C. L. (1988). *J. Am. Chem. Soc.* **110**, 2687–2688.

Ijames, C. F., and Wilkins, C. L. (1990). *Anal. Chem.* **62**, 1295–1299.

Johlman, C. L., Wilkins, C. L., Hogan, J. D., Donavan, T. L., Laude, D. A., and Youssefi, M. J. (1990). *Anal. Chem.* **62**, 1167–1172.

Karas, M., Bachmann, D., and Hillenkamp, F. (1985). *Anal. Chem.* **57**, 2935–2939.

Karas, M., Bachmann, D., Bahr, U., and Hillenkamp, F. (1987). *Int. J. Mass Spectrom. Ion Processes* **78**, 53–68.

Karas, M., Bahr, U., Ingendoh, A., and Hillenkamp, F. (1989). *Angew. Chem., Int. Ed. Engl.* **6**, 760–761.

Kelland, D., Wallach, E. R., Cottee, F. H., and Williamson, F. A. (1987). *Anal. Chim. Acta* **197**, 81–88.

Kistemaker, P. G., Lens, M. M. J., van der Peyl, G. J. Q., and Boerboom, A. J. H. (1979). *Adv. Mass Spectrom.* **8A**, 928–934.

Kolaitis, L., and Lubman, D. M. (1986). *Anal. Chem.* **58**, 2137–2142.

Köster, C., Grotemeyer, J., Rohwer, E. R., Lindner, J., and Schlag, E. W. (1989). *Adv. Mass Spectrom.* **11**, 1192–1193.

Krauss, J. H., Seydel, U., Weckesser, J., and Mayer, H. (1989). *Eur. J. Biochem.* **180**, 519–526.

Krier, G., and Muller, J. F. (1986). In *Proceedings of the Third Laser Microprobe Mass Spectrometry Workshop, 1986, Antwerp* (F. Adams and L. Van Vaeck, eds.), pp. 133–137. University of Antwerp, Antwerp.

Krier, G., Pelletier, M., Muller, J. F., Lazare, S., Granier, V., and Lutgen, P. (1989). In *Mircobeam Analysis—1989* (P. E. Russell, ed.), pp. 347–349. San Francisco Press, San Francisco.

Krueger, F. R., and Schueler, B. (1980). *Adv. Mass Spectrom.* **8A**, 978–927.

Kubis, A. J., Somayajula, K. V., Sharkey, A. G., and Hercules, D. M. (1989). *Anal. Chem.* **61**, 2516–2523.

Kupka, K.-D., Hillenkamp, F., and Schiller, C. (1980). *Adv. Mass Spectrom.* **8A**, 935–941.

Lam, Z., Comisarow, M. B., Dutton, G. G. S., Weil, D. A., and Bjarnason, A. (1987). *Rapid Commun. Mass Spectrom.* **1**, 83–86.

Lam, Z., Comisarow, M. B., and Dutton, G. G. S. (1988). *Anal. Chem.* **60**, 2306–2309.

Land, D. P., Tai, T. L., Lindquist, J. M., Hemminger, J. C., and McIver, R. T., Jr. (1987). *Anal. Chem.* **59**, 2927–2930.

Li, L., and Lubman, D. M. (1988). *Appl. Spectrosc.* **42**, 411–417.

Li, L., and Lubman, D. M. (1989a). *Int. J. Mass Spectrom. Ion Processes* **88**, 197–210.

Li, L., and Lubman, D. M. (1989b). *Rapid Commun. Mass Spectrom.* **3**, 12–16.

Lindner, B., and Seydel, U. (1985). *Anal. Chem.* **57**, 895–899.

Lindner, B., Helander, I., and Seydel, U. (1988). *Adv. Mass Spectrom.* **11B**, 1484–1485.

Lindner, J., Grotemeyer, J., and Schlag, E. W. (1990). *Int. J. Mass Spectrom. Ion Processes* **100**, 267–285.

Lineman, D. N., Viswanadham, S. K., Sharkey, A. G., and Hercules, D. M. (1989). In *Microbeam Analysis—1989* (P. E. Russell, ed.), pp. 297–298. San Francisco Press, San Francisco.

Lohmann, W., Hillenkamp, F., Rosmarinowski, J., Bachmann, D., and Karas, M. (1984). *Fresenius' Z. Anal. Chem.* **317**, 129–130.

Lub, J., van Velzen, P. N. T., van Leyen, D., Hagenhoff, B., and Benninghoven, A. (1988). *Surf. Interface Anal.* **12**, 53–57.

Lub, J., van Leyen, D., and Benninghoven, A. (1989). *Polym. Commun.* **30**, 74–77.

Lubman, D. M., and Li, L. (1990). In *Lasers and Mass Spectrometry* (D. M. Lubman, ed.), pp. 353–382. Oxford University Press, New York.

Marchon, J. C., Wilkins, C. L., and Yang, L.-C. (1989). *Org. Mass Spectrom.* **24**, 197–200.

Martin, W. B., Silly, L., Murphy, C. M., Raley, T. J., and Cotter, R. J. (1989). *Int. J. Mass Spectrom. Ion Processes* **92**, 243–265.

Mathey, A., Van Vaeck, L., and Steglich, W. (1987). *Anal. Chim. Acta* **195**, 89–96.

Mathey, A., Van Vaeck, L., and Ricci, P. (1989). In *Microbeam Analysis—1989* (P. E. Russell, ed.), pp. 350–352. San Francisco Press, San Francisco.

Mattern, D. E., and Hercules, D. M. (1985). *Anal. Chem.* **57**, 2041–2046.

Mauney, T. (1987). *Anal. Chim. Acta* **195**, 337–341.

Mauney, T., and Adams, F. (1984a). *Int. J. Mass Spectrom. Ion Processes* **59**, 103–119.

Mauney, T., and Adams, F. (1984b). *Sci. Total Environ.* **36**, 215–224.

McCrery, D. A., and Gross, M. L. (1985a). *Anal. Chim. Acta* **178**, 91–103.

McCrery, D. A., and Gross, M. L. (1985b). *Anal. Chim. Acta* **178**, 105–116.

McCrery, D. A., Ledford, E. B., and Gross, M. L. (1982). *Anal. Chem.* **54**, 1437–1439.

McCrery, D. A., Peake, D. A., and Gross, M. L. (1985). *Anal. Chem.* **57**, 1181–1186.

McLafferty, F. W., and Stauffer, B. (1989). *The Wiley/NBS Registry of Mass Spectral Data*, Vols. 1–7. Wiley, New York.

Miller, L. L., Thomas, A. D., Wilkins, C. L., and Weil, D. A. (1986). *J. Chem. Soc., Chem. Commun.*, pp. 661–663.

Moesta, P., Seydel, U., Lindner, B., and Grisebach, H. (1982). *Z. Naturforsch.* **37C**, 748–751.

Monaghan, J. J., Barber, M., Bordoli, R. S., Sedgwick, R. D., and Tyler, A. N. (1982). *Org. Mass Spectrom.* **17**, 569–574.

Morelli, J. J., and Hercules, D. M. (1984). In *Microbeam Analysis—1984* (A. D. Romig and J. I. Goldstein, eds.), pp. 15–18. San Francisco Press, San Francisco.

Morelli, J. J., and Hercules, D. M. (1986). *Anal. Chem.* **58**, 1294–1298.

Morelli, J. J., and Hercules, D. M. (1989). *Appl. Spectrosc.* **43**, 1073–1081.

Morelli, J. J., Visnawadham, S. K., Sharkey, A. G., Jr., and Hercules, D. M. (1987). *Int. J. Environ. Anal. Chem.* **31**, 295–323.

Muller, J. F., Berthé, C., and Magar, J. M. (1981). *Fresenius' Z. Anal. Chem.* **308**, 312–320.

Muller, J. F., Ricard, A., and Ricard, M. (1984). *Int. J. Mass Spectrom. Ion Processes* **62**, 125–136.

Muller, J. F., Krier, G., Verdun, F., Lamboule, M., and Muller, D. (1985). *Int. J. Mass Spectrom. Ion Processes* **64**, 127–138.

Neussner, H. J. (1987). *Int. J. Mass Spectrom Ion Processes* **79**, 141–181.

Niessner, R., Klockow, D., Bruynseels, F., and Van Grieken, R. (1985). *Int. J. Environ. Anal. Chem.* **23**, 281–295.

Nuwaysir, L. M., and Wilkins, C. L. (1988). *Anal. Chem.* **60**, 279–282.

Nuwaysir, L. M., and Wilkins, C. L. (1989). *Anal. Chem.* **61**, 689–694.

Odom, R. W., Radicati di Brozoli, F., Harrington, P. B., and Voorhees, K. J. (1989). In *Microbeam Analysis—1989* (P. E. Russell, ed.), pp. 283–285. San Francisco Press, San Francisco.

Pachuta, S. J., and Cooks, R. G. (1987). *Chem. Rev.* 647–669.

Parker, C. D., and Hercules, D. M. (1985). *Anal. Chem.* **57**, 698–704.

Parker, C. D., and Hercules, D. M. (1986). *Anal. Chem.* **58**, 25–30.

Pierce, J. L., Busch, K. L., Cooks, R. G., and Walton, R. A. (1982). *Inorg. Chem.* **21**, 2597–2602.

Posthumus, M. A., Kistemaker, P. G., Meuzelaar, H. L. C., and Ten Noever de Brauw, M. C. (1978). *Anal. Chem.* **50**, 985–991.

Prokai, L. (1990). *Field Desorption Mass Spectrometry*, Pract. Spectrosc. Ser. Vol. 9. Dekker, New York.

Qureshi, N., Honovich, J. P., Hara, H., Cotter, R. J., and Takayama, K. (1988). *J. Biol. Chem.* **263**, 5502–5504.

Roczko, A. W., Visnawadham, S. K., Sharkey, A. G., and Hercules, D. M. (1989). *Fresenius' Z. Anal. Chem.* **334**, 521–526.

Rogers, D. E. C., and Derrick, P. J. (1984). *Org. Mass Spectrom.* **19**, 490.

Rohly, K. E., Heffren, J. S., and Douglas, B. E. (1984). *Org. Mass Spectrom.* **19**, 398–402.

Rosmarinowsky, J., Karas, M., and Hillenkamp, F. (1985). *Int. J. Mass Spectrom. Ion Processes* **67**, 109–119.

Rosnack, K. J., Somayajula, K. V., Sharkey, A. G., Jensen, N. J., and Hercules, D. M. (1989). *Appl. Spectrosc.* **43**, 1087–1092.

Sammons, M. C., Bursey, M. M., and White, C. K. (1975). *Anal. Chem.* **47**, 1165–1166.

Scanlan, F. P., Muller, J. F., and Fiquet, J. M. (1989). In *Microbeam Analysis—1989* (P. E. Russell, ed.), pp. 345–346. San Francisco Press, San Francisco.

Schiebel, H. M., and Schulten, H.-R. (1986). *Mass Spectrom. Rev.* **5**, 249–311.

Schiller, C., Kupka, K.-D. and Hillenkamp, F. (1981). *Fresenius' Z. Anal. Chem.* **308**, 304–308.

Schmidt, P. F., and Brinkmann, B. (1989). In *Microbeam Analysis—1989* (P. E. Russell, ed.), pp. 330–331. San Francisco Press, San Francisco.

Schueler, B., and Krueger, F. R. (1979). *Org. Mass Spectrom.* **14**, 439–441.

Schueler, B., and Krueger, F. R. (1980). *Org. Mass Spectrom.* **15**, 295–301.

Schueler, B., Feigl, P., Krueger, F. R., and Hillenkamp, F. (1981). *Org. Mass Spectrom.* **16**, 502–506.

Schulten, H.-R., and Lattimer, R. P. (1984). *Mass Spectrom. Rev.* **3**, 231–315.

Segar, K. R., and Johnston, M. V. (1989). *Org. Mass Spectrom.* **24**, 176–182.

Seydel, U., and Lindner, B. (1983). In *Ion Formation from Organic Solids* (A. Benninghoven, ed.), Chem. Phys. No. 25, pp. 240–244. Springer-Verlag, Berlin.

Seydel, U., Lindner, B., Wollenweber, H. W., and Rietschel, E. T. (1984). *Eur. J. Biochem.* **145**, 505–509.

Shabanowitz, J., and Hunt, D. F. (1990). In *Lasers and Mass Spectrometry* (D. M. Lubman, ed.), pp. 340–352. Oxford University Press, New York.

Shomo, R. E., II, Marshall, A. G., and Weisenberger, C. R. (1985). *Anal. Chem.* **57**, 2940–2944.

Shomo, R. E., II, Marshall, A. G., and Lattimer, R. P. (1986). *Int. J. Mass Spectrom. Ion Processes* **72**, 209–217.

Shomo, R. E., II, Marshall, A. G., Reuning, R. H., and Robertson, L. W. (1988). *Biomed. Environ. Mass Spectrom.* **15**, 295–302.

Southon, M. J., Witt, M. C., Harris, A., Wallach, E. R., and Myatt, J. (1984). *Vacuum* **34**, 903–909.

Speir, J. P., Gorman, G. S., Cornett, D. S., and Amster, I. J. (1991). *Anal. Chem.* **63**, 65–69.

Spengler, B., and Cotter, R. J. (1990). *Anal. Chem.* **62**, 793–795.

Spengler, B., Karas, M., Bahr, U., and Hillenkamp, F. (1987). *J. Phys. Chem.* **91**, 6502–6506.

Spengler, B., Pan, Y., Cotter, R. J., and Kan, L. S. (1990). *Rapid Commun. Mass Spectrom.* **4**, 99–102.

Stoll, R., and Röllgen, F. W. (1979). *Org. Mass Spectrom.* **14**, 642–645.

Stoll, R., and Röllgen, F. W. (1980). *J. Chem. Soc., Chem. Commun.*, pp. 789–789.

Stoll, R., and Röllgen, F. W. (1982). *Z. Naturforsch.* **37A**, 9–14.

Sunner, J., Ikonomou, M. G., and Kebarle, P. (1988a). *Int. J. Mass spectrom. Ion Processes* **82**, 221–257.

Sunner, J., Morales, A., and Kebarle, P. (1988b). *Int. J. Mass Spectrom Ion Processes* **86**, 169–186.

Tabet, J.-C., and Cotter, R. J. (1984). *Anal. Chem.* **56**, 1662–1667.

Tabet, J. C., Jablonski, M. J., Cotter, R. J., and Hunt, J. E. (1985). *Int. J. Mass Spectrom. Ion Processes* **65**, 105–117.

Tanaka, K., Waki, H., Ido, Y., Akita, S., Yoshida, Y., and Yoshida, T. (1988). *Rapid Commun. Mass Spectrom.* **2**, 151–153.

Tembreull, R., and Lubman, D. M. (1987). *Anal. Chem.* **59**, 1003–1006.

Van Breemen, R. B., Snow, M., and Cotter, R. J. (1973). *Int. J. Mass Spectrom. Ion Phys.* **49**, 35–50.

van der Peyl, G. J. Q., Isa, K., Haverkamp, J., and Kistemaker, P. G. (1981). *Org. Mass Spectrom.* **18**, 416–420.

van der Peyl, G. J. Q., Haverkamp, J., and Kistemaker, P. G. (1982). *Int. J. Mass Spectrom. Ion Phys.* **47**, 125–141.

van der Peyl, G. J. Q., Isa, K., Haverkamp, J., and Kistemaker, P. G. (1983). *Int. J. Mass Spectrom. Ion. Phys.* **47**, 11–14.

van der Peyl, G. J. Q., Van der Zande, W. J., and Kistemaker, P. G. (1984). *Int. J. Mass Spectrom. Ion Processes* **62**, 51–71.

van Leyen, D., Hagenhoff, B., Niehuis, E., Benninghoven, A., Bletsos, I. V., and Hercules, D. M. (1989). *J. Vac. Sci. Technol.* **A7**, 1790–1794.

Van Vaeck, L., and Gijbels, R. (1989). In *Microbeam Analysis—1989* (P. E. Russell. ed.), pp. xvii–xxv. San Francisco Press, San Francisco.

Van Vaeck, L., and Gijbels, R. (1990a). *Fresenius' Z. Anal. Chem.* **337**, 743–754.

Van Vaeck, L., and Gijbels, R. (1990b). *Fresenius' Z. Anal. Chem.* **337**, 755–765.

Van Vaeck, L., and Lauwers, W. (1990). *Adv. Mass Spectrom.* **11A**, 316–317.

Van Vaeck, L., De Waele, J., and Gijbels, R. (1984). *Mikrochim. Acta* **3**, 237–257.

Van Vaeck, L., Claereboudt, J., De Waele, J., Esmans, E., and Gijbels, R. (1985a). *Anal. Chem.* **57**, 2194–2951.

Van Vaeck, L., Claereboudt, J., Van Espen, P., Adams, F., Gijbels, R., and Cautreels, W. (1985b). *Adv. Mass Spectrom.* **9B**, 957–958.

Van Vaeck, L., Claereboudt, J., Veldeman, E., Vermeulen, M., and Gijbels, R. (1986a). *Bull. Soc. Chim. Belge* **95**, 351–372.

Van Vaeck, L., Claereboudt, J., De Waele, J., and Gijbels, R. (1986b). *Adv. Mass Spectrom.* **10B**, 991–992.

Van Vaeck, L., Van Espen, P., Jacob, W., Gijbels, R., Cautreels, W. (1988a). *Biomed. Environ. Mass Spectrom.* **16**, 113–119.

Van Vaeck, L., Van Espen, P., Gijbels, R., and Lauwers, W. (1988b). *Biomed. Environ. Mass Spectrom.* **16**, 121–130.

Van Vaeck, L., Gijbels, R., and Lauwers, W. (1989a). *Adv. Mass Spectrom.* **11A**, 348–349.

Van Vaeck, L., Bennett, J., Van Espen, P., Schweikert, E., Gijbels, R., Adams, F., and Lauwers, W. (1989b). *Org. Mass Spectrom.* **24**, 787–796.

Van Vaeck, L., Bennett, J., Van Espen, P., Schweikert, E., Gijbels, R., Adams, F., and Lauwers, W. (1989c). *Org. Mass Spectrom.* **24**, 797–806.

Van Vaeck, L., Van Espen, P., Adams, F., Gijbels, R., Lauwers, W., and Esmans, E. (1989d). *Biomed. Environ. Mass Spectrom.* **18**, 581–591.

Van Vaeck, L., Bennett, J., Lauwers, W., Vertes, A., and Gijbels, R. (1990). *Mikrochim. Acta* **3**, 283–303.

Vastola, F. J., and Pirone, A. J. (1968). *Adv. Mass Spectrom.* **4**, 107–111.

Vastola, F. J., Mumma, R. O., and Pirone, A. J. (1970). *Org. Mass Spectrom.* **3**, 101–104.

Veith, H. J. (1978). *Org. Mass Spectrom.* **13**, 280–283.

Veith, H. J. and Röllgen, F. W. (1985). *Org. Mass Spectrom.* **20**, 689.

Verbueken, A. H., Van Grieken, R. E., De Broe, M. E., and Wedeen, R. P. (1987). *Anal. Chim. Acta* **195**, 97–115.

Vertes, A., Balazs, L., and Gijbels, R. (1990). *Rapid Commun. Mass Spectrom.* **4**, 263–266.

Viswanadham, S. K., Hercules, D. M., Schreiber, E. M., Weller, R. R., and Giam, C. S. (1988). *Anal. Chem.* **60**, 2346–2353.

Watson, C. H., Baykut, G., and Eyler, J. R. (1987). *Anal. Chem.* **59**, 1133–1138.

Weber, R., Borchers, F., Levsen, K., and Rollgen, F. W. (1978). *Z. Naturforsch.* **33A**, 540–548.

Weintraub, A., Zähringer, U. Wollenweber, H.-W., Seydel, U., and Rietschel, E. T. (1989). *Eur. J. Biochem.* **183**, 425–431.

Wieser, P., and Wurster, R. (1983). In *Ion Formation from Organic Solids* (A. Benninghoven, ed.), Chem. Phys. No. 25, pp. 235–239. Springer-Verlag, Berlin.

Wilk, Z. A., Viswanadham, S. K., Sharkey, A. G., and Hercules, D. M. (1988). *Anal. Chem.* **60**, 2328–2346.

Wilkins, C. L., and Tang, C. L. C. (1986). *Int. J. Mass Spectrom. Ion Processes.* **72**, 195–208.

Wilkins, C. L., Weil, D. A., Yang, C. L. C., and Ijames, C. F. (1985). *Anal. Chem.* **57**, 520–524.

Wollenweber, H.-W., Seydel, U., Lindner, B., Lüderitz. O., and Rietschel, E. T. (1984). *Eur. J. Biochem.* **145**, 265–272.

Wuepper, J. J. (1979). *Anal. Chem.* **51**, 997–1000.

Wünsche, C., Benninghoven, A., Eicke, A., Heinen, H. J., Ritter, H. P., Taylor, L. C. E. and Veith, J. (1984). *Org. Mass Spectrom.* **19**, 176–182.

Yang, L. C., and Wilkins, C. L. (1989). *Org. Mass Spectrom.* **24**, 409–414.

Zakett, D., Schoen, A. E., Cooks, R. G., and Hemberger, P. H. (1981). *J. Am. Chem. Soc.* **103**, 1295–1297.

CHAPTER

3

METHODS UTILIZING LOW AND MEDIUM LASER IRRADIANCE

C. TWO-STEP METHODS FOR SEPARATED VOLATILIZATION AND LASER-INDUCED MULTIPHOTON IONIZATION OF SMALL BIOLOGICAL MOLECULES IN SUPERSONIC JETS

DAVID M. LUBMAN

Department of Chemistry
The University of Michigan
Ann Arbor, Michigan

3C.1. INTRODUCTION

Resonance-enhanced multiphoton ionization (REMPI) has been shown to be an ionization source with unique properties for chemical analysis. The general multiphoton ionization technique (MPI) depends upon the absorption of several photons by a molecule on irradiation with an intense visible or ultraviolet (UV) light source (Antonov et al., 1978; Bernstein, 1982; Boesl et al., 1978, 1980, 1981; Brophy and Rettner, 1979; Carlin and Freiser, 1983; Cooper et al., 1980; Dietz et al., 1980a,b; Duncan et al., 1979; Engelke et al., 1987; Fisanick et al., 1980; Frueholz et al., 1980; Gobeli et al., 1985; Grotemeyer et al., 1986a,b,c; Hager and Wallace, 1988; Hahn et al., 1987; Irion et al., 1982; Johnson, 1975, 1980; Johnson et al., 1975; Krogh-Jespersen et al., 1979; Lubman 1986; Lubman and Kronick, 1982; Lubman et al., 1980; Nieman and Colson, 1978; Parker et al., 1978; Petty et al., 1975; Rava and Goodman, 1982; Reilly and Kompa, 1980; Rettner and Brophy, 1981; Rhodes et al., 1983; Robin and Kuebler, 1978; Sack et al., 1985; Seaver et al., 1980; Tembreull and Lubman, 1984, 1986, 1987a,b,c; Tembreull et al., 1985a,b, 1986; Wessel et al., 1981; Zandee and Bernstein, 1979a,b, Zandee et al., 1978). When the laser frequency is tuned to a real intermediate electronic state, the cross section for

Laser Ionization Mass Analysis, Edited by Akos Vertes, Renaat Gijbels, and Fred Adams. Chemical Analysis Series, Vol. 124.
ISBN 0-471-53673-3

ionization is greatly enhanced and REMPI is produced. A molecule will ionize only if the sum of the energy of the photons absorbed exceeds the ionization potential of the molecule. When the laser wavelength is not tuned to a real electronic state, the probability for MPI is negligible. Thus, although ions are produced as the final product that can be detected in a mass spectrometer, the ion current obtained depends on the ionization cross section of the molecule at the selected laser frequency. Thus, the ionization cross section reflects the absorption–excitation spectrum of the intermediate state. The truly unique property of MPI spectroscopy is that it is an optical technique that can be used as a means of achieving spectral selection for a compound prior to mass analysis.

3C.2. PRINCIPLES OF RESONANCE-ENHANCED MULTIPHOTON IONIZATION

There are several means by which MPI can be induced, as shown in Figure 3C.1. The MPI method that has found most extensive application in analytical

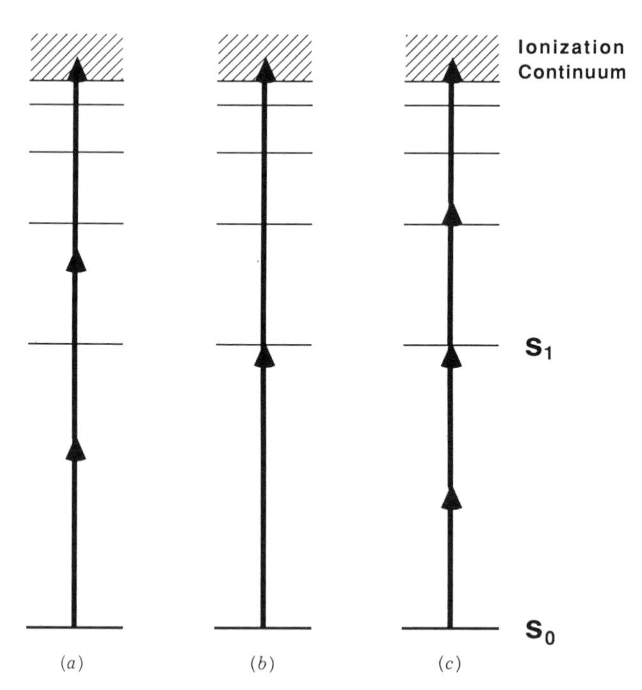

Figure 3C.1. Energy-level diagram showing MPI transitions: (a) nonresonant MPI; (b) resonant two-photon ionization; (c) two-photon REMPI.

chemistry and that is used throughout most of our work is resonant two-photon ionization (R2PI) (Boesl et al., 1980, 1981; Brophy and Rettner, 1979; Cable et al., 1988; Carlin and Freiser, 1983; Dietz et al., 1980a,b; Engelke et al., 1987; Gobeli et al., 1985; Grotemeyer and Schlag, 1988a, 1989; Grotemeyer et al., 1986a,b,c; Hager and Wallace, 1988; Hahn et al., 1987, 1988; Irion et al., 1982; Lubman, 1986, 1988b,c, 1990; Lubman and Kronick, 1982; Rettner and Brophy, 1981; Rhodes et al., 1983; Sack et al., 1985; Syage and Wessel, 1988; Tembreull and Lubman 1986, 1987a,b,c; Tembreull et al., 1985a,b, 1986; Wessel et al., 1981). In this process, one photon excites a molecule to an excited electronic state (i.e., $S_0 \rightarrow S_1$), and a second photon ionizes the molecule (see Figure 3C.1). Thus, the sum of the two-photon energies must be greater than the ionization potential of the molecule for R2PI, although the two photons may have either the same or different frequencies. Since most organic species have ionization potentials between 7 and 13 eV, R2PI can be achieved using near-UV pulsed laser sources. Thus, high peak power and broadly tunable Nd:YAG and excimer-pumped frequency-doubled dye lasers in the UV region serve as versatile sources for selective resonant two-photon ionization and detection of molecules. The selectivity afforded by this method has been used to solve problems such as the discrimination of isomers and isobars, which are often difficult to distinguish using conventional mass spectrometric sources (Tembreull and Lubman, 1984). A further discussion on the principles of resonance ionization can be found in Chapter 4, Part C (Section 4C.2.2) of this volume.

Other MPI processes such as two-photon resonant ionization (TPRI) or, more generally, n-photon resonant ionization are also possible. In these processes, at least two or more photons are needed to reach the first resonance (see Figure 3C.1). Because at least one photon is nonresonant and interacts with a very short-lived virtual state ($< 10^{-15}$ s), the efficiency for ionization in these processes is far less than that achieved in R2PI. Totally nonresonant ionization is also possible; however, very high laser power is needed to drive this very inefficient process (Figure 3C.1). In addition, the wavelength selectivity that is unique to the MPI process as an ionization source for mass spectrometry is lost.

It should be noted that direct photoionization has been used as an ionization source in mass spectrometry for some time (Ditchburn and Arnot, 1929; Lossing and Tanaka, 1956). Ionization results if the single-photon energy is greater than the ionization potential of the molecule. This requires the use of vacuum ultraviolet (VUV) radiation (~ 100 nm). Because VUV lamps are low intensity and are noncoherent sources, the ion yields, and consequently the sensitivity for mass spectrometry, are also often low. The recent development of VUV laser sources may yet greatly increase the possible photoionization yield, although this emerging field of technology needs further development (Herman et al., 1985). It should be noted that in MPI the light source is generally in the visible or near UV range and multiple photon absorption is used to reach the IP. Thus, wavelength selectivity is achieved in MPI when the excitation frequency is in resonance with a real electronic state, whereas in direct photoionization no such selectivity is possible since this process does not proceed through an intermediate state. Also, the laser light used in MPI can be transmitted in air as opposed to VUV sources, which must operate in vacuum. In addition, the laser can be located external to the mass spectrometer source, and because laser light is a coherent form of radiation, the laser can be a conveniently long distance from the ion source without significant loss of intensity. In general, MPI obviates the need for large pumping systems and the VUV monochromator required for direct photoionization. Nevertheless, laser-induced VUV radiation has recently proven a valuable alternate means of softly ionizing and detecting organic molecules in a mass spectrometer (Pallix et al., 1989; Steenvoorden et al., 1991; van Bramer and Johnston, 1990).

R2PI has several important attributes for mass spectrometry in addition to its potential for selectivity. In particular, R2PI can be used to control the fragmentation pattern obtained. This method can provide very efficient soft ionization of molecules yielding the molecular ion with little or no fragmentation (Bernstein, 1982; Boesl et al., 1981; Dietz et al., 1980a,b; Gobeli et al., 1985; Lubman and Kronick, 1982; Lubman et al., 1980; Rettner and Brophy, 1981; Rhodes et al., 1983; Sack et al., 1985; Tembreull and Lubman, 1984; Tembreull et al., 1985a,b, 1986; Wessel et al., 1981; Zandee and Bernstein, 1979a,b; Zandee et al., 1978. Soft ionization generally appears to occur at modest laser energies ($<10^6$ W/cm^2) in a wide range of organic species. In small aromatics such as aniline (Dietz et al., 1980a,b; Lubman and Kronick, 1982; Rettner and Brophy, 1981), benzene (Boesl et al., 1978; Lubman et al., 1980; Reilly and Kompa, 1980), naphthalene (Lubman and Kronick, 1982; Lubman et al., 1980; Rhodes et al., 1983), substitued aromatics (Tembreull and Lubman, 1987b), N, S, and O-heterocycles (Tembreull et al., 1985c), and other larger but rigid aromatics such as polynuclear aromatic hydrocarbons (Lubman and Kronick, 1982; Rhodes et al., 1983; Sack et al., 1985; Tembreull and Lubman, 1986), strong molecular ion signals with no fragmentation can be routinely produced. In more recent studies, relatively soft ionization has resulted from R2PI of fragile biological and thermally labile molecules as discussed in more detail later in this chapter (Engelke et al., 1987; Grotemeyer et al., 1986a,b,c; Hager and Wallace, 1988; Hahn et al., 1987; Lubman, 1986; Tembreull and Lubman, 1986, 1987a,b,c). This process simplifies the mass spectrum considerably and allows for identification based upon molecular weight. Of course, some compounds, such as certain inorganic complexes [e.g., $Fe(CO)_5$] and some organometallics, are quite fragile and fragment even at low power owing to photodissociation by the first photon before MPI can occur (Duncan et al., 1979); however, these systems are the exception rather than the rule.

The R2PI technique is an extremely efficient means of producing molecular ions. Often ionization of up to several percent of the seed molecules present in a molecular beam can be produced without fragmentation within the intersection of the laser beam and the molecular beam while the laser is on (Frueholz et al., 1980; Lubman et al., 1980; Rhodes et al., 1983; Sack et al., 1985; Seaver et al., 1980; Tembreull and Lubman, 1984). In several cases such as aniline and naphthalene, ionization efficiencies ranging from 10 to 100% have been estimated (Boesl et al., 1981; Brophy and Rettner, 1979; Dietz et al., 1980a,b; Lubman et al., 1980). The high efficiency of R2PI contrasts sharply with that of the electron impact source, which typically ionizes less than $1/10^4$ of those molecules that enter the ionization chamber of a mass spectrometer. Also, soft ionization with an electron beam can be achieved only at low beam energies at which the ionization cross section is small, resulting in a loss in ionization efficiency of several orders of magnitude.

The ultimate limits to the efficiency of laser R2PI are fundamental considerations, such as the absorption cross section of the molecule at a particular wavelength and the radiationless transition rate (generally due to internal conversion), that is, the rate at which energy leaks out of S_1 before the second photon can induce ionization. Thus, molecules with groups that induce radiationless transitions, such as chlorinated and brominated groups in aromatic rings, will generally exhibit less efficient ionization than that of their unsubstituted counterparts. However, when R2PI is used, even molecules with very short lifetimes will ionize efficiently. Indeed, molecules that have picosecond or subpicosecond lifetimes can be easily ionized by increasing the rate of up-pumping by increasing the laser output power, thereby overcoming the radiationless transition rate. More recently, picosecond laser pulses have been used to enhance the ionization efficiency for species with short excited state lifetimes (Wilkerson and Reilly, 1990; Wilkerson et al., 1989). Theories that model the competing processes in MPI and discuss the effect of substitution on the efficiency of R2PI have appeared in the literature (Dietz et al., 1980b; Tembreull et al., 1985a,b).

A key feature of MPI is that it is a versatile ionization source where either soft or hard ionization can be produced. Extensive fragmentation can be produced either by increasing the laser power density or as a function of the wavelength chosen for ionization. Zandee and Bernstein, for example, have demonstrated that C^+ is the predominant ion obtained from REMPI fragmentation of benzene when a laser power density of $\sim 10^9$ W/cm^2 is used at a wavelength of 391.4 nm (Bernstein, 1982; Zandee and Bernstein, 1979a,b; Zandee et al., 1978). As the power density is varied, the fragmentation pattern will correspondingly alter where different ratios of carbon fragments including $C_6H_n^+$, $C_5H_n^+$, $C_4H_n^+$, $C_3H_n^+$, $C_2H_n^+$, and CH_n^+ are observed. Similar results have been observed in MPI as a function of laser power in the near-UV (300–260 nm) by numerous other investigators (Boesl et al., 1981; Carlin and Freiser, 1983; Fisanick et al., 1980; Gobeli et al., 1985; Irion et al., 1982; Lubman and Kronick, 1982; Lubman et al., 1980; Rettner and Brophy, 1981; Rhodes et al., 1983; Sack et al., 1985; Seaver et al., 1980; Tembreull and Lubman, 1984, 1986, 1987a,b,c; Tembreull et al., 1985a,b, 1986; Wessel et al., 1981). The fragmentation process generally occurs through the ladder switching mechanism for most organic molecules using a 10 ns laser pulse (Boesl et al., 1980). In this mechanism the molecular ion is initially produced by R2PI or by one of the other MPI processes outlined previously. As the laser power is increased, subsequent absorption of additional photons may occur, resulting in excitation to a state that dissociates, producing ionic (and neutral) fragments. The ionic fragments may absorb subsequent photons, producing yet smaller ionic fragments.

The fragmentation produced by R2PI/MPI is often similar to EI-MS rather than to particle bombardment techniques, since it involves fragmentation of the radical cation, $M^{+\cdot}$. Although the fragmentation produced by R2PI/MPI is similar to EI, there are several fundamental reasons why they may be different. Fragmentation patterns produced by laser ionization are initiated by light typically $\sim$300 nm in wavelength (i.e., much larger than the dimension of the molecule), so that the transitions occurring are Franck–Condon controlled. Thus, light-induced transitions occur vertically without a change in the internuclear distance and only transitions to certain states will be readily allowed. Also, because of spin conservation from molecule to ion, fewer states are accessible than by EI. This is true because ionization by 70 eV electron impact corresponds to radiation of $\sim$0.15 nm energy, so that strong perturbation of the molecule occurs and many more states are accessible with the potential for much more interesting fragmentation for analysis.

The use of laser photon up-pumping or the ability to vary the wavelength of light may lead to several unique properties of R2PI/MPI for producing fragmentation relative to EI. In particular, it has been shown that an important consequence of the ladder switching process in MPI/dissociation is that it leads to a sharp energy distribution in the parent ion (Boesl et al., 1980, 1982; Kuhlewind et al., 1985). Any further photon absorption occurs within the parent ion and produces parent ions excited at a specific energy. Thus molecular ions can be excited to definite energy levels with UV or visible light, thus producing very specific intermediate states—producing in turn very specific products. In addition, the products have a well-defined energy that will change as a function of photon energy. This is in contrast to EI or conventional photoionization, where a range of energies is placed into a molecule and a broad range of fragmentations is observed. This is especially significant in using laser radiation of a given energy to enhance the production of metastables that may not be readily observed in EI owing to an intensity that is too low to be easily detected (Boesl et al., 1982; Kuehlwind et al., 1985). In addition, the ability to use high-powered lasers to induce MPI can result in the easy formation of fragments of high appearance potential (AP) not readily observed in EI (Tembreull and Lubman, 1987a,b,c).

An important examples of the use of discrete photon selective excitation/fragmentation is given in the case of the isomers of alkyl iodides performed by groups in Bernstein's and Schlag's laboratories (see Kuehlwind et al., 1985; Parker and Bernstein, 1982). These groups found that the structural isomers of various alkyl iodides gave distinctly different fragmentation patterns in the MPI–dissociation (MPI-D) mass spectrum, although those of the EI-MS were very similar. The implication here was that isomerization occurs before dissociation for EI, but not for MPI-D. Experiments by the Schlag group showed that as the energy placed into the ion by rapid photon absorption increased, the rates of isomerization and dissociation crossed over so that the rate of dissociation exceeded that of isomerization. Thus, MPI-D allows for structural sensitivity in the mass spectra based upon the total photon energy placed in the parent ion. This, of course, can be varied by changing the photon energy or power density to increase the up-pumping rate.

The R2PI method is used to best effect in combination with supersonic jet introduction. The theory of jet expansions has been reviewed extensively in the literature and appropriate references are included therein (Hayes, 1987; Levy, 1984; Miller, 1984; Smalley et al., 1977). In this technique, the analyte molecule is seeded into a monoatomic carrier gas at typically 1 atm and expanded through a pinhole orifice into vacuum at 10^{-5} torr. In the expansion, the internal energy of the analyte molecules is relaxed through two-body collisions with the carrier gas. The internal energy of the molecules is converted ultimately into the translational energy of the carrier, resulting in ultracold molecules. At room temperature, large polyatomic molecules exhibit broad unresolvable absorption spectra. This phenomenon is due to the thermal population of a large manifold of rovibronic states in the ground electronic state, which results in spectral congestion in the electronic transition. However, in a jet expansion the cold molecules that result exhibit sharp features in R2PI spectroscopy. These sharp spectral absorptions can be used for selective ionization and identification and in providing well-defined internal energy distribution for excitation in MPI-D experiments. In addition, the narrowing of the translational energy distribution that occurs in the jet provides enhanced resolution in a time-of-flight (TOF) mass spectrometer.

The R2PI method is particularly compatible even with high-density jets because the near-UV wavelengths are transparent to the expansion gas and only interact with the analyte species.

3C.3. EXPERIMENTAL METHODOLOGY

3C.3.1. Apparatus

The strategy behind the experiment described herein involves entraining the neutral species produced by pulsed laser desorption into a supersonic jet expansion. These species are subsequently ionized by laser R2PI in a TOF-MS. The two different experimental setups used in this work are shown in Figures 3C.2 and 3C.3. The basic experiment consists of a supersonic beam TOF-MS (Tembreull and Lubman, 1987b). The experiment is oriented such that a pulsed supersonic molecular beam crosses the acceleration region of the TOF device perpendicular to the flight tube. A laser beam orthogonal to both of these ionizes the sample within the acceleration region of the mass spectrometer.

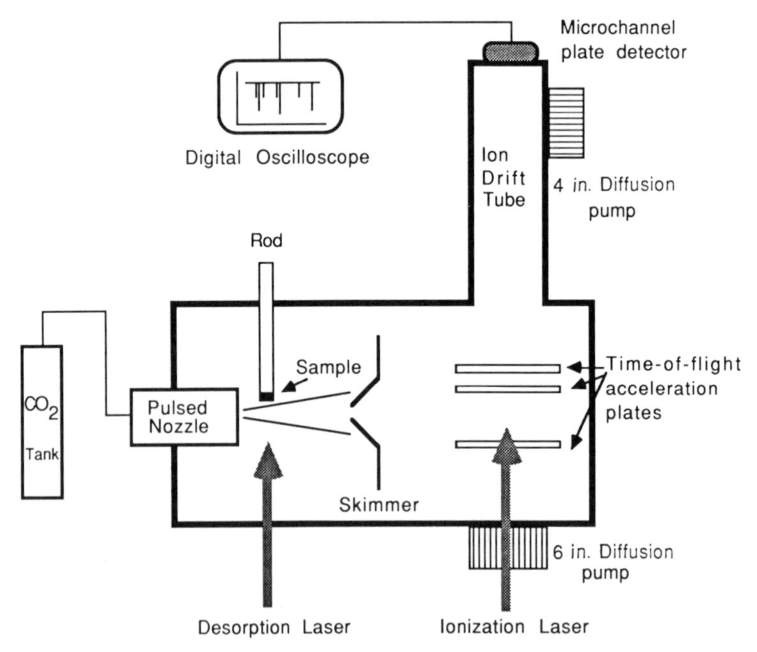

Figure 3C.2. Schematic of the experimental apparatus for the linear TOF-MS setup.

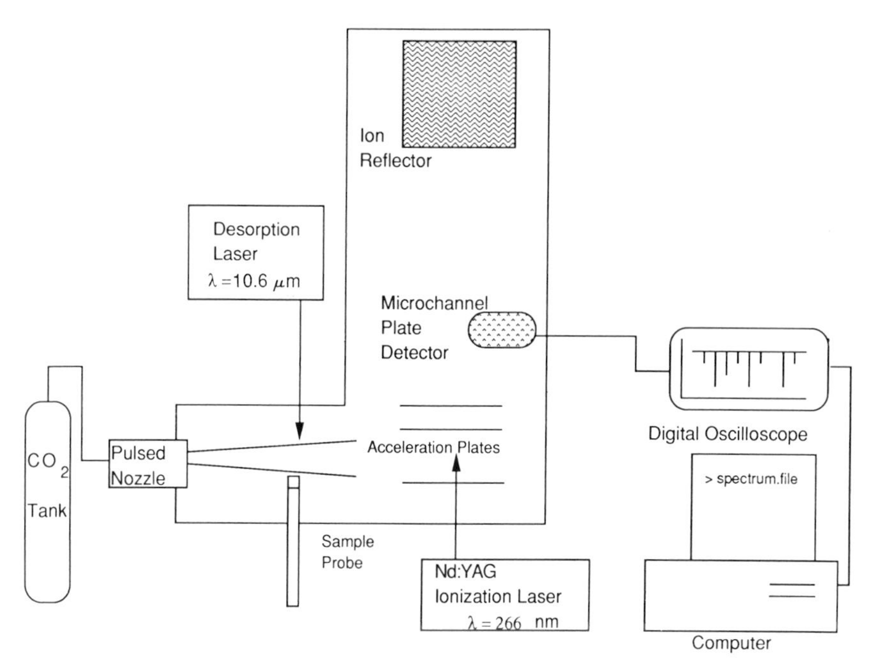

Figure 3C.3. Schematic of the experimental apparatus for the reflectron TOF-MS setup.

In most of the experiments described here, we used a simple linear TOF-MS designed after that of Wiley and McLaren (1955) which has been modified for supersonic beam injection and had sufficient resolution for the small biological molecules (< 500 u) studied in this work. The TOF-MS is mounted vertically in a 6 in. stainless-steel cross that is pumped by a 6 in. diffusion pump. The acceleration region is enclosed by a liquid N_2-cooled cryoshield that contains a conical skimmer to skim the molecular beam and ports to allow the laser beam to enter this region. This enclosure introduces a conductance limit to pumping by the main pump so that a 4 in. liquid N_2-baffled diffusion pump provides a pressure of $< 10^{-5}$ torr in the flight tube. The combination of the cryoshield and differential pumping significantly reduces the background ionization over that observed in a single-stage system (Tembreull and Lubman, 1986, 1987b).

In more recent work, a reflectron TOF-MS (reTOF-MS) has been used in order to obtain increased resolution for higher-molecular-weight compounds (Boesl et al., 1982; Mamyrin et al., 1973). The reTOF (manufactured by R. M. Jordan Co., Grass Valley, California) (see Figure 3C.3) is constructed of a 6 in. diameter stainless-steel cross that serves as an ionization chamber and an 8 in. diameter stainless-steel flight tube. Ions created in the chamber are accelerated into the flight tube at 2–3 kV. At the reflector the ions are turned back toward the source with an angular displacement from the initial flight axis. The ions are then detected by a two-stage 40 mm microchannel plate detector. The reTOF-MS is pumped by two diffusion pump stacks. A 6 in. pump is used on the ionization chamber, and a 4 in. pump is used on the flight tube. Additional pumping is provided to the chamber via a liquid N_2 cold baffle mounted on the cross. This baffle pumps away the CO_2 carrier gas from the jet. Typically, the pressure in the flight tube is 2×10^{-6} torr.

The supersonic beam source uses a stainless-steel pulsed valve (R. M. Jordan Co., Grass Valley, California) that is based on the magnetic repulsion principle (Byer and Duncan, 1981; Liverman et al., 1979; Rorden and Lubman, 1983). This nozzle provides gas pulses of $\sim 55\,\mu s$ fwhm (full width at half-maximum) at "choked flow" so that at a 10-Hz laser repetition rate the valve is on only 550 μs/s. The essential part of this design is the "hairpin loop" of highly conductive metal (Byer and Duncan, 1981; Gentry and Giese, 1978; Liverman et al., 1979; Rorden and Lubman, 1983), originally published by Gentry and Giese (1978). This loop is clamped on both ends, and an insulator

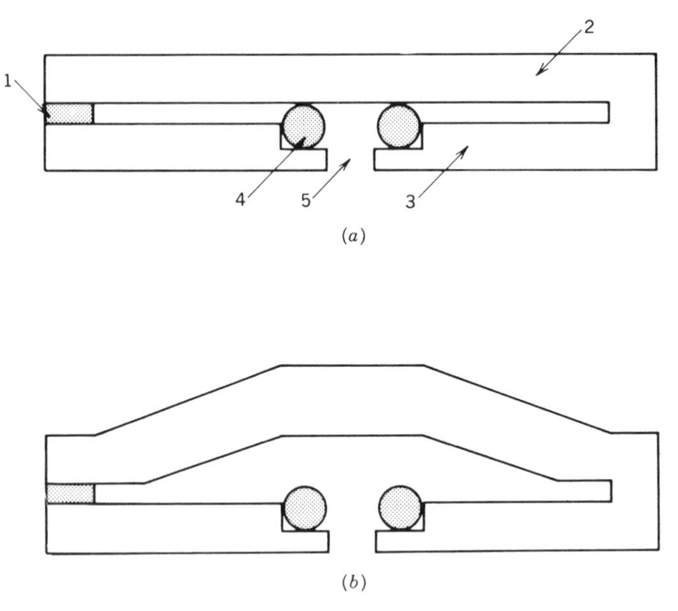

Figure 3C.4. (a) Hairpin loop in a position where the top spring is sealed against the O-ring so that the valve is in the closed position: 1, top plate; 2, bottom plate; 3, insulator; 4, O-ring; 5, orifice. (b) Hairpin loop with current flowing in opposite directions in the top and bottom plates so that the resulting forces generated by the opposing magnetic fields deflect the top plate and the O-ring seal is broken. Reprinted from Lubman (1988b) with permission.

is placed between the top and bottom plates of the hairpin (Figure 3C.4). When a high-current pulse on the order of several thousand amperes flows through the two plates in opposite directions, the mutual repulsion forces generated by the magnetic fields resulting from the antiparallel currents cause a displacement of the top plate. The displacement of the top plate breaks an O-ring seal that normally seals the top plate against an orifice drilled into the bottom plate. The gas is allowed to flow through the orifice and nozzle until the upper plate mechanically returns to its original position and seals the orifice.

The fast pulse produced from the commercial version of this valve is such that the duty cycle needed for pumping is reduced by a factor of ~2000 over a comparable continuous source. More critical, however, is the percentage of molecules actually sampled by the laser. Because the beam velocity is $\sim 4 \times 10^4$ cm/s and the pulse width is 50 μs (Rorden and Lubman, 1983), the beam spread in space is ~2 cm so that a 4 mm laser beam probes 20% of the available molecules on-axis. The shortest gas pulse consistent with achieving full hydrodynamic flow through the orifice is therefore desirable. Although gas pulses of 20 μs fwhm can be achieved with this valve, the transmission does not reach "choked flow," which is maximum flow through the orifice under the expansion conditions, until a 55 μs pulse is obtained. In addition, it should be noted that the throughput of the orifice increases as D^2 (where D = diameter of the orifice) and the ultimate cooling attained also increases with the orifice diameter, so that the largest aperture is desirable—limited, of course, by the ultimate pumping speed available (Lubman et al., 1982). If a continuous wave (CW) expansion were used under the same conditions in our experiment, only a 45 μm orifice could be used (instead of a 500 μm orifice), with a corresponding decrease in sensitivity proportional to D^2. The nozzle-to-excitation distance is 16 cm, which is limited by the constraints of our particular apparatus. The shortest distance that places the jet expansion in its "free-flow" or collisionless region by the time the skimmer is reached is preferable since the density decreases on-axis as $1/x^2$ (x = distance from the orifice). The jet must reach the free-flow region before the skimmer; otherwise shock waves will destroy the jet expansion.

One important feature of the use of supersonic jets is a marked improvement in the resolution of the TOF device. The Boltzmann energy spread of molecules in the acceleration region generally limits the width of the ion packets at the detector so that a mass resolution of ~150–200 may be observed in a TOF device. However, in an ideal supersonic jet all the molecules are traveling in he same direction with the same velocity so that the energy spread is minimized, and a resolution of up to 1600 has been obtained, limited only by our 5 ns laser pulse (Lubman and Kronick, 1982). This resolution, of course, holds only in the absence of large energy release or if severe fragmentation does not occur where spatial effects may limit the mass resolution. If higher resolution is required, then the reTOF provides a resolution of 2300 at *m/z* 93 and the resolution increases to 4100 at *m/z* 306.

3C.3.2. Desorption Method

In our experiments desorption is produced using a CO_2 laser at ~10–40 mJ/pulse (Quanta-Ray Exc-1, Mountain View, California). This contrasts to the direct laser desorption (LD) experiments in which generally a thin layer of material several monolayers thick is rapidly desorbed and ionized by using a high-powered pulsed CO_2 laser ~1 J in energy (Coates and Wilkins, 1987; Lubman et al., 1982; McCrery and Gross, 1985; Wilkins et al., 1985). Because we are interested in neutral species rather than ions, significantly lower desorption energies can be used. The sample introduction into the supersonic expansion was performed by desorbing the compound of interest from the surface of an $\frac{1}{8}$ in. diameter rod of machineable Macor™ that was found to produce the least amount of decomposition in the desorption process. In our initial laser R2PI-MS experiments, the sample was deposited on the face of the rod by dissolving the compound in benzene or methanol and coating the surface with the use of a spatula. Rather large (milligram) samples were employed in our preliminary studies to demonstrate the general feasibility of the method rather than to demonstrate quantitative analysis. A typical sample may consist of many hundreds of monolayers. By controlling the laser energy, several monolayers can be desorbed on each laser pulse.

In more recent work, glycerol or various other viscous fluids have been used as matrices from which to desorb the material (Li and Lubman, 1989a). It has been found that in thick neat samples that the CO_2 laser radiation is absorbed by the sample and may result in pyrolysis. Thus, the glycerol serves as a heat sink to prevent thermal decomposition. Alternatively, thin films of sample can be used to prevent decomposition in the desorption process.

The key to the desorption method used for the R2PI jet-cooled spectroscopy experiments is the ability to obtain shot-to-shot desorption stability. This is achieved by dissolving the material of interest in a high-viscosity fluid such as glycerol or silicone diffusion pump fluid. Generally ~100 μg of sample is used in these experiments and the glycerol matrix causes the sample to form a very even thin layer from which pulse-to-pulse stability appears quite good (i.e., somewhat less than $\pm 5\%$). The rate of desorption can then be controlled by adjusting the laser power on the surface so that total desorption can be produced in seconds or over an extended length of time (i.e., more than a half hour).

The desorption process in these experiments is actually a matrix-assisted desorption effect that produces neutral species rather than ions (Beavis and Chait, 1989; Karas and Hillenkamp, 1988). The CO_2 laser radiation at 10.6 μm is strongly absorbed by the glycerol matrix. The laser wavelength matrix-assisted desorption of the glycerol enhances the desorption of the neutral species dissolved in the glycerol. If a Nd:YAG laser at 1.06 μm, which is not absorbed by the glycerol, is used for this experiment, then the radiation is absorbed by the Macor probe surface. The result is heating of the probe surface and the resulting thermal decomposition of the sample. A further more detailed discussion on the principles of laser desorption and volatilization is found in Chapter 3, Part A, of this volume.

3C.3.3. Supersonic Jet Cooling

In order to perform R2PI jet-cooled spectroscopy by using the pulsed laser desorption method for entrainment into the supersonic jet expansion, the cooling must be optimized because the sample is no longer expanded with the carrier gas through the nozzle. In this case the sample is introduced into the jet expansion outside the nozzle orifice. Thus, a critical parameter for obtaining optimal rovibronic cooling is the position of the ceramic rod with respect to the jet expansion. There are two dimensions to consider here: (1) the distance of the rod from the molecular beam axis, and (2) the distance of the rod from the nozzle orifice. In the first case, the best cooling was observed when the rod was as close to the supersonic beam as possible without destroying the jet. This was performed in our work by including a volatile sample such as aniline in the jet and moving the rod toward the beam until the molecular ion peak intensity started to decrease. As the rod was moved further, generally a dramatic decrease occurred as the rod began to interfere with the jet. In the second case, the distance between the center of the rod and the nozzle in our experiments was ~5.5 mm. This distance provided excellent cooling; however, if this distance was decreased, a significant decrease in signal was observed with no noticeable increase in cooling. This appeared to occur because, at shorter distances at which the carrier density was very high, it was difficult to efficiently penetrate the beam without causing shock waves. At a longer expansion distance at which the carrier density was lower, sample penetration appears to be much more efficient. The use of CO_2 vs. Ar or He provides an increased number of collisions at a longer expansion distance for enhanced collisional cooling (Lubman et al., 1982).

The desorption was performed using a pulsed CO_2 laser at 10.6 μm. The IR beam was softly focused with a 10 cm focal length biconvex germanium lens to a ~2–3 mm spot for desorption, although the focus was adjusted for each sample molecule to optimize our results. The amount of sample desorbed per pulse depends on the power density on the surface and on the properties of the sample—in particular, the melting point. An estimate of the desorption power density of the surface can be made by measuring the beam energy using a power meter (Scientech Model 365) and from the known beam temporal profile (Li and Lubman, 1988c). In the case of catechol, a low melting point compound (mp = 106 °C), the desorption was so efficient that a low laser power was used in the experiments ($\sim 1 \times 10^5$ W/cm^2), whereas for tyrosine (mp = 325 °C) a power density of nearly 5×10^6 W/cm^2 was used. By carefully controlling the desorption laser power the sample can be made to last for an extended period of time for spectroscopic scans. In several spectroscopic scans in our work a 100 μg sample has lasted for ~0.5 h at a repetition rate of 10 Hz. This is equivalent to $\sim 3 \times 10^{13}$ molecules or ~0.1 monolayers desorbed per CO_2 laser pulse.

3C.3.4. Laser Ionization

The laser ionization (R2PI) was performed using the frequency-doubled dye output from a Quanta-Ray DCR-2A Nd:YAG pumped dye laser system. For the R2PI-MS experiments, the fourth harmonic of the Nd:YAG laser at 266 nm can also be used as the ionization source. The 6 mm output beam was collimated with a combination of 30 cm focal length positive lens and a 10 cm focal length negative lens to produce a laser beam 2–3 mm in diameter. The power density was adjusted to give the desired signal level, which depended on the amount of material entrained in the jet and the efficiency of ionization of each particular compound.

The actual sequence of events was controlled by several delay generators in which the pulsed CO_2 laser fires first to produce desorption, followed by the pulsing of the valve. The two events were synchronized in time so that the desorbed plume was entrapped into the jet expansion of CO_2 and carried into the acceleration region of the TOF-MS. The flight time of the jet from the pulsed valve to the ionization region was ~300 μs, and the laser was set to pulse as the gas pulse arrived at this point. Laser R2PI was produced and a LeCroy 9400 digital oscilloscope was used to record the mass spectrum. The wavelength spectrum was obtained by using an SRS 250 gated integrator to monitor only the molecular ion as the dye laser was scanned.

3C.4. APPLICATIONS

3C.4.1. Spectroscopy and Mass Spectrometry of Neurotransmitters and Other Small Biological Molecules

We have used the pulsed laser desorption/laser R2PI technique to study the detection of catecholamines, indoleamines, and their metabolites. The selective detection of catecholamines would be of great utility in the investigation of neuroblastomas and other neurogenic tumors, Parkinson's disease, and psychological stress (Boniforti et al., 1983; Robinson, 1980). For example, neuroblastomas affect nerve structures involved in synthesizing catecholamines, so this condition can be diagnosed by the concentration of metabolites in the urine. The quantitative and qualitative determination of these compounds is of importance for cases in which high concentration values for the metabolites homovanillic acid and vanilmandelic acid in urine are indicative of an unfavorable prognosis and rapid metastasis. At present, simple chemical tests are used to determine total catecholamine concentration or the presence of several metabolites, but the degree of selectivity available is unsatisfactory for the type of exact identification needed for the analytes present. A more selective means of specifically monitoring each catecholamine metabolite would be highly desirable (Robinson, 1980). In addition, the lability of these compounds even under mild temperature conditions makes this an important test of the capabilities of any given technique for analyzing other fragile biological species.

Alternatively, we have also been interested in the selective detection of indoleamines and their metabolites. This group of compounds include OH- and CH_3O-substituted indoleacetic acid, tryptamine, tryptophol, and *N*,*N*-dimethyltryptamine. Each of these biogenic indoles serves as important metabolites of biological processes in the human body (Hooper, 1981). The concentration of each of these metabolites in blood or urine reflects various conditions that clinicians are interested in monitoring. Serotonin (5-hydroxytryptamine) (Veca and Breisbach, 1988), for example, is a neurotransmitter found in high concentrations in the blood stream. Depression is thought to be caused by a deficiency of serotonin (Veca and Breisbach, 1988). The pineal specific methoxyindoles exert important effects upon reproductive activity (Hooper, 1981), and the *N*,*N*-dimethyltryptamines are possible causative agents of psychotic illnesses (Hooper, 1981). Gas chromatography–mass spectrometry (GC-MS) using derivatization to enhance the volatility of these labile compounds has been employed for analysis of these compounds (Hooper, 1981). The potential of LD-R2PI supersonic jet spectroscopy is that it allows volatilization and detection with minimal sample preparation and enhanced optical selectivity prior to mass analysis. The importance of this group of compounds has led us to investigate R2PI as a soft ionization tool for detection via the molecular ion and highly controlled fragmentation.

The catecholamines and related species were examined using the pulsed laser desorption–laser R2PI method (Figure 3C.5). The key result is that generally molecular ions with little or no fragmentation were observed at modest laser input energy at 280 nm ($< 7 \times 10^5$ W/cm^2). The ion peak observed is the radical cation, $M^{+\cdot}$, as in EI. There are some exceptions such as epinephrine and norepinephrine, which undergo facile cleavage even at the

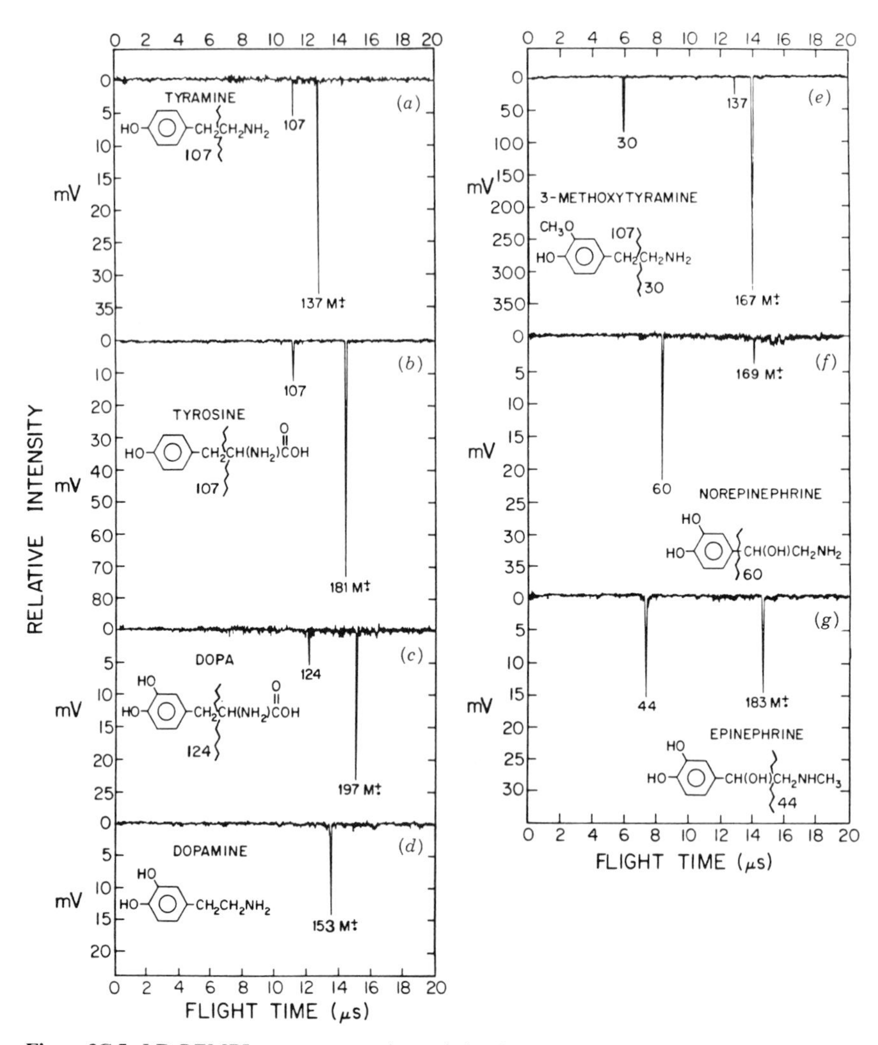

Figure 3C.5. LD-REMPI mass spectra of catecholamines and related derivatives performed at (a) 280 nm, (b) 280 nm, (c) 266 nm, (d) 280 nm, (e) 280 nm, (f) 266 nm (g) 280 nm. Reprinted from Tembreull and Lubman (1987b) with permission.

threshold laser ionization energy used to obtain an observable signal, so that production of only the molecular ion is not observed. R2PI of 3-methoxytyramine, dopamine, DOPA, and tyrosine produced only the molecular ion at 280 nm with laser beam energies below 2–3 mJ. However, even a small increase in energy of the 280 nm ionizing beam, such as was used to obtain the results in Figure 3C.5, significantly enhances the intensity of the molecular ion while only moderately promoting bond scission in the alkylamine fragment. The 280 nm wavelength was chosen since *p*-cresol, which is the spectroscopic precursor to tyrosine and tyramine, has its origin absorption at 282.97 nm. As will be shown later, it would be expected that these compounds absorb in a very similar region of the spectrum, where the absorption is only shifted by the effects of the different substituent groups (Tembreull et al., 1985a,b).

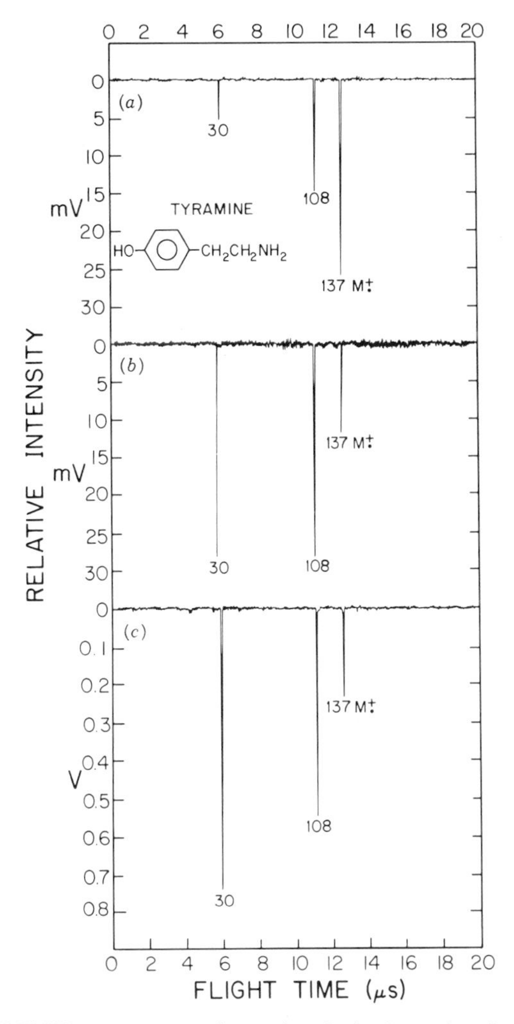

Figure 3C.6. LD-REMPI mass spectra of tyramine obtained as a function of wavelength: (a) 266 nm, (b) 245 nm, (c) 222 nm. Reprinted from Tembreull and Lubman (1987b) with permission.

The ionization wavelength is a key parameter in these experiments that can have important effects on the mass spectrum observed. The effect of ionization wavelength on mass spectral fragmentation patterns was studied for several of the catecholamines and indoleamines, as illustrated in Figure 3C.6 for tyramine. In the cases studied here, there is no significant change in the type of fragments observed as the ionization wavelength decreases. In fact, most of the fragment ions can be produced at any wavelength at which the molecule absorbs by simply increasing the ionizing laser power (Tembreull and Lubman, 1987b). However, the threshold power density required for the appearance of fragments with a relatively high appearance potential decreases as the lasing frequency increases. In addition, substantial changes in the relative fragment ion abundance are observed as a function of wavelength at the same laser beam intensity. For example, as the laser frequency increases in Figure 3C.6, one notes the appearance of a small ionic fragment at mass 30 (CNH_4^+) that was not observed at the same input energy at 280 nm (Figure 3C.5a). As the frequency increases, the abundance of the mass 30 peak continues to increase relative to the other peaks.

If the frequency of the ionizing laser light is increased to 245 or 222 nm from 280 or 266 nm, respectively, the relative abundance of the ionic fragments also increases as shown in Figure 3C.6. In the case of the catecholamines and related precursors, we found that the trend is for benzylic cleavage to dominate at low-photon energy, whereas charge migration followed by cleavage β to the aromatic amine nitrogen becomes more important at high-photon frequency given the same laser flux. In the case of tyramine, for example, at 266 nm the mass 107 peak due to simple β cleavage is large relative to the mass 30 peak, whereas at $\lambda = 222$ nm the mass 30 peak due to the second mechanism becomes dominant. At 222 nm norepinephrine yields a prominent aromatic fragment (m/z 109) due to simple cleavage α to the benzene ring. This compound also undergoes the familiar cleavage between the bonded carbons α and β to the nitrogen atom at this wavelength, resulting in a peak at mass 30.

An important point common to all spectra observed throughout this work is that the laser-induced mass spectra exhibit fragment ions that are similar to those obtained in conventional electron impact spectra. This result was not unexpected, as other investigators have reported this to be the case (Budzikiewicz et al., 1964). However, the great advantage of laser MPI is that it provides a greater degree of control over the number of fragments observed and their relative abundance. By simple variation of the laser energy or wavelength, it is often possible to obtain only molecular ions with high efficiency or diagnostic fragments of relatively high appearance potential. The mass spectra are frequently simpler than those obtained in EI because the selection rules involved are different. However, with MPI, fragments whose production are energetically unfavorable can be produced in higher abundance by successive photon pumping as the laser power is increased.

Although the main fragments observed in the laser-induced mass spectra shown here are those that are expected, generally fewer fragments are observed than in EI and these fragments are characteristic of the structure of the molecule. This is demonstrated for the catecholamines at 280 or 266 nm (see Figure 3C.5), the fragmentation pattern of which is dominated by two competing mechanisms when a relatively low energy ($< 10^6$ W/cm^2) is used. These mechanisms, which are similar to those of electron beam ionization, involve either simple cleavage of the benzylic C_α—C_β bond to form a stable even-electron fragment ion or rapid charge migration to the amine nitrogen and subsequent β-cleavage to this atom (Budzikiewicz et al., 1964). Following charge migration, bond scission probably occurs via expulsion of an aralkyl radical, again resulting in formation of an even-electron cation.

Another important point common to all the spectra observed in these experiments is the absence of cationization. This phenomenon appears to be ubiquitous in laser desorption experiments in which ionization and desorption are performed in a single step with the use of high laser power density ($> 10^8$ W/cm^2) (Coates and Wilkins, 1987; McCrery and Gross, 1985; Wilkins et al., 1985). However, in the experiments performed here cationization was never observed. In previous work (Tembreull and Lubman, 1986) pulsed laser desorption with a CO_2 laser followed by R2PI using a UV laser beam was performed within the acceleration region of the TOF-MS. The use of the pulsed CO_2 laser alone at the power densities used in these experiments produced essentially no ions but only neutrals that could be subsequently ionized with the UV source. The power density of the pulsed CO_2 laser used here is typically at least 1 or 2 orders of magnitude lower than that used in LD-MS experiments, to enhance the production of neutrals over ions.

In the case of the indoleamines, relatively soft ionization could be obtained at 286 nm, near the indole origin. At 280 or 266 nm, as the laser intensity is increased, two basic fragmentation mechanisms are observed that are similar to those known from EI work (Powers, 1968). One channel similar to that observed in the catecholamines and their precursors involves the migration of a positive charge from the indole ring system to the amine nitrogen in the side chain, followed by expulsion of an indole radical. This accounts for ions at low mass, such as m/z 30 in Figure 3C.7, which are produced by rupture of the bond β to the positively charged nitrogen atom. The second fragmentation pathway involves loss of a substituted alkyl radical, which results in breaking of the highly activated bond β to the indole nucleus. The m/z 130 peak in Figure 3C.7 results from expansion of the indole ring, resulting in formation of the quinolinium ion, and sequential loss of HCN and acetylene from this ion results in the ions observed at m/z 103 and 77, respectively. Although at 266 nm mainly the molecular ion is produced at modest laser energy for tryptamine (Figure 3C.7), as the frequency increases, smaller ionic fragments are produced and there is a shift in the relative importance of the two fragmentation pathways. Thus, as in the case of the catecholamines, ions of greater appearance potential are more easily produced by using photons of higher frequency. As the photon frequency increases, it appears that the molecule is excited to ionic states that dissociate to produce ionic fragments. The excited ionic state

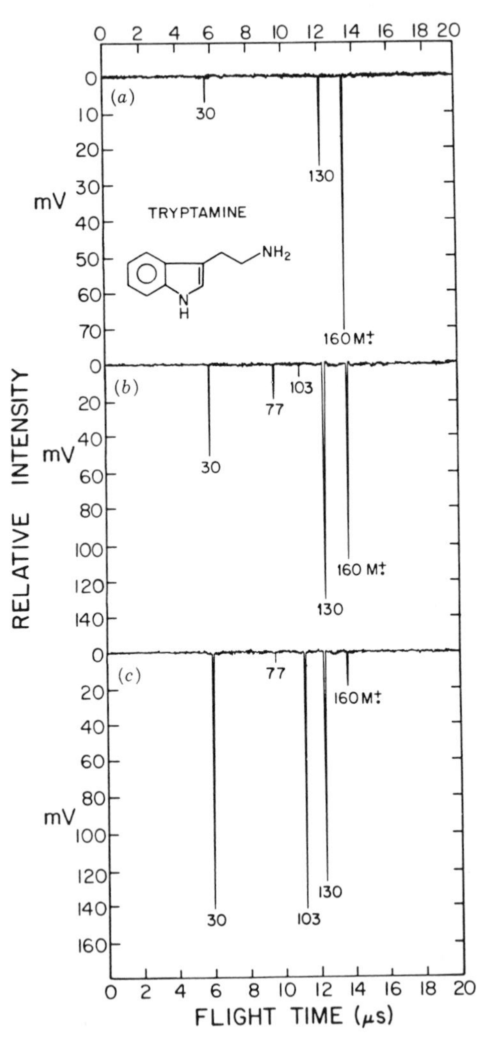

Figure 3C.7. LD-REMPI mass spectra of tryptamine obtained as a function of wavelength: (a) 266 nm, (b) 245 nm, (c) 222 nm. Reprinted from Tembreull and Lubman (1987b) with permission.

reached will in turn determine the degree and mechanism of fragmentation observed.

Spectroscopic Analysis. The identification of these compounds by soft or hard ionization in MS can be accomplished by other ionization sources and tandem MS methods besides REMPI. The truly unique feature of REMPI methods is the selectivity that can be achieved in ionization based upon the optical properties of the method. In combination with supersonic jet introduction, various problems in the discrimination of isomeric and isobaric compounds have been demonstrated. In the past, the optical selectivity of REMPI was applied mainly to relatively volatile species that had a reasonable vapor pressure at $<100\,^{\circ}C$. However, the development of the pulsed laser desorption method with entrainment into jet expansions now allows the extension of this method to thermally labile biological species.

In Figure 3C.8 are shown jet-cooled spectra of tryptamine and its substituted analogues (Li and Lubman, 1988b). These spectra were obtained by using pulsed laser desorption with subsequent entrainment in a jet expansion of CO_2 carrier. The molecular ion with either no or minimal fragmentation results from R2PI in each case. In these spectra the molecular ion only is monitored using the gate of a gated integrator. The wavelength spectra obtained exhibit sharp spectral features (<0.1 nm fwhm) that can be used to uniquely identify each species.

The spectra of tryptamine obtained in our work by using the pulsed laser desorption–volatilization method are very comparable to those obtained in previous work (Bersohn et al., 1984; Hayes et al., 1983; Park et al., 1986; Rizzo et al., 1986). Serotonin though is too labile for a spectrum to be obtained by heating methods. The pulsed laser desorption method thus serves to minimize decomposition due to direct heating methods, and no pyrolysis products were observed in the R2PI mass spectrum. The method serves as a means of repetitively and reproducibly introducing nonvolatile compounds into a jet expansion. It should be noted that there are numerous bands observed in the origin region. It has been shown that these can be attributed to the presence of multiple rotational conformers that differ by rotation of part of the extended alkyl chain in space or to the presence of vibrational progressions. Each of these different configurations are frozen out and are stable in the jet expansion. This phenomenon has been reported by a number of previous studies (e.g., Dunn et al., 1985; Oikawa et al., 1984) and has been observed in a jet spectroscopy study of tyrosine analogues (Li and Lubman, 1988d).

The key point of this work is the ability to obtain selectivity in analysis. Several experiments were performed in order to evaluate the analytical potential of this method in regard to the discrimination capabilities and sensitivity. In previous experiments using supersonic jet cooling in conjunc-

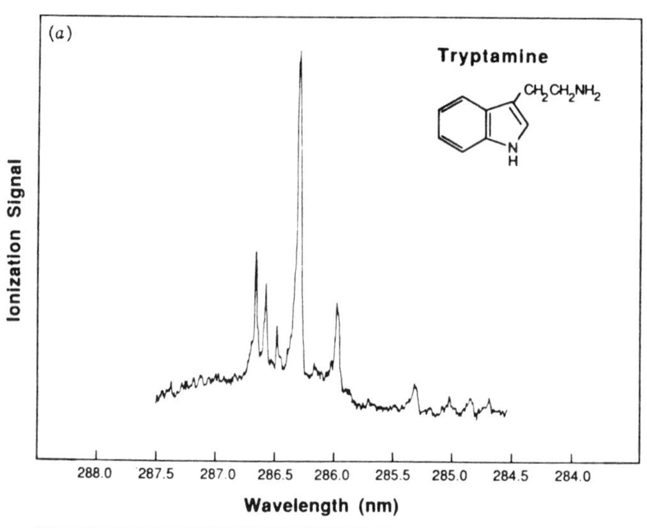
(a)
Tryptamine
CH2CH2NH2
N
H
Ionization Signal
288.0
287.5
287.0
286.5
286.0
285.5
285.0
284.5
284.0
Wavelength (nm)

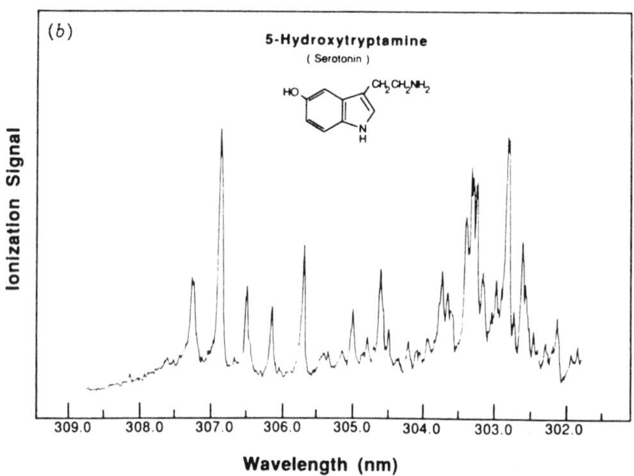
(b)
5-Hydroxytryptamine
(Serotonin)
HO
CH2CH2NH2
N
H
Ionization Signal
309.0
308.0
307.0
306.0
305.0
304.0
303.0
302.0
Wavelength (nm)

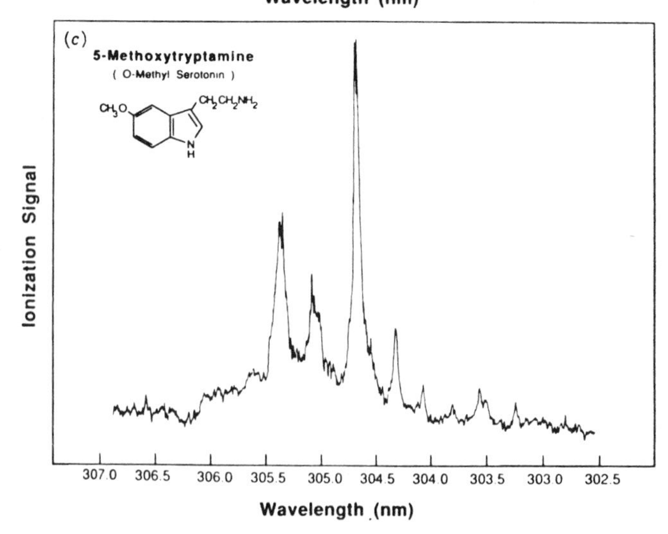
(c)
5-Methoxytryptamine
(O-Methyl Serotonin)
CH3O
CH2CH2NH2
N
H
Ionization Signal
307.0
306.5
306.0
305.5
305.0
304.5
304.0
303.5
303.0
302.5
Wavelength (nm)

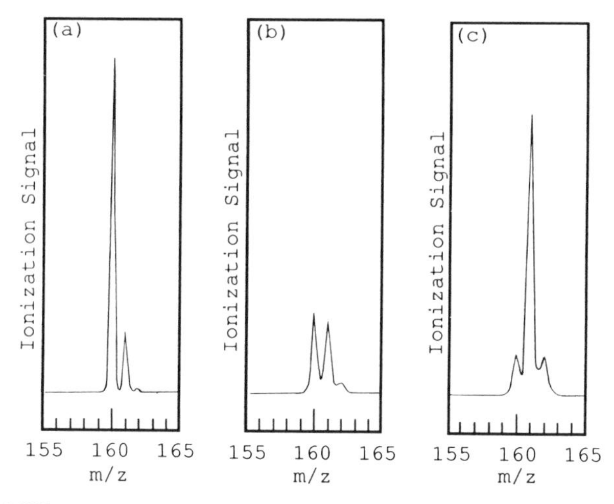

Figure 3C.9. R2PI mass spectra of 1:1 mixture of tryptophol (*m/z* 161)/tryptamine (*m/z* 160) obtained as a function of wavelength: (a) 286.30 nm, (b) 285.27 nm, and (c) 285.59 nm. The laser ionization power density was 5×10^5 W/cm^2. Reprinted from Li and Lubman (1988b) with permission.

tion with R2PI-MS, enhanced discrimination based upon optical selectivity was demonstrated for various disubstituted benzenes under different conditions (Sin et al., 1984). However, in these systems there is generally one sharp origin band observed with few other bands until considerably higher frequency. In the indole- and catechol-based compounds studied, the situation becomes more complex because of the presence of multiple origin bands resulting in congestion in the spectrum. Several types of discrimination situations may arise, as described in the following paragraphs:

One situation, demonstrated in Figure 3C.9, shows wavelength selective ionization mass spectra in a 1:1 mixture of tryptophol/tryptamine dissolved in methanol and placed on a glycerol matrix. In this case, the two molecules have multiple bands located in closely overlapping regions of their wavelength spectra. In Figure 3C.9, the relative abundances of the mass peaks corresponding to these compounds are monitored as a function of wavelength. The mass difference between tryptamine (MW 160) and tryptophol (MW 161) is easily resolvable in the supersonic jet TOF device. When the laser is tuned to 286.30 nm (Figure 3C.9a), the wavelength is in resonance with the strongest tryptamine band in the wavelength spectrum (see Figure 3C.8a) and the tryptamine peak dominates in the mass spectrum while the tryptophol peak is of relatively low

Figure 3C.8. R2PI jet-cooled spectra of (a) tryptamine, (b) 5-hydroxytryptamine, and (c) 5-methoxytryptamine. Reprinted from Li and Lubman (1988b) with permission.

abundance. If the laser wavelength is shifted to 285.27 nm (Figure 3C.9b), there is an absorption band in both tryptamine and tryptophol of almost equal intensity, which results in two peaks of almost equal intensity. When the wavelength is tuned to 285.59 nm (Figure 3C.9c), which is the second strongest absorption band in the tryptophol absorption spectrum (but removed from the tryptamine absorption band), a strong tryptophol molecular ion results at the expense of the tryptamine.

The ultimate selectivity will depend upon the degree of cooling, which will minimize interference from background resonant ionization of molecules that have not been completely cooled in the jet. It will also be limited by the presence of the ^{13}C peak of tryptamine at MW 161, which will overlap with the mass peak of tryptophol in our TOF device. Nevertheless, using this method we have been able to optically discriminate 1:100 of tryptamine in tryptophol at 286.30 nm. The term *optical discrimination* refers to the ability to distinguish two compounds based upon their unique spectral absorption bands. A formula for calculating this term has been defined in previous work (Sin et al., 1984) as $D = 3.3\delta(C_a/C_b)IS$, where δ is the distance between neighboring peaks/fwhm peak width, C_a and C_b are the concentrations of components a and b, I is the correction in laser intensity at different wavelengths, and S is a correction factor for the different relative peak heights between the two absorption lines under the same conditions. When $\delta = 1$, two adjacent peaks of equal height can be resolved from each other such that the peaks overlap at 10% of the peak height from the baseline. The above definition has been formulated so that discrimination is achieved when the peaks overlap at 33% of the peak height from the baseline. Note that the discrimination figure obtained from the experiment does not include the additional discrimination capability provided by the mass spectrometer.

A second case that may arise is if the target molecule for detection has an origin band absorption at a longer wavelength than the origin region of other components in the mixture. An example of this situation is the detection of tryptamine in a mixture with tyramine. The various origin conformer transitions of tyramine occur around 280 nm (Li and Lubman, 1988d), whereas the strongest origin band of tryptamine is at 286.30 nm. At 286.30 nm (1×10^6 W/cm^2), only the tryptamine could be observed in the mass spectrometer whereas the tyramine was transparent to the laser radiation. A series of successive dilution mixtures were made of tryptamine in tyramine, and at 286.30 nm an optical discrimination of at least $1{:}10^5$ was achieved. Tyramine can be ionized at 286.30 nm, but only at relatively high power (10^7 W/cm^2), and the signal observed is very weak. This may be due to nonresonant MPI or resonant ionization of a small number of molecules that have not been completely cooled in the jet. In either case, almost no molecular ion is observed and the major peaks are smaller fragments.

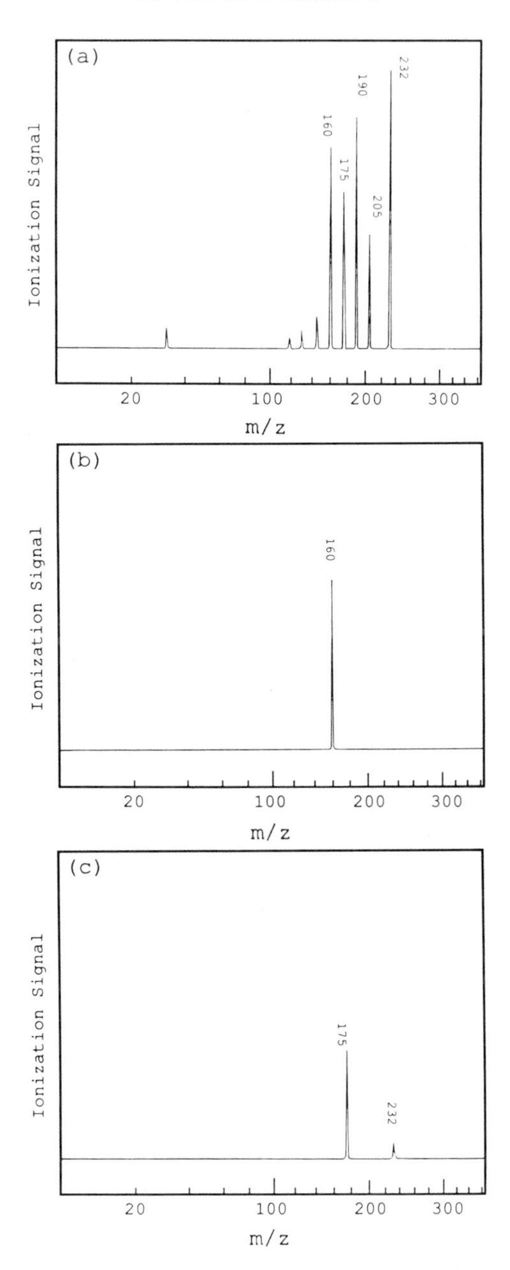

Figure 3C.10. R2PI mass spectra of a mixture of an equal amount of (1) tryptamine (*m/z* 160), (2) indole-3-acetic acid (*m/z* 175), (3) 5-methoxytryptamine (*m/z* 190), (4) 5-methoxyindole-3-acetic acid (*m/z* 205), and (5) melatonin (*m/z* 232), obtained as a function of wavelength: (a) 285.75 nm ($P = 2 \times 10^6$ W/cm^2); (b) 286.30 nm (P 5×10^5 W/cm^2); (c) 285.30 nm ($P = 5 \times 10^5$ W/cm^2). Reprinted from Li and Lubman (1988b) with permission.

A third case that may arise is if the target molecule of interest has an origin region at a shorter wavelength than the origin region of other components in the mixture. An example of this case is illustrated in Figure 3C.10 for a mixture of five compounds in a glycerol matrix that were desorbed using pulsed laser desorption and subsequently ionized by R2PI at various selected wavelengths. The mixture in Figure 3C.10 consists of an equal amount of tryptamine (m/z 160), indole-3-acetic acid (m/z 175), 5-methoxytryptamine (m/z 190), 5-methoxyindole-3-acetic acid (m/z 205), and melatonin (m/z 232).

With the laser tuned to 285.75 nm, all five components are observed in the mass spectrum (Figure 3C.10a). At a power density of 2×10^6 W/cm^2, the molecular ion of each compound is observed accompanied by only minor fragments of relatively low abundance. When the laser wavelength is tuned to 286.30 nm, tryptamine is selectively ionized as shown in Figure 3C.10b. If the ionization laser power density is carefully adjusted to 5×10^5 W/cm^2, the molecular ion peak of tryptamine is observed with no other interfering components. When the laser is tuned to 285.30 nm, the indole-3-acetic acid peak appears whereas the tryptamine is no longer observed (see Figure 3C.10c). In these experiments, the laser power must be carefully controlled. Since the other components have origin bands at around 305 nm, there will be some background continuum absorption observed in these molecules at higher frequency. Although this continuum absorption is small, the resulting ionization signal at 285.75 nm will become significant as the laser power is increased. Note that in Figure 3C.10c, although indole-3-acetic acid is selectively ionized, there is some background ionization of melatonin (N-acetyl-5-methoxytryptamine) even at the relatively low power being used. With selective laser ionization at 286.30 nm for this five-component mixture, the optical discrimination ranged from 1:200 for tryptamine in melatonin to 1:1000 for tryptamine in 5-methoxyindole-3-acetic acid. In the case of indole-3-acetic acid at 285.30 nm, similar optical discrimination limits were also observed, although the limit for indole-3-acetic acid in melatonin was only 1:70.

A measurement of the sensitivity of this method was made by preparing a series of successive dilutions in methanol/benzene (1:1), which were then dissolved in the glycerol matrix. Using a microliter syringe, we placed the sample into a 1 mm diameter hole bored into the rod surface. The desorption laser was focused to a $\sim$1 mm beam, and the power density was raised to approximately $\sim 10^7$ W/cm^2, so that desorption was complete within several laser pulses. The resulting R2PI molecular ion signal was monitored at the strongest transition over the desorption period, and a plot of signal vs. concentration provided an estimate of the lower limit of detection. The detection limit observed at a single wavelength with a signal-to-noise ratio (S/N) of 3 was as follows: tryptophol, 15.0 pg; indole-3-acetic acid, 10.0 pg; and tryptamine, 20.0 pg. By use of conventional derivatization techniques in GC-MS, a typical detection limit of $\sim$40 pg ($S/N = 2$) was reported for tryptamine and indole-3-acetic acid (Artigas and Gelpi, 1979). Similar detection limits (50–100 pg) have been reported for the indoleamines using thermospray liquid chromatography–MS (Artigas and Gelpi, 1987). Thus, the method described in this work compares favorably in terms of the sensitivity with available MS methods.

3C.4.2. Mass Spectrometry and Spectroscopy of Peptides

The laser desorption–evaporation method with entrainment into jet expansions has recently been used to systematically study small peptides by R2PI.

Table 3C.1. R2PI/MPI Mass Spectra of Tyrosine-Containing Dipeptides

Compound	MW	Wavelength (nm)	Major Fragments [m/z (rel. int. %)]	
			Soft[a]	Hard[b]
Tyr-Gly	238	281.40	238(100)	238(100), 221(7), 147(5), 136(62), 131(45), 107(13), 85(9), 74(6), 57(17), 30(54)
Gly-Tyr	238	280.70	238(100)	238(87), 221(6), 180(5), 165(35), 131(7), 107(41), 74(12), 73(65), 47(7), 30(100)
Tyr-Ala	252	281.60	252(100)	252(100), 145(38), 136(67), 107(13), 44(100)
Ala-Tyr	252	280.40	252(100)	252(97), 164(16), 107(17), 89(32), 44(100)
Tyr-Val	280	281.50	280(100)	280(66), 173(30), 156(11), 147(7), 136(100), 127(10), 116(5), 107(20), 85(18), 72(20)
Val-Tyr	280	280.60	280(100), 72(100)	280(27), 107(5), 72(100)
Tyr-Leu	294	281.25	294(100)	294(100), 277(5), 187(41), 136(52), 107(5), 86(10), 43(3)
Leu-Tyr	294	280.80	294(100) 86(35)	294(64), 132(7), 107(5), 86(100), 43(5)
Pro-Tyr	278	280.60	278(100)	278(45), 70(100)
Ile-Tyr	294	280.80	294(100)	294(47), 164(5), 132(6), 107(9), 86(100)

[a]Soft ionization at $P = \sim 5 \times 10^5$ W/cm^2.
[b]Hard ionization at $P = \sim 4 \times 10^6$ W/cm^2.

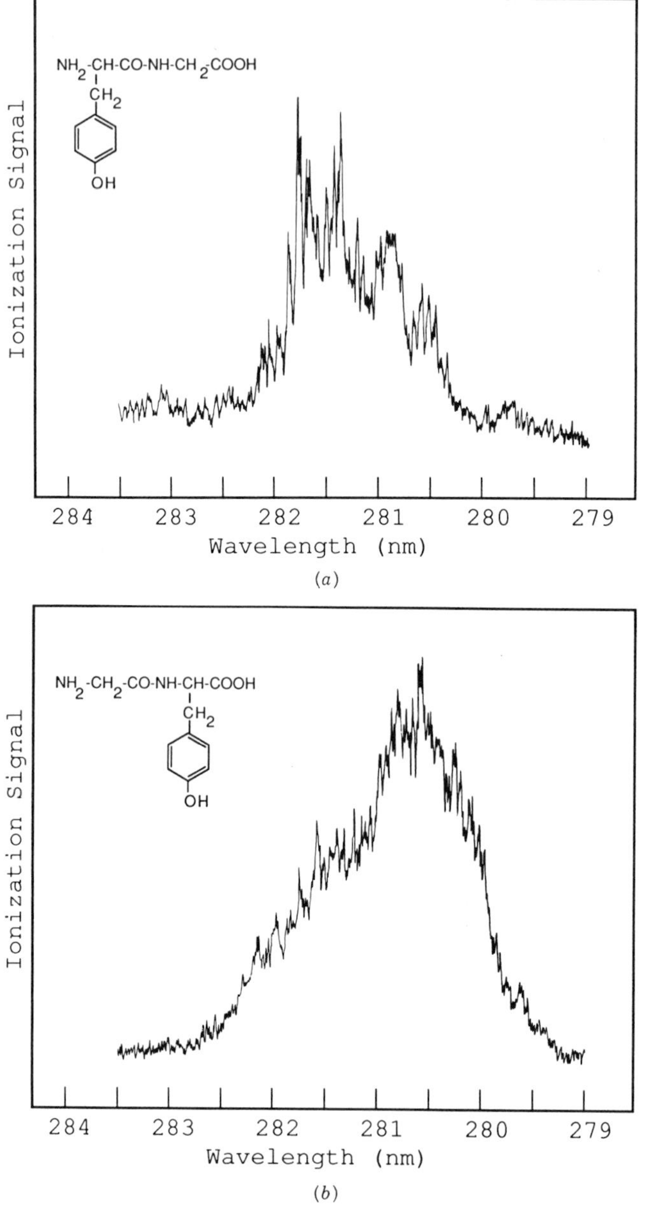

Figure 3C.11. Supersonic jet-cooled resonant two-photon ionization spectra of tyrosine-containing dipeptides obtained with the use of pulsed laser desorption and subsequent entrainment in the jet expansion of CO_2 carrier: (a) Tyr-Gly; (b) Gly-Tyr; (c) Tyr-Ala; (d) Ala-Tyr. Reprinted from Li and Lubman (1989a) with permission.

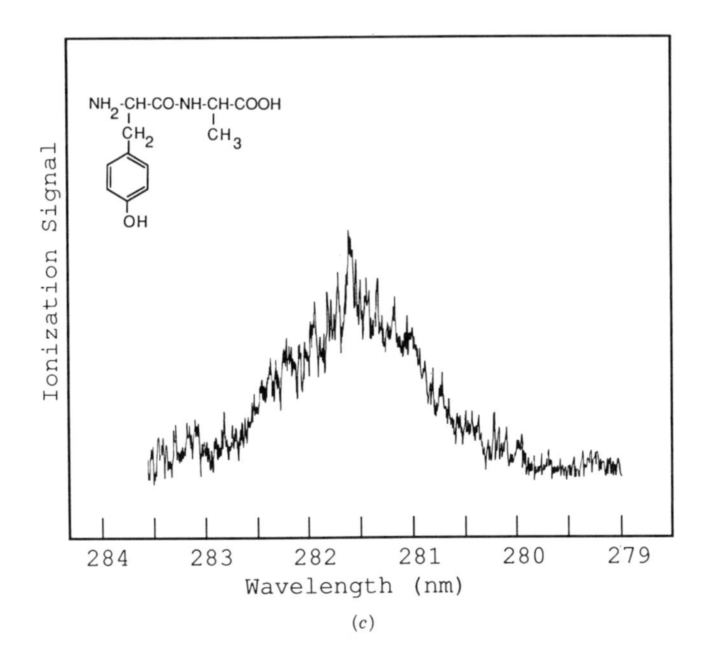

(*c*)

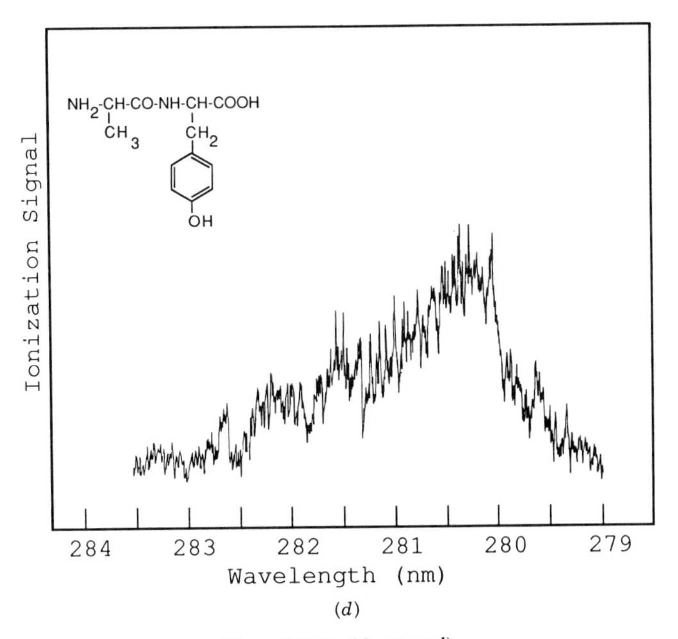

(*d*)

Figure 3C.11. (*Continued*)

In particular, di- and tripeptides with an absorbing aromatic group such as tryptophan, tyrosine and phenylalamine have been studied (Li and Lubman, 1989a; Tembreull and Lubman, 1987a). These compounds absorb radiation in the near-UV through their $\pi \rightarrow \pi^*$ transitions of the aromatic moiety and subsequently are ionized by R2PI. Tyrosine-containing dipeptides were chosen for study since the gas-phase jet-cooled spectroscopy of tyrosine and many of its derivatives have previously been studied in our laboratory and are known to absorb strongly between 280–282 nm (Li and Lubman, 1988d). It must thus be emphasized here that R2PI is a spectroscopic ionization tool since it depends on the presence of an intermediate absorbing electronic state.

In the tyrosine-containing peptides studied, the M^{+} ion is generally observed with no fragmentation if the conditions of the experiment are adjusted properly, as shown in Table 3C.1 (Li and Lubman, 1989a). These mass spectra were obtained under conditions where the molecules were cooled by the effects of the supersonic jet. In Figure 3C.11 are shown the corresponding jet-cooled R2PI spectra of two pairs of tyrosine-containing dipeptide isomers obtained with the use of pulsed laser desorption with subsequent entrainment in the jet expansion (Li and Lubman, 1989a). Each of these spectra were obtained by monitoring only the molecular ion signal intensity as a function of wavelength using a gated integrator at a power of $P = 5 \times 10^5\ \mathrm{W/cm^2}$. The spectra shown in Figure 3C.11a–d are relatively broad and featureless compared to those obtained for tyrosine- and indole-based compounds in previous work (Li and Lubman, 1988d), where sharp spectral features were clearly evident. The jet-cooled spectra of the dipeptides are typically 1.0–1.5 nm fwhm. No sharp spectral features were observed, although the shape of the contour is quite distinctive for each dipeptide. This result appears to be due to the large number of possible conformers of these dipeptides, which are not resolved under the conditions of this experiment. There appear to be two distinct sets of dipeptides in our experiments in terms of the spectral shifts. The dipeptides with tyrosine on the C-terminal end have contours that peak at $\sim$280.5 nm, which essentially corresponds to the strongest conformeric origin band of tyrosine. The dipeptides with tyrosine on the N-terminal end generally have contours that peak at $\sim$281.5 nm or have shifted $\sim$1 nm to longer wavelength than that of the dipeptides with tyrosine on the C-terminal end. Thus, dipeptide isomers such as Gly-Tyr and Tyr-Gly can easily be distinguished by their different optical absorption spectra. It should be noted that these tyrosine-containing dipeptides absorb in a region of the spectrum very similar to the tyrosine precursor, whose origin band region consists of several conformers between 280 and 281 nm (Li and Lubman, 1988d). The nonaromatic amino acids in these dipeptides just act as substituent groups that shift the absorption of the tyrosine aromatic center.

In Table 3C.1 are shown the mass spectra corresponding to the wavelength spectra of each of the dipeptides shown in Figure 3C.11 (Li and Lubman, 1989a). In this table only fragments with an abundance of 5% or greater have been included. Each of these mass spectra has been obtained by optimizing the cooling in the jet expansion following desorption and entrainment. The frequency of the doubled dye laser output is then tuned to the optimal absorption in the spectral contour. By optimizing the jet entrainment and cooling and exciting at the specific optimal wavelength for absorption, one maximizes the signal and so relatively low laser power can be used for ionization. Soft ionization, where only the molecular ion is observed, can be obtained in almost every case under these conditions at a laser energy of 250 μJ ($P = 5 \times 10^5$ W/cm^2). In earlier work where only incomplete cooling was obtained and nonspecific wavelength ionization was used (Tembreull and Lubman, 1987a), a laser energy of about 5 mJ, or nearly 20 times that of the present studies, was required for efficient ionization. Under these conditions, it was difficult to prevent fragmentation. By optimizing the jet cooling and exciting on-resonance in these studies, one enhances the partition function for absorption so that a relatively low ionization laser power density can be used, resulting in minimal fragmentation. These results thus illustrate the importance of the spectroscopic aspects of R2PI in optimizing the mass spectrometric results in addition to the advantages of selectivity.

As the laser power is increased, extensive fragmentation can be produced for structural studies (Table 3C.1). This fragmentation is induced at the given wavelength that corresponds to the peak of the dipeptide absorption at a laser energy of 2 mJ ($P = 4 \times 10^6$ W/cm^2). This laser power is relatively low compared with that used in other studies (Tembreull and Lubman, 1987a) for inducing fragmentation and is a result of obtaining efficient jet cooling in the expansion, followed by excitation at the optimal absorption. It should be noted that M^{+} is still present in each case, although it may or may not be the dominant ion peak, depending on the laser power. Also isomeric dipeptides exhibit different fragmentation, by which they may be uniquely identified. For example, the mass spectra of Tyr-Val and Val-Tyr can be compared in Table 3C.1. In both cases, a strong molecular ion peak is observed in the mass spectrum. The base peak in each case is the A_1 fragment (in the Roepstorff–Fohlman notation) (Roepstorff and Fohlman, 1984), which has generally been observed to be a dominant fragment product in many of the dipeptides. If the tyrosine is on the N-terminal end, such as in Tyr-Val, then the A_1 fragment results in an ion at m/z 136, whereas if the tyrosine is on the C-terminal end, as in Val-Tyr, the fragments ion is at m/z 72. In Val-Tyr the other cleavages result in only minor fragment at the given laser energy. In Tyr-Val, simple β-cleavage in the tyrosine moiety results in a m/z 107, which corresponds to a hydroxybenzyl ion, and m/z 173, which corresponds to the dipeptide minus this species. There is also a strong m/z 72 ion, which arises from Y_1-cleavage followed by loss of the COOH group with an accompanying H migration. Such differences in sequence ions are observed in all the isomeric dipeptides upon laser-induced fragmentation by R2PI/MPI. Note also that Leu-Tyr and Ile-Tyr have different fragmentation patterns, with the appearance of a m/z 43 peak for Leu-containing peptides but not for Ile-containing peptides. This is significant because the presence of Ile or Leu is difficult to establish with present MS sources.

The effect of jet cooling on the laser-induced mass spectrum can be demonstrated by monitoring the ion peaks obtained as the laser frequency is scanned. As the laser frequency is tuned off the optimal resonance in the jet-cooled wavelength spectrum (see Figure 3C.11), the absorption intensity decreases. Increased laser power is required in order to obtain the same $M^{\ddagger}$ signal level off-resonance, as compared to when the signal is optimized on-resonance. The result is that increased fragmentation is obtained off-resonance in order to obtain an $M^{\ddagger}$ signal comparable to that obtained on-resonance. Alternatively, soft ionization can be obtained off-resonance but at the expense of sensitivity by a factor typically 10–100 times.

The discrimination of isomeric peptides is a generally important problem in biomedical analysis. This is demonstrated using R2PI at 266 nm ($P = 1.5 \times 10^7$ W/cm^2) for the tripeptide Gly-Gly-Trp, Gly-Trp-Gly, and Trp-Gly-Gly in Figure 3C.12 (Li and Lubman, 1988a). In each case a strong $M^{\ddagger}$ ion is clearly observed as well as an ion at $m/z = 130$ u due to simple β-cleavage at the tryptophan moiety. Although the preferred cleavage under the conditions of the experiment appears to be β-cleavage with a resulting ion fragment, in each isomer a different fragment pattern is obtained depending on the initial structure of the molecule. For example, the ion at $M - 74$ (244 u) is present in configurations b and c due to the cleavage as indicated; however, the same cleavage will not provide this ion when, as in configuration a in Figure 3C.12, the tryptophan is on the other side of the bond. In general, the fragments obtained from the three species are quite different.

One truly unique feature of the fragmentation induced by laser R2PI/MPI is the ability to distinguish the isomeric amino acid pair leucine and isoleucine that cannot be easily distinguished based upon the fragmentation induced by classical methods (Biemann and Martin, 1987). This had been demonstrated previously by Zare's group workers for the PTH-leucine and PTH-isoleucine amino acids (see Engelke et al., 1987). In our work (Li and Lubman, 1988c) this has been illustrated for several carbobenzoxy (CBZ)-oligopeptides containing leucine and isoleucine, as shown in Figure 3C.13, where the CBZ group provides an absorbing center at 266 nm for this linear peptide. At above $m/z = 100$ the mass spectra for Leu-containing and Ile-containing peptides are indistinguishable. However, at m/z below 100 these peptides can be distinguished by observing the intensity differences between the $m/z = 43$ ion

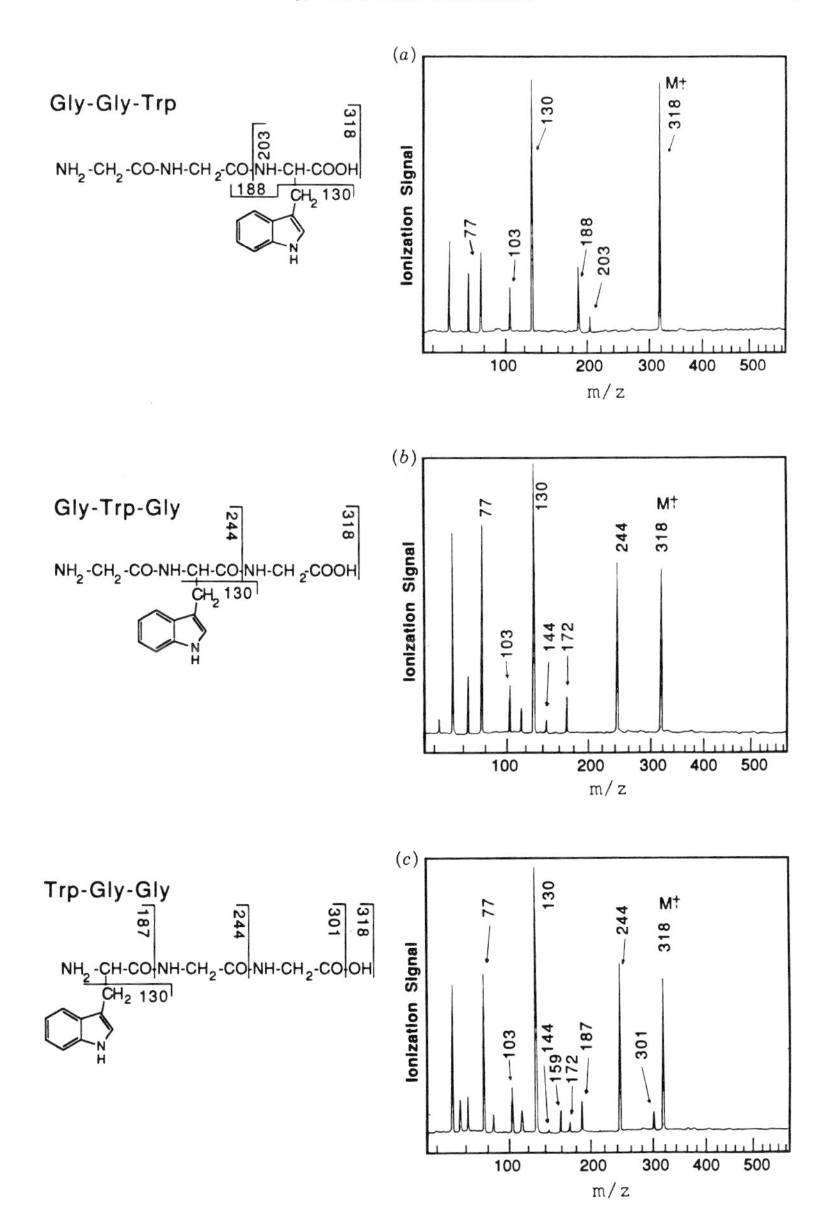

Figure 3C.12. LD–R2PI/MPI mass spectra of (a) Gly-Gly-Trp, (b) Gly-Trp-Gly, and (c) Trp-Gly-Gly at 266 nm ($P = 1.5 \times 10^7$ W/cm^2). Reprinted from Li and Lubman (1988a) with permission.

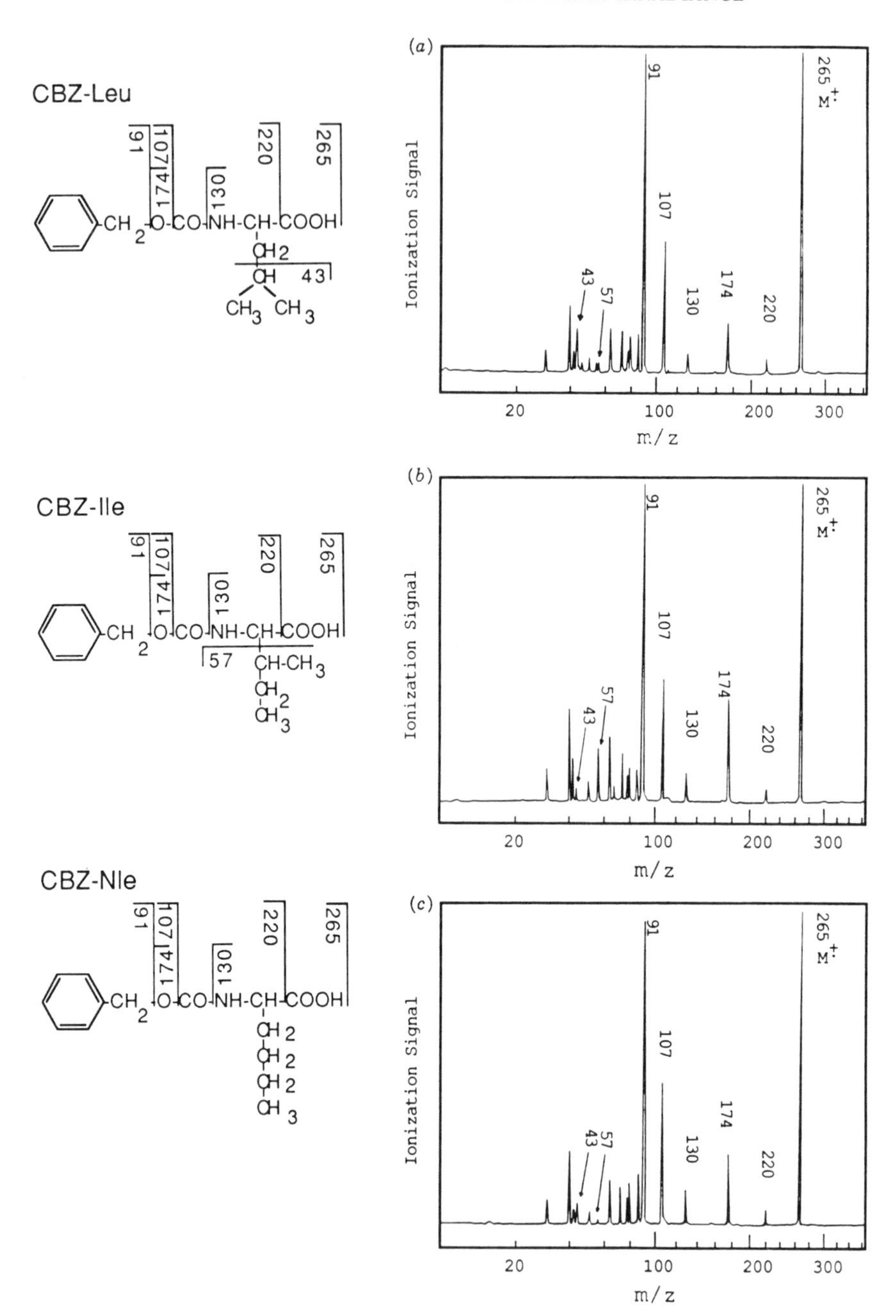

Figure 3C.13. Laser-induced mass spectra and fragmentation patterns of the three isomers CBZ-Leu, CBZ-Ile, and CBZ-Nle at 266 nm ($P = 1.5 \times 10^7$ W/cm^2). Reprinted from Li and Lubman (1988c) with permission.

and the $m/z = 57$ ion. It was found that the $m/z = 43$ ion is of much greater relative abundance than the $m/z = 57$ ion for the Leu-containing peptides, whereas the opposite is true for the Ile-containing peptides. This is the result of different cleavages induced in the side chains of these amino acids. The —$CH(CH_3)_2$ fragment is more stable for leucine based upon the structure of the side chain, whereas —$CH(CH_3)CH_2CH_3$ is the stable fragment produced from the isoleucine side chain. This is also demonstrated for the case of the isomers of CBZ-Leu, CBZ-Ile, and CBZ-Nle. For CBZ-Leu an $m/z = 43$ ion of much greater relative abundance is observed compared to the $m/z = 57$ ion; for CBZ-Ile an $m/z = 57$ ion is observed that is of relatively greater abundance than the $m/z = 43$ ion. In either case, the differences involve the presence of minor features in the mass spectrum that can be observed since there is no matrix background that would interfere at these low masses as would occur in fast atom bombardment mass spectrometry (FAB-MS). In the case of CBZ-Nle, ions are observed at $m/z = 43$ and $m/z = 57$, which are of relatively low abundance compared to other ions in the spectrum. Such isomeric discrimination has also been demonstrated for CBZ-(α)-lysine, CBZ-(ε)-lysine, and CBZ-glutamine (Li and Lubman, 1988c).

In other recent work (Li and Lubman, 1988a) laser desorption with jet entrainment has been extended to larger species such as pentapeptides including Met- and Leu-enkephalin, which are important neurotransmitters in the brain. By using high laser power ($P = 1.5 \times 10^7$ W/cm^2) at 266 nm, extensive fragmentation could be induced in these compounds for structural analysis. The fragmentation induced under these conditions resulted in all the classical *A*, *B*, *C* and *X*, *Y*, *Z* cleavage products with formation of characteristic acylium and aldimine ions as well as β-cleavage at the aromatic moiety. In addition, a substantial molecular ion, $M^{+\cdot}$ is observed in every case and $(M - OH)^+$ or $(M - COOH)^+$ ions are often also observed. In every case only $M^{+\cdot}$, and not MH^+ as in FAB-MS, is observed. The result is that the fragmentation generated by R2PI/MPI would be expected to be very different from that generated by FAB-MS as indeed is observed. The laser-induced mass spectrum of Leu-enkephalin is shown in Figure 3C.14.

In the case of Leu-enkephalin (Figure 3C.14a) cleavage occurs at the —CO—NH peptide bond, with a resulting series of acylium ions at m/z values 278, 221, and 164. In addition, at 28 mass units lower than the acylium ions the corresponding acyl-immonium ions are observed owing to loss of CO at m/z values of 397 and 136. A second degradation pathyway begins from the N-terminal end and results in formation of aldimine ions at m/z values of 391, 335, and 277 in the laser-induced fragmentation of Leu-enkephalin. Such aldimine ions are often difficult to produce using electron impact ionization. In the R2PI/MPI-induced mass spectra these ions are formed by bond cleavage of the —CO—NH bond in the molecular cation with a resulting even-electron ion. No hydrogen migration is observed as in FAB-MS (Biemann and Martin, 1987) except for *m/z* 335 in Leu-enkephalin, in which a hydrogen has been added to the *Y*-cleavage product to form an odd-electron ion.

A series of fragments due to cleavage at every N—C bond is observed for Leu-enkephalin in which charge retention can occur on the N or C fragment. Charge retention on the N results in a series of fragments at *m/z* values of 179, 236, and 441, whereas charge retention on the C results in ions at *m/z* 376, 319, and 262. The formation of the fragment at *m/z* 441 formed by cleavage of the N—C bond with rentention of the charge at the N-containing fragment involves the addition of one H to form an odd-electron fragment so that C′ is formed. However, the fragments at *m/z* 179 and 236 result in even-electron fragments with no migration of the H. In addition, simple β-cleavage of the

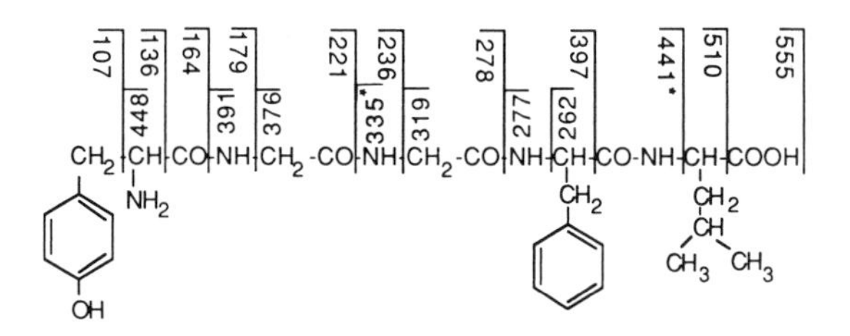

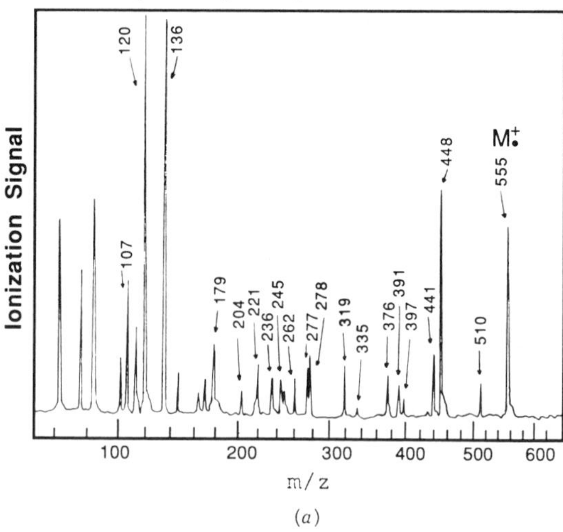

(*a*)

Figure 3C.14. LD–R2PI/MPI ionization fragmentation patterns at $\lambda = 266$ nm of Leu-enkephalin at (a) high laser ionization energy ($P = 1.5 \times 10^7$ W/cm^2) and (b) low ionization energy ($P = 1.0 \times 10^6$ W/cm^2). Reprinted from Li and Lubman (1988a) with permission.

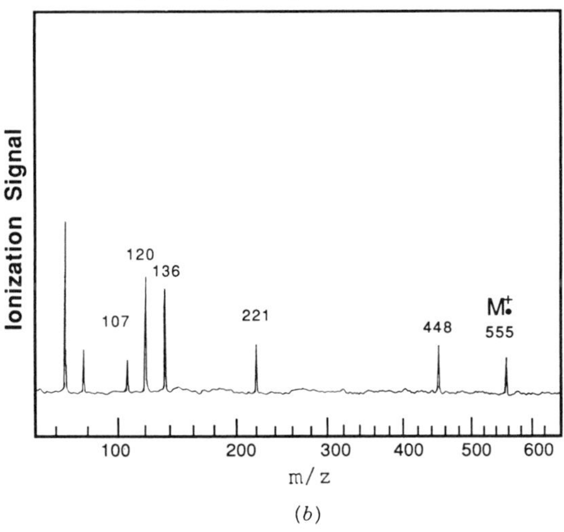

(*b*)

Figure 3C.14. (*Continued*)

aromatic side chain in tyrosine occurs, with charge retention on either fragment resulting in an *m/z* of 107 if the charge remains on the hydroxybenzyl ion or in an *m/z* of 448 if the charge remains on the remaining fragment. Also, a strong peak is observed at an *m/z* of 120 due to an internal fragment resulting from the formation of an immonium ion characteristic of the phenylalanine group (Biemann and Martin, 1987).

It should be noted that when the laser energy is lowered to $\sim 1\text{–}2 \times 10^6$ W/cm^2 the molecular ion can be obtained with reduced fragmentation, as demonstrated in Figure 3C.14b for Leu-enkephalin. The ability to reduce fragmentation will be of importance in analysis of complicated mixtures. However, this occurs at the expense of sensitivity by an order of magnitude or more.

Moreover, note that the R2PI results obtained may also depend on the

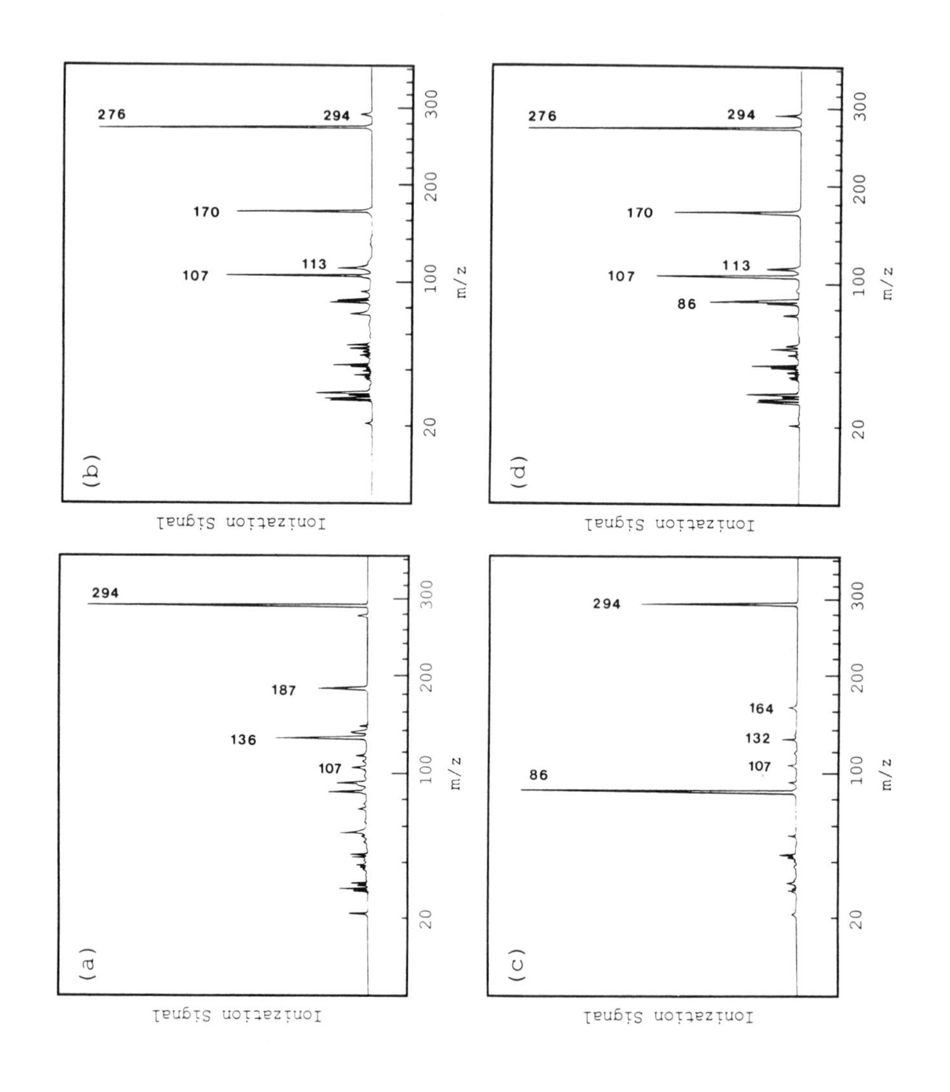
(a)
294
187
136
107
(b)
276
294
170
113
107
(c)
294
164
132
107
86
(d)
276
294
170
113
107
86
Ionization Signal
m/z
20
100
200
300

desorption process and whether thermal decomposition is induced. It has been shown that the laser-induced thermal decomposition product of small linear oligopeptides is a 2,5-diketopiperazine which is formed by cyclization of the amino acid moieties in position 1 and 2 from these linear oligopeptides (Li and Lubman, 1989b). This is shown in Figure 3C.15, which shows the R2PI/MPI-induced mass spectra of the two isomeric dipeptides Tyr-Leu and Leu-Tyr desorbed from thin layer samples by pulsed infrared radiation (10.6 μm). These spectra were taken at an ionization laser power of $\sim 5 \times 10^6$ W/cm^2, which induced some fragmentation that is useful for structural studies. The molecular ion (*m/z* 294) is observed as a strong peak in each case, and the fragmentation observed is distinctly different for the two isomers. As the laser power is decreased, the molecular ion only is observed without fragmentation. Thermal decomposition is avoided in these thin films where minimal absorption and bulk heating is possible. Figure 3C.15b,d shows the R2PI/MPI-induced mass spectra of the same two peptides desorbed from thick samples on the probe surface. The mass spectra obtained are almost identical for the isomeric dipeptides. The molecular ion is visible only as a very minor peak, whereas the $(M-18)^{\ddagger}$ ion (*m/z* 276) forms the base peak in the mass spectrum. The ratio between the signal intensities of the $M^{\ddagger}$ and $(M-18)^{\ddagger}$ peaks can be varied by changing the position of the IR beam on the surface. It is assumed that the decomposition is thermally induced as the thick sample absorbs the radiation. The Fourier-transform infrared (FTIR) spectra of the dipeptides such as Tyr-Leu remaining on the probe surface after laser irradiation are different from those obtained from the dipeptides before desorption or from the pure dipeptides. The FTIR spectrum of Tyr-Leu after irradiation appears to be a combination of that observed for pure Tyr-Leu and pure cyclo-(Leu-Tyr). In fact, a laser-induced R2PI/MPI mass spectrum of a synthetic sample of cyclo-(Leu-Tyr) yielded the same results as Figure 3C.15b,d (minus the $M^{\ddagger}$ peak). Thus, we conclude that the species formed from thermal decomposition during the laser desorption process is the cyclic dipeptide and that the same fragments in Figure 3C.15b,d are obtained from this common precursor.

Table 3C.2 shows that 2,5-diketopiperazines can be formed in the laser-induced thermal decomposition process for oligopeptides and their derivatives whose N-terminal group is not blocked (Li and Lubman, 1989b). The mechanism of formation of 2,5-diketopiperazines or cyclic dipeptides by the desorption laser beam is shown in Figure 3C.16 and can be summarized for

Figure 3C.15. R2PI/MPI mass spectra of (a) Tyr-Leu and (c) Leu-Tyr obtained using a thin film of sample found on the desorption surface. R2PI/MPI mass spectra of the thermally decomposed species induced during the desorption of (b) Tyr-Leu and (d) Leu-Tyr obtained using a thick sample. Ionization was produced using $\lambda = 266$ nm at a power density of 5×10^6 W/cm^2. Reprinted from Li and Lubman (1989b) with permission.

Table 3C.2. The Characteristic Fragment Ions of the 2,5-Diketopiperazines Obtained by R2PI/MPI[a] of the Laser-Induced Thermal Decomposition Products of the Linear Peptides

Original Linear Peptides	Thermal Decomposition Product	MW	Major Fragments [m/z (rel. int. %)] of cyclo-(Dipeptide)
Gly-Tyr	cyclo-(Gly-Tyr)	220	220(100), 114(24), 107(77)
Tyr-Gly	cyclo-(Gly-Tyr)	220	220(100), 114(25), 107(73)
Tyr-Gly-Gly	cyclo-(Gly-Tyr)	220	220(100), 114(24), 107(76)
Tyr-Gly-Gly-Phe-Met	cyclo-(Gly-Tyr)	220	220(100), 114(24), 107(76)
Ala-Tyr	cyclo-(Ala-Tyr)	234	234(100), 128(68), 107(100)
Tyr-Ala	cyclo-(Ala-Tyr)	234	234(100), 128(68), 107(100)
Trp-Gly	cyclo-(Gly-Trp)	243	243(100), 130(72), 103(8), 77(12)
Gly-Trp	cyclo-(Gly-Trp)	243	243(100), 130(75), 103(9), 77(18)
Trp-Gly-Gly	cyclo-(Gly-Trp)	243	243(100), 130(76), 103(10), 77(17)
Gly-Trp-Gly	cyclo-(Gly-Trp)	243	243(100), 130(75), 103(11), 77(22)
Pro-Tyr	cyclo-(Pro-Tyr)	260	260(100), 154(85), 114(8), 107(42)
Val-Tyr	cyclo-(Val-Tyr)	262	262(100), 156(53), 114(15), 107(87)
Tyr-Val	cyclo-(Val-Tyr)	262	262(100), 156(53), 114(16), 107(87)
Leu-Tyr	cyclo-(Leu-Tyr)	276	276(100), 170(48), 113(13), 107(47)
Tyr-Leu	cyclo-(Leu-Tyr)	276	276(100), 170(47), 113(15), 107(47)
Ile-Tyr	cyclo-(Ile-Tyr)	276	276(100), 170(45), 113(21), 107(58)
Trp-Leu	cyclo-(Leu-Trp)	299	299(86), 170(5), 130(100), 103(5), 77(15)
Leu-Trp-Met	cyclo-(Leu-Trp)	299	299(88), 170(8), 130(100), 103(5), 77(18)
Leu-Trp-Met-Arg	cyclo-(Leu-Trp)	299	299(80), 170(10), 130(100), 103(5), 77(20)
Leu-Trp-Met-Arg-Phe	cyclo-(Leu-Trp)	299	299(86), 170(10), 130(100), 103(9), 77(15)

[a]The ionization wavelength was 266 nm and the power density was 5×10^6 W/cm^2.

several cases as follows: (a) X can be a hydroxy group, as is the case in dipeptides, where the cyclic dipeptide is formed with the elimination of water; (b) X can be an amino group, as in Gly-Tyr-amide, where the cyclic dipeptide is formed with elimination of NH_3; or (c) X can be an amino acid or peptide, as in peptides with three or more amino acids. In this last case, the displacement of the amino acid or peptide forms the cyclic dipeptide.

Other important features of this method include sensitivity and the ability to perform quantitative measurements. To provide an estimate of the sensitivity of this method, successive dilutions of several different peptides in methanol/benzene (1:1) were prepared. A microliter syringe was used to place the sample into a 1 mm diameter hole bored into the rod surface. The desorption laser was focused to a ~1 mm beam and the power density raised to

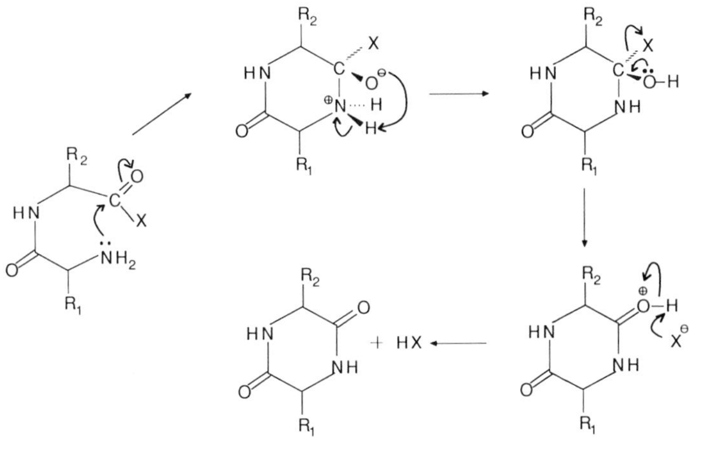

HX = H_2O, NH_3, amino acid, peptide, etc.

Figure 3C.16. Mechanism for thermally induced cyclic dipeptide formation. Reprinted from Li and Lubman (1989b) with permission.

Table 3C.3. Detection Limit for CBZ-Derivatized Compounds Using R2RI at 266 nm[a]

Compound	Detection Limit (pg)
CBZ-Ala	125
CBZ-Leu	56
CBZ-Pro	125
CBZ-Gly-Ala	333
CBZ-Leu-Gly	200
CBZ-Pro-Leu	250
CBZ-Gly-Gly-Ile	500
CBZ-Gly-Gly-Ala	300
CBZ-Ile-Gly-Gly	1000

[a] $P = 1.5 \times 10^7$ W/cm^2.

$\sim 10^7$ W/cm^2 so that desorption was complete within several laser pulses. The resulting R2PI molecular ion signal was monitored over the desorption period, and a plot of signal vs. concentration provided an estimate of the lower limit of detection with a signal-to-noise (S/N) ratio of 2 as shown in Table 3C.3. The detection limit achieved in our work, which is typically tens of picograms, is not nearly as low as the picogram detection limit achieved for phenylthiohydantoin(PTH)–amino acids by Zare's group (see Engelke et al., 1987). The differences probably reflects the loss of molecules during the entrainment process, as compared to the work of Engelke et al., 1987, in which no beam entrainment is used.

The apparent detection limit appears to increase as the size of the peptide increases. A detection limit for CBZ–amino acids was typically 50–100 pg, whereas for CBZ-tripeptides a detection limit of 500 pg–1 ng resulted. As the peptide size increases further, the detection limit becomes higher—as in Met-enkephalin, where only 250 ng of sample can be detected. The detection limit appears to be affected mainly by the laser-induced ionization efficiency. The latter will be influenced by the absorption coefficient of the molecule and the competing radiationless processes. These will depend on the structure of the particular molecule, and thus sensitivity in this technique may be expected to vary significantly with the sample. However, in nonrigid peptide chains there are many active modes that can promote internal conversion, thus resulting in a decreased ionization efficiency. The entrainment efficiency of the sample into the jet and the resulting jet cooling may also affect the detection limit, although the relative contribution of these processes in limiting the sensitivity is still unknown. Nevertheless, for small aromatic-based species the sensitivity is excellent, often reaching into the low-picogram range.

It should be noted that similar work has been carried out by Grotemeyer et al. (1986a,b,c, 1988a,b, 1989), who are using a high-resolution V-shaped reflectron TOF device. They have desorbed neutrals into a supersonic jet of Ar by using a pulsed CO_2 laser and performed REMPI with an excimer-pumped dye laser. In their configuration a gridless reflectron TOF-MS is used and the jet expansion proceeds down the flight tube, as opposed to expanding perpendicular to the TOF as in our work. The advantage of their method is that it is more compatible for work at high mass. The supersonic expansion causes a transverse velocity component which carries the ions off the axis of the flight tube as it expands at right angles to the TOF direction. Thus, ions need to be steered back onto the axis of the flight tube via a deflection plate voltage that will be different for each m/z. In the design of Grotemeyer et al. the expansion down the axis of the flight tube eliminates this problem. This group has been able to obtain a mass-resolving power of at least 6500 at $m/z = 96$ by using this technique. In one study, they were able to ionize the decapeptide angiotensin I and obtain only the molecular ion at $m/z = 1295$ by using 271.23 nm radiation at a power density of 1×10^6 W/cm^2. By increasing the laser power to 1×10^8 W/cm^2, they were able to generate fragments characteristic of the structure of the peptide (see Scheme I of Grotemeyer et al., 1986c). Fragmentation mainly due to bond breaking around the carbonyl groups of the peptide chain resulted in efficient formation of acylium ions. In more recent work this group has also studied several angiotensins (Grotemeyer and Schlag, 1988a,b), gramicidin D (Kinsel et al., 1991), substance P (Grotemeyer and Schlag, 1989), and oligonucleotides (Lindner et al., 1990) using this methodology.

3C.5. Conclusion

Thus, in conclusion, laser desorption serves as a powerful means of volatilizing thermally labile molecules for entrainment into jet expansions. The supersonic jet technique provides ultracold, sharp R2PI spectra of various small biological molecules such as indole-, catechol-, and tyrosine-based compounds that can be used to uniquely identify each of these species in a mass spectrometer. This optical method can provide for identification and enhanced discrimination of these compounds without extensive sample preparation or derivatization. In these experiments, even isomeric dipeptides can be optically distinguished by the spectral shifts that result, depending on which amino acid is on the N- or C-terminal end. In the case of compounds whose absorptions are shifted by only several nanometers, an optical discrimination on the order of $1{:}10^5$ can be obtained. In addition, the sensitivity for detection of these small aromatic biological compounds by R2PI can reach the low-picogram level in many cases.

A second important aspect of this work is that, since ions are produced by the R2PI process, further identification can be obtained using the soft ionization capabilities with detection in a TOFMS. In particular, soft ionization can be enhanced at low laser power by using the cooling properties of the jet expansion and exciting on-resonance at the maximum intensity of each specific origin transition. However, extensive fragmentation can also be produced by increasing the laser power or frequency. This fragmentation can be used for structrual analysis and identification as demonstrated herein for various di- and tripeptides, where even the isomeric forms of these compounds could be uniquely distinguished. Further, even isomeric amino acids such as leucine and isoleucine can be distinguished, which is generally very difficult by conventional mass spectrometric methods. In addition, the effect of thermal decomposition in the desorption process upon laser-induced mass spectra has been studied. The laser-induced thermal decomposition process results in formation of a cyclic dipeptide and a linear peptide whose fragmentation can also provide important structural information. The thermal decomposition process can usually be prevented by the use of thin films or a heat-absorbing matrix such as glycerol in order to study the mass spectral products due solely to the R2PI/MPI process. Thus, R2PI/MPI can provide important capabilities for detection, identification, and structural analysis of molecules not readily available by other mass spectrometric methods.

The limitations of the method at present involve the range of molecules that can be effectively probed by these techniques. At present we are limited to molecules that will absorb near-UV and visible radiation and ionize in a two-photon process at these wavelengths. This generally restricts us to small aromatic and N-, S-, and O-heterocyclic compounds and their substituted counterparts. A multicolor wavelength scheme can extend this methodology to larger extended ring systems such as polynuclear aromatic hydrocarbons. However, this will increase the complexity and expense of the experiment. In addition, many N-heterocyclic and related systems such as purines and pyrimidine bases may not exhibit sharp spectral features in their $\pi \rightarrow \pi^*$ absorptions but may have sharp features for the $n \rightarrow \pi^*$ transitions. Thus the selectivity aspect of this experiment may be limited in its applicability to certain classes of molecules or certain transitions within molecules. Making the spectroscopic aspect of this method a widely applicable tool will require the development of broadly tunable solid state sources to replace the present dye laser systems. When these broadly tunable sources exist, the real potential of the selective capabilities of jet methodology in combination with R2PI will be developed. In addition, most nonaromatic and linear molecules absorb in the VUV region of the spectrum below 180 nm. At present tunable VUV sources are at an early stage of development and can only generate low-energy pulses, which are insufficient for highly efficient R2PI. The future development of such VUV sources will undoubtedly see the further extension of R2PI to a wide range of species for molecular detection.

A second limitation of the R2PI method appears to be the size of the sample that can be efficiently ionized. At the time of this writing, few molecules with MW > 1000 have been ionized by R2PI. Most of the larger molecules that can be effectively ionized either have rigid structures (i.e., porphyrins) or have strongly absorbing tryptophan moieties. In addition, the molar sensitivity may drop rapidly as the molecular weight increases, especially for nonrigid molecules. A fundamental limitation here is the radiationless transition rate, which increases ever more rapidly as the number of active modes increase. As energy is placed in the molecule, the energy can be rapidly degraded via radiationless processes before ionization proceeds, thus limiting the sensitivity. This may yet be partly overcome with the use of picosecond and femtosecond laser pulses that can produce ionization more rapidly than the input energy can be degraded. Another limitation in detecting large molecules is the ability to efficiently vaporize these samples intact as free molecules. Laser desorption can often result in thermal degradation or large clusters of sample desorbed from the surface. The recent development of laser matrix-assisted desorption allows for desorption of large molecules, i.e., 250,000 amu, as ions into the gas phase. As matrices are developed that can be used for efficient desorption of large neutral molecules into the gas phase, the capabilities of R2PI for ionization of large species can be examined.

Despite the present limitations of the R2PI method, especially in terms of its limited applicability, the development of new technology should continue to extend the unique capabilities of this technique for chemical analysis. R2PI still remains a unique ionization source for mass spectrometry based upon its selectivity in trace and mixture analysis, its high sensitivity, and its ability to effectively control the amount of energy placed in an ion and thus control the fragmentation process. R2PI will no doubt continue to play a central role in spectroscopic chemical analysis and mass spectrometry based upon these properties.

REFERENCES

Antonov, V. S., Knyazev, I. N., Letokhov, V. S., Matjiuk, V. M., Moshev, B. G., and Potapov, V. K. (1978). *Opt. Lett.* **3**, 37–39.

Artigas, F., and Gelpi, E. (1979). *Anal. Biochem.* **92**, 233–242.

Artigas, F., and Gelpi, E. (1987). *J. Chromatogr.* **394**, 123–134.

Beavis, R. C., and Chait, B. T. (1989). *Rapid Commun. Mass Spectrom.* **3**, 436–439.

Bernstein, R. B. (1982). *J. Phys. Chem.* **86**, 1178–1184.

Bersohn, R., Even, U., and Jortner, J. (1984). *J. Chem. Phys.* **80**, 1050–1058.

Biemann, K., and Martin, S. A. (1987). *Mass Spectrom. Rev.* **6**, 1–76.

Boesl, U., Neusser, H. J., and Schlag, E. W. (1978). *Z. Naturforsch.* **33A**, 1546–1548.

Boesl, U., Neusser, H. J., and Schlag, E. W. (1980). *J. Chem. Phys.* **72**, 4327–4333.

Boesl, U., Neusser, H. J., and Schlag, E. W. (1981). *Chem. Phys.* **55**, 193–204.

Boesl, U., Neusser, H. J., Weinkauf, R., and Schlag, E. W. (1982). *J. Phys. Chem.* **86**, 4857–4863.

Boniforti, L., Citti, G., Lostia, O., and Lucarelli, C. (1983). In *Recent Developments in Mass Specrtrometry in Biochemistry, Medicine and Environmental Research* (A. Frigerio, ed.), Vol. 12, p. 25. Elsevier, Amsterdam.

Brophy, J. H., and Rettner, C. T. (1979). *Opt. Lett.* **4**, 337–339.

Budzikiewicz, H., Djerassi, C., and Williams, D. H. (1964). *Interpretation of Mass Spectra of Organic Compounds*, pp. 63–65. Holden-Day, San Francisco.

Byer, R. L., and Duncan, M. D. (1981). *J. Chem. Phys.* **74**, 2174–2179.

Cable, J. R., Tubergen, M. J., and Levy, D. H. (1988). *J. Am. Chem. Soc.* **110**, 7349–7355.

Carlin, T. J., and Freiser, B. S. (1983). *Anal. Chem.* **55**, 955–958.

Coates, M. L., and Wilkins, C. L. (1987). *Anal. Chem.* **59**, 197–200.

Cooper, C. D., Williamson, A. D., Jr., Miller, J. C., and Compton, R. N. (1980). *J. Chem. Phys.* **73**, 1527–1537.

Dietz, T. G., Duncan, M. A., Liverman, M. G., and Smalley, R. E. (1980a). *Chem. Phys. Lett.* **70**, 246–250.

Dietz, T. G., Duncan, M. A., Liverman, M. G., and Smalley, R. E. (1980b). *Chem. Phys. Lett.* **73**, 4816–4821.

Ditchburn, R. W., and Arnot, F. L. (1929). *Proc. R. Soc. London* **A123**, 516–536.

Duncan, M. A., Dietz, T. G., and Smalley, R. E. (1979). *Chem. Phys.* **44**, 415–419.

Dunn, T. M., Tembreull, R., and Lubman, D. M. (1985). *Chem. Phys. Lett.* **121**, 453–457.

Engelke, F., Hahn, J. H., Henke, W., and Zare, R. N. (1987). *Anal. Chem.* **59**, 909–912.

Fisanick, G. J., Eichelberger, T. S., IV, Heath, B. A., and Robin, M. B. (1980). *J. Chem. Phys.* **72**, 5571–5580.

Frueholz, R., Wessel, J., and Wheatley, E. (1980). *Anal. Chem.* **52**, 281–284.

Gentry, W. R., and Giese, C. F. (1978). *Rev. Sci. Instrum.* **49**, 595–600.

Gobeli, D. A., Yang, J. J., and El-Sayed, M. A. (1985). *Chem. Rev.* **85**, 529–554.

Grotemeyer, J., and Schlag, E. W. (1988a). *Angew. Chem.* **100**(4), 461–474.

Grotemeyer, J., and Schlag, E. W. (1988b). *Org. Mass Spectrum.* **23**, 388–396.

Grotemeyer, J., and Schlag, E. W. (1989). *Acc. Chem. Res.* **22**, 399–406.

Grotemeyer, J., Boesl, U., Walter, K., and Schlag, E. W. (1986a). *Org. Mass Spectrom.* **21**, 595–597.

Grotemeyer, J., Boesl, U., Walter, K., and Schlag, E. W. (1986b). *Org. Mass Spectrom.* **21**, 645–653.

Grotemeyer, J., Boesl, U., Walter, K., and Schlag, E. W. (1986c). *J. Am. Chem. Soc.* **108**, 4233–4239.

Hager, J. W., and Wallace, S. C. (1988). *Anal. Chem.* **60**, 5–10.

Hahn, J. H., Zenobi, R., and Zare, R. N. (1987). *J. Am. Chem. Soc.* **109**, 2842–2843.

Hahn, J. H., Zenobi, R., Bada, J. L., and Zare, R. N. (1988). *Science* **239**, 1523–1525.

Hayes, J. M. (1987). *Chem. Rev.* **87**, 745–760.

Hayes, J. M., Henke, W. E., Selzle, H. L., and Schlag, E. W. (1983). *Chem. Phys. Lett.* **97**, 347–351.

Herman, P. R., LaRocque, P. E., Lipson, R. H., Janroz, W., and Stoicheff, B. P. (1985). *Can. J. Phys.* **63**, 1581–1588.

Hooper, R. J. L. (1981). In *Recent Developments in Mass Spectrometry in Biochemistry, Medicine and Environmental Research* (A. Frigerio, ed.), Vol. 7, p. 69. Elsevier, Amsterdam.

Irion, M. P., Bowers, W. D., Hunter, R. L., Rowland, F. S., and McIver, R. T., Jr. (1982). *Chem. Phys. Lett.* **93**, 375–379.

Johnson, P. M. (1975). *J. Chem. Phys.* **62**, 4562–4563.

Johnson, P. M. (1980). *Acc. Chem. Res.* **13**, 20–26.

Johnson, P. M., Berman, M. R., and Zakheim, D. (1975). *J. Chem. Phys.* **62**, 2500–2502.

Karas, M., and Hillenkamp, F. (1988). *Anal. Chem.* **60**, 2299–2301.

Kinsel, G. R., Lindner, J., Grotemeyer, J., and Schlag, E. W. (1991). *J. Phys. Chem.* **95**, 7824–7830.

Krogh-Jespersen, K., Rava, R. P., and Goodman, L. (1979). *Chem. Phys.* **44**, 295–302.

Kuehlwind, H., Neusser, H. J., and Schlag, E. W. (1985). *J. Phys. Chem.* **89**, 5593–5599.

Levy, D. H. (1984). *Sci. Am.* **250**, 96–109.

Li, L., and Lubman, D. M. (1988a). *Anal. Chem.* **60**, 1409–1415.

Li, L., and Lubman, D. M. (1988b). *Anal. Chem.* **60**, 2591–2598.

Li, L., and Lubman, D. M. (1988c). *Appl. Spectrosc.* **42**, 411–417.

Li, L., and Lubman, D. M. (1988d). *Appl. Spectrosc.* **42**, 418–424.

Li, L., and Lubman, D. M. (1989a). *Appl. Spectrosc.* **43**, 543–549.

Li, L., and Lubman, D. M. (1989b). *Rapid Commun. Mass Spectrom.* **3**, 12–16.

Lindner, J., Grotemeyer, J., and Schlag, E. W. (1990). *Int. J. Mass Spectrom. Ion Processes* **100**, 267–285.

Liverman, M. G., Beck, S. M., Monts, D. L., and Smalley, R. E. (1979). *J. Chem. Phys.* **70**, 192–198.

Lossing, F. P., and Tanaka, I. (1956). *J. Chem. Phys.* **25**, 1031–1034.

Lubman, D. M. (1986). *Anal. Chem.* **59**, 31A–40A.

Lubman, D. M. (1988b). *Mass Spectrom. Rev.* **7**, 535–554.

Lubman, D. M. (1988c). *Mass Spectrom. Rev.* **7**, 559–592.

Lubman, D. M., ed. (1990). *Lasers and Mass Spectrometry.* Oxford University Press, New York.

Lubman, D. M., and Kronick, M. N. (1982). *Anal. Chem.* **54**, 660–665.

Lubman, D. M., and Naaman R., and Zare, R. N. (1980). *J. Chem. Phys.* **72**, 3034–3040.

Lubman, D. M., Rettner, C. T., and Zare, R. N. (1982). *J. Phys. Chem.* **86**, 1129–1135.

Mamyrin, B. A., Karataev, V. I., Shmik, D. V., and Zagulin, V. A. (1973). *Sov. Phys.—JETP (Engl. Transl.)* **37**, 4–48.

McCrery, D. A., and Gross, M. L. (1985). *Anal. Chim. Acta* **91**, 178–183.

Miller, T. A. (1984). *Science* **223**, 545–553.

Nieman, G. C., and Colson, S. D. (1978). *J. Chem. Phys.* **68**, 5656–5657.

Oikawa, A., Abe, H., Mikami, N., and Ito, M. (1984). *J. Phys. Chem.* **88**, 5181–5186.

Pallix, J. B., Schuhle, U., Becker, C. H., and Huestis, D. L. (1989). *Anal. Chem.* **61**, 805–811.

Park, Y. D., Rizzo, T. R., Peteanu, L. A., and Levy, D. H. (1986). *J. Chem. Phys.* **84**, 6539–6549.

Parker, D. H., and Bernstein, R. B. (1982). *J. Phys. Chem.* **86**, 60–66.

Parker, D. H., Berg. J. O., and El-Sayed, M. A. (1978). *Advances in Laser Chemistry, Springer Ser. Chem. Phys.*, p. 320. Springer-Verlag, New York.

Petty, G., Tai, C., and Dalby, F. W. (1975). *Phys. Rev. Lett.* **34**, 1207–1208.

Powers, J. C. (1968). *J. Org. Chem.* **33**, 2044–2050.

Rava, R. P., and Goodman, L. (1982). *J. Am. Chem. Soc.* **104**, 3815–3822.

Reilly, J. P., and Kompa, K. L. (1980). *J. Chem. Phys.* **73**, 5468–5476.

Rettner, C. T., and Brophy, J. H. (1981). *Chem. Phys.* **25**, 53–61.

Rhodes, G., Opsal, R. B., Meek, J. T., and Reilly, J. P. (1983). *Anal. Chem.* **55**, 280–286.

Rizzo, T. R., Park, Y. D., Peteanu, L. A., and Levy, D. H. (1986). *J. Chem. Phys.* **84**, 2534–2541.

Robin, M. B., and Kuebler, N. A. (1978). *J. Chem. Phys.* **69**, 806–810.

Robinson, R. (1980). *Tumors that Secrete Catecholamines*, p. 148. Wiley, Chichester.

Roepstorff, P., and Fohlman, J. (1984). *Biomed. Mass Spectrom.* **11**, 242–257.

Rorden, R. J., and Lubman, D. M. (1983). *Rev. Sci. Instrum.* **54**, 641–643.

Sack, T. M., McCrery, D. A., and Gross, M. L. (1985). *Anal. Chem.* **57**, 1290–1295.

Seaver, M., Hudgens, J. W., and DeCorpo, J. J. (1980). *Int. J. Mass Spectrom. Ion Processes* **34**, 159–173.

Sin, C. H., Tembreull, R., and Lubman, D. M. (1984). *Anal. Chem.* **56**, 2776–2781.

Smalley, R. E., Wharton, L., and Levy, D. H. (1977). *Acc. Chem. Res.* **10**, 139–145.

Steenvoorden, R. J. M., Kistemaker, P. G., de Vries, A. E., Michalak, L., and Nibbering, N. M. M. (1991). *Int. J. Mass Spectrom. Ion Processes* **107** (3), 475–489.

Syage, J. A., and Wessel, J. E. (1988). *Appl. Spectrosc.* **24**, 1–79.

Tembreull, R., and Lubman, D. M. (1984). *Anal. Chem.* **56**, 1962–1967.

Tembreull, R., and Lubman, D. M. (1986). *Anal. Chem.* **58**, 1299–1303.

Tembreull, R., and Lubman, D. M. (1987a). *Anal. Chem.* **59**, 1003–1006.

Tembreull, R., and Lubman, D. M. (1987b). *Anal. Chem.* **59**, 1082–1088.

Tembreull, R., and Lubman, D. M. (1987c). *Appl. Spectrosc.* **41**, 431–436.

Tembreull, R., Sin, C. H., Li, P., Pang, H. M., and Lubman, D. M. (1985a). *Anal. Chem.* **57**, 1186–1192.

Tembreull, R., Sin, C. H., Pang, H. M., and Lubman, D. M. (1985c). *Anal. Chem.* **57**, 2911–2917.

Tembreull, R., and Dunn, T. M., and Lubman, D. M. (1986). *Spectrochim. Acta* **42A**, 899–906.

van Bramer, S. E., and Johnston, M. V. (1990). *J. Am. Soc. Mass Spectrom.* **1**, 419–426.

Veca, A., and Breisbach, J. H. (1988). *J. Chem. Educ.* **65**, 108–111.

Wessel, J. E., Cooper, D. E., and Klimcak, C. M. (1981). *Proc. Soc. Photo-Opt. Instrum. Eng.* **286**, 48–55.

Wiley, W. C., and McLaren, I. H. (1955). *Rev. Sci. Instrum.* **26**, 1150–1157.

Wilkerson, C. W., Jr., and Reilly, J. P. (1990). *Anal. Chem.* **62**, 1804–1808.

Wilkerson, C. W., Jr., Colby, S. M., and Reilly, J. P. (1989). *Anal. Chem.* **61**, 2669–2673.

Wilkins, C. L., Weil, D. A., Yang, C. L. C., and Ijames, C. F. (1985). *Anal. Chem.* **57**, 520–524.

Zandee, L., and Bernstein, R. B. (1979a). *J. Chem. Phys.* **70**, 2574–2575.

Zandee, L., and Bernstein, R. B. (1979b). *J. Chem. Phys.* **71**, 1359–1371.

Zandee, L., Bernstein, R. B., and Lichtin, D. A. (1978). *J. Chem. Phys.* **69**, 3427–3429.

CHAPTER

4

THE HIGH LASER IRRADIANCE REGIME

*A. LASER ABLATION AND PLASMA FORMATION**

CLAUDE R. PHIPPS

Chemical and Laser Sciences Division
Los Alamos National Laboratory
Los Alamos, New Mexico

RUSSELL W. DREYFUS

IBM Research Division
T. J. Watson Research Center
Yorktown Heights, New York

4A.1. PLASMA-MEDIATED LASER–TARGET COUPLING AT IRRADIANCE GREATER THAN 1 GW/cm^2

4A.1.1. Introduction

The present section treats the specialized physics of high-irradiance vacuum target coupling. The reader is cautioned that this specialization restricts our attention to situations that meet our definition of "high irradiance" outlined in Section 4A.1.3, below. This restriction prevents us from reviewing experiments in air, in materials that absorb in depth rather than on the surface, and those that occur below the plasma threshold. Our focus on surface absorbers is consistent with the limitation to high irradiance, since it can be shown that *all* absorbers in vacuum coalesce to the behavior of surface absorbers at high irradiance. We also avoid discussing work with picosecond and shorter pulses, because the interaction physics is still poorly understood. Often, we have had to exclude data where the irradiance was not clearly specified. In return for these restrictions, our specialized purview has the advantage that it permits

* Dedicated to Professor Heinrich Hora on his sixtieth birthday.

Laser Ionization Mass Analysis, Edited by Akos Vertes, Renaat Gijbels, and Fred Adams. Chemical Analysis Series, Vol. 124.
ISBN 0-471-53673-3

us to theoretically predict all the main features of the majority of high-irradiance ablation data available at this time.

For consideration of the vast store of data taken in physical regimes outside the range we consider here, the reader is referred to other chapters in this volume. Discussion of RIMS elemental analysis appears in Dietze and Becker's contribution (Part C of this chapter, Section 4C.2.2).

By definition, the high-irradiance regime features a laser–surface interaction dominated by a plasma layer that forms just above the surface. For this reason, the observables of the interaction depend less on the chemical or molecular composition of the preirradiation solid material and more on the simple atomic abundances and ion charge states present in the ablation plume. The photoinitiated plasma layer near the target surface is mainly responsible for laser absorption and quickly redistributes the incident energy into thermal conduction and reradiation (with the radiation spectrum of the plasma, rather than of the laser), which actually communicate energy to the solid surface.

The present discussion aims at showing that most aspects of high-irradiance laser–surface interaction in vacuum—such as pressure, plasma density, and ion velocity—can be predicted quite well from plasma physics theory. The physical and chemical properties of the underlying solid material enter as average ionic mass number, A, and average ionic charge, Z, in the ablation plume, evaluated close to the region where laser absorption occurs. As an example, for KrF lasers with $\tau = 10$ ns and target irradiance $\approx 10\,\mathrm{GW/cm^2}$, vacuum ablation pressures can be tens of kilobars, and temperatures on the order of 50,000 K. It is obvious that molecular and even chemical structure in the ablated target material are destroyed early in such interactions.

Here, in Section 4A.1, we define the boundary to this case in terms of the range of laser irradiation parameters which occur in laser ionization mass analysis, and outline the theory for the limiting case of high irradiance. In Section 4A.2, we compare predictions to experimental results. Section 4A.3 treats ion acceleration, and Section 4A.4 considers measurement techniques. Section 4A.5 presents our conclusions.

Why Does High Irradiance in LIMA Work? Above $5\,\mathrm{GW/cm^2}$ irradiance, the ionization fraction approaches unity (e.g., Bingham and Salter, 1976a,b), whereas it can be 10^{-2}–10^{-3}, or even lower, with 1 decade less irradiance. A high source flux is a benefit for laser ionization mass analysis (LIMA) work. High irradiance around $1\,\mathrm{GW/cm^2}$ ($20\,\mathrm{TW/m^2}$) also makes for more uniform relative sensitivity coefficients (RSC) (Conzemius and Capellen, 1980; Ramendik et al., 1987). Mass resolution $\geqslant 300$ together with a 10^{-3} transmission coefficient are conditions commonly achieved at $10\,\mathrm{GW/cm^2}$ irradiance ($100\,\mathrm{TW/m^2}$) (Eloy, 1974), and values up to 1500 at this irradiance are indicated (Kovalev et al., 1978). For elemental analysis, higher resolution is not required.

Table 4A.1. Basic Constants and Units Used

Quantity	Symbol	cgs	SI
Absorption coefficient	α	cm^{-1}, $\times$ 100 =	m^{-1}
Light speed	c	2.998×10^{10} cm/s	2.998×10^{8} m/s
Specific heat (practical)	C	$J \cdot g^{-1} \cdot K^{-1}$, $\times 10^3$ =	$J \cdot kg^{-1} \cdot K^{-1}$
Electric field strength (physical)	E	esu/cm, $\times$ 29,979 =	V/m
Electric field strength (practical)	E	V/cm, $\times$ 100 =	V/m
Electron charge	e	4.803×10^{-10} esu	1.602×10^{-19} C
Free-space permittivity	ε_0	1 esu	8.854×10^{-12} F/m
Fluence (physical)	Φ	erg/cm^2, $\times 10^{-13}$ =	J/m^2
Fluence (practical)	Φ	J/cm^2, $\times 10^4$ =	J/m^2
Planck's constant	h	6.6256×10^{-27} erg·s	6.6256×10^{-34} J·s
Planck's constant	$\hbar$	$h/2\pi$	$h/2\pi$
Viscosity	η	cm^2/s, $\times 10^4$ =	m^2/s
Volumetric irradiance (practical)	i	W/cm^3, $\times 10^6$ =	W/m^3
Irradiance (physical)	I	erg $cm^{-2}s^{-1}$, $\times 10^{-3}$ =	W/m^2
Irradiance (practical)	I	W/cm^2, $\times 10^4$ =	W/m^2
Momentum	J	dyn·s, $\times 10^{-5}$ =	n·s
Boltzmann's constant	k	1.3805×10^{-16} erg/K	1.3805×10^{-23} J/K
Thermal conductivity (practical)	K	W/cm, $\times$ 100 =	W/m
Thermal diffusivity	κ	cm^2/s, $\times 10^{-4}$ =	m^2/s
Wavelength	λ	cm, $\times$ 0.01 =	m
Mach number	M	Dimensionless	Dimensionless
Electron mass	m_e	9.1091×10^{-28} g	9.1091×10^{-31} kg
Proton mass	m_p	1.673×10^{-24} g	1.673×10^{-27} kg
Particle density	n	cm^{-3}, $\times 10^6$ =	m^{-3}
Potential	ϕ	esu, $\times$ 299.79 =	V
Pressure (physical)	p	dyn/cm^2, $\times$ 0.1 =	Pa
Pressure (practical)	p	bar, $\times 10^{-5}$ =	Pa
Mass density	ρ	g/cm^3, $\times 10^3$ =	kg/m^3
Stefan–Boltzmann constant	σ	5.67×10^{-12} $W \cdot cm^{-2} \cdot K^{-4}$	5.67×10^{-8} $W \cdot m^{-2} \cdot K^{-4}$
Temperature (physical)	T	K	K
Temperature (practical)	T	eV, $\times$ 11,605 =	K
Energy	W	erg, $\times 10^{-7}$ =	J (joule)

Units. Equations are written alternately in practical cgs and SI units throughout this work, to conform to the broadly accepted SI convention while maintaining connection with the laser and plasma literature, which almost exclusively uses practical cgs. In the latter convention, physics is done in cgs but some common parameters are expressed in practical form. For example, temperature will be shown in kelvin units (K), and the electron charge will be in electostatic units (esu), but irradiance will be given in W/cm^2 rather than $erg \cdot cm^{-2} \cdot s^{-1}$ (see Table 4A.1). Occasional exceptions designed to make a specific point [e.g., electron volts (eV) in Eq. (18)] are clearly noted. Power-of-ten prefixes we use are f (10^{-15}), p (10^{-12}), n (10^{-9}), μ (10^{-6}), m (10^{-3}), k (10^{3}), M (10^{6}), G (10^{9}), T (10^{12}), and P (10^{15}).

4A.1.2. Basic Concepts: Survey of the Four Interaction Regions

Figure 4A.1 outlines the extremely wide range of plasma physics variables that are important in a pulsed laser–surface irradiation event in vacuum. The notation "not drawn to scale" cautions the reader that it is impossible to draw this illustration in a universally applicable way. The horizontal axis is approximately logarithmic. The distances and other parameters indicated across the top are dependent on scale lengths in a particular problem, and may vary by some orders of magnitude, depending on parameters such as laser pulse duration, τ, and irradiance. The specific numbers chosen for the illustration are typical of 100 μm diameter irradiation on aluminum, using 10–100 ns excimer laser pulses at 5–10 J/cm^2 (50–100 kJ/m^2).

Figure 4A.1 is meant to illustrate a sequence for plasma expansion into the vacuum and hence does not necessarily represent an achievable "snapshot" at any particular instant. Rather, it represents phenomena that are dominant near the plume density maximum, as the ablated material expands through many decades of decreasing density. For example, the laser may be "off" before the hot ion beam reaches Region IV, but such a situation does not refute any arguments we shall present.

Region I. Most of the laser absorption occurs in a thin plasma layer close to the target in Region I. Steady expansion away from the target in vacuum reduces the plasma density far below n_{ec} (see below) as distance from the target increases, leaving a narrow high-pressure zone close to the target where most absorption occurs. "Thin" in this context means a layer that is significantly smaller than the laser spot diameter [see Section 4A.1.5 (*Inverse Bremsstrahlung Absorption...*), below].

The absorption zone is close to the target because that is where the electron density first reaches the so-called critical density, n_{ec}. This density is called "critical" in plasma physics because, as it is approached, the complex magnitude

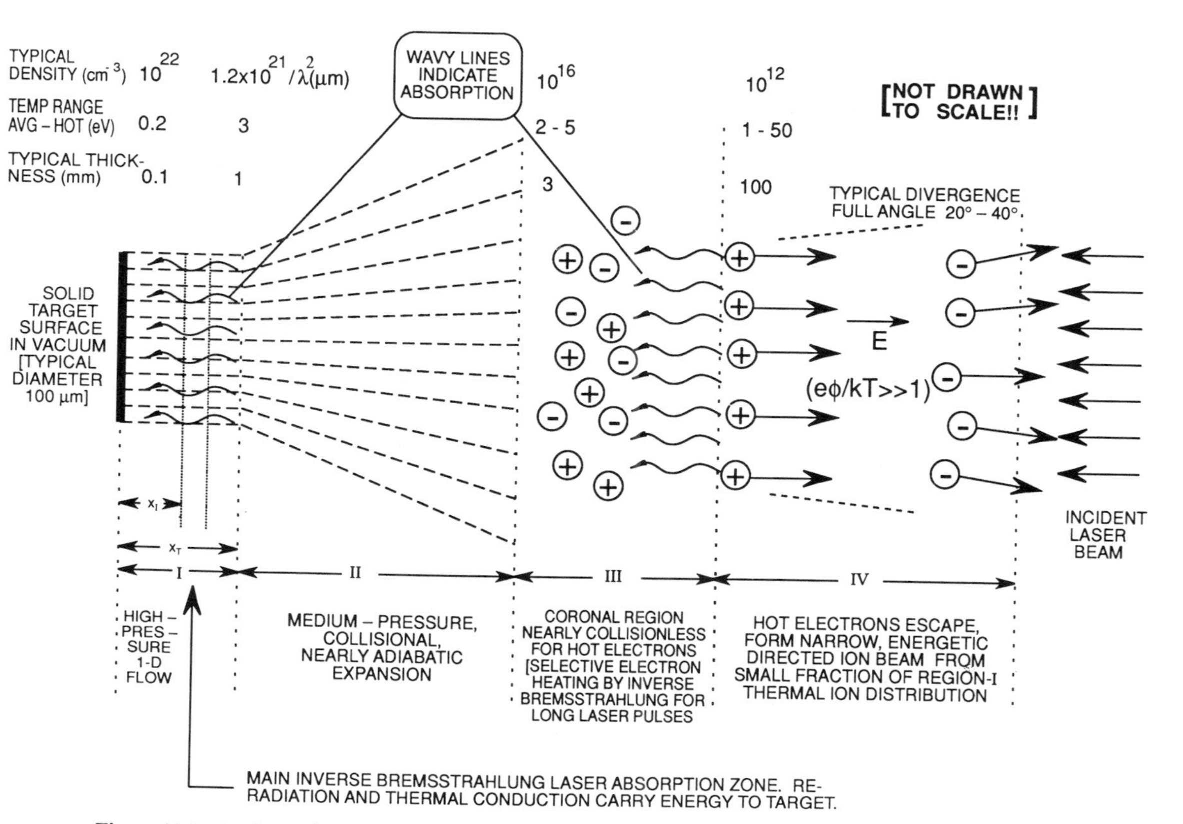

Figure 4A.1. A schematic representation of the four distinct physical regions in a vacuum laser–surface interaction at high irradiance.

of the plasma index of refraction decreases from near 1 to 0, first causing strong absorption, then total reflection of the incident light. The critical density has a very large, near-solid value (10^{21}–10^{23} cm^{-3} or 10^{27}–10^{29} m^{-3}) for visible laser wavelengths and is high enough to create target surface pressures on the order of 100 kbar for KrF lasers (248 nm wavelength) and 2 eV temperature. It is related to other parameters by

$$n_{ec} = m_e \boldsymbol{n}^2 \omega^2 / 4\pi e^2 = 1.115 \times 10^{13} / \lambda^2 \qquad cm^{-3}$$

or

$$n_{ec} = m_e \varepsilon_0 \boldsymbol{n}^2 \omega^2 / e^2 = 1.115 \times 10^{15} / \lambda^2 \qquad m^{-3} \tag{1}$$

In Eq. (1), $\boldsymbol{n}$ is the refractive index of the surrounding medium, which is taken as 1.0 on the right-hand side of Eq. (1) and throughout this work, and ω is the laser frequency (in radians).

It can be shown by integration from infinity [see Section 4A.1.5 (*Inverse Bremsstrahlung Absorption*...), below] and, with proper assumptions about the variation $n_e(x)$ with distance x, from the target surface that the effective thickness of the absorption zone is about $1/\alpha$, where α is the inverse bremsstrahlung absorption coefficient discussed in subsequent sections [i.e., absorption due to inelastic scattering of photons by free electrons (see Spitzer, 1967)]. The equivalent thickness of the absorption zone can be as small as a laser wavelength. For example, with a KrF laser, 2 eV plasma temperature, and $n_e = 0.2\, n_{ec} = 3.3 \times 10^{21}$ cm^{-3} (3.3×10^{27} m^{-3}), $1/\alpha = \lambda = 248$ nm.

Even with very small laser spots, the physics governing *laser coupling* can be "one-dimensional" [see Section 4A.1.6 (*Heat Conduction and Reradiation*), below], despite the fact that, in free space, all expansions *eventually* become multidimensional. We call the distance at which the transition to multidimensional flow occurs x_T. It is the ratio of x_T to the irradiated diameter d_s that determines the "dimensionality" of the expansion after laser coupling has occurred.

Region II and the Knudsen Layer. In Region II, the expansion inevitably takes on a three-dimensional character. Naively, one might expect a hemispherical, bubble-like expansion into vacuum. However, even below the plasma threshold, the expansion of vapor from a half-plane has more of the character of a beam than a hemisphere, as was shown by Kelly and Dreyfus (1988a). This is because, in a highly collisional regime like Region I/II, a so-called *Knudsen layer* will form. In this layer, an explicit center-of-mass velocity develops, producing a beam-like expansion. It also produces a *reduced temperature* and *number density*, a consequence of collisional expansion of a Maxwellian velocity distribution of particles from an infinite half-plane (traveling with the center of mass).

The Kelly and Dreyfus derivation slightly underestimates the short-pulse collimation, as an additional cosine factor does not appear in such a steady-state derivation, i.e., a cosine term that resembles the case of collisionless vaporization into a vacuum or Knudsen nozzle expansion—the effective difference being, however, that the "vaporization" is from the dense medium in Region I.

The Knudsen layer will always be present in laser–target interactions that we consider, since sufficient (3–5) collisions to cause inverse bremsstrahlung absorption are also sufficient to form the Knudsen layer. The beam-like qualities of a Knudsen-layer expansion do not by themselves explain the features of *ion* beam formation in laser ablation, which we shall discuss in Section 4A.3.

Kelly and Dreyfus (1988a) show that the beamlike character of the expansion as it leaves the target in Region II is at least as sharply peaked in the forward direction as $\cos^4\theta$ and that an unsteady adiabatic expansion (such as in Region II) will produce a forward peaking more like $\cos^9\theta$. The latter distribution has $\theta_d = 44^\circ$ full width at half-maximum (fwhm), and not $\theta_d = 90^\circ$, as one finds for a "free" expansion.

Many examples of this forward peaking have been reported. Viswanathan and Hussla (1986) irradiated copper with a KrF laser in vacuum at 450 MW/cm^2, and saw forward peaking with an fwhm of about 45°, which was fit with a $\cos^8\theta$ distribution. At much higher laser fusion intensities, Mulser et al. (1973) irradiated a solid deuterium target with a Nd:glass laser at about 1 TW/cm^2 (10 PW/m^2) and obtained an angular flux dependence varying approximately as $\cos^5\theta$.

In any case, we take

$$(90/\theta_d)d_s = x_T \tag{2}$$

as the boundary beyond which the dimensionality of the free expansion is assumed to be >1. In Region II, an adiabatic expansion occurs, with a predicted variation

$$n(x)/n(x_T) = (x/x_T)^{-20/9} \tag{3}$$

for the ion or electron density with distance, x, from the target [see Puell, 1970; and Section 4A.1.6 (*Adiabatic Expansion Theory*...), below]. We note that von Gutfeld and Dreyfus (1989) observed an ion current variation

$$I_i(x) \propto (x/x_T)^{-2.5} \tag{4}$$

giving an experimentally determined exponent that agrees with the ion density profile predicted by adiabatic expansion theory to within 10%.

Three-body recombination (see Section 4A.1.5, below) is usually important in Regions I and II but ceases to be an important electron energy loss and atom-heating mechanism beyond the boundary between Regions II and III. Where ion densities n_i reach about $10^{18}\,\mathrm{cm}^{-3}$ ($10^{24}\,\mathrm{m}^{-3}$), this recombination time is about 10 ns, a rough upper limit for achieving equilibrium during a typical ablation pulse via this mechanism.

Region III. In Region III, also called the coronal region by analogy with solar physics, collisions are infrequent for thermal electrons and virtually nonexistent for energetic electrons. Hot electrons that developed in previous regions begin to decouple from the Maxwell–Boltzmann distribution, violating local thermodynamic equilibrium (LTE) [see Figure 4A.1; also the next subsection (*Region IV*) and Section 4A.1.5 (*Coulombic Equilibration Times...*), below]. Here, three-body recombination ceases to be a factor, except for *very* long (millisecond-duration) laser pulses.

Heating by inverse bremsstrahlung still exists (see the wavy lines in Figure 4A.1) and, although small compared to the heating rate in Region I, may counterbalance adiabatic cooling, making it difficult to estimate the variation $T_e(x)/T_{eT}$ in a universally applicable way. Boland et al. (1968) did measure $T_e(x)$ and $n_e(x)$ in a ruby laser experiment with poly(ethylene) targets at irradiance levels around 0.3 TW/cm² (3 PW/m²) and found

$$T_e(x)/T_e(x_T) = [n_e(x)/n_e(x_T)]^{2/3} \tag{5}$$

which is the prediction of ideal-gas adiabatic expansion theory (see Section 4A.1.6, below). Taken together, Eqs. (3) and (5) predict a $(x/x_T)^{-1.5}$ variation for the (thermal) plasma temperature.

Region IV. In Region IV, a minority population of hot electrons completely escapes the plasma plume, owing to the lack of energy-exchanging collisions while still in the experiment volume, causing a high positive potential $e\phi \gg kT_e$ to develop, which in turn accelerates a few ions that are still in the plume region to high energy. Because $e\phi \gg kT_e$, LTE no longer applies to these hot electrons [see the discussion of TE and LTE in Section 4A.1.5 (*Coulombic Equilibration Times...*) and of acceleration in Section 4A.3, below]. These hot electrons depart from the majority Maxwell–Boltzmann distribution and cease to obey the physics derived from the assumptions of LTE. Because of the strong electric field due to the separation Δx [which may be much greater than a Debye length (D_e—see Section 4A.1.5 (*The Debye Shielding Distance...*), below],

$$E = -4\pi e n_e \Delta x \qquad \text{esu/cm}$$

or

$$E = -e n_e \Delta x/\varepsilon_0 \qquad \text{V/m} \tag{6}$$

the ions rapidly reach the same average velocity as the electrons. Given a sufficiently large volume, the electron and ion clouds may actually oscillate through each other's position because of the low collision rate, at a continually decreasing plasma frequency corresponding to the decreasing electron cloud density as expansion proceeds. The plasma frequency is

$$f_p \sim \frac{\omega_p}{2\pi} = \left(\frac{n_e e^2}{\pi m_e}\right)^{1/2} = 8978(n_e^{1/2}) \qquad \text{Hz}$$

or, in SI, (7)

$$f_p = \frac{\omega_p}{2\pi} = \frac{1}{2\pi}\left(\frac{n_e e^2}{\varepsilon_0 m_e}\right)^{1/2} = 8.978(n_e^{1/2}) \qquad \text{Hz}$$

In a time-of-flight experiment where $n_e = 10^6\ \text{cm}^{-3}$ ($10^{12}\ \text{m}^{-3}$), plasma oscillations might be seen with a period $\tau_p = 1/f_p \approx 100$ ns.

After the interaction, the co-motional cloud of ions has nearly all the hot electrons' initial drift energy, because of the electrons' much smaller mass. Depending on Z, this energy may be distributed among a smaller number of particles (see Section 4A.3.2, below), but the drift velocity of the two velocity-decoupled clouds should ultimately be equal.

Plasma physics is normally treated by classical approximations because the quanta are normally quite small compared to, e.g., thermonuclear values of kT_e. However, it is important to realize that

$$\hbar\omega_p = 3.71 \times 10^{-11}(n_e^{1/2}) \qquad \text{eV}$$

or, in SI, (8)

$$\hbar\omega_p = 3.71 \times 10^{-14}(n_e^{1/2}) \qquad \text{eV}$$

Thus, the energy of a plasma quantum is already 1 eV when the plasma electron density is $7.2 \times 10^{20}\ \text{cm}^{-3}$ ($7.2 \times 10^{26}\ \text{m}^{-3}$), a density which is equal to the critical density for 1.25 μm light. Plasma quanta are on the order of 10 eV in metals where n_e is on the order of $10^{23}\ \text{cm}^{-3}$ ($10^{29}\ \text{m}^{-3}$). For $n(x)$ given by Eq. (3), f_p continually decreases with the expansion, nearly linearly, according to $(x/x_T)^{-10/9}$.

4A.1.3. Applicability Conditions for High-Irradiance Theory

In the heading of Section 4A.1, we picked the irradiance value 1 GW/cm^2 to illustrate the boundary to the high-irradiance regime. This choice is a convenient oversimplification. In fact, there is no single intensity that can define the boundary. For the theory of high irradiance to clearly apply, the following four conditions must be met, at least approximately.

Condition I: Vacuum Environment. Many ablation experiments (as members of the general class in which LIMA work is found) are done in air, and in some cases the ablation depth is not too different in air or in vacuum (see, e.g., Kelly et al., 1985; and Section 4A. 2, below). Nevertheless, the underlying physics of an ablation jet in air is so much more complex than that in vacuum that simple analytical predictive capability is lost. In terms of observables, the main effect of adding more than about 1 torr of air to the problem is to *increase* target momentum coupling due to "tamping" by the air (Beverley and Walters, 1976), to decrease ablation velocity due to the "snowplow" effect, and to limit the laser irradiance that can be delivered to the target to a value equal to the air breakdown threshold irradiance (Figueira et al., 1981; Golub' et al., 1981). Further, interaction energy is devoted to the launching of so-called laser-supported detonation (LSD) waves and, for longer pulses, laser-supported combustion (LSC) waves, which can totally decouple the interaction from the original target (Beverley and Walters, 1976). This is in addition to obvious gas phase chemistry (see, e.g., Nogar, 1991; Otis and Dreyfus, 1991).

Condition II: Surface Absorber. Surface absorbers are defined (Phipps et al., 1988) as those for which the thermal penetration depth during a laser pulse of duration τ,

$$x_{\text{th}} = 0.969[(\kappa\tau)^{1/2}] \gg 1/\alpha \tag{9}$$

is much larger than the laser penetration depth. Here, $\kappa = K/\rho C$ is the thermal diffusivity of the material (units L^2/t), which among common metals ranges from 1.15 (Cu) to 0.068 cm^2/s (Bi) (Batanov et al., 1972); K is the thermal conductivity; ρ, the mass density; and C, the specific heat. This condition has also been expressed as (Kelly et al., 1985)

$$\tau_{\text{thermal}} \approx L_0^2/\kappa \ll \tau \tag{10}$$

where $L_0 = 1/\alpha$ is the laser penetration depth. In any case, the result is a problem in which the laser interaction can be modeled as a continuous plane heat source at the target surface.

"Volume absorbers" such as undoped poly(methyl methacrylate) (PMMA) at 248 nm [at low laser irradiance near plasma threshold (see Phipps et al., 1990)] show momentum coupling and other properties that increase dramatically from what is obtained for surface absorbers. One obtains momentum coupling coefficients on the order of 100 $\text{dyn}\cdot\text{s}\cdot\text{J}^{-1}$ ($10^7\ \text{n}\cdot\text{s}\cdot\text{J}^{-1}$) at irradiance near the plasma threshold, rather than the 1–10 $\text{dyn}\cdot\text{s}\cdot\text{J}^{-1}$ (10^5–$10^6\ \text{n}\cdot\text{s}\cdot\text{J}^{-1}$) experienced by surface absorbers.

This coupling enhancement is not yet predictable in simple closed form from just the laser irradiance I, wavelength λ, and pulse duration τ, but depends strongly on material properties such as specific heat, ionization energy, thermal conductivity, and equation of state.

As a consequence, vaporization depth x_v in volume absorbers is very material dependent and depends mainly on laser "fluence" $\Phi \equiv \int_{-\infty}^{\infty} d\tau\, I(\tau)$ (see Taylor et al., 1987). As an example of the importance of fluence, the very short pulse laser ablation work ($\tau = 160$–$300\,\text{fs}$) of Küper and Stuke (1989) and Srinivasan et al. (1987) shows little difference in ablation depth obtained from work done with 10^4 times longer pulse widths $\tau \approx 10\,\text{ns}$ at the same fluence (though the sharpness of the ablation features obtained was quite different). In this connection, see Condition IV, below.

Volume absorbers benefit from trapped ablation either through absorption at depth within the material or through deliberate imposition of a heterogeneous transparent tamping layer on top of the absorbing medium. Although their experiments were done in air rather than vacuum, and the significant "tamping" effect of air (tending also to improve coupling of the detonation) was ignored in their analysis, interesting results on the order of $700\,\text{dyn}\cdot\text{s}\cdot\text{J}^{-1}$ were obtained by Fabbro et al. (1990), using 3 and 30 ns pulses from an Nd:glass laser (1.06 μm) and $I \approx 300\,\text{MW/cm}^2$ ($3\,\text{TW/m}^2$).

High irradiance in itself ensures the achievement of condition (9). This is because the ambient spectrophotometric absorption depth $1/\alpha_a$ is dramatically decreased by high-irradiance temperature and pressure effects, including free-electron absorption, to the point where the interaction with a target that was "volume absorbing" at low irradiance is indistinguishable in its behavior from that of a metallic target.

Condition III: Plasma Ignition. For surface absorbers in vacuum, an estimate of the onset of plasma dominance of the laser–surface interaction is given by the relationship (Berchenko et al., 1981; Phipps et al., 1988)

$$I(\tau^{1/2}) \geqslant B \tag{11}$$

where τ is the laser pulse length. Over at least a 5 orders of magnitude range in pulse duration from 1 ms to 10 ns, and laser wavelength in the range 0.25–10 μm, the constant $B_{\max} \approx 8 \times 10^4\,\text{W}\cdot\text{s}^{1/2}\cdot\text{cm}^{-2}$ ($8 \times 10^8\,\text{W}\cdot\text{s}^{1/2}\cdot\text{m}^{-2}$) when plasma dominates the interaction. The plasma formation threshold occurs at $B_p \approx \frac{1}{2} B_{\max} = 4 \times 10^4\,\text{W}\cdot\text{s}^{1/2}\cdot\text{cm}^{-2}$ ($4 \times 10^8\,\text{W}\cdot\text{s}^{1/2}\cdot\text{m}^{-2}$). This point is illustrated in Figure 4A.2.

This value for B_p is accurate within a factor of 2 for most metals in vacuum. It is also a good upper limit value for dielectrics but can be lower in the presence of strongly absorbing surfaces (Phipps et al., 1990). For very

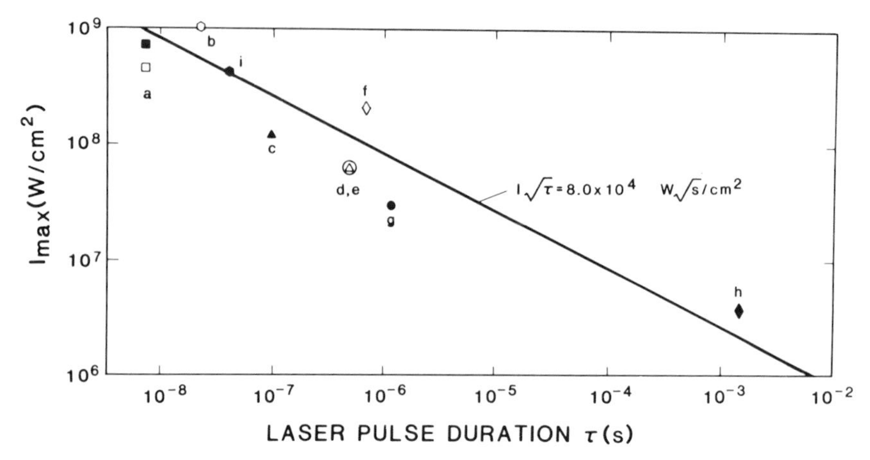

Figure 4A.2. A survey of intensities for plasma dominance of the laser–surface interaction in vacuum on metals at several wavelengths from Phipps et al. (1988), used by permission: (a) Gregg and Thomas (1966a,b), $\lambda = 694$ nm, $\tau = 7.5$ ns, Be; (b) Phipps et al. (1988), $\lambda = 248$ nm, $\tau = 22$ ns, Al; (c) Ursu et al. (1981), $\lambda = 10.6\,\mu$m, $\tau = 100$ ns, stainless steel; (d) Rosen et al. (1982b), $\lambda = 350$ nm, $\tau = 500$ ns, Ti alloy; (e) Rosen et al. (1982a), $\lambda = 350$ nm, $\tau = 500$ ns, Al alloy; (f) Phipps et al. (1988), $\lambda = 10.6\,\mu$m, $\tau = 700$ ns, Al; (g) Duzy et al. (1980), $\lambda = 350$ nm, $\tau = 1.2\,\mu$s, Al; (h) Afanas'ev et al. (1969), $\lambda = 1.06\,\mu$m, $\tau = 1.5$ ms, Al; and (i) Phipps et al. (1988), $\lambda = 248$ nm, $\tau = 37$ ns, Al.

short pulses and transparent materials, B is somewhat larger. For pulses in the range 10 ps–10 ns and for transparent dielectrics such as NaCl, B_p increases to about $5B_{max}$ (see Smith, 1978). Further details of plasma ignition are discussed below in Section 4A.1.4.

Condition IV: Laser Pulse Not Too Short. For high-irradiance scaling theory to predict the depth, x_v, of material removed as well as it does the impulse, we have one more requirement: the duration of the pressure pulse on the target surface must be of the same order as or smaller than the laser pulse duration. In high-irradiance laser irradiation experiments, this condition is normally considered to be satisfied. Quantitatively, the condition is satisfied if the laser-absorption region near the target surface is able to expand much farther than the standoff distance x_I [see Section 4A.1.6 (*Adiabatic Expansion Theory*...), below], allowing the surface pressure to drop substantially after the laser pulse ceases, in a time much shorter than the laser pulse. In other words, a validity limit for our theory is

$$\tau \gg t_I = x_I/v_a \qquad \text{s} \tag{12}$$

where the velocity of the absorption front, v_a, is approximately equal to c_s

for purposes of comparison. Note that the standoff distance x_I is not necessarily equal to the flow transition distance x_T.

To understand this limitation, we need to compute x_I and t_I. We will do this later [in Section 4A.1.6 (*Adiabatic Expansion Theory*...)], after development of additional concepts. Here, we will simply state the result:

$$t_I = 8.6 \times 10^{-5} A^{7/8} I^{1/2} \lambda^{3/2} \tau^{3/4} \qquad \text{s}$$

or, in *SI*, (13)

$$t_I = 8.6 \times 10^{-4} A^{7/8} I^{1/2} \lambda^{3/2} \tau^{3/4} \qquad \text{s}$$

For example, for 10 ns KrF experiments on a silicon target ($A = 28$) at $10\,\text{GW/cm}^2$ ($100\,\text{TW/m}^2$) irradiance, the theory should predict impulse well, because $t_I = 20$ ps. For a 10 ns, 10 μm laser, the same conditions would require $\tau \gg 5$ ns, so that the prediction might not be as good. As a third example, with $I = 10\,\text{PW/cm}^2$ ($10^{20}\,\text{W/m}^2$), $\lambda = 248$ nm, $\tau = 1$ ps, and $A = 63.5$ (copper), we find $t_I = 40$ ps, and $\tau \gg t_I$ is clearly not satisfied.

For pulses of picosecond and shorter duration, special considerations such as shock heating come into play and for example, very-high-irradiance 100 fs ablation experiments will not be modeled by our theory [see Section 4A.1.6 (*Adiabatic Expansion Theory*...).

4A.1.4. High-Irradiance Absorption and Ignition at the Solid Surface

Under most experimental conditions, awaiting the few initial electrons does not appear to be a hindrance to breakdown. The first few electrons are probably present owing to one or more of the following phenomena:

- Multiphoton ionization of atomic or molecular species in the plume
- Penning ionization from excited states in the plume
- Fracture ejection of ions and electrons from the surface
- Thermionic emission from hot surfaces (including those of ejected particulates), usually at temperatures >1000 K, even for polymers; we will emphasize this mechanism

While the last-named mechanism could supply large currents, space charge effects over the surface choke off the supply of electrons at a flux about $10^9\,\text{cm}^{-2}$, as charge-compensating ions are not present.

If these electron sources were not present, then ablation-plasma thresholds would be in the $100\,\text{GW/cm}^2$ region, instead of $1\,\text{GW/cm}^2$.

In the general case, after vapor is present, the physical mechanisms that produce plasma are either collisional ionization or photoionization, depending on the ratio of incident photon energy to the atomic energy levels available in the vapor (Raizer, 1977).

Initiation. The functional form of Eq. (11) results from analysis of uniform-rate heating of a semi-infinite slab, where the heated layer thickness is given by (see Carslow and Jaeger, 1959)

$$x_{\text{th}} = 0.969[(\kappa\tau)^{1/2}] \tag{14}$$

The resulting surface temperature is

$$\Delta T = \frac{2I_{\text{abs}}(\tau^{1/2})}{(\pi K\rho C)^{1/2}} \tag{15}$$

In Eqs. (14) and (15), I_{abs} is the absorbed irradiance; K, the thermal conductivity; C, the specific heat; and $\kappa = K/\rho C$, the thermal diffusivity.

In the simplest analysis, the temperature of the irradiated solid surface rises until the vapor pressure above the surface is sufficient for the laser irradiance to cause plasma formation. Even if the surface absorptivity I_{abs}/I is known, the microscopic state of the surface may depress the ignition threshold below what would be predicted by Eq. (10). See, for example, the insulated surface defect model of Glickler et al. (1980), which supposes that microscopic flakes on the material surface ignite plasma in the vapor via thermionic emission. The particle current density for the thermionic emission is given by the Richardson–Dushman equation ($\text{e}\phi_w$ is the surface work function energy (Coles, 1976):

$$\Gamma_{\text{e}} = \frac{4\pi m_{\text{e}}(kT)^2}{h^3}\exp(-\text{e}\phi_w/kT) \qquad \text{e}\cdot\text{s}^{-1}\cdot\text{cm}^{-2} \tag{16}$$

Collisional Breakdown. In the infrared (IR), e.g., with CO_2 lasers, the mechanism for initial plasma formation is (collisional) ac avalanche breakdown in the target vapor. The small initial background density of electrons is accelerated by the electric field of the laser light, repeatedly colliding with neutral atoms until some of them reach their ionization threshold. This mechanism is inverse bremsstrahlung heating. With I in W/cm^2 (W/m^2) and E in V/cm (V/m), the magnitude of the electric field of the light wave is

$$E = 27(I^{1/2}) \tag{17}$$

With sufficient dephasing collisions per optical cycle, this field is available for electron acceleration. As an example, the available accelerating field is

27 MV/cm at 1 TW/cm^2 (10 PW/m^2) irradiance. With sufficient irradiance and time, this process exponentiates as the collision frequency increases with ionization density, until complete ionization occurs. Avalanche ionization can be quite efficient and rapid. Even if $\hbar\omega = 117$ meV and $E_I/\hbar\omega \approx 100$, as with CO_2 laser-induced ionization, the critical density $n_{ec} = 9.9 \times 10^{18}$ cm^{-3} can be reached in picoseconds through this mechanism, starting from perhaps $n_{e0} = 10^3$ cm^{-3}. Enough atoms are available in air at 150 torr to create n_{ec} with full ionization. This process is particularly efficient in the IR, as collision and laser frequencies are similar in magnitude.

The (collision-dominated) IR optical intensity breakdown threshold for neutral gases is inversely proportional to vapor pressure up to several atmospheres (see Gill and Dougal, 1965; Grey Morgan, 1978). Therefore, above a certain threshold laser intensity, laser vaporization simply increases the pressure at the surface until the breakdown threshold is achieved.

Photoionizaton. On the other hand, for ultraviolet (UV) wavelengths, direct photoionization of the vapor atoms is often the predominant ionization mechanism. For example, the photon energy $\hbar\omega = 5$ eV of KrF is sufficient to photoionize many metal vapors via two or three steps involving low-energy, intermediate excited states [e.g., titanium (see Rosen et al., 1982a, and associated references) and copper (see Dreyfus, 1991)].

Between these extremes of wavelength, one might expect that the dominant ionization mechanism during the laser pulse would depend on the atomic physics of the target material. After the initial formation stage, direct photoionization can play an important role with IR laser irradiation as well, since a dense plasma, once formed, is an efficient IR-to-UV converter. This is because one is dealing with a hot-electron gas. Since the peak of the blackbody spectrum (per unit frequency interval) is given by

$$T\lambda_{max} = 440 \qquad \text{nm}\cdot\text{eV} \tag{18}$$

an optically thick hot-electron gas radiates a principal color that is well into the hard UV for the expected temperature at $\geqslant 2$ eV.

Collisionless Ionization. Multiphoton, collisionless, photoionization is a very-high-irradiance, low-pressure phenomenon. One necessary precondition is (Grey Morgan, 1978)

$$(p\tau)_{mp} < 1 \times 10^{-7} \qquad \text{torr}\cdot\text{s}$$

or, in SI, (19)

$$(p\tau)_{mp} < 1.33 \times 10^{-5} \qquad \text{Pa}\cdot\text{s}$$

The other is adequate optical electric field to actually strip electrons off individual atoms without collisions. The transition probability for the process is

$$\mathbf{P} = C_i \omega \tau N_{min}^{2/3} \left[\frac{W_{osc}}{2W_i} \right]^{N_{min}} \tag{20}$$

(see Johann et al., 1986), where $\mathbf{P}$ is the probability (per atom); τ is the laser pulse width; $C_i \approx 1$; W_i is the ionization potential of the atom; N_{min} is the minimum number of laser photons of energy $\hbar\omega$ sufficient to overcome the modified ionization threshold $W_i(I) = W_i + W_{osc}$; and W_{osc} is the "quiver energy" of an electron in the laser field,

$$W_{osc} = eE^2/4m\omega^2 = 8.4 \times 10^{-13} I \qquad \text{eV} \tag{21}$$

with I in W/cm^2 (the coefficient is 8.4×10^{-17} for I in W/m^2 to yield eV).

In other words,

$$N_{min} = [W_i + W_{osc}]/\hbar\omega \tag{22}$$

As an example, for argon at 0.2 torr, subjected to a 50 ns, 3 μm IR laser, the collisionless ionization threshold is about 25 TW/cm^2, and $N_{min} = 89$. The conditions $n_0\mathbf{P} = 10^{10}\ cm^{-3}$ can be used as the ionization threshold.

At vapor pressures above about 0.2 torr (27 Pa), collisional processes will come into play at lower irradiance levels than collisionless ones. Therefore, we expect this mechanism to have small importance for most LIMA work, considering the very high irradiance and low pressure required for it to be the dominant source of ionization.

Saha Equilibrium. Exact ignition scenarios must be calculated using a specific computer model that takes detailed account of material properties and laser parameters (see, e.g., Vertes et al., 1989). However, a good understanding of the relation between temperature and ionization density in LTE can be obtained from the simplified Saha equation (see Vertes and Juhasz, 1989):

$$\frac{\eta^2}{1-\eta} = \frac{1}{n_\Sigma} \left(\frac{2\pi m_e k T_e}{h^2} \right)^{3/2} \exp(-W_i/kT_e) \tag{23}$$

In this expression, W_i is the first ionization potential of the neutral vapor; T_e, the electron temperature; $n_\Sigma = n_0 + n_e$ (cm^{-3}), the total number density; and $\eta = n_e/n_\Sigma$, the ionization fraction obtained. It should be pointed out that, in the high-irradiance cases we consider in this section, we have $\eta \rightarrow 1$ early in the laser pulse, the precise value of η being unimportant.

4A.1.5. Absorption and Energy Redistribution Processes in Dense LTE Plasma

Local Thermodynamic Equilibrium (LTE). Most systems never achieve complete thermodynamic equilibrium (TE). Although the departure may be small, a system to which a pulse of energy is added or subtracted is, at least momentarily, out of TE. Equilibria exist that are not thermal equilibria. Collisions and collision-like particle-field damping processes restore TE. Because relaxation time increases with increasing size of a system, separate small parts of a system will reach a state of internal equilibrium long before they equilibrate with each other (see Landau and Lifschitz, (1958). Therefore, TE is first restored locally, hence LTE.

In electrolytes, these times are very fast because the density is high, and departures from TE in chemistry are often short-lived. In plasmas, equilibration times can be short enough, in some experimental conditions, for energetic particles to escape the experiment before TE is achieved.

Coulombic Equilibration Times and Mean Free Paths. For temperatures of a few electron volts and modest densities one would expect that, in Regions III or IV (see Figure 4A.1), dephasing collisions that produce LTE are dominated by coulombic interactions among the plasma particles. Since these are of quite different mass—in an electron–ion plasma, $m_i/m_e = 1836A$, where A is the atomic mass number—separate LTEs for each set of particles may coexist for a time because of the coulombic equilibration-time hierarchy (Montgomery and Tidman, 1964):

$$\tau\begin{bmatrix}\text{electrons become}\\ \text{isotropic and}\\ \text{Maxwellian}\end{bmatrix}:\tau\begin{bmatrix}\text{ions become}\\ \text{isotropic and}\\ \text{Maxwellian}\end{bmatrix}:\tau\begin{bmatrix}\text{electrons \& ions}\\ \text{reach}\\ \text{equilibrium}\end{bmatrix}$$

$$= \left(\frac{m_e}{m_i}\right):\left(\frac{m_e}{m_i}\right)^{1/2}:1 \qquad (24)$$

$$= \frac{1}{1836A}:\frac{1}{(1836A)^{1/2}}:1$$

As an example, for copper, these times are in the ratio 10^5:300:1. The electrons reach LTE quite rapidly because of their high thermal velocity, whereas the ions do so more slowly owing to their smaller velocity and higher mass, Because at most m_e/m_i values the kinetic energy can be transferred in an inefficient electron–ion collision, the population of electrons take still longer to share energy with the ion population.

The expression for the coulombic electron–ion energy exchange time is (Spitzer and Härm, 1958)

$$\tau_{eiE} = \frac{2^{5/2} A m_p (kT_e)^{3/2}}{\pi^{3/2} (m_e)^{1/2} Z n_e e^4 \ln\Lambda} \tag{25}$$

while the corresponding expression for the mean free path (MFP) for electron–ion energy exchange is

$$\begin{aligned} \lambda_{eiE} = v_{the}\tau_{eiE} &= \frac{2^{5/2}}{\pi^{3/2}} \frac{A m_p}{m_e} \frac{(kT_e)^2}{Z n_e e^4 \ln\Lambda} \\ &\approx 10^8 A T_e^2 / Z n_e \qquad \text{cm} \end{aligned} \tag{26}$$

taking $\ln\Lambda = 7$. The quantity Λ is defined in a later subsection [see Eq. (31)]. The coefficient in Eq. (26) becomes 10^{12}, rather than 10^8, when SI units are used. The quantity $v_{the} = (kT_e/m_e)^{1/2}$ is the electron thermal velocity (rigorously, the speed corresponding to the most probable kinetic energy in a Maxwellian velocity distribution).

Physically, the form of Eq. (26) reflects the fact that the cross section σ_{sc} for coulombic scattering of an electron by an ion is inversely proportional to the fourth power of electron velocity and thus the second power of electron temperature (see, e.g., Spitzer, 1967), so that a coulombic MFP has the dependence $\lambda_{eiE} = 1/(n_e\sigma_{sc}) \propto T_e^2/n_e$ on density and temperature.

Figures 4A.3 and 4A.4 show a calculation of the MFP for electrons in energy exchange, $\lambda_{eiE} = v_{the}\tau_{eiE}$, for a Maxwellian plasma formed from copper vapor, and of τ_{eiE}, respectively, with three-body recombination effects (see the following subsection) included. It is seen that the electron MFP for energy loss, even for a temperature of 1 eV, exceeds the size of a typical experiment when $n_e(x)$ has dropped to $10^{16}\,\text{cm}^{-3}$. This situation, which can occur in a Region II thickness as small as 300 μm in a KrF experiment with target spot size $d_s = 1\ \mu$m, is the definition of Region III. For a typical initial laser-induced plasma expansion velocity (in Region I) of 10 μm/ns, the plasma can expand into Region III within 30 ns.

Effects of Three-Body Recombination on Equilibration. Most of the electron–ion relaxation times in Table 4A.2 are quite long. However, for low temperatures and high densities such as one expects in Region I or II (see Figure 4A.1), three-body recombination may dominate coulombic interactions. Three-body recombination and its inverse, collisional ionization, must take

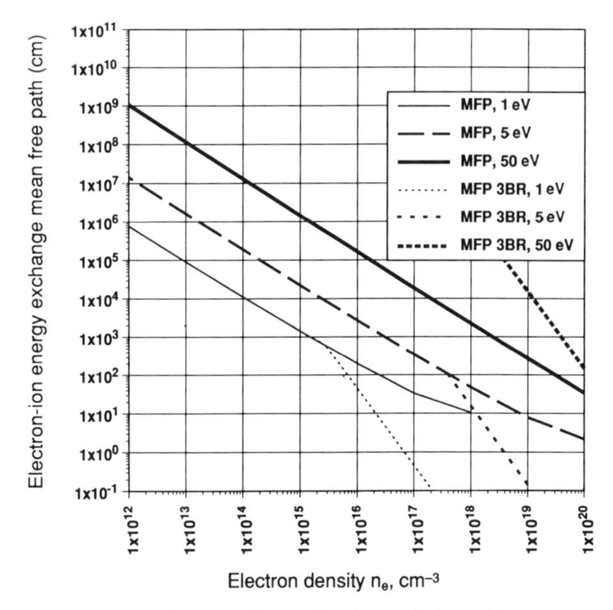

Figure 4A. 3. Graph showing the mean free paths λ_{eiE} and λ_{ei3BR} (electrons on cold ions) for copper with $Z = 1$ vs. electron density for three electron temperatures.

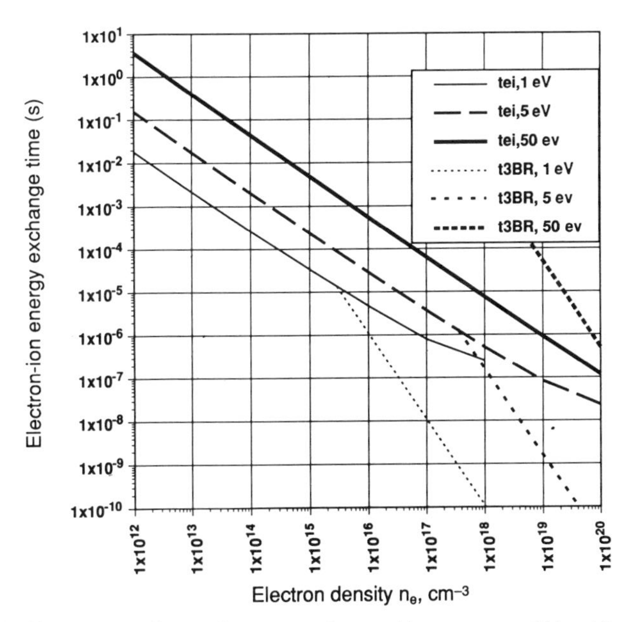

Figure 4A. 4. The energy exchange times τ_{eiE} and τ_{ei3BR} (electrons on cold ions) for copper with $Z = 1$ vs. electron density for three electron temperatures.

Table 4A.2. Illustrative Values for Singly Ionized Copper Plasma

Region	n_e (cm^{-3})	T_e (eV)	T_e (K)	D_e	f_p	$\Lambda = n_e D_e^3$ (cm)	τ_{ee}	τ_{ii}	τ_{ei}	τ_{ei3BR}
I ($\lambda = 1\ \mu m$, IB absorption zone)	10^{21}	3	3.48×10^4	4.1 Å	284 THz	0.28	—[a]	—[a]	—[a]	16 fs
II (mid-expansion near target)	10^{18}	5	5.80×10^4	17 nm	9 THz	19	4.4 ps	1.5 ns	520 ns	160 ns
III (mid-corona, hot electrons)	10^{14}	10	1.17×10^5	2.4 μm	90 GHz	5.4×10^3	43 ns	15 μs	5 ms	0.5 yr
IV (ion beam, hot electrons)	10^8	50	5.80×10^5	5.3 mm	90 MHz	6.1×10^7	230 ms	79 s	7.5 h	16 Gyr
Interstellar gas	1	1	1.16×10^4	740 cm	9 kHz	1.7×10^9	15 h	220 days	200 yr	3.6×10^{18} yr
Stellar interior	10^{24}	2000	2.32×10^7	3.3 Å	9 PHz	150	21 fs	7.1 ps	2.4 ns	80 ns

[a] Concept not valid because $\Lambda \neq \gg 1$.

place at equal rates in LTE:

$$2e + n_i(Z+1) \leftrightarrow n_i(Z) + e \tag{27}$$

It is a nonradiative process.

When conditions are appropriate in laser–plasma interactions, three-body recombination becomes dominant very rapidly, since it depends on the square of the electron density and the 4.5 power of temperature (see, e.g., Book, 1989):

$$\tau_{ei3BR} = 5.85 \times 10^7 T_e^{4.5}/n_e^2 \tag{28}$$

The corresponding electron MFP has an even stronger temperature dependence:

$$\lambda_{ei3BR} = v_{the}\tau_{ei3BR} = 2.28 \times 10^{13} T_e^5/n_e^2 \qquad \text{cm} \tag{29}$$

[for SI units, the correct prefatory constant in Eq. (28) is 5.85×10^{19}, and in Eq. (29) is 2.28×10^{23} to yield MFP in meters].

The decay of both τ_{ei3BR} and τ_{eiE} with distance can be quite rapid, depending sensitively on the actual evolution of the plasma temperature during the expansion. For example, if the density varies with distance from the target according to $(x/x_T)^{-2.5}$ [see Eq. (4)] while the temperature stays constant, the recombination time τ_{ei3BR} varies at least as rapidly as

$$\tau_{ei3BR} \propto (x/x_T)^{-5} \tag{30}$$

while the coulombic τ_{eiE} for the thermal plasma, under the same conditions, varies only as $(x/x_T)^{-1}$. Using electron density and temperature implied by 1 GW/cm^2 of 248 nm radiation striking a Cu target for 10 ns, one estimates an ablation flux of 2.2×10^{18} Cu^0 cm^{-2} and 3×10^{17} Cu^+ cm^{-2} (see Vertes and Juhasz, 1989). For the sake of discussion, let us take the distribution of ion expansion velocity to be a "top hat" extending from 1.5 to 3 cm/μs, i.e., from one to two times the Boltzmann velocity, and the laser spot diameter on target to be 1 mm. Because the Cu^+ density is about 2×10^{19} cm^{-3}, three-body recombination is extremely rapid during the 10 ns ablation pulse ($\tau_{ei3BR} \approx$ 20 ps), while the ablation front achieves a thickness of about 200 μm.

The success of models based on LTE, especially for short times and expansion distances, is explained by the fact that collisional processes can be so rapid near the target.

Assuming an expansion cone angle $\theta_d = 24°$, we have $x_T = 3.7$ mm. We take the initial electron temperature under these conditions to be 2.6 eV, based on the work of von Gutfeld and Dreyfus (1989). Applying Eq. (28) to this experiment, we find that τ_{ei3BR} will have increased from 20 ps to 10 ns in a drift distance $x = 4x_T = 1.5\,\text{cm}$ ($\tau \approx 0.7\,\mu\text{s}$) and to $10\,\mu\text{s}$ in $x \approx 7\,\text{cm}$ ($\tau \approx 3.5\,\mu\text{s}$). Meanwhile, the coulombic τ_{eiE} in the thermal plasma will have increased from 750 ns to $10\,\mu\text{s}$, so that coulombic processes will begin dominating three-body processes, insofar as equilibration between the electron and ion populations is concerned, only after expansion has proceeded 7 cm from the target. Note that the dominant collisional processes act to rapidly equilibrate the low-velocity component of the particle velocity distributions, which then equilibrate within themselves much more rapidly than τ_{eiE} [by the factors given by Eq. (24)].

Now, let us consider a LIMA-type ablation, in which the laser spot on target is as small as $10\,\mu\text{m}$ in diameter. With a cone angle $\theta_d = 24^\circ$, we have $x_T \approx 40\,\mu\text{m}$, a distance that can be attained by the plume at $2\,\text{cm}/\mu\text{s}$ in $\approx 2\,\text{ns}$, well before the laser pulse ends. Again, $\tau_{ei3BR} \approx 20\,\text{ps}$, whereas $\tau_{eiE} \approx 750\,\text{ns}$. Here, the crossover to dominance by coulombic interactions occurs much sooner because of the smaller scale. Applying Eq. (28) to this experiment, we find that τ_{ei3BR} will have increased from 20 ps to 10 ns in a drift distance $x = 160\,\mu\text{m}$ ($\tau \approx 8\,\text{ns}$), and to $10\,\mu\text{s}$ in $x \approx 0.8\,\text{mm}$ ($\tau \approx 40\,\text{ns}$). Coulombic processes will become more important than three-body processes after expansion has proceeded 0.8 mm from the target, although they will still be as infrequent as before and, when dominant, support the generation of hot electrons.

Choosing $\sigma_{3BR} = 3\,\text{Å}^2$ as a typical recombination cross section, we see that at a density $1 \times 10^{19}\,\text{cm}^{-3}$, the ion MFP for recombination $1/(n\sigma_{3BR}) = 3\,\mu\text{m}$, a significant fraction of the ablation region radius in this case. This may explain why diatomics are commonly dissociated with large (1 mm diameter) ablations whereas they often survive the plasma in a $10\,\mu\text{m}$ diameter interaction.

Experimentally, the lack of communication back to the slower neutrals has been observed, in the sense that it is the faster diatomics that are dissociated. The more intense the plasma, the fewer energetic diatomic species are observed (see Dreyfus et al., 1986c; Pappas et al., 1992).

The Debye Shielding Distance and Debye Sphere. The quantity Λ is the number of particles in a Debye sphere,

$$\Lambda = \frac{4\pi}{3} n_e D_e^3 \qquad (31)$$

The electron Debye length, D_e, is the distance within which ion charge fluctuations cannot be neutralized by electron motions. The significance of Λ

is that, if $\Lambda \neq \gg 1$, the plasma may no longer be regarded as a gas, and the Boltzmann distribution is inapplicable; D_e and Λ are among the most important LTE concepts in plasma physics.

The Debye–Hückel theory, originally developed to treat charge motion in electrolytes, also describes a plasma with a distribution of electrons in thermal equilibrium. Their spacial density will then be given by the Boltzmann factor $\exp(-e\phi/kT)$, where $\phi(r)$ is the electrical potential, or

$$n_e(r) = n_0 e^{-e\phi/kT} \tag{32}$$

If now a single test ion Ze is placed at the origin in this distribution of electrons (assuming charge neutrality, but considering the other ions to be a smeared-out, positive-charge background), Poisson's equation is

$$\begin{aligned}\nabla^2\phi &= -4\pi e\{Z\delta(r) - n_0[e^{-(e\phi/kT} - 1]\} \\ &\approx -4\pi e\{Z\delta(r) + en_0\phi/kT\}\end{aligned} \tag{33}$$

with the linearizing approximation $e\phi \ll kT$. The solution to Eq. (33) is

$$\phi = \phi_0 \exp(-r/D_e) \tag{34}$$

where $\phi_0 = e/r$ ($e/4\pi\varepsilon_0 r$ in SI) is the electric potential of a single ion. The Debye length, D_e, in a plasma depends on electron density, n_e, and temperature, T_e, according to

$$D_e = \frac{v_{the}}{\omega_p} = \left(\frac{kT_e}{4\pi n_e e^2}\right)^{1/2} = 6.9\left[\left(\frac{T_e}{n_e}\right)^{1/2}\right] \quad \text{cm} \tag{35a}$$

or

$$D_e = \frac{v_{the}}{\omega_p} = \left(\frac{\varepsilon_0 kT_e}{n_e e^2}\right)^{1/2} = 69\left[\left(\frac{T_e}{n_e}\right)^{1/2}\right] \quad \text{m} \tag{35b}$$

Table 4A.2 gives Debye lengths from Eq. (35), electron plasma frequencies from Eq. (7), and the times τ_{ee}, τ_{ii}, and τ_{ei} from Eq. (25) for typical densities and temperatures illustrative of the regions of Figure 4A.1 and for other conditions. Note that, if the ions are *not* assumed to be a smeared-out positive-charge background, but rather discrete point sources, one gets a coefficient $(1/8\pi)^{1/2}$ vs. $(1/4\pi)^{1/2}$ in Eq. (35a) (see Ginzburg, 1970). However, the convention is as stated in Eq. (35a,b). Normally D_e is much smaller than the plume thickness for nanosecond laser ablation work; hence electric and magnetic fields do not penetrate the plume except in the outermost coronal region.

Ambipolar Diffusion and Debye Sheaths. In this section, we assume that the Debye length D_e is less than d_s, the diameter of the laser interaction region. Free rather than ambipolar diffusion occurs in the unusual event that this condition is violated (Weyl, 1989). In LIMA work, such a violation would occur for laser spot sizes on the order of 1 Å (see Section 4A.3), an extremely unlikely occurrence.

Where Γ is the particle current (units: $L^{-2} \cdot s^{-1}$), the electron and ion particle current equations are

$$\Gamma_e = -\mathscr{D}_e \nabla n_e - \mu_e E n_e$$

and (36)

$$\Gamma_i = -\mathscr{D}_i \nabla n_i + \mu_i E n_i$$

The mobility μ_i of species j is given by its collision time $\tau_j = \tau_{jkp}$ for momentum exchange on the other species:

$$\mu_j = e\tau_j / m_j \tag{37}$$

and the diffusion coefficient is

$$\mathscr{D}_j = (kT/e)\mu_j \tag{38}$$

in the absence of magnetic fields (see Rose and Clark, 1961).

We take

$$n_e = Z n_i$$

and (39)

$$\Gamma_e = Z\Gamma_i$$

assuming charge neutrality. When both species flow freely, limited only by self-generated fields, an electric field will develop,

$$E = \frac{(\mathscr{D}_i - \mathscr{D}_e)\nabla n_i}{(\mu_i + \mu_e) n_i} \tag{40}$$

in a so-called sheath of thickness about equal to D_e. However, since $\mathscr{D}_e \gg \mathscr{D}_j$ and $\mu_e \gg \mu_i$, we find

$$-eE/kT_e \approx \nabla n_e / n_e \tag{41}$$

The electric field E can be very large. Eliezer and Hora (1989) predict values on the order of 100 MV/cm (10 GV/m) for the plasma electric field generated by 0.5 ps Nd:glass laser irradiation at 10 PW/cm^2 (10^{20} W/m^2).

However, under LTE, the *distance* over which such a field exists is correspondingly small. What normally happens in response to this field is the immediate formation of a so-called double layer (Hora et al., 1989), composed of just enough charge separation dipole moment

$$en_e \Delta x = -E/4\pi = eD_e^2 \nabla n_e \tag{42}$$

to balance this field and prevent the flow of current. For $n(x)$ given by Eq. (3), the density gradient scale length in the laser absorption region,

$$L_n = |n_e/\nabla n_e| = 0.45x \tag{43}$$

is on the order of 0.1 μm in typical KrF laser ablation experiments.

Thus,

$$(\Delta x/D_e) = D_e/L_n \tag{44}$$

depends on the ratio of the Debye length to the density gradient scale length and is very small near the absorption layer, where D_e is itself on the order of angstroms. If $n(x)$ is given by Eq. (3), then (near the absorption layer $x = x_I$), $\Delta x/D_e = D_e/L_n = 2.22D_e/x_I = 0.01$, for the conditions of typical KrF ablation experiments ($I = 10$ GW/cm^2, $\tau = 10$ ns, $\lambda = 248$ nm, $A = 28$, and $Z = 1$). This means that, under LTE conditions, a large field operates over a small distance, and the product

$$e\phi = E\,\Delta x = (D_e/L_n)^2 kT_e \tag{45}$$

is approximately equal to $10^{-4}\,kT_e$ at the absorption layer, small enough to be insignificant.

Of course, at extremely high irradiance levels sufficient to develop plasma ponderomotive forces, density gradient steepening by the light pressure can drastically reduce L_n, creating larger potentials than this. At irradiance levels above 10–100 TW/cm^2 (100–1000 PW/m^2), suitable for laser fusion, a copious suprathermal "hot" electron flux is generated. These electrons can bounce off the sheath repeatedly to give extra energy to ions that do escape the DL (Hauer et al., 1989). We shall return to the subject of charge acceleration in Section 4A.3.

The Plasma Dielectric Function. Defining

$$\tilde{n} = n + i\chi \tag{46}$$

for the plasma complex refractive index and

$$\varepsilon = \tilde{\boldsymbol{n}}^2 = (\boldsymbol{n}^2 - \chi^2) + \mathrm{i}2\boldsymbol{n}\chi \tag{47}$$

as the plasma complex dielectric function, the real (refractive) and imaginary (absorptive) parts of the dielectric constant, ε, of a nonmagnetized plasma are given by

$$\mathrm{Re}(\varepsilon) = \boldsymbol{n}^2 - \frac{\omega_p^2}{\omega^2(1 + \nu^2/\omega^2)} \tag{48}$$

and

$$\mathrm{Im}(\varepsilon) = \frac{\nu}{\omega}\left[\frac{\omega_p^2}{\omega^2(1 + \nu^2/\omega^2)}\right] \tag{49}$$

so that the optical absorption coefficient is

$$\alpha = \frac{2\omega}{c}\chi = \frac{\nu}{\boldsymbol{n}c}\left[\frac{\omega^2{}_p}{\omega^2(1 + \nu^2/\omega^2)}\right] \tag{50}$$

The damping frequency, ν, is provided by long-distance collision-like electron–ion coulombic, momentum-exchanging interactions in the plasma,

$$\begin{aligned} \nu = \nu_{\mathrm{ei}p} &= 1/\tau_{\mathrm{ei}p} \\ &= 7.49 Z n_{\mathrm{e}}/T_{\mathrm{e}}^{3/2} \qquad \mathrm{Hz} \end{aligned} \tag{51}$$

in cgs units. For SI, the coefficient is 7.49×10^{-6} in Eq. (51). The parameter $\tau_{\mathrm{ei}p}$ is the time for electron–ion momentum exchange. Although ν_{ei} is often large in the visible and UV range, it is also often smaller than the laser frequency itself. Then, in the limits $\nu_{\mathrm{ei}}/\omega \ll 1$, and $\boldsymbol{n} \approx 1$ (valid for vacuum propagation), we have—

absorption coefficient:

$$\alpha \cong (\nu_{\mathrm{ei}}/c)(n_{\mathrm{e}}/n_{\mathrm{ec}}) \tag{52}$$

optical cross section:

$$\sigma_{\mathrm{opt}} \equiv \alpha/n_{\mathrm{e}} \cong [\nu_{\mathrm{ei}}/(c n_{\mathrm{ec}})] \tag{53}$$

and

$$\varepsilon \approx \boldsymbol{n}^2 \approx [1 - n_{\mathrm{e}}/n_{\mathrm{ec}}] \tag{54}$$

Inverse Bremsstrahlung Absorption and Related Terms. Inverse bremsstrahlung is so named because it is physically the inverse of the process by which electron collisions with ions generate radiation in a plasma. The quantum mechanically correct formula for the inverse bremsstrahlung absorption coefficient in a plasma where LTE applies is (Kidder, 1971)

$$\alpha = \frac{4}{3}\left(\frac{2\pi}{3\,kT_e}\right)^{1/2}(2\pi)^3 Z^2 n_e n_i g_{ff}\, e^6 \frac{[1-\exp(-\hbar\omega/kT_e)]}{hcm_e^{3/2}\omega^3} \qquad \mathrm{cm}^{-1}$$

or, with SI quantities, (55)

$$\alpha = \frac{2\pi}{3\times 10^{23}}\left(\frac{2\pi}{3\,kT_e}\right)^{1/2}(2\pi)^3 Z^2 n_e n_i g_{ff}\, e^6 c^5 \frac{[1-\exp(-\hbar\omega/kT_e)]}{hm_e^{3/2}\omega^3} \qquad \mathrm{m}^{-1}$$

Here, $Z = n_e/n_i$ is the average ion charge state, assuming charge neutrality. The Gaunt factor, g_{ff}, is a quantum mechanical correction with magnitude very close to unity for most laser–plasma interactions (see Gaunt, 1930). In the case of longer wavelengths, e.g., radio wavelengths in the ionosphere, g_{ff} is related to the so-called Coulomb logarithm by

$$g_{ff} = (\sqrt{3}/\pi)\ln\Lambda \tag{56}$$

(see Allen, 1973; Spitzer, 1967). The quantity Λ was defined in Eq. (31).

It should be noted that this formula (55) treats electromagnetic-wave heating of plasma by assuming electron dephasing is produced exclusively by long-range coulombic interactions with ions. Electron-neutral collisions can also be important as a source of dephasing for partially ionized plasmas, but we use the full ionization approximation.

The expression $[1-\exp(-\hbar\omega/kT_e)]$ in Eq. (55) may be taken as approximately equal to $\hbar\omega/kT_e$ for $\geqslant 5\,\mathrm{eV}$ plasmas and IR or visible lasers. These simplifications produce the familiar formula laser absorption:

$$\alpha = 1.97\times 10^{-23} Z^3 n_i^2 \lambda^2 / T_e^{3/2} \qquad \mathrm{cm}^{-1}$$

or, in SI, (57)

$$\alpha = 1.97\times 10^{-29} Z^3 n_i^2 \lambda^2 / T_e^{3/2} \qquad \mathrm{m}^{-1}$$

The optical absorption cross section $\sigma = \alpha/n_e$ for a plasma electron can be many orders of magnitude larger than its geometrical size πr_e^2, as can be seen by reexpressing Eq. (57) in the form

$$\sigma_{\mathrm{opt}} = 250\left(\frac{Zn_e\lambda^2}{T_e^{3/2}}\right) r_e^2 \qquad \mathrm{cm}^2 \tag{58a}$$

or, in SI,

$$\sigma_{\text{opt}} = 2.50\left(\frac{Zn_e\lambda^2}{T_e^{3/2}}\right)r_e^2 \qquad \text{m}^2 \tag{58b}$$

For example, if $\lambda = 10.6\,\mu$m, $T_e = 11{,}605$ K (1 eV), $n_e = 10^{19}\,\text{cm}^{-3}$, and $Z = 1$, then $\sigma_{\text{opt}} = 2.25 \times 19^9\, r_e^2 = 1.8 \times 10^{-16}\,\text{cm}^2$. A plasma with these parameters has an absorption length for 10.6 μm light given by $1/\alpha = 5\,\mu$m and is 99.9% opaque in 40 μm thickness. The functional form of Eqs. (57) and (58a,b) was predicted by the preceding analysis of the plasma dielectric function, Eqs. (52) and (53).

These equations describe a laser-produced plasma after its creation. If $n_e = 10^{19}\,\text{cm}^{-3}$, as in the preceding example, and $T_e = 3$ eV, the plasma energy density $w_p = n_e kT_e = 4.8\,\text{J/cm}^3$ (4.8 MJ/m^3). It should be kept in mind that the more one retreats toward the plasma formation threshold irradiance value, the more important become the conditions *during* the plasma creation, when this energy was deposited. At threshold irradiance, it becomes important that photon–plasma coupling mechanisms other than inverse bremsstrahlung were necessary early in the laser pulse to deposit this energy density w_p. Such processes include local density fluctuations that enhance ν_{ei}, and three-body recombination when $n_i > 10^{18}\,\text{cm}^{-3}$.

4A.1.6. What Can Be Learned About Ablation Parameters from a General Theory?

Analysis of high-irradiance ($I \geqslant 10^{11}\,\text{W/cm}^2$), short-pulse ($\tau \sim 10$ ns) surface ablation has existed for years. In this section, we will show that the resulting dense ($> 10^{19}\,\text{cm}^{-3}$), thin ($\geqslant 30\,\mu$m) surface plasma can be described down to a low-power limit ($\sim 10^9\,\text{W/cm}^2$ for UV) determined primarily by the inelastic scattering cross section of plasma electrons, i.e., inverse bremsstrahlung. The result (for Cu, but virtually *material independent*) is that the ablation depth is approximately equal to $40(I\tau^{1.5}/\lambda)^{1/2}$ nm, where λ is the wavelength (in centimeters). The plasma temperature is approximately equal to $3(I\lambda\sqrt{\tau})^{1/2}$ eV, a number that is useful for calculating X-ray emission (for photolithography). The directed velocity of neutral atoms is $0.6(I\lambda\sqrt{\tau})^{1/4}$ cm/μs; however, ions are ejected at a $\geqslant$ 3-times-higher velocity due to plasma voltages. In summary, one now has an extension of earlier results *down* into the $\geqslant 1\,\text{GW/cm}^2$ range, below which plasma ablation converts into (*material-independent*) thermal or photochemical vaporization (see Phipps and Dreyfus, 1992).

Basic Relationships. Surface absorber ablation theory (Phipps et al., 1988) is developed from the work of Kidder (1968, 1971) and others (e.g., Basov et al.,

1968; Krokhin, 1971; Caruso et al., 1966; Nemchinov, 1967), which formed the foundation of our understanding of laser-fusion hydrodynamics.

The principal assumptions used in the surface-absorber theory are as follows:

- One-dimensional flow
- Inverse bremsstrahlung heating mechanism
- All incident light employed to support the ablation flow
- Mach-1 flow (Chapman–Jouguet condition, $M = 1$)
- Ideal gas ($\gamma = 5/3$)
- Macroscopic charge neutrality ($n_e = Zn_i$)
- Self-regulating plasma opacity, giving unity optical thickness $S = 1$ at the laser wavelength, after a brief transition time τ_1

The assumption that inverse bremsstrahlung is the dominant plasma heating mechanism implies irradiance I is below the level (0.1–$1\ \mathrm{PW/cm^2}$, or 10^3–$10^4\ \mathrm{PW/m^2}$) required for dominance by anomalous (i.e., noncollisional) absorption processes (see Hughes, 1979).

The following three equations describe the physical problem:

$$S = \alpha c_s \tau \tag{59}$$

$$I = [M(M^2 + 3)/2]\rho_a c_s^3 \tag{60}$$

and

$$\alpha = \frac{CZ^3(Z+1)^{3/2}\lambda^2}{A^{7/2}} \frac{\rho_a^2}{c_s^3} \tag{61}$$

Equation (61) reexpresses the inverse bremsstrahlung absorption formula (57) in terms of the hydrodynamic variables ρ_a and c_s by employing the additional expressions:

$$\rho_a = (Am_p/Z)n_e \tag{62}$$

and

$$c_s = \left[\frac{\gamma(Z+1)kT_e}{Am_p}\right]^{1/2} \tag{63}$$

In Eq. (60), I is in physical rather than practical units, ρ_a is the ablation plasma mass density, and constant C has units $\mathrm{L^6 \cdot M^{-2} \cdot T^{-3}}$, with numerical value 1.136×10^{37} (cgs) or 1.136×10^{43} (SI).

When Eqs. (59)–(61) are solved, there results a fully transient theory in which the ablation pressure varies slowly with time during the evolution of the laser pulse, according to

$$p_a = p_0, \qquad \tau < \tau_1$$
$$p_a = (\tau_1/\tau)^{1/8} p_0, \qquad \tau \geqslant \tau_1 \tag{64}$$

where

$$p_0 = \frac{(1+\gamma M^2)}{\gamma} \frac{[(Am_p/Z)^{1/3} n_{ec}] I^{2/3}}{[M(M^2+3)/2]^{2/3}} \tag{65}$$

is the maximum ablation pressure that results from achievement of the critical density for $\tau < \tau_1$ early in the pulse, during the ablation.

The time τ_1 is given by

$$\tau_1 = \frac{I^{2/3} A^{7/2}}{2^{2/3} C Z^3 (Z+1)^{3/2} \lambda^2 \rho_a^{8/3}} \tag{66}$$

For convenience, we reexpress the transition time τ_1 using practical units for I:

$$\tau_1 = C_1 \frac{I^{2/3} A^{5/6} \lambda^{10/3}}{Z^{1/3}(Z+1)^{3/2}} \tag{67}$$

where the constant $C_1 = 1.05 \times 10^{-4}$ (cgs) or 1.05 (SI).

Two different sets of predictions for ablation variables (e.g., pressure, temperature, velocity, and mass ablation rate) result depending on whether $\tau \gtrless \tau_1$. The time τ_1 is so short compared to laser pulses normally used in LIMA work (e.g., pulse durations greater than 1 ps for KrF lasers, or 100 ps for Nd lasers) that the added complication of carrying along both sets of predictions for $\tau \gtrless \tau_1$ in the present work is not justified. We will just carry the relationships for $\tau \geqslant \tau_1$, keeping in mind that $\tau < \tau_1$ governs the field of laser fusion [for both sets of relationships, see Table II of Phipps et al. (1988)].

These results may be used to predict many of the interesting ablation parameters with good accuracy over a broad range of laser wavelengths (0.25–10 μm), pulse durations (up to 1 ms), and irradiances. Employing the conditions $S = 1$, $M = 1$, etc. listed at the beginning of this subsection gives the following set of relationships, the basis of our work in this chapter:

pessure $$p_a = 5.83 \frac{\Psi^{9/16}}{A^{1/8}} \frac{I^{3/4}}{(\lambda\sqrt{\tau})^{1/4}} \quad \text{dyn/cm}^2 \tag{68}$$

(with SI variables, the correct coefficient to obtain Pa is 1.84×10^{-4});

temperature $$T_e = 2.98 \times 10^4 \frac{A^{1/8}Z^{3/4}}{(Z+1)^{5/8}}(I\lambda\sqrt{\tau})^{1/2} \quad \text{K} \tag{69}$$

(with SI variables, the correct coefficient is 2.98×10^3);

density $$n_e = 3.59 \times 10^{11} \frac{A^{5/16}}{Z^{1/8}(Z+1)^{9/16}} \frac{I^{1/4}}{(\lambda\sqrt{\tau})^{3/4}} \quad \text{cm}^{-3} \tag{70}$$

(with SI variables, the correct coefficient to obtain m^{-3} is 1.135×10^{15});

ablation rate $$\dot{m} = 2.66 \frac{\Psi^{9/8}I^{1/2}}{A^{1/4}\lambda^{1/2}\tau^{1/4}} \quad \mu\text{g}\cdot\text{cm}^{-2}\cdot\text{s}^{-1} \tag{71}$$

(with SI variables, the correct coefficient to obtain $\mu\text{g}\cdot\text{m}^{-2}\cdot\text{s}^{-1}$ is 26.6); and

velocity $$c_s = 1.37 \times 10^6 \frac{A^{1/8}}{\Psi^{9/16}}(I\lambda\sqrt{\tau})^{1/4} \quad \text{cm/s} \tag{72}$$

(with SI variables, the correct coefficient to obtain m/s is 4.33×10^7). The ablation pressure p_a, electron temperature T_e, and sound speed c_s are evaluated in the Region I absorption zone, where laser light is converted into plasma pressure.

The parameter

$$\Psi = \frac{A}{2[Z^2(Z+1)]^{1/3}} \tag{73}$$

is a function of A and Z, and its magnitude is often ≈ 1. Among these quantities, the impulse coupling coefficient,

$$C_m = \int_{-\infty}^{\infty} d\tau p_a(\tau)/\Phi$$

is predicted to within a factor of 1.5 for a broad range of experimental data above plasma threshold in vacuum.

Data for $\dot{m}$ are predicted quite well in the high irradiance limit, but ablation depth $x_v = \dot{m}\tau/\rho$ is not predicted so well, especially at low intensities, as we discussed in developing conditions II and IV, because the heated surface continues to emit material after the laser pulse has ceased (see Kelly and Rothenberg, 1985). Many laser pulses (e.g., CO_2) have a long, low-irradiance tail, which extends the vaporization time and complicates modeling. With a correction due to charge acceleration, ion velocities seem to be well predicted (see Phipps, 1989; also Section 4A.3, below).

Heat Conduction and Reradiation. In the high-irradiance regime, heat is mainly transported to the target via plasma blackbody radiation. Depending on density, temperature, and laser irradiance, bremsstrahlung and/or electron thermal conduction may also be important or even dominant. Because "high irradiance" in the context of laser ablation is still well below the values where nonclassical transport applies [approximately, $I \geqslant 100\,\mathrm{TW/cm^2}$ (or $10^{18}\,\mathrm{W/m^2}$) on target (see Pearlman and Anthes, 1975)], we consider only classical transport here. At irradiance levels appropriate for laser fusion, a "flux limitation" adjustment must be applied to the classical heat conduction formula (see, e.g., Max et al., 1980). We now derive typical magnitudes for heat transport variables in order to assess their importance in typical situations.

Electron Thermal Conduction. Heat transfer by thermal conduction in a plasma is mainly due to electrons because of their small mass and high velocity, and is given by

$$\mathbf{Q} = -K\nabla T \tag{74}$$

The classical electron thermal conductivity is given by (see Manheimer and Colombant, 1982; Spitzer, 1967):

$$K = K_0 T_e^{5/2} \tag{75}$$

watts per unit length, where (in cgs)

$$K_0 = 1.95 \times 10^{-11} \delta_t / Z \ln\Lambda$$

or (SI) (76)

$$K_0 = 1.95 \times 10^{-9} \delta_t / Z \ln\Lambda.$$

In Eq. (76), δ_t is a tabulated function of charge state Z that varies from 0.23 to 0.79 as Z varies from 1 to 16 (Spitzer, 1967), and $\ln\Lambda$ is the Coulomb logarithm defined in Eqs. (31) and (56). The heat flux transferred over a dimension equal to the laser focal spot radius is about (Phipps et al., 1988)

$$Q = 1.78 \times 10^4 A^{7/16} f(Z)(I\lambda\sqrt{\tau})^{7/4}/d_s \qquad \mathrm{W/cm^2} \tag{77}$$

For SI units yielding W/m^2, the correct coefficient in Eq. (77) is 563. The function

$$f(Z) = \delta_t Z^{13/8}/(Z+1)^{35/16} \tag{78}$$

is nearly a constant, varying from 0.05 to 0.18 as Z varies from 1 to 16, and returning to a value of 0.07 when $Z = 100$. Taking $I = 1\,\mathrm{GW/cm^2}$ ($10\,\mathrm{TW/m^2}$) and $I\lambda\sqrt{\tau} = 5\,\mathrm{W\cdot s^{1/2}\cdot cm^{-1}}$ ($500\,\mathrm{W\cdot s^{1/2}\cdot m^{-1}}$) for typical laser ionization microprobe conditions, $A = 12$ (carbon), $f(Z) = 0.1$, and $d_s = 2\,\mu\mathrm{m}$, the heat flux Q due to electron thermal conduction will be on the order of $0.44\,\mathrm{GW/cm^2}$ ($4.4\,\mathrm{TW/m^2}$). It is readily appreciated that the size of this flux approaches that of the absorbed laser flux.

Blackbody Irradiance. For an optically thick plasma (that is, one which is dense enough to be opaque to its own radiation), the emitted flux will be

$$\mathbf{Q} = \sigma\varepsilon T^4 \tag{79}$$

where ε is the emissivity. In the case of a perfect blackbody at 7 eV ($T = 11{,}605\,\mathrm{K/eV}$), $Q = 0.25\,\mathrm{GW/cm^2}$ ($2.5\,\mathrm{TW/m^2}$). This situation will occur, for example, in the high-pressure region very near the target surface. It is seen that the magnitude of this flux also approaches that of the absorbed laser flux, so that radiation and thermal conduction can both be important means of conducting heat to the target in high-irradiance LIMA work.

Bremsstrahlung Irradiance. For wavelengths or densities for which the plasma is *not* optically thick, classical bremsstrahlung emission describes the plasma's self-irradiance, rather than the blackbody formula. Appropriately, this is characterized by a volumetric rather than a surface source density (Allen, 1973):

$$\mathrm{i_B} = \left(\frac{64\pi}{3}\right)\left(\frac{\pi}{6}\right)^{1/2}\left(\frac{\mathrm{e}^6 Z^2}{hc^3 m_\mathrm{e}}\right)\left(\frac{kT_\mathrm{e}}{m_\mathrm{e}}\right)^{1/2} n_\mathrm{e} n_\mathrm{i} \qquad \mathrm{erg\cdot cm^{-3}\cdot s^{-1}} \tag{80}$$

or

$$\mathrm{i_B} = 1.425 \times 10^{-34} Z(T_\mathrm{e}^{1/2}) n_\mathrm{e}^2 \qquad \mathrm{W/cm^3} \tag{81}$$

The coefficient in Eq. (81) should be 1.425×10^{-40} for SI parameters, to yield $\mathrm{W/m^3}$. For a 7 eV plasma with electron density $n_\mathrm{e} = 10^{20}\,\mathrm{cm^{-3}}$ ($10^{26}\,\mathrm{m^{-3}}$), $\mathrm{i_B} = 0.4\,\mathrm{GW/cm^3}$ ($0.4\,\mathrm{PW/m^3}$). To compare these results, we note that the chief blackbody self-emission color per unit frequency interval for this plasma [see Eq. (18)] is at 63 nm wavelength, that Eq. (57) gives $1/\alpha = 8\,\mathrm{nm}$ absorption length for this color, and that $Q/i_B \approx 6$ nm. The difference between these values is a proper integration over the bremsstrahlung spectrum. Further, it is important to realize that LTE may not always apply well enough to establish the concept of T_e. Bremsstrahlung is the dominant type of plasma self-emission in Region III (see Figure 4A.1), yet it is small in comparison with the laser irradiance.

Adiabatic Expansion Theory: Thickness and Dimensionality of the Absorption Zone. The theory that led to Eqs. (68)–(72) is derived for one-dimensional expansions. However, it is intuitively clear that LIMA work sometimes involves two-dimensional expansions because of the small laser spot diameters used in this work. As mentioned previously, the phenomena concerning *laser coupling* to the target are considered to be correctly treated by one-dimensional theory when the standoff distance, x_I, of the absorption layer from the target is small compared to the laser spot diameter d_s on the target.

In the rare case when this is not so, the dimensionality corrections involved (Mora, 1982; Phipps et al., 1988) are very small, being proportional to $(x_I/d_s)^{1/9}$, and would amount to reduction of the pressure predicted by the one-dimensional laser coupling theory by at most a factor of 2 even in the extreme case $x_I/d_s = 300$.

In order to determine the standoff distance x_I, we follow the analysis of Manheimer and Colombant (1982), taking the distance between the absorption layer and the target surface x_I to be the thermal gradient scale length in Region I over which thermal convection and thermal conduction are balanced.

$$\tfrac{5}{2}\rho v C T = -K\frac{dT}{dx} \tag{82}$$

Since [see Eq. (75)] $K = K_0 T^{5/2}$, and using $v = c_s = \sqrt{(\gamma k T/A m_p)}$, we can integrate the resulting equation to find

$$x_I = 0.032\frac{K_0(A^{1/2})}{k^{3/2}[(\gamma/m_p)^{1/2}]}\frac{T^2}{n} \tag{83}$$

or, with $\ln\Lambda = 7$,

$$x_I = 5.5 \times 10^5[\delta_t(A^{1/2})/Z]T^2/n_e \quad \text{cm}$$

or (SI) (84)

$$x_I = 5.5 \times 10^9[\delta_t(A^{1/2})/Z]T^2/n_e \quad \text{m}$$

Note that Eqs. (84) have the form $x \propto T^2/n$, which is characteristic of a coulombic mean free path. Now, it only remains to substitute Eqs. (69) and (70) for T and n into Eq. (84) to obtain (in cgs).

$$x_I = 1365\left[\frac{\delta_t A^{7/16} Z^{5/8}}{(Z+1)^{11/16}}\right](I^{3/4}\lambda^{7/4}\tau^{7/8}) \qquad \text{cm}$$

or (SI) (85)

$$x_I = 43\left[\frac{\delta_t A^{7/16} Z^{5/8}}{(Z+1)^{11/16}}\right](I^{3/4}\lambda^{7/4}\tau^{7/8}) \qquad \text{m}$$

By taking the function $g(z) = \delta_t Z^{1/4}/(Z+1)^{7/8} = 0.13$ [it varies only slightly from 0.12 to 0.13 as Z varies from 1 to 16 (see Spitzer, 1967)] and substituting Eq. (72) for c_s, the time for collapse of the absorption zone pressure after the laser pulse, $t_I = x_I/c_s$, stated in Eq. (13), results.

It is clear that the dimensionality that applies to laser coupling in Region I can be assessed from the ratio x_I/d_s, which should be $\ll 1$ for a one-dimensional expansion. That this is normally the case can be seen by substitution of typical irradiation parameters. For example, substitution of $\lambda = 248$ nm, $I = 10\,\text{GW/cm}^2$, $\tau = 10$ ns, $A = 30$, and $Z = 1$ gives $x_I = 200$ nm, much smaller than laser spots.

Adiabatic Expansion Theory. At some distance $x = x_T$, plume expansion eventually makes a transition to multidimensional flow, even when laser coupling can be described as one-dimensional. In order to better understand the plasma plume expansion, we will ignore heat input and reradiation outside the absorption zone in Region I, and consider the predictions of adiabatic expansion theory. In such an expansion, the volume available to a fixed total number of particles expanding from the original perimeter at the flow-transition distance x_I depends only on the flow dimensionality a:

$$1 < a < 3 \tag{86}$$

Then the density

$$n(x)/n(x_I) = (x/x_T)^{-a} \tag{87}$$

We showed in the discussion surrounding Eq. (3) that theory predicts $a = 20/9 = 2.22$ whereas experiment shows $a \approx 2.5$. For an ideal gas, $\gamma = 5/3$, so

$$(p/p_T) = (n/n_T)^{\gamma} = (x/x_T)^{-5a/3} \tag{88}$$

whence

$$T(x)/T(x_T) = (n/n_T)^{2/3} = (x/x_T)^{-2a/3} \tag{89}$$

Since $\alpha \propto n^2/T^{3/2}$,

$$(\alpha/\alpha_T) = (x/x_T)^{-a} \tag{90}$$

Integrating along the laser beam path from infinity (where $x = \infty$ and $I = I_\infty$) to position x then gives

$$\frac{I(x)}{I_\infty} = \exp\left\{-\left[\frac{\alpha_T x_T}{(a-1)}\right]\left(\frac{x_T}{x}\right)^{[a-1]}\right\} \tag{91}$$

a function that has a scale thickness

$$x_{\text{abs}} = \left[\frac{\alpha_T}{a-1}\right]^{1/(a-1)} x_T^{a(a-1)} \tag{92}$$

If we take $x_T = (1/\alpha_T)$,

$$x_{\text{abs}} = \left(\frac{1}{a-1}\right)^{1/(a-1)} \frac{1}{\alpha_T} \tag{93}$$

Choosing $a = 2.5$ for the dimensionality of the flow in Regions II–IV to agree with experiment, we find $x_{\text{abs}} = 0.77\,(1/\alpha_T)$, where α_T is the inverse bremsstrahlung absorption coefficient for the laser wavelength in Region I. This was stated earlier without proof.

Shock Heating and Other Redistribution Processes. When picosecond duration or shorter pulses are used, one may ask whether the surface-absorber ablation theory we have presented applies or not. One notices, for example, that mass ablation rates are not predicted well. Several effects that have not been considered in deriving the basic theory may come into play to extend the heated region and dissipate incident heat flux. These include:

- Thermal conduction
- UV and X-ray heating of the target
- Hot-electron heating with front velocity, v_{hot}
- Shock heating

Of these, the last is the most important. Thermal conduction can develop a high velocity for picosecond pulses, but not high enough to penetrate a significant amount of material. X-ray heating can be shown to be less significant than shocks, and the hot-electron source term is weak until irradiance reaches the laser fusion regime [10 TW/cm^2 (100 PW/m^2) or so, outside the range of interest for most laser ablation work.

Shocks are important energy sinks for short pulses even at modest laser fluence levels. For such short pulses, the acoustic intensity will be $I_a = p^2\tau/\rho$, and mechanical coupling efficiency I_a/I can be quite good. We assume the sound wave is launched into the target material with intensity I_a and an attenuation coefficient α_a, propagating a distance $c_s\tau$ during the laser pulse. The fraction of the acoustic intensity that goes into solid material heating is given by

$$Q_s/I_a = c_s\tau\alpha_a \tag{94}$$

where

$$\alpha_a \equiv 1/(c_s\tau_B) \tag{95}$$

so that

$$Q_s/I_a = (\tau/\tau_B) \tag{96}$$

We have used the symbol τ_B for the acoustic relaxation time of the medium at the acoustic frequency $\omega_a = 2\pi/\tau$ induced by the laser pulse, because it can be found in the literature of stimulated Brillouin scattering (SBS) (see e.g., Heiman et al., 1978). The relaxation time

$$\tau_B = (2\pi/\eta)(c/\omega_a)^m \tag{97}$$

is a function of the material viscosity η and the acoustic frequency ω_a, which we take as determined by the laser pulse width that launches the shock, $\omega_a = 2\pi/\tau$. The coefficient m also varies with frequency:

$$m = 2/[1 + (\omega_a\tau_0)^2] \tag{98}$$

Based on the results of Heiman et al. (1978), we estimate that, for glasses as an example, where $\tau_0 \approx 50$ ps, $m \rightarrow 0$ at $\tau \approx 5$ ps, and that $\tau_{Bmin} \approx 50$ ps. Values for other materials will vary. Thus, a condition for inapplicability of our analysis is:

$$\tau \ll \tau_B \tag{99}$$

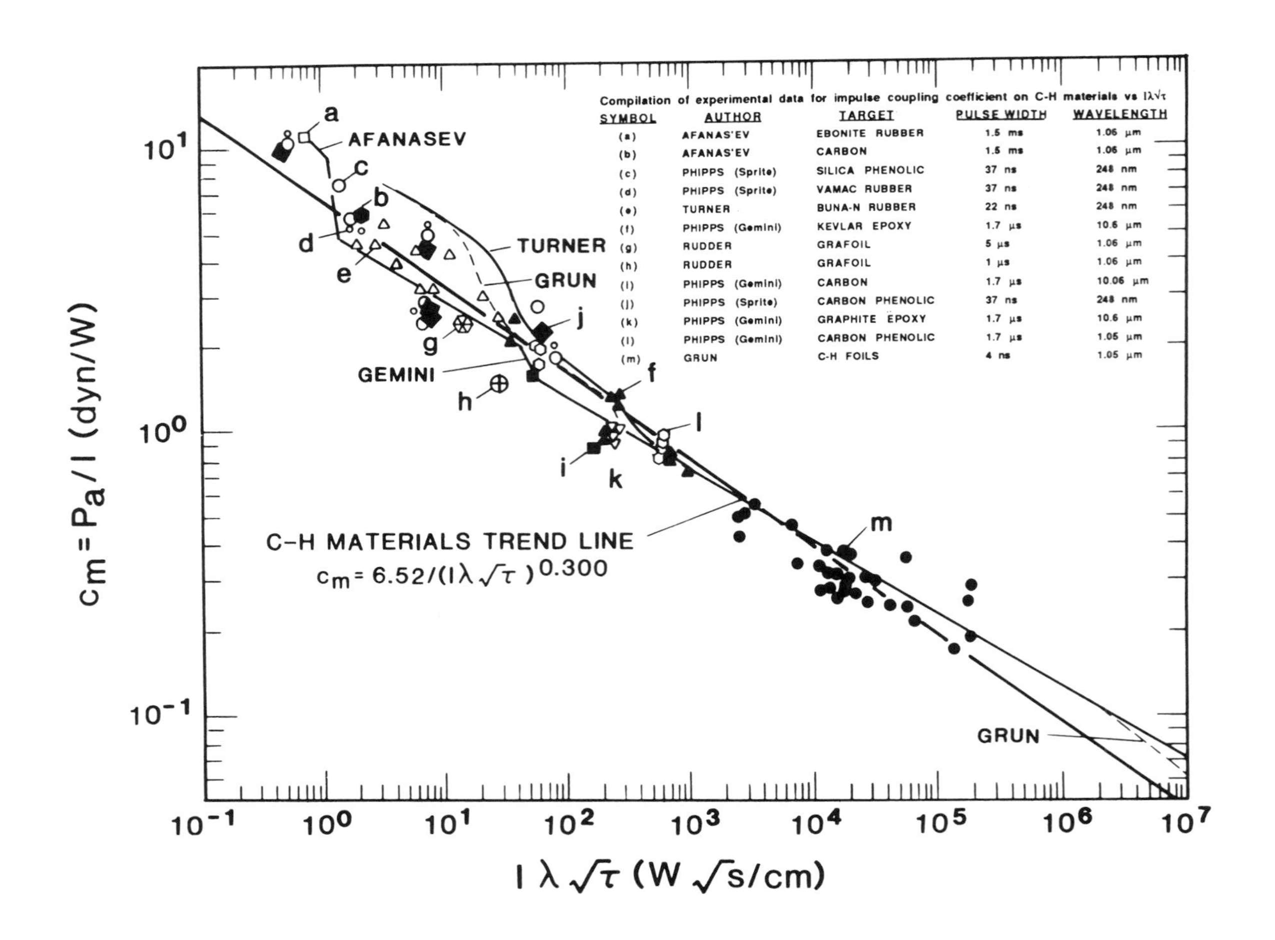

Compilation of experimental data for impulse coupling coefficient on C-H materials vs IλVτ
SYMBOL AUTHOR TARGET PULSE WIDTH WAVELENGTH
(a) AFANAS'EV EBONITE RUBBER 1.5 ms 1.06 μm
(b) AFANAS'EV CARBON 1.5 ms 1.06 μm
(c) PHIPPS (Sprite) SILICA PHENOLIC 37 ns 248 nm
(d) PHIPPS (Sprite) VAMAC RUBBER 37 ns 248 nm
(e) TURNER BUNA-N RUBBER 22 ns 248 nm
(f) PHIPPS (Gemini) KEVLAR EPOXY 1.7 μs 10.6 μm
(g) RUDDER GRAFOIL 5 μs 1.06 μm
(h) RUDDER GRAFOIL 1 μs 1.06 μm
(i) PHIPPS (Gemini) CARBON 1.7 μs 10.06 μm
(j) PHIPPS (Sprite) CARBON PHENOLIC 37 ns 248 nm
(k) PHIPPS (Gemini) GRAPHITE EPOXY 1.7 μs 10.6 μm
(l) PHIPPS (Gemini) CARBON PHENOLIC 1.7 μs 1.05 μm
(m) GRUN C-H FOILS 4 ns 1.05 μm
AFANASEV
TURNER
GRUN
GEMINI
GRUN
C-H MATERIALS TREND LINE
$c_m = 6.52/(I\lambda\sqrt{\tau})^{0.300}$
a
b
c
d
e
f
g
h
i
j
k
l
m
$c_m = P_a/I$ (dyn/W)
$I\lambda\sqrt{\tau}$ (W $\sqrt{s}$/cm)
10^1
10^0
10^{-1}
10^{-1}
10^0
10^1
10^2
10^3
10^4
10^5
10^6
10^7

in the target material. The second condition, which must occur simultaneously to invalidate the analysis, is that the thickness $c_{\text{solid}}\tau$ be significant compared to the ablation depth during the pulse, $x_v = \dot{m}\tau/\rho$. They may be stated as

$$\dot{m} \ll \rho c_{\text{solid}} \tag{100}$$

For many solids, $\rho c_{\text{solid}} \approx 10^6\ \text{g}\cdot\text{cm}^{-2}\cdot\text{s}^{-1}$ ($10^7\ \text{kg}\cdot\text{m}^{-2}\cdot\text{s}^{-1}$). Since we nearly always deal with ablation rates that are smaller than this value, condition (99) is the limiting criterion. The foregoing can then be summarized as saying that we do not expect our theory to model picosecond and shorter pulses.

4A.2. EXPERIMENTAL RESULTS FOR HIGH IRRADIANCE

Figure 4A.5, reproduced from Phipps et al. (1988), shows how well the theory predicts vacuum surface impulse at high irradiance. Laser wavelengths represented in the figure are 10.6, 3, and 1.06 μm and 248 nm. Also represented are pulse lengths from 500 ps to 1.5 ms, and irradiances from 3.5 MW/cm^2 to 30 TW/cm^2. The data shown describe interaction with nonmetallic materials, but data from metals are fit equally well. The steps in the data fit show how changes in ionization state Z affect Ψ and thus C_m, which occur as laser irradiance increases. We can estimate the charge state Z with sufficient accuracy for calculating Ψ from, e.g., Eq. (23), if it is not otherwise known.

Note that we have plotted the impulse data vs. the scaling parameter $I\lambda\sqrt{\tau}$. This particular choice of parameter is suggested by Eq. (68) and the instantaneous relationship $C_m = p_a/I$.

4A.2.1. Mass Loss Rate

Figure 4A.6 shows the excellent agreement of the high-irradiance theory with measurements of $\dot{m}$ from the work of several authors (Yamanaka et al., 1986; Shirsat, et al., 1989; Gupta and Kumbhare, 1984). In the figure, the data of Yamanaka et al. are taken as the norm. Normalization adjustments are on the order of 30%.

The *lowest* irradiance represented on the graph is 1×10^{11} W/cm^2 in the data of Gupta and Kumbhare, and the highest 4×10^{14} W/cm^2 in the data

Figure 4A.5. Graph illustrating good agreement between an extensive compilation of experimental data for high-irradiance vacuum impulse coupling coefficient on C—H type materials vs. the parameter $I\lambda\sqrt{\tau}$. All Phipps, Rudder, and Turner references may be found in Phipps et al. (1988). Other references are Afanas'ev et al. (1969) and Grun et al. (1983).

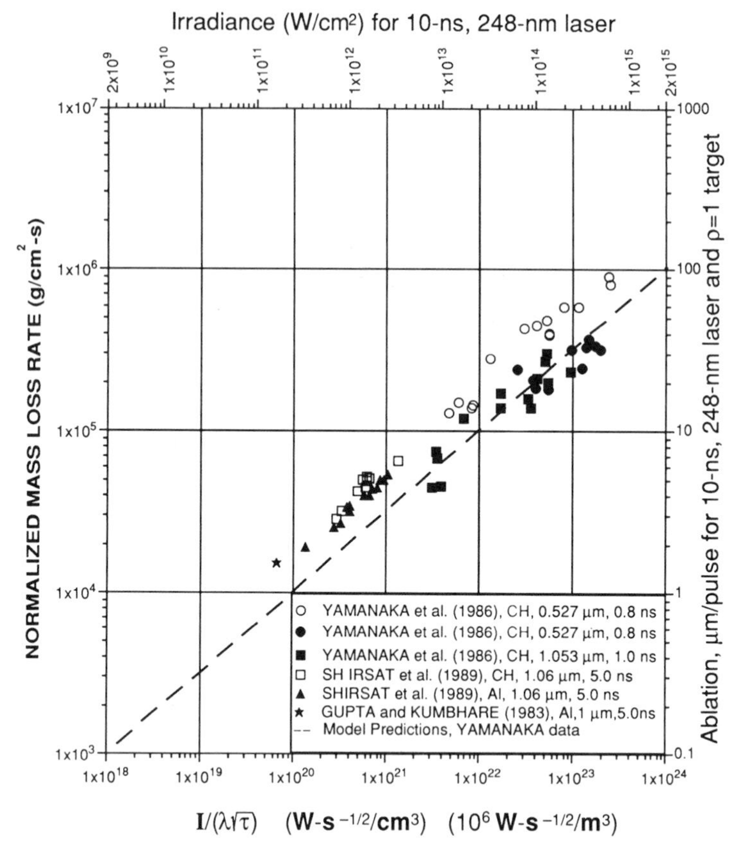

Figure 4A.6. Compilation of mass loss rate for high irradiance, compared to the Eq. (71) prediction. Mass loss is adjusted slightly to compensate for differences in atomic mass A and ion charge number Z, e.g., $C_{\mathrm{Yamanaka}} = \Psi^{9/8}/A^{1/4} = 0.384$; see Eq. (71). Other data are adjusted for different conditions of atomic mass A and charge state Z by multiplying published $\dot{m}$ by $C_{\mathrm{Yamanaka}}/C_{\mathrm{data}}$ in order to make a comparison of data from widely different experimental conditions possible.

of Yamanaka et al. Note that we have plotted the mass loss data vs. the scaling parameter $I/(\lambda\sqrt{\tau})$, as suggested by Eq. (71).

4A.2.2. Ablation Depth

Our success in modeling $\dot{m}$ at high irradiance gives us confidence that we might also be able to approximately model high-irradiance measurements of ablation depth:

$$x_v = \int_{-\infty}^{\infty} \frac{dt}{\rho} \dot{m} \approx \dot{m}\tau/\rho \qquad \text{cm} \tag{101}$$

Here, because $\dot{m}\tau \propto (I\tau^{3/2}/\lambda)^{1/2}$ [see Eq. (71)], the suggested plotting variable is $I\tau^{3/2}/\lambda$. Of course, exact modeling requires careful numerical calculations specific to each circumstance (see, e.g., Balazs et al., 1991).

We exclude data that do not at least approximately satisfy each of the conditions I–IV. The most frequent reason for excluding data from Figures

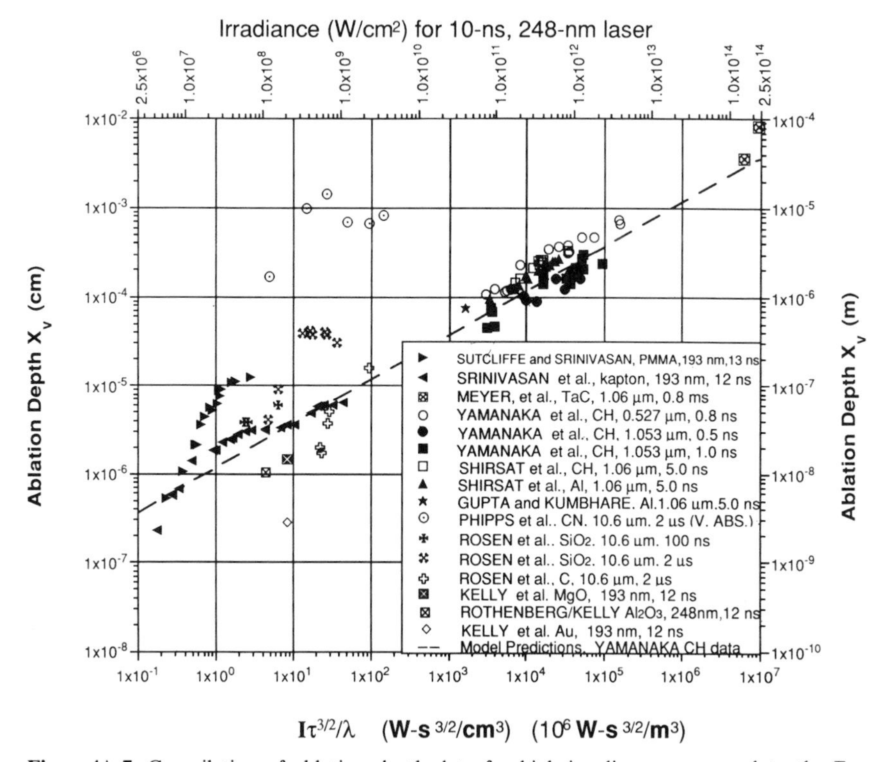

Figure 4A.7. Compilation of ablation depth data for high irradiance, compared to the Eq. (101) prediction. Ablation depth is adjusted slightly to compensate for differences in atomic mass A and ion charge number Z. The normalization is made to the conditions of Yamanaka as a standard by multiplying the published value of x_v by the factor $C2_{\text{Yamanaka}}/C2_{\text{data}}$, where $C2 = C/\rho$, ρ being the target solid mass density. One volume absorber data set is included for comparison, that of Phipps et al. (1990) [symbol ⊙], and one surface absorber set taken in air, that of Srinivasan et al. (1986a). Other data are the same as in Figure 4A.6, except for the addition of Rosen et al. (1988), Rothenberg and Kelly (1984), and Kelly et al. (1985).

4A.6 or 4A.7 was that the data were taken at insufficient irradiance to guarantee plasma dominance of the interaction (condition III) (e.g., Chuang et al., 1988). The next most frequent reasons were that the data were taken in air (condition I) or that the target was a volume absorber rather than a surface absorber (condition II).

Figure 4A.7 shows the results of this effort. In addition to the $\dot{m}$ data references given previously and the exceptional data discussed below, data are taken from Meyer et al. (1973), Kelly et al. (1985), and Rothenberg and Kelly (1984), for a total of 16 data sets. In cases where at least one point (typically the highest intensity point) of a data set crosses over into the parameter space that satisfies all the conditions, we have plotted the entire set.

The data of Meyer et al. (1973) [symbol ⊠] that we used shown in Figure 4A.7 were taken in vacuum, but no difference was seen when background pressure was as high as 0.2 torr (27 Pa) of O_2 or H_2. Meyer and coworkers' 25 ns carbon data were also modeled well, but they were not plotted in the figure because the additional points could not be seen in the data cluster at midrange of $I\tau^{3/2}/\lambda$.

We have also included three examples in Figure 4A.7 that violate one of our conditions strongly. The first, a volume-absorber data set [symbols ⊙] (see Phipps et al., 1990; Rosen et al., 1988) that violates condition II, shows the extent of ablation enhancement that is possible at low irradiance near plasma threshold in volume absorbers. These data represent the largest ablation depth per joule yet reported at any laser wavelength, as well as the largest momentum coupling coefficient C_m in a homogeneous target material.

The second, which is the polyimide set at the very short 193 nm wavelength, from Srinivasan et al. (1986b) [symbol ◀], gives excellent agreement with the model and is included to make the following point. While an air environment changes the interaction physics so much that the analysis in this chapter should be irrelevant—and, indeed, some impulse measurements in air show strong disagreement with measurements in vacuum—at irradiance above plasma threshold, but low enough to avoid forming laser-supported detonation waves (LSD waves) in the atmosphere above the target, very strong absorbers like kapton at 193 nm (surface absorbers) still seem to give the right answer. The LSD wave threshold at 1 bar for 10.6 μm lasers on aluminum is just under 1 GW/cm^2 (Beverley and Walters, 1976). The spectrophotometric absorption depth $1/\alpha$ for kapton at the 193 nm UV wavelength and standard temperature and pressure (STP) is 230 Å (Sutcliffe and Srinivasan, 1986).

The third set (from Sutcliffe and Srinivasan, 1986) [symbol ▶] is identical to the first, except that the target material is PMMA. This data set also violates our plasma ignition limit by about a factor of 10 at its lowest irradiance level. However, at irradiances where the data can be compared, it is interesting to note how small the region of volumetric absorption need be to change the character of the ablation toward "volume" from "surface." The 193 nm absorption depth $1/\alpha$ is about 100 times larger with PMMA as compared to kapton—but still only 2.2 μm. However, the resulting ablation depth for the same irradiance is noticeably larger with PMMA than with kapton.

At high irradiance, *all* absorbers in vacuum coalesce to the behavior predicted by the dashed line in Figure 4A.7. As was explained in Section 4A.1.3, under the high temperature and pressure associated with high irradiance, the relatively long spectrophotometric absorption depths characteristic of volume absorbers shrink to resemble metallic absorption depths.

A lot of the scatter in Figure 4A.7 at low irradiance could well be due to modeling uncertainty: it is difficult to guess the values of $\langle A \rangle$ and $\langle Z \rangle$ that are required to apply the theory, and they are not given.

In summary, Figure 4A.7 shows it is possible to model x_v for surface absorbers over a 7 orders of magnitude range in the parameter $I\tau^{3/2}/\lambda$, including about a factor of 50 in laser wavelength, and at least 6 orders of magnitude in pulse duration from nanosecond to millisecond duration pulses, using a simple scaling formula. The analysis predicts ablation depth within a factor of 2 when $I\tau^{3/2}/\lambda > 50\ \mathrm{W \cdot s^{3/2} \cdot cm^{-3}}$ ($500\ \mathrm{kW \cdot s^{3/2} \cdot m^{-3}}$), corresponding to irradiances above about 1 GW/cm^2 for 12 ns KrF lasers. This explains the choice of irradiance in the heading of Section 4A.1. For lower irradiance, the prediction becomes progressively poorer, as plasma physics no longer dominates surface chemistry. The x_v modeling may apply in air in some cases.

4A.3. ION ACCELERATION IN THE LOW-DENSITY PLUME

4A.3.1 Experiments

This section concerns what happens in Region IV of Figure 4A.1. Figure 4A.8, based on the ion velocity measurements of von Gutfeld and Dreyfus (1989), Eidmann et al. (1984), Gregg and Thomas (1966a,b), and Dreyfus (1991), demonstrates that there is good agreement with the velocity trend predicted by the model, but it is consistently found that a multiplier amounting to about $\sqrt{10}$ must be used to shift the prediction upward and match the measurements. The falloff at $I\lambda\sqrt{\tau} \approx 1$ is forseeable, as the area of validity for the present theory has been violated, e.g., power density is too low for inverse bremsstrahlung to be the major energy input ($\leqslant 1\ \mathrm{GW/cm^2}$).

It is concluded that, over a wide range of high-irradiance values in vacuum, something accelerates the ions, increasing their energy by a factor of 10 between the time of their creation in the ablation layer and their detection

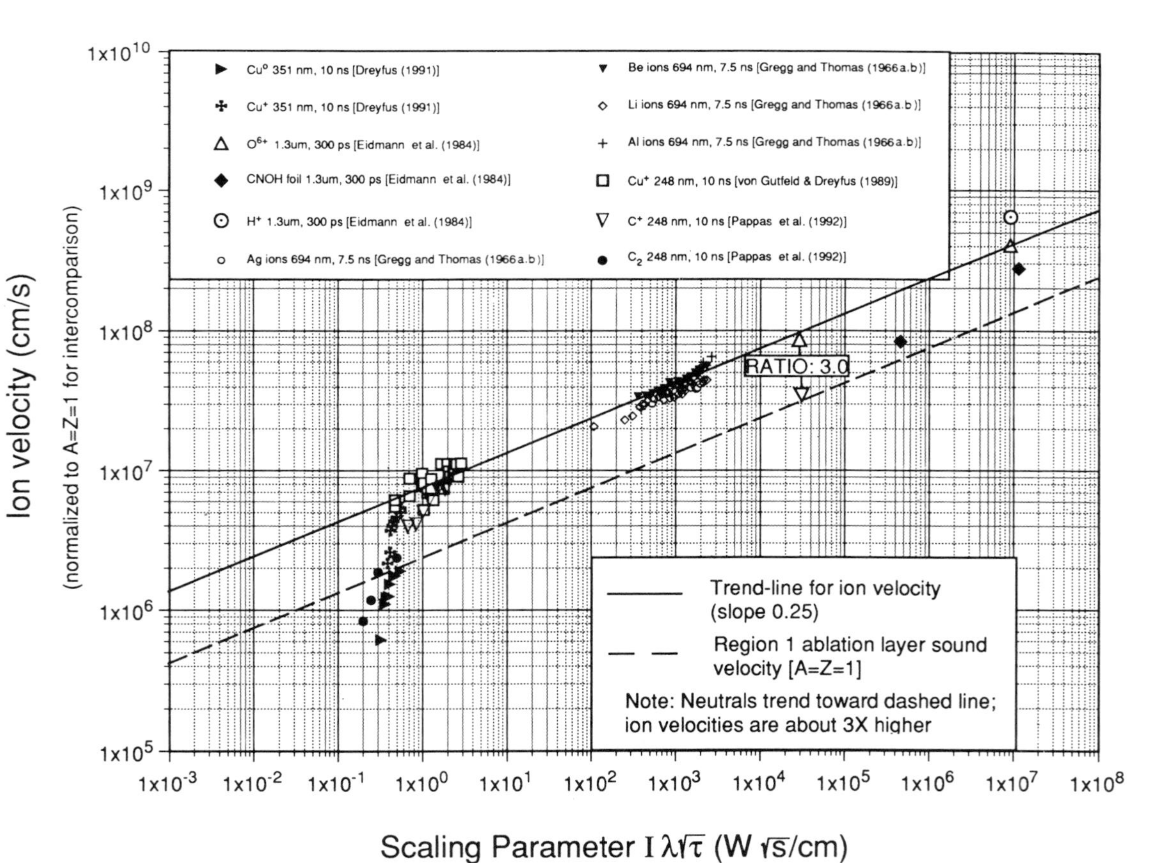

Figure 4A.8. Compilation of ion velocities in the high-irradiance regime, illustrating the point that velocity predicted by Eq. (72) for Region I seems to be consistently augmented by a multiplicative factor on the order of $\sqrt{10}$ in the plasma plume.

in the plasma plume. The loss of this acceleration at the lower boundary of the high-irradiance region is clearly seen in the Dreyfus (1991) copper ion data.

We note that the velocity of some neutral species falls on the same (accelerated) trend line as the ions [e.g., CN from polyimide at 248 nm wavelength (see Srinivasan and Dreyfus, 1985)]. This result suggests that these species gain energy, presumably due to photochemical ablation rather than a thermal one. We have plotted the CN data to illustrate this point, even though the species are not ions and are below the high-irradiance boundary.

If one appeals only to expansion work to find this energy, one finds from Kelly and Dreyfus (1988a) that the expected drift velocity u_K is close to the sound velocity that, from Eq. (5), is expected to show a $(x/x_T)^{-3/4}$ variation, i.e., a deceleration rather than an acceleration. Along with the approach of the experimental points to the theoretical prediction, note that only the neutral species approach the dashed line.

It is well known that very large ion-accelerating electric fields are created in high-irradiance laser–plasma interaction experiments. Langer et al. (1966) were possibly the first to notice suprathermal ion kinetic energy in vacuum laser irradiation experiments. They obtained C^{2+} ions with energies from 100 to 1000 eV from 10 GW/cm^2 laser irradiance (100 TW/m^2). They also recorded a threshold for formation of multiply charged C ions of 1.5 GW/cm^2 (15 TW/m^2). Multiply charged ions were always seen to have a narrower, higher energy spectrum. At 60 GW/cm^2 (0.6 PW/m^2), for example, C^{4+} ions with 1 keV energy were seen. The threshold for C^{4+} formation was 20 GW/cm^2. This corresponds to a *temperature* at their creation of only 64 eV (Allen, 1973).

Many other workers have also measured large laser-induced plasma voltages; e.g., Silfvast and Szeto (1977), Bergmann et al. (1980), and Cook and Dyer (1983) obtained 0.1–1 kV pulses from CO_2-laser-induced plasmas.

Bykovskii et al. (1972) observed typical Li^+ and D^+ ion drift-kinetic energies of 100 eV with 1 GW/cm^2 (10 TW/m^2) irradiation of LiD targets in vacuum. At 10 GW/cm^2 (100 TW/m^2), they saw Li^{2+} ions for the first time with a characteristic energy of 270 eV, whereas the Li^+ and D^+ ions showed a double-peaked distribution at about 120 and 300 eV, respectively, with tails extending up to 700 eV. With heavier ZrH targets at the same irradiance, Zr^{5+} ions were created with typical energy 2.5 keV, and the Zr^+ spectrum was essentially flat, rising slightly between 100 eV and 2.5 keV. Bykovskii et al. (1972) characterized these energy spectra as being much higher than thermal energy (about 15 eV for the LiD experiment). They explained these results by appealing to an accelerating field arising from hot electrons leaving the plasma. Their data is reproduced in Figure 4A.9.

David and Weichel (1969) noticed that $T_e \approx 10\, T_{\text{plasma}}$, in experiments with 1 GW/cm^2 irradiance on carbon.

Bykovskii et al. (1971) earlier observed clear evidence for separate

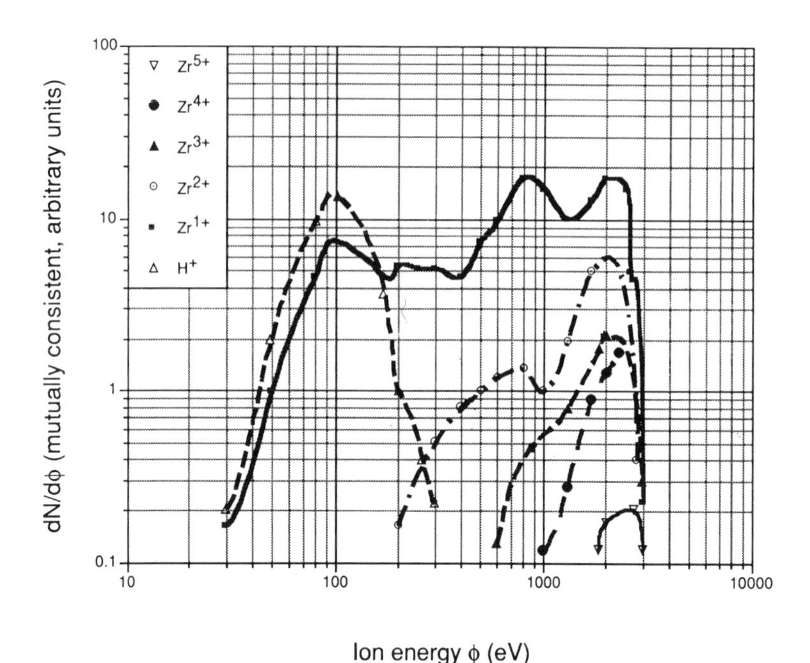

Figure 4A.9. Reproduction of Figure 3 of Bykovskii et al. (1972) showing the ion energy spectrum obtained in 1 μm wavelength laser ablation of zirconium hydride at 10 GW/cm^2 irradiance (100 TW/m^2). Where one would expect a temperature of about 15 eV for these irradiation parameters, very substantial suprathermal behavior is observed (used by permission).

"thermal" ($\approx$ 10 eV) and "accelerated" ($\approx$ 100 eV) time-of-flight (TOF) ion peaks at $I = 8\,\text{GW/cm}^2$ (80 TW/cm^2) irradiance ($\lambda = 1.06\,\mu$m; $\tau = 15$ ns) with aluminium targets. The hot peak tended to be multiply charged ions. With 10 TW/cm^2 irradiance, these workers created 35 keV Co^{25+} ions. In their experiments, the threshold for this acceleration is clearly an irradiance of $I_{\text{thresh}} = 500\,\text{MW/cm}^2$ (5 TW/m^2). Bykovskii et al. (1972) estimate that the fraction represented by the accelerated ions n_{i*} is

$$f_i = n_{i*}/n_i \approx D_e/d_s \tag{102}$$

This ratio is about 10^{-5} in the laser absorption zone, for KrF lasers and 100 μm diameter spots. Bykovskii et al. (1969) observed certain random anisotropies in the angular distribution of the ejected ions.

Inoue et al. (1970) were perhaps the first to observe beamlike behavior in the accelerated ions, in experiments producing 70 eV C^{2+} ions from poly(ethylene) with a ruby laser ($\lambda = 694.3$ nm) and irradiance around 2 TW/cm^2.

Bykovskii et al. (1971) also saw beamlike behavior, with ions ejected progressively closer to the target normal at $I > I_{\text{thresh}}$ in laser–target interactions. The measurements of Gupta et al. (1986) add further support to the picture of acceleration and ion beam narrowing. At 268 nm irradiances of 0.5–10 TW/cm^2, fast and slow aluminum ion peaks were clearly seen. Flux measurements showed the slow peak to be nearly isotropic, whereas the fast peak ions were clearly confined to a narrow cone around the target-normal.

Workers in laser fusion—where irradiance levels exceed 10 TW/cm^2 (100 PW/m^2)—are accustomed to seeing a ratio $f_T = T_*/T \approx 8$–20 (Haines 1979), with a population of "hot electrons" at least as large as

$$f_e = \frac{n_{e*}}{n_e} = \frac{(e\phi/kT_e)^{1/2}\exp(-e\phi/kT_e)}{\pi^{1/2}}\left(1 + \frac{1}{2e\phi/kT_e}\right) \tag{103}$$

In Eq. (103), $e\phi/kT_e$ is the ratio of the charge separation potential to the thermal energy in consistent units; Eq. (103) is derived by integrating a single Maxwellian energy distribution from $e\phi/kT_e$ to ∞ (Pearlman and Dahlbacka, 1977). At irradiance levels we often utilize in LIMA work, Eq. (103) provides a broadly applicable estimate of the hot fraction.

The work of Akhsakhalyan et al. (1982), one of the few instances in which the energy distribution was measured with irradiance as low as 1 GW/cm^2 at $\lambda = 1\ \mu$m, clearly shows $f_T = T_*/T \approx 10$ with a total hot particle flux about 2×10^{-3} of the total particle flux.

4A.3.2. Theory and Modeling

The best general theoretical treatment is given by Hora's theory of so-called double layers (DLs) (see Hora et al., 1989; Eliezer and Hora, 1989). DLs become important in the corona or outer region of the laser–plasma system.

In LTE, one expects (as in our earlier discussion of Debye sheaths) these layers of positive and negative charges to generate a potential jump that is on the order of kT_e, depending on the magnitude of the density gradient scale length $L_n = n_e/\nabla n_e$.

At high irradiance, the electron velocity distribution that develops from laser–plasma heating is no longer described by LTE concepts. Instead of a single Maxwellian velocity distribution, one finds that separate "thermal" (cold) and "hot" components with temperatures T and T_*, respectively, can briefly coexist. The simultaneous existence of "hot" and "thermal" electrons in the corona region results in charge separation sufficient to strongly accelerate positive ions.

Physical mechanisms that can create this departure from LTE, causing the production of "hot" electrons and ions, include resonance absorption (Denisov, 1957), relativistic self-focusing, at very high irradiance (Hora, 1975), and magnetic field effects; Afanas'ev and Kanavin (1983) predict generation of magnetic fields on the order of 10 kG (1 T) at $5\,\mathrm{GW/cm^2}$ ($50\,\mathrm{TW/m^2}$) irradiance on a $100\,\mu\mathrm{m}$ diameter spot.

An idea of the predicted ratio of "hot" to "normal" particle number at laser fusion irradiance levels is obtained from the work of Forslund et al. (1977). Their theory is based partly on the theory of resonant absorption in laser fusion (beyond our scope here) and is partly empirically derived from two-dimensional relativistic particle simulations at five different wavelengths spanning the range from 0.7 to $3\,\mu\mathrm{m}$. They found:

$$f_e = \frac{n_{e*}}{n_e} = 0.24\left[\left(\frac{I\lambda^2}{T}\right)^{1/2}\right] \tag{104}$$

For SI unit parameters, the coefficient is unchanged. If applied to the Akhsakhalyan et al. (1982) case, Eq. (103) would predict $f_e = n_{e*}/n \approx 8 \times 10^{-3}$, in rough agreement with what they observed.

However, the prediction of Forslund et al. for the temperature ratio $f_T = T_*/T$ is not correct for irradiance levels below about $100\,\mathrm{TW/cm^2}$.

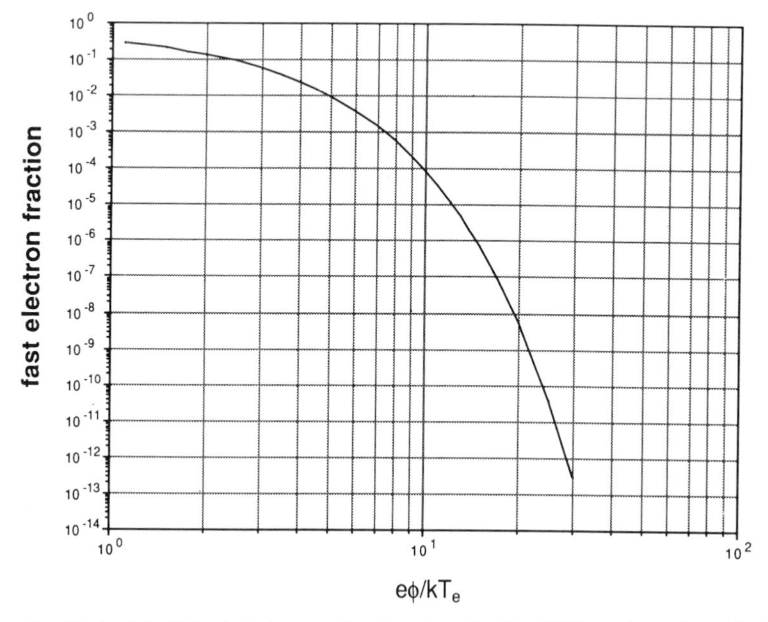

Figure 4A.10. A plot of the hot electron fraction given by Eq. (103) vs. the ratio $e\phi/kT_e$ defined in the text, for a Maxwellian velocity distribution of electrons in LTE.

Instead, one may use the theory of Eliezer and Hora (1989), which gives

$$(6f_T - f_T^2 - 1)^2 - 16f_T^2 \geqslant 0 \tag{105}$$

requiring that $f_T \geqslant 5 + \sqrt{24} = 9.9$.

The function f_e in Eq. (103) is plotted in Figure 4A.10. In the figure, we can see that a potential $e\phi/kT_e = 10$ times the thermal potential can be reached by $f_e = 10^{-4}$ of the total electron population of a single Maxwellian velocity distribution, about 20 times less than observed by Akhsakhalyan et al. (1982), reinforcing the idea that a complete non-LTE treatment would be better.

Afanas'ev et al. (1988) show that the electric field created should have a logarithmic dependence on D_e/d_s. Mulser (1971) proved that, while there is no question that plasma electric fields in high-irradiance experiments are capable of causing the ion acceleration that we see, the velocity *difference* between ions of different mass observed in TOF experiments *cannot* be explained by different forces acting on them in the accelerating field but must instead be due to irradiance inhomogeneities in the laser spot on target. Langer et al. (1966) showed that irradiance inhomogeneities could be deliberately manipulated via astigmatic focus to create anisotropic ion ejection patterns.

We now address acceleration in the experiment of von Gutfeld and Dreyfus (1989), as most relevant to problems that might arise in LIMA work. This problem has been previously addressed by Phipps (1989). We begin with the paper of Pearlman and Dahlbacka (1977), in which they reported observing laser-produced electric potentials on the order of 10 kV persisting for up to 200 times the laser pulse duration in a 50 ps laser–target irradiation experiment at $\lambda = 1.06\,\mu$m wavelength. The development of a 30° fwhm ion beam with energy as large as 1 MeV was inferred from the data.

We believe that the physical mechanism for charge acceleration in the ablation experiment of von Gutfeld and Dreyfus (1989) is the same as in Pearlman and Dahlbacka's case. A suprathermal accelerating electric potential is set up between a separate "hot" component of the electron velocity distribution and the slower-moving ions, which are initially left behind—a topic that we introduced in our discussion of Region IV. The "hot" electron fraction escapes LTE simply by escaping the experiment before interactions with other particles have a chance to operate. The most energetic ions (of equal and opposite total charge) then accelerate up to the hot electrons, taking most of their energy as already described in Section 4A.1.2.

The experimental parameters of von Gutfeld and Dreyfus (see Table 4A.3) set this hot electron fraction at $f_e \approx 8 \times 10^{-4}$. If we insert this value into Eq. (102), we obtain a single Maxwellian acceleration factor $e\phi/kT_e = 9$ and

Table 4A.3. Self-Consistent Values for von Gutfeld and Dreyfus Experimental Parameters

Fluence Φ	$6\,\mathrm{J/cm^2}$
Material	Cu, $\rho = 8\,\mathrm{g/cm^3}$
Enthalpy	$H_v = 5300\,\mathrm{J/g}$
Emitted neutrals	$N_0 = 3.7 \times 10^{18}\,\mathrm{cm^{-2}}$
Escape fraction	$f_{esc} = 8 \times 10^{-4}$
Ionization fraction	$\eta = 1$
Target beam area	$\mathscr{A}_b = 0.02\,\mathrm{cm^2}$
Total emitted electrons	$N_e \mathscr{A}_b = 7.4 \times 10^{16}$
Escaping electrons	$f_{esc} N_e \mathscr{A}_b = 6 \times 10^{13}$
Ion divergence angle	$\phi_d = 40°$ fwhm
Ion burst duration	$\tau = 3\,\mu\mathrm{s}$
Escaping electron current	$\mathscr{J} = 3\,\mathrm{A}$
Ion potential	$e\phi = 38\,\mathrm{eV}$
Beam impedance	$Z = 160\,\boldsymbol{\Omega}$ (ohms)
Probe distance	$x = 10\,\mathrm{cm}$
$n_i(x)$	$1 \times 10^{12}\,\mathrm{cm^{-3}}$
Velocity enhancement	$(e\phi/kT_e)^{1/2} = 3$
Temperature	$T_e = 4\,\mathrm{eV}$

note that $(e\phi/kT_e)^{1/2} \approx 3$ agrees with the ion velocity enhancement factor shown by the von Gutfeld and Dreyfus data (discussed in Section 4A.3.1). We also find that this value is consistent with the measurements of Bykovskii et al. (1972). We find that the von Gutfeld and Dreyfus irradiance, wavelength, and temperature values predict $f_e \approx 5 \times 10^{-4}$ via Eq. (103); Eq. (104) gives $e\phi/kT_e = 9.9$, also in excellent agreement with observation.

We can give a heuristic argument that allows prediction of the quantity $e\phi/kT_e$ from the particle mass ratio. First, we apply the constraint of current equality $[j_e = j_i]$, necessary for a steady potential, taking the energetic electron current to be limited by the ion thermal current:

$$n_{e*} v_{e*} = Z n_i v_i \tag{106}$$

Then, applying the twin conditions of LTE $[v_e = (m_i/m_e)^{1/2} v_i]$ and overall charge neutrality $[Z n_i = n_e]$ within the electron and ion thermal populations of Eq. (106) gives

$$(e\phi/kT_e)^{1/2} = (v_{e*}/v_e = n_e/n_{e*})\,[(m_e/m_i)^{1/2}] \tag{107}$$

or

$$f_e^2(e\phi/kT_e) = m_e/m_i \tag{108}$$

Now, applying (103), we find

$$\exp(e\phi/kT_e) = (m_i/\pi m_e)^{1/2}(e\phi/kT_e + \tfrac{1}{2}) \tag{109}$$

or

$$e\phi/kT_e = \tfrac{1}{2}\ln(m_i/\pi m_e) + \ln(e\phi/kT_e + \tfrac{1}{2}) \tag{110}$$

Because of the slow variation of the first logarithmic term, $e\phi/kT_e \approx 7$ within 15% for atoms ranging from copper to carbon. As an example, for a carbon plasma, Eq. (109) gives $e\phi/kT_e = 6.26$ and, for a copper plasma, $e\phi/kT_e = 7.32$. Therefore, the velocity ratio $(e\phi/kT_e)^{1/2} \approx 2.5$ over the same range of atoms.

These energetic electrons then attract and accelerate a cloud of ions from the thermal plasma. Given enough space, it is important to realize that these ultimately move at the same velocity as the electrons,

$$v_{e*}|_{\text{after}} = v_{i*} \tag{111}$$

slowing them down until the ions have effectively captured all the electrons' kinetic energy.

We further find that we can satisfy the requirements for energy conservation between the two energetic clouds,

$$m_e n_{e*} v_{e*}^2|_{\text{before}} = m_i n_{i*} v_{i*}^2|_{\text{after}} \tag{112}$$

in addition to Eqs. (106) and (108) as well as charge neutrality $[Zn_i = n_e]$, by requiring that

$$f_i = \frac{n_{i*}}{n_i} = \left(\frac{Z}{e\phi/kT_e}\right) f_e \tag{113}$$

For copper ions, Eq. (113) gives $f_i = 1.5 \times 10^{-4}$ and $f_e = 1.1 \times 10^{-3}$, very close to observations and earlier estimates in this section.

The acceleration process further narrows the emitted ion beam—down to 20° fwhm in the von Gutfeld and Dreyfus experiment on copper targets, using KrF laser irradiation. At the same time, as the coulomb potential accelerates the ions, there exists a competing dilution process in which resonant charge exchange as well as classical ion–molecular collisions transfer the energy into the neutral cloud, i.e., there is a distinct energy gain by neutrals as well. This is noted in Figure 4A.8 for the case of CN from polyimide (though this artifact may be explained instead by photochemical processes) and has been noted for Al from Al_2O_3 (Dreyfus et al., 1986c) and C_2 from PMMA (Srinivasan et al., 1986b). While these systems did not involve intense plasmas dissociating diatomics, there does exist a conflicting process in which plasma-induced dissociation eliminates many of the more energetic diatomics. The result of this selective dissociation is an apparent *slowing* (loss of kinetic energy) for the diatomics surviving (see Dreyfus et al., 1986c).

The foregoing considerations are capable of reproducing the key elements of the von Gutfeld and Dreyfus experiment. In Table 4A.3 we give the results of a self-consistent calculation that agrees with the reported aspects of that experiment (Phipps, 1989).

4A.4. MEASUREMENT TECHNIQUES IN THE LOW-DENSITY PLUME CREATED BY HIGH LASER IRRADIANCE

4A.4.1. Langmuir Probes

The Langmuir probe is one of the best known and—in general—best understood of plasma diagnostic techniques. Von Gutfeld and Dreyfus (1989) were among the first to apply this tool to the *quantitative* measurement of coronal plasma parameters in vacuum ablation experiments.

A Langmuir wire probe is deceptively simple, consisting of a dc-biased current probe inserted into the plasma [see Figure 4A.11 (from von Gutfeld and Dreyfus, 1989)]. It is a tool that is useful for exploring the instantaneous electron temperature and ion density in a thermal (Maxwellian, LTE) plasma. Ideally, the collected electron current I_e depends on the potential ϕ and the electron temperature according to (Chen, 1965)

$$I_e/I_0 = \exp(e\phi/kT_e) \tag{114}$$

when $\phi = (\phi_{probe} - \phi_{space})$ is positive so as to repel ions and collect electrons. Then, a plot of $\ln(I_e/I_0)$ vs. ϕ gives the electron temperature through the slope e/kT_e. The so-called space potential, ϕ_s, is the voltage assumed by the probe when it is allowed to float. In the work of von Gutfeld and Dreyfus, a value of $T_e \approx 2\,eV$ was determined at $x = 10\,cm$. Even though Eq. (69) predicts a value of $4\,eV$, ionization after the initial pulse and effects of the adiabatic expansion could easily have decreased T_e (T_e commonly falls to $\sim 1/3$ of the average ionization energy in a collisional plasma).

When the probe is biased negative, so as to collect ions, the ion density

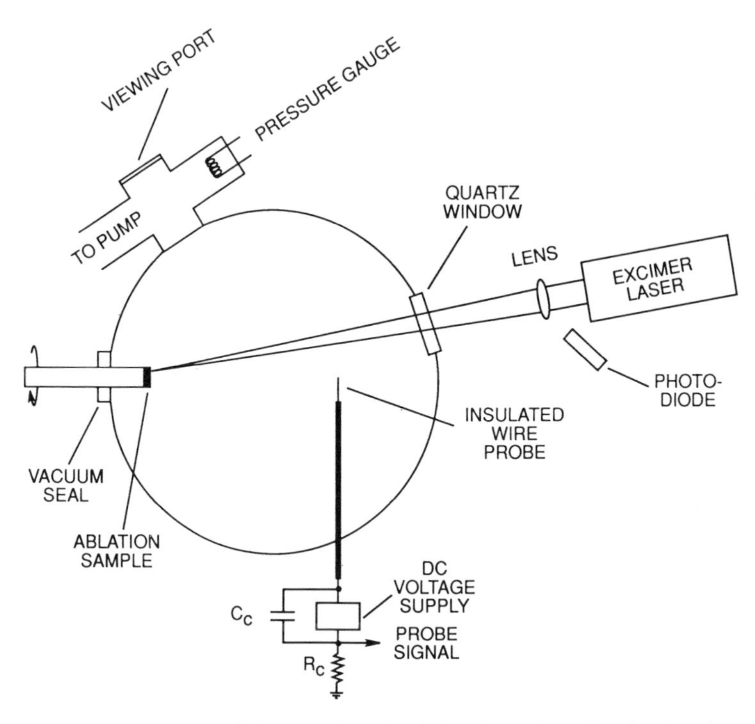

Figure 4A. 11. Reproduction of Figure 1 of von Gutfeld and Dreyfus (1989) showing the arrangement for the Langmuir probe experiment (used by permission).

can be determined reliably from

$$n_i = \frac{1}{\mathscr{A}} \left\{ \left[\frac{-4\pi m_i}{3_{e^3}} \frac{d(I_i^2)}{d\phi_p} \right]^{1/2} \right\} \qquad \text{cm}^{-3} \tag{115}$$

where $\mathscr{A}$ is the probe area exposed to the plasma; I_i is the ion current; ϕ_p is the probe voltage; and m_i is the ion mass. Note that T_i is usually difficult to measure from the ion current characteristic, owing to the overlapping large electron currents at lower negative potentials.

TOF data are also available from the probe signals and offer valuable information about the nature and drift velocity of charged particles, ion charge states, etc.

Experimental Precautions:

Nondisturbance. In order not to disturb the thermal energy distributions being measured, care must be taken to limit the maximum instantaneous electron or ion current withdrawn from the plasma, and not to cool the plasma by using an unduly large probe. These considerations lead to the use of a very-small-diameter wire for the probe, e.g., 75 μm in the work of von Gutfeld and Dreyfus. The electron current limitation can be observed experimentally when the apparent slope obtained from Eq. (114) starts to change with the current.

Probe Heating. One has to be careful not to draw such a large ion current as to cause thermionic emission from the probe. This regime is commonly only a problem for dc plasmas, but not for the few microseconds of a laser-generated plasma.

Sheath Thickness. It is important to calculate the thickness of the plasma sheath that forms around the probe: the collecting area $\mathscr{A}$ in Eq. (115) is the sheath area, which is greater than the geometric area of the probe by an amount that depends on plasma density and temperature. This correction acts to reduce the apparent ion density.

Termination Impedance. The Langmuir probe is a very delicate charge collection device with a small capacitance, so that a too-large termination resistor (used to convert current into voltage at the oscilloscope) will cause the probe voltage to vary during the measurement, generating faulty data. A simple Ohm's law calculation avoids this problem.

Negative Ion Species. Unfortunately, not all ions are positive. In ablation plasma measurements, TOF will usually allow discrimination between the negative ions and the electrons. It is important to locate the probe far enough from the ablation surface to take advantage of this discrimination capability.

4A.4.2. Time of Flight

Probably the earliest instances of kinetic energy measurements of laser-produced ions is the work of Linlor (1963) and that of Gregg and Thomas (1966a,b) with a 7.5 ns ruby laser (694.3 nm). In the latter work, irradiances from about 5 GW/cm^2 to 0.4 TW/cm^2 (50 TW/m^2–4 PW/m^2) were incident on a 220 μm diameter illumination spot, producing ion velocities from 7×10^6 to 3×10^7 cm/s (70–300 km/s) and ion kinetic energy as large as 2 keV. Materials investigated were Li, LiH, Be, C, Al, S, Zn, and Ag. These data are reproduced, in part, in Figure 4A.8. It was found that ion velocity varied with irradiance according to

$$v_{\rm i} \propto I^b \tag{116}$$

where the coefficient $0.19 \leqslant b \leqslant 0.43$, the smallest value applying to Ag, and the largest to Al. We note that Eq. (70) predicts $b = 0.250$, in the middle of this range. In this experiment, no reflecting potentials were used.

One of the most important features about the expected signal in TOF experiments is the result of Kelly and Dreyfus (1988a,b). Since

$$n_i \propto 1/t \tag{117}$$

where

$$v_i = x/t \tag{118}$$

and

$$|dv_i| = x\,dt/t^2 \tag{119}$$

then, for a density-sensitive detection system such as LIF, one obtains a counterintuitive result that the detected signal will be proportional to v^4:

$$\begin{aligned} ds \propto n_i v_i f_i(v)\,dv &\propto (x^2/t^4)\exp\{-[mx^2/(2kT_i t^2)]\}\,dt \\ &= (v^4/x^2)\exp(-mv_i^2/2kT_i)\,dt \end{aligned} \tag{120}$$

For a flux-sensitive detector, the response even has a v^5 dependence.

In the high-intensity regime, sophisticated TOF experiments have augmented simple TOF with a retarding potential electrostatic filter, pulsed deflection voltage, and extraction field (Vertes et al., 1988).

4A.4.3. Laser-Induced Fluorescence

Laser-induced fluorescence (LIF) is also an extremely useful diagnostic for laser ablation plasmas and is the only way to get certain kinds of information. In the pioneering work of Dreyfus et al. (1986a), this technique made it possible to detect and measure the energy distributions of the species AlO, C_2, and CN during KrF irradiation of aluminum oxide and polyimide (kapton).

The technique has also been used to advantage in measuring velocity distributions and relative populations of excited-state and fine-structure levels of atoms sputtered from surface samples (Zare, 1984).

Table 4A.4 shows examples of laser wavelengths that have been used to detect particular species by selectively inducing fluorescence. Observables with this technique include drift velocity (kinetic energy), rotational temperature, and vibrational temperature. As an example of the latter work, Dreyfus et al. (1986b) were able to use LIF to determine both the rotational and vibrational energies of AlO diatomics, as well as the translational kinetic energy of AlO and Al.

Table 4A.4. Laser-Induced Fluorescence (LIF) at High Irradiance: Species and Wavelengths

Species	Pump Wavelength (nm)	Detection Wavelength (nm)	Reference
CN	385–388.3 (dye)	422	Dreyfus et al (1986a)
C	230.1–230.2	230	Dreyfus et al. (1987)
C_2	230 (doubled dye)	230	Dreyfus et al. (1986c)
C_2	248 (KrF)	470	Walkup et al. (1985)
C_3	230	230	Dreyfus et al. (1987)
Al	248	396.2	Rothenberg and Kelly (1984)
AlO	447–450	—	Dreyfus et al (1986)

Koppman et al. (1986) used the LIF technique to assess the velocity, energy, and density of Li and Al ions created by ruby laser irradiation of an Li/Al sandwich at 1.5 GW/cm^2 irradiance. The purpose of the experiment was to create short ion beam bursts for injection into magnetic fusion plasma machines.

4A.4.4. Other Ion Diagnostics

Emission spectroscopy (Venugoplan, 1971), resonance ionization (Bonnie et al., 1987), and laser beam deflection by optical inhomogeneities (Chen and Yeung, 1988; Dreyfus et al., 1986b) have been used in ion source plasma diagnostics. Vertes et al. (1990) has given an excellent review of the subject of measuring and modeling plasma processes in ion sources.

PPTD. Dreyfus et al. (1987) developed the pulsed photothermal deformation (PPTD) technique for measuring energy deposited in surfaces by focused laser light. This technique, which utilizes the thermal deformation of the irradiated surface as a diagnostic, should be as useful for ablation work in vacuum as it was for their work in air. PPTD is extremely sensitive, with a noise level corresponding to 20 nJ absorbed.

"Laser Microprobe" Spacial Mapper. As regards this instrument, see Huie and Yeung (1985), Steenhoek and Yeung (1981), and Yappert et al. (1987). Spacially and temporally resolved maps of ablated atom concentrations over the target surface were obtained by interrogating the laser-generated plume with a second laser beam of appropriate wavelength chosen for selective absorption by the ablated species and recording the transmitted beam on a vidicon. Time gating at 50 μs intervals was provided by a Bragg cell. For maximum tunability to match different species, the interrogating laser was an argon-ion-laser-pumped pulsed dye laser.

Thomson Parabola. The so-called Thomson parabola instrument (Olsen et al., 1973) seems not to have been used yet in laser ablation experiments, although it has been extremely useful in inertial confinement fusion. In this instrument, parallel electric and magnetic fields oriented perpendicular to the ion velocity give combined deflections that, combined with TOF information, are able to uniquely define the charged particles charge, mass, and energy without changing the particle energy. The device gets its name from the fact that the electrostatic deflection $x_e \propto Z/(Av_i^2)$, while the magnetic deflection $x_m \propto Z/(Av_i)$, so that each charged particle traces out a distinct parabola moving toward the origin as time evolves on the instrument screen,

$$x_e/x_m^2 = A/Z \tag{121}$$

with only mass-to-charge ratio A/Z determining the shape of the parabola.

Faraday Cage. As a closing note, we mention the very interesting and novel ion diagnostics work of Siekhaus et al. (1986). They used a Faraday cage (with very small openings for the laser beam) surrounding the ablation experiment to measure the total electron charge emitted during 1.06 and 0.35 μm vacuum irradiation of insulator and semiconductor surfaces at pulse widths from 1 to 40 ns. Irradiance ranged from about 15 MW/cm^2 up to 15 GW/cm^2—at the upper end of their data, definitely within the high-irradiance regime. The sample was held by a grounded metal holder inside a collector can, surrounded by an outer shield can. Both of the latter cans were maintained at a common variable bias potential in the range ± 300 V. Materials studied were CdTe, ZnS, NaCl, SiO_2, CeF_3, W, and GeO_2. In the 1 ns work at the Nd:YAG third harmonic on ZnS and SiO_2, the total charge measured at 10 J/cm^2 fluence was about 1 μC, or 6.2×10^{12} electrons . We see that this figure is about 10^{-4} of $N_e \mathscr{A}_b$, which we calculated in Table 4A.3. Siekhaus et al. found the emitted charge increased strongly with laser target fluence Φ. They found

$$Q/Q_0 = (\Phi/\Phi_0)^m \tag{122}$$

with $m = 9$ for SiO_2, suggesting that the factor-of-3 more irradiance would have achieved the emission we calculate for the von Gutfeld and Dreyfus experiments on copper at $\Phi = 6$ J/cm^2. Unfortunately, Siekhaus and colleagues did no measurements on metals at high irradiance. Space charge limitation may also have affected their results.

4A.5. SUMMARY AND CONCLUSIONS

Laser ablation and plasma formation are discussed in the high-irradiance regime, as it applies to LIMA work. Four regions of the laser–target interaction are visualized to occur as one moves away from the target, mainly distinguished by progressively longer collision times.

Four conditions were listed and explained, all of which must be satisfied for the high-irradiance analysis to apply. These included irradiance above the plasma formation threshold, a vacuum environment, surface absorption, and sufficiently small targets. The concept of local thermodynamic equilibrium (LTE) was discussed as it applies to the physics of pulsed laser–plasma interaction.

A straightforward analysis based on plasma physics was shown to give the expected mass ablation rate and total ablation depth. We reviewed experimental data on ablation rate, ablation depth, and ion velocity from a wide range of lasers and target materials and found that the analysis presented in this chapter provided a very convenient way of organizing such data.

We further found that vacuum laser ablation rate and ablation depth for many materials that are surface absorbers could be predicted quite well over a very broad range of laser parameters.

In addition, we found that many ion experiments could be predicted quantitatively if the velocity given by LTE plasma physics theory were enhanced by a constant factor $\sqrt{g} \approx \sqrt{10}$, derived from measurements of ion flux at a distance from the target in one instance.

Finally, we reviewed a number of measurement techniques for the low-density plasma plume created in a laser ablation experiment, including PPTD, Langmuir probes, Faraday cages, laser microprobes, TOF, and laser-induced fluorescence (LIF), and summarized results obtained with them.

ACKNOWLEDGMENTS

The authors gratefully acknowledge very helpful discussions with Prof. E. Matthias of the Free University, Berlin, and the travel support of NATO Research Grant SA.5-2-05(CRG.880020)1133/90/AHJ-514.

REFERENCES

Afanas'ev, Yu. V., and Kanavin, A. P. (1983). *Kvantovaya Elektron.* (*Moscow*) **11**, 423–426; *Sov. J. Quantum Electron.* (*Engl. Transl.*) **14**, 292–294 (1984).

Afanas'ev, Yu. V., Basov, N. G., Krokhin, O. N., Morachevskii, N. V., and Sklizkov,

G. V. (1969). *Zh. Tekh. Fiz.* **39**, 894–905; *Sov. Phys.—Tech. Phys.* (*Engl. Transl.*) **14**, 669–676 (1969).

Afanas'ev, Yu. V., But, S. M., and Kanavin, A. P. (1988). *Kvantovaya Elektron.* (*Moscow*) **15**, 744–746; *Sov. J. Quantum Electron.* (*Engl. Trans.*) **18**, 474–475 (1988).

Akhsakhalyan, A. D., Bityurin, Yu. A., Gaponov, S. V., Gudkov, A. A., and Luchin, V. I. (1982). *Zh. Tekh. Fiz.* **52**, 1584–1589; *Sov. Phys.—Tech. Phys.* (*Engl. Transl.*) **27**, 969–973 (1982).

Allen, C. W. (1973). *Astrophysical Quantities*, 3rd ed., p. 33. Athlone Press, London.

Balazs, L., Gijbels, R., and Vertes, A. (1991). *Anal. Chem.* **63**, 314–320.

Basov, N. G., Gribkov, V. A., Krokhin, O. N., and Sklizkov, G. V., (1968). *Zh. Eksp. Teor. Fiz.* **54**, 1073–1080; *Sov. Phys.—JETP* (*Engl. Transl.*) **27**, 575–582.

Batanov, V. A., Bunkin, F. V., Prokhorov, A. M., and Fedorov, V. B. (1972). *Zh. Eksp. Teor. Fiz.* **63**, 586–608; *Sov. Phys.—JETP* (*Engl. Transl.*) **36**, 311–322 (1973).

Berchenko, E. A., Koshkin, A. V., Sobolev, A. P., and Fedyushin, B. T. (1981). *Kvantovaya Elektron.* (*Moscow*) **8**, 1582–1584; *Sov. J. Quantum Electron.* (*Engl. Trans.*) **11**, 953–955 (1981).

Bergmann, E. E., McLellan, E. J., and Webb, J. A. (1980). *Appl. Phys. Lett.* **37**, 18–19.

Beverley, R. E., III, and Walters, C. T. (1976). *J. Appl. Phys.* **47**, 3485–3495.

Bingham, R. A., and Salter, P. L. (1976a). *Int. J. Mass Spectrom. Ion Phys.* **21**, 133–140.

Bingham, R. A., and Salter, P. L. (1976b). *Anal. Chem.* **48**, 1735–1750.

Boland, B. C., Irons, F. E., and McWhirter, R. W. P. (1968). *J. Phys.* **B1**, 1180–1191.

Bonnie, J. H. M., Eenshuistra, P. J., and Hopman, H. J. (1987). *AIP Conf. Proc.* **158**, 133–142.

Book, D. L. (1989). In *AIP 50th Anniversary Physics Vade Mecum* (H. Anderson, ed.), p. 278. Am. Inst. Phys., New York.

Bykovskii, Yu. A., Dudoladov, A. G., Degtyarenko, N. N., Elesin, V. F., Kozyrev, Yu. P., and Nikolaev, I. N. (1969). *Zh. Eksp. Teor. Fiz.* **56**, 1819–1822; *Sov. Phys.—JETP* (*Engl. Transl.*) **29**, 977–978 (1969).

Bykovskii, Yu. A., Degtyarenko N. N., Elesin, V. F., Kozyrev, Yu. P., and Sil'nov, S. M. (1971). *Zh. Eksp. Teor. Fiz.* **60**, 1306–1319; *Sov. Phys.—JETP* (*Engl. Transl.*) **33**, 706–712.

Bykovskii, Yu. A., Vasil'ev, N. N., Degtyarenko, N. N., Elesin, V. F., Laptev, I. D., and Nevolin V. N. (1972). *Zh. Eksp. Teor. Fiz., Pis'ma. Red.* **15**, 308–311; *JETP Lett.* (*Engl. Transl.*) **15**, 217–220 (1972).

Carslaw, H. S., and Jaeger, J. C. (1959). *Conduction of Heat in Solids*, 2nd ed., p. 75. Oxford University Press (Clarendon), New York.

Caruso, A., Bertotti, B., and Giupponi, P. (1966). *Nuovo Cimento* **45**, 176–189.

Chen, F. F. (1965). In *Plasma Diagnostic Techniques* (R. H. Huddlestone and S. L. Leonard, eds.) Chapter 4, pp. 113–199. Academic Press, New York.

Chen, G., and Yeung, E. S. (1988). *Anal. Chem.* **60**, 864–868.

Chuang, T. J., Hiraoka, H., and Mödl, A. (1988). *Appl. Phys.* **A45**, 277–288.

Coles, J. N. (1976). *Surf. Sci.* **55**, 721–724.

Conzemius, R. J., and Capellen, J. M. (1980). *Int. J. Mass Spectrom. Ion Phys.* **34**, 197–271.

Cook, G., and Dyer, P. E. (1983). *J. Phys.* **D16**, 889–896.

David, C. D., and Weichel, H. (1969). *J. Appl. Phys.* **40**, 3764–3770.

Denisov, N. G. (1957). *Sov. Phys.—JETP* (*Engl. Transl.*) **4**, 544–550.

Dreyfus, R. W. (1991). *J. Appl. Phys.* **69**, 1721–1729.

Dreyfus, R. W., Kelly, R., and Walkup, R. E. (1986a). *Appl. Phys. Lett.* **49**, 1478–1480.

Dreyfus, R. W., Walkup, R. E., and Kelly, R. (1986b). *Radiat. Eff.* **99**, 199–211.

Dreyfus, R. W., Kelly, R., Walkup, R. E., and Srinivasan, R. (1986c). *Proc. Soc. Photo-opt. Instrum. Engi.* **710**, 46–54.

Dreyfus, R. W., McDonald, F. A., and von Gutfeld, R. J. (1987). *J. Vac. Sci. Technol.* **B5**, 1521–1527.

Duzy, C., Woodroffe, J. A., Hsia, J. C., and Ballantyne, A. (1980). *Appl. Phys. Lett.* **37**, 542–544.

Eidmann, K., Amiranoff, F., Fedosejevs, R., Maaswinkel, A. G. M., Petsch, R., Sigel, R., Spindler, G., Teng, Y., Tsakiris, G., and Witkowski, S. (1984). *Phys. Rev.* **A30**, 2568–2589.

Eliezer, S., and Hora, H. (1989). *Phys. Rep.* **172**, 339–458.

Eloy, J. F. (1974). *Bull. Inf. Sci. Tech., Commis. Energ. At. (Fr.)* **192**, 71–75.

Fabbro, R., Fournier, J., Ballard, P., Devaux, D., and Virmont, J. (1990). *J. Appl. Phys.* **68**, 775–784.

Figueira, J. F., Czuchlewski, S. J., Phipps, C. R., and Thomas, S. J. (1981). *Appl. Opt.* **20**, 838–841.

Forslund, D. W., Kindel, J. M., and Lee, K. (1977). *Phys. Rev. Lett.* **39**, 284–290.

Gaunt, J. A. (1930). *Proc. R. Soc. London* **A126**, 654–690.

Gill, D. H., and Dougal, A. A. (1965). *Phys. Rev. Lett.* **15**, 845–849.

Ginzburg, V. L. (1970). *The Propagation of Electromagnetic Waves in Plasmas*, 2nd ed., p. 36. Pergamon, New York.

Glickler, S. L., Shraiman, B. J., Woodroffe, J. A., and Smith, M. J. (1980). *Am. Inst. Aeronaut. Astronaut., 13th Fluid Plasma Dyn. Conf.*, Pap. 80–1320.

Golub', A. P., Nemchinov, I. V., Petrukhin, A. I., Pleshanov, Yu. E., and Rybakov, V. A. (1981). *Zh. Tekh. Fiz.* **51**, 316–323; *Sov. Phys.—Tech. Phys.* (*Engl. Transl.*) **26**, 191–196 (1981).

Gregg, D. W., and Thomas, S. J. (1966a). *J. Appl. Phys.* **37**, 2787–2789.

Gregg, D. W., and Thomas, S. J. (1966b). *J. Appl. Phys.* **37**, 4313–4316.

Grey Morgan, C. (1978). *Sci. Prog.* (*Oxford*) **65**, 31–50.

Grun, J., Obenschain, S. P., Ripin, B. H., Whitlock, R. R., McLean, E. A., Gardner, J., Herbst, M. J., and Stamper, J. A. (1983). *Phys. Fluids* **26**, 588–597.

Gupta, P. D., and Kumbhare, S. R. (1984). *J. Appl. Phys.* **55**, 120–124.

Gupta, P. D., Tsui, Y. Y., Popil, R., Fedosejevs, R., and Offenberger, A. A. (1986). *Phys. Rev.* **A33**, 3531–3534.

Haines, M. G. (1979). *Proc. Scott. Univ. Summer Sch. Phys.* **20**, 145–218.

Hauer, A. A., Forslund, D. W., McKinstrie, C. J., Wark, J. S., Hargis, P. J., Hamill, R. A., and Kindel, J. M. (1989). In *Laser Induced Plasmas and Applications* (L. J. Radziemski and D. A. Cremers, eds.), pp. 385–436. Dekker, New York.

Heiman, D., Hamilton, D. S., and Hellwarth, R. W. (1978). *Phys. Rev.* **B19**, 6583–6592.

Hora, H. (1975). *J. Opt. Soc. Am.* **65**, 882–890.

Hora, H., Min, G., Eliezer, S., Lalousis, P., Pease, R. S., and Szichman, H. (1989). *IEEE Trans. Plasma Sci.* **17**, 284–289.

Hughes, T. P. (1979). *Proc. Scott. Univ. Summer Sch. Phys.* **20**, 1–90.

Huie, C. W., and Yeung, E. S. (1985). *Spectrochim. Acta* **40B**, 1255–1258.

Inoue, N., Kawasumi, Y., and Miyamoto, K. (1970). *Plasma Phys.* **13**, 84–87.

Johann, J., Luk, T. S., Egger, H., and Rhodes, C. K. (1986). *Phys. Rev.* **A34**, 1084–1102.

Kelly, R., and Dreyfus, R. W. (1988a). *Nucl. Instrum. Methods* **B32**, 341–348.

Kelly, R., and Dreyfus, R. W. (1988b). *Surf. Sci.* **198**, 263–276.

Kelly, R., and Rothenberg, J. E. (1985). *Nucl. Instrum. Methods* **B7**, 755–763.

Kelly, R., Cuomo, J. J., Leary, P. A., Rothenberg, J. E., Braren, B. E., and Aliotta, C. F. (1985). *Nucl. Instrum. Methods* **B9**, 329–340.

Kidder, R. E. (1968). *Nucl. Fusion* **8**, 3–12.

Kidder, R. E. (1971). In *Proceedings of the International School of Physics, Course 48* (P. Caldirola and H. Knoepfel, eds.), pp. 306–352. Academic Press, New York.

Koppman, R., Refaei, S. M., and Pospieszczyk, A. (1986). *J. Vac. Sci. Technol.* **A4**, 79–85.

Kovalev, I. D., Maksimov, G. A., Suchkov, A. I., and Larin, N. V. (1978). *Int. J. Mass Spectrom. Ion Phys.* **27**, 101–137.

Krokhin, O. N. (1971). In *Proceedings of the International School of Physics, Course 48* (P. Caldirola and H. Knoepfel, eds.), pp. 278–305. Academic Press, New York.

Küper, S., and Stuke, M. (1989). *Appl. Phys. Lett.* **54**, 4–6.

Landau, L. D., and Lifschitz, E. M. (1958). *Statistical Physics*, p. 13. Pergamon, London.

Langer, P., Tonon, G., Floux, F., and Ducauze, A. (1966). *IEEE J. Quantum Electron.* **QE-2**, 499–506.

Linlor, W. I. (1963). *Appl. Phys. Lett.* **3**, 210–213.

Manheimer, W. M., and Colombant, D. G. (1982). *Phys. Fluids* **25**, 1644–1652.

Max, C. E., McKee, C. F., and Mead, W. C. (1980). *Phys. Rev. Lett.* **45**, 28–31.

Meyer, R. T., Lynch, A. W., and Freese, J. M. (1973). *J. Phys. Chem.* **77**, 1083–1092.

Montgomery, D. C., and Tidman, D. A. (1964). *Plasma Kinetic Theory*, pp. 33–37. McGraw-Hill, New York.

Mora, P. (1982). *Phys. Fluids* **25**, 1051–1056.

Mulser, P. (1971). *Plasma Phys.* **13**, 1007–1012.

Mulser, P., Sigel, R., and Witkowski, S. (1973). *Phys. Rep.* **6**, 187–239.

Nemchinov, I. V. (1967). *Prikl. Mat. Mekh.* **31**, 300–319; *J. Appl. Math. Mech.* (*Engl. Transl.*) **31**, 320–338 (1967).

Nogar, N. S. (1991). *Workshop Laser Albation Mech. Appl., Oak Ridge, Tennessee, 1991.* **389**, 3–11.

Olsen, N., Kuswa, G. W., and Jones, E. D. (1973). *J. Appl. Phys.* **44**, 2275–2283.

Otis, C. E., and Dreyfus, R. W. (1991). *Phys. Rev. Lett.* **67**, 2102–2105.

Pappas, D., Sanger, K., Cuomo, J., and Dreyfus, R. (1992). *J. Appl. Phys.* **72,** 3966–3970.

Pearlman, J. S., and Anthes, J. P. (1975). *Appl. Phys. Lett.* **27**, 581–585.

Pearlman, J. S., and Dahlbacka, G. H. (1977). *Appl. Phys. Lett.* **31**, 414–417.

Phipps, C. R. (1989). *Mechanical Effects Induced by Laser*, Gen. Lect. I, Euromech. 257 Conf. University of Aix-Marseille (unpublished).

Phipps, C. R., and Dreyfus, R. W. (1992). *Bull. Am. Phys. Soc.* [2] **37**, 83.

Phipps, C. R., Turner, T. P., Harrison, R. F., York, G. W., Osborne, W. Z., Anderson, G. K., Corlis, X. F., Haynes, L. C., Steele, H. S., and Spicochi, K. C. (1988). *J. Appl. Phys.* **64**, 1083–1096.

Phipps, C. R., Harrison, R. F., Shimada, T., York, G. W., Turner, T. P., Corlis, X. F., Steele, H. S., Haynes, L. C., and King, T. R. (1990). *Laser Part. Beams* **8**, 281–295.

Puell, H. (1970). *Z. Naturoforsch.* **25**, 1807–1815.

Raizer, Yu. P. (1977). *Laser-Induced Discharge Phenomena*, p. 9. Consultants Bureau, New York.

Ramendik, G. I., Manzon, B. M., and Tyurin, D. A. (1987). *Talanta* **34**, 61–67.

Rose, D. J., and Clark, M. (1961). *Plasmas and Controlled Fusion*, p. 67. Wiley, New York.

Rosen, D. J., Mitteldorf, J., Kothandaraman, G., Pirri, A. N., and Pugh, E. R. (1982a). *J. Appl. Phys.* **53**, 3190–3200.

Rosen, D. J., Hastings, D. E., and Weyl, G. M. (1982b). *J. Appl. Phys.* **53**, 5882–5890.

Rosen, D. J., Rollins, C. R., and Chen, J. (1988). Report SR-334. "Issues in Laser Propulsion." Physical Science, Inc., Andover, Mass. (unpublished).

Rothenberg, J. E., and Kelly, R. (1984). *Nucl. Instrum. Methods* **229**(B1), 291–300.

Shirsat, T. S., Parab, H. D., and Pant, H. C. (1989). *Laser Part. Beams* **7**, 795–805.

Siekhaus, W. J., Kinney, J. H., Milam, D., and Chase, L. L. (1986). *Appl. Phys.* **A39**, 163–166.

Silfvast, W. T., and Szeto, L. H. (1977). *Appl. Phys. Lett.* **31**, 726–728.

Smith, W. L. (1978). *Opt. Eng.* **17**, 489–503.

Spitzer, L. (1967). *Physics of Fully ionized Gases*, 2nd ed., pp. 120–153. Wiley (Interscience), New York.

Spitzer, L., and Härm, R. (1953). *Phys. Rev.* **89**, 977–982.

Srinivasan, R., and Dreyfus, R. W. (1985). *Laser Spectrosc.* **7**, 396–400.

Srinivasan, R., Braren, B., and Dreyfus, R. W. (1986a). *J. Appl. Phys.* **61**, 372–376.

Srinivasan, R., Braren, B., Dreyfus, R. W., Hadel, L., and Seeger, D. E. (1986b). *J. Opt. Soc. Am.* **B7**, 785–791.

Srinivasan, R., Sutcliffe, E., and Braren, B. (1987). *Appl. Phys. Lett.* **51**, 1285–1287.

Steenhoek, L. E., and Yeung, E. S. (1981). *Anal. Chem.* **53**, 528–532.

Sutcliffe, E., and Srinivasan, R. (1986). *J. Appl. Phys.* **60**, 3315–3322.

Taylor, R. S., Singleton, D. L., and Paraskevopoulos, G. (1987). *Appl. Phys. Lett.* **50**, 1779–1781.

Ursu, I., Apostol, I., Barbulescu, D., Mihailescu, I. N., and Moldovan, M. (1981). *Opt. Commun.* **39**, 180–184.

Venugopalan, M., ed. (1971). *Reactions under Plasma Conditions*, Vol. 1, 367–542. Wiley, New York.

Vertes, A., and Juhasz, P. (1986). *Int. J. Mass Spectrom. Ion Process* **94**, 63–85.

Vertes, A., Juhasz, P., Jani, P., and Czitrovszky, A. (1988). *Int. J. Mass Spectrom. Ion Processes* **83**, 45–70.

Vertes, A., DeWolf, M., Juhasz, P., and Gijbels, R. (1989). *Anal. Chem.* **61**, 1029–1035.

Vertes, A., Gijbels, R., and Adams, F. (1990). *Mass Spectrom. Rev.* **9**, 71–113.

Viswanathan, R., and Hussla, I. (1986). *J. Opt. Soc. Am.* **B3**, 796–800.

von Gutfeld, R. J., and Dreyfus, R. J. (1989). *Appl. Phys. Lett.* **54**, 1212–1213.

Weyl, G. (1989). In *Laser Induced Plasmas and Applications* (L. J. Radziemski and D. A. Cremers, eds.), p. 25. Dekker, New York.

Yamanaka, C., Nakai, S., Yamanaka, T., Izawa, Y., Mima, K., Nishihara, K., Kato, Y., Mochizuki, T., Yamanaka, M., Nakatskuka, M., and Yabe, T. (1986). In *Laser Interaction and Related Phenomena* (H. Hora and G. Miley, eds.), pp. 395–419. Plenum, New York.

Yappert, M. C., Kimbrell, S. M., and Yeung, E. S. (1987). *Appl. Opt.* **26**, 3536–3541.

Zare, R. N. (1984). *Science* **226**, 298–303.

CHAPTER

4

THE HIGH LASER IRRADIANCE REGIME

B. SOLID SAMPLING FOR ANALYSIS BY LASER ABLATION

LIESELOTTE MOENKE-BLANKENBURG

Department of Chemistry
Martin-Luther-University Halle-Wittenberg
Halle, Germany

4B.1. INTRODUTION

Laser ablation (LA) offers significant potential for the direct analysis of solids.

There has been considerable interest in recent years in reducing sample preparation time by the use of solid sampling techniques. These techniques have additional advantages when the sample is difficult, hazardous, or tedious to digest and when errors caused by contamination or losses may arise during dissolution of the solid. The principle is simple: When a laser beam of sufficient power density hits a solid surface, material is ejected into the gas phase. In this way the laser ablates and vaporizes localized sections of solids. The vapor contains significant populations of excited and ionized atoms, which suggests the use of a laser source for elemental analysis by atomic emission spectrometry (AES) and mass spectrometry (MS). The first use of a solid-state laser for laser micro mass spectrography was reported by Honig and Woolston as early as 1963. Since the demands made on both the performance of the laser and on the focusing optics were similar to those in laser micro emission analysis, it was natural to supplement a laser microscope with an ion source and to couple it with a suitable mass spectrograph. The starting point for these experiments was the interest in extending local elemental analysis in the micro region to the determination of isotopes in order, among other things, to open up new possibilities for tracer technique in solid-state research and in isotope analysis in geology and biology.

Laser Ionization Mass Analysis, Edited by Akos Vertes, Renaat Gijbels, and Fred Adams.
Chemical Analysis Series, Vol. 124.
ISBN 0-471-53673-3

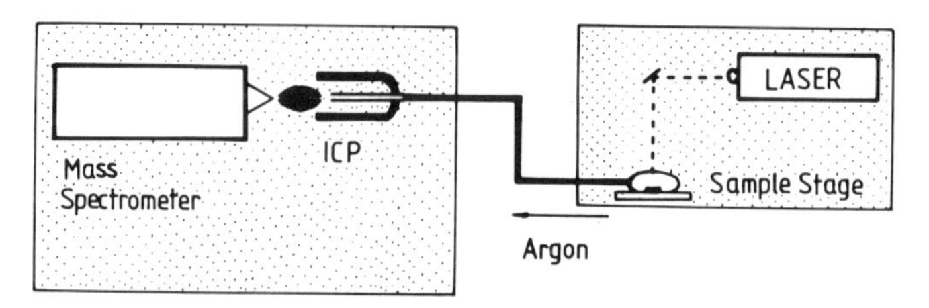

Figure 4B.1. Schematic view of the most commonly used laser ablation sampling setup with ICP detection.

The disadvantage of a direct one-step method has been the hardly controllable laser–target interactions, which caused low reproducibility and therefore bad presumptions for quantitative analysis. To overcome this disadvantage, the idea was to insert an additional step between the laser operation and the spectrometry. The laser-ablated material could be introduced in a separate hot plasma with the aim of reaching full excitation of atoms (for AES) or full ionization of atoms (for MS).

First attempts to combine LA with inductively coupled plasma (ICP) excitation were made by Abercrombie et al. (1977a,b), Salin et al. (1979), Carr and Horlick (1980), and Thompson et al. (1981a,b) using atomic emission spectrometry. First results with the tandem technique LA-ICP-MS were obtained in the mid-1980s by Gray (1985), Houk (1986), and Arrowsmith (1987).

The principle most commonly used is shown in Figure 4B.1. Solid samples of the appropriate size are placed in a glass-capped chamber and orientated such that their surface is at the focus of the laser beam. The radiation of a solid state laser with an output energy of about 0.1–1 J is focused by an optical system on a selected region of the sample and ablates nanogram, microgram, or milligram amounts of analytical material. The carrier gas flow is passed over the sample surface and then through a tube to the ICP torch. The gas transports the majority of the sample vapor, small amounts of microscopic liquid droplets, solidified and solid particles of sizes up to about 2 μm and more.

Requirements of an ideal technique for solid sampling by LA are an analyte aerosol of low particle size and representative composition as well as an analyte transport process of high removal efficiency and with low memory effects.

4B.2. MICROPLUME GENERATION

Well-known properties of laser beams are high intensity, directionality, and monochromaticity. The main feature of interest of LA is the ability to sample, vaporize, atomize, excite, and ionize both conducting and nonconducting solids in micro and macro regions.

4B.2.1. Laser

Most frequently used lasers are solid-state types such as ruby (694 nm), Nd: glass (1064 nm), and Nd:YAG (1064 nm, frequency-doubled: 532 nm; frequency-quadrupled: 266 nm), but also gas lasers, such as CO_2 (10.6 μm) and N_2 (337 nm), dye lasers (220–740 nm), and excimer lasers [ArF (193 nm), KrCl (222 nm), and XeCl (308 nm)].

In laser microanalysis, the following values are common, giving rise to plasma temperatures between 3000 and about 25,000 K (see Table 4B.1): power densities in the range of approximately 10 MW cm^{-2} to 10 GW cm^{-2} for a free-running laser of 1 J energy; about 1 ms time duration; 10 kW power, focused to a spot diameter of about 250 and 10 μm, to approximately 1 GW·cm^{-2} to 1 TW·cm^{-2} for a *Q*-switched pulse of 0.1 J energy; 50 ns time duration of one giant pulse; 2 MW power, focused to the same spot diameter (Piepmeier and Malmstadt, 1969).

4B.2.2. Laser–Target Interaction

In times on the order of the duration of a laser pulse, the electrons, which absorb the photons, will produce many collisions, both among themselves and with lattice phonons. The energy absorbed by an electron will be distributed and passed on to the lattice. Ready (1971) therefore regards the optical energy as being turned into heat instantaneously at the point at which the light was absorbed. The intense local heating experienced by the target causes a rapid rise in the surface temperature of the material. Heat is conducted into the interior of the target and a thin molten layer forms below the surface.

Table 4B.1. Solid-State (Ruby) Laser Parameters for Laser Microanalysis

	Free-Running	*Q*-Switched Giant Pulse
Energy	1 J	0.1 J
Number of spikes	$\geqslant 100$	1
Time duration	≈ 1 ms	≈ 50 ns
Power	≈ 10 kW	≈ 2 MW
Power density		
$r \approx 125\ \mu$m	2×10^7 W·cm^{-2}	4×10^9 W·cm^{-2}
$r \approx 5\ \mu$m	10^{10} W·cm^{-2}	3×10^{12} W·cm^{-2}

As the thermal energy deposited at the surface increases, a point is reached where the deposited energy exceeds the latent heat of vaporization. When this happens, heat cannot be conducted away from the point of irradiation fast enough to prevent the surface from reaching its boiling temperature and evaporation occurs from the surface. The distribution occurs so rapidly in the time scale of *Q*-switched and free-running laser pulses that Ready (1971) assumes a local equilibrium to have been established during the pulse. This assumption may break down for the case of very short pulses.

The energy density deposited at the surface of the target by a power density F is Ft_e; hence the average energy density per unit mass acquired by the thin layer of molten metal is $Ft_e/d(at_e)^{1/2}$, where d is the mass density of the target, a the thermal diffusivity, and t_e, the duration of the laser pulse. For evaporation to occur, the energy deposited in this layer must exceed the latent heat of vaporization of the target, L_v. Thus the following threshold condition is obtained for the minimum absorbed power density ($F_{\min}$), below which no evaporation will occur:

$$F_{\min} = dL_v a^{1/2} t_e^{-1/2} \tag{1}$$

Houk (1986) described the effect of power density as follows: At low power density ($\leqslant 10^8\ \mathrm{W \cdot cm^{-2}}$) positive ions are readily observed from volatile elements of low ionization energy, for example, Na, K, and Pb. These ions are likely formed by ordinary thermal ionization in a relatively cool microplasma. The efficiency of atomization and ionization under these conditions varies widely between different elements and will also depend on the particular sample matrix investigated. Thus, low power densities are in general undesirable for quantitative element analysis.

At higher power density ($\geqslant 10^9\ \mathrm{W \cdot cm^{-2}}$) the laser transfers considerably more energy to the sample surface. The vaporization process typically leaves a crater in the sample surface. A more or less dense microplume forms from ablated material above the sample surface. The laser pulse transfers energy directly to free electrons in the microplume. The conditions in the microplume are sufficiently energetic for atomization and ionization to be quite efficient. On the other side, these ionization conditions for elemental analysis yield some multiply charged ions.

The processes generated by a low-power-density laser pulse are usually classified as *laser desorption*, whereas the high-power-density limit is referred to as *laser plasma ionization*.

Vertes et al. (1988, 1989a,b,c, 1990) developed a one-component one-dimensional hydrodynamic model to describe the expansion of laser-generated plumes in ion sources. A comprehensive study of the model covered three laser types (ruby, CO_2, and frequency-quadrupled Nd:YAG) and three classes of solid targets (metals, transparent insulators, and opaque insulators).

The full real-time characterization of a plasma is often a hopeless task, but approximate methods for plasma diagnostics and modeling have been used to obtain a better understanding of the processes involved and to increase the reliability of the analytical results (Adams and Vertes, 1990).

Beyond the power density of a laser the spiking behavior will also influence the vaporization (Moenke-Blankenburg, 1989). With the operation mode of a semi-Q-switch by saturable absorbers the number of spikes and their duration can be chosen. Fabbro et al. (1980) studied the effect of laser wavelengths on the mass ablation rate and postulated the following equation:

$$m = 110 \frac{\phi_a^{1/3}}{10^{14}} \lambda^{-4/3} \tag{2}$$

where m is the mass ablation rate (kg/s cm^2), and ϕ_a the absorbed power density. They concluded that the mass ablation rate would increase strongly at shorter wavelengths and subsequently give higher sensitivity. Mitchell et al. (1987) varied the pulse repetition rate at 1–20 Hz of a low-energy laser and concluded that the higher the repetition rate, the higher the ablation rate and the better the precision of the analysis.

4B.3. LASER MICROPROBE MASS SPECTROMETRY (LMMS)

The first laser of the first-generation apparatus used by Honig and Woolston (1963) delivered an output energy of 1 J and, with a pulse duration of 200 μs, a power of 5 kW. It was therefore suitable for ionizing all elements with an ionizing potential of $\leqslant 10$ eV in the microplasma (i.e., more than 50% of all elements of the periodic table). Trace analysis in the ppm range and isotope analysis can be performed in micro regions of solid surfaces down to 0.004 cm^2. The new method was appreciated because no special preparation of the sample was needed, nor were transformation of the sample into measurable compounds and separation of disturbing elements necessary, in contrast to the traditional analysis with ordinary thermionic sources.

In the period 1963–1972, the development of laser ion mass spectrometers for the characterization of solids proceeded simultaneously in several laboratories in different countries (Table 4B.2). Some excellent reviews, doctoral theses, and chapters of books provide a broad view of the applications of LMMS between 1970 and 1980 (see Capellen and Conzemius, 1980).

Table. 4B.2. LMMS Studies Between 1963 and 1972

	Solid-State Laser Characteristics					
References	Energy (J)	Pulse Length (μs)	Power (W)	Power Density ($W \cdot cm^{-2}$)	Beam Area (cm^2)	Analyzer[a]
Honig and Woolston (1963)	0.2–1	200	5×10^3	5×10^7	10^{-4}	DF
Giori et al. (1963)	5–13	500	2×10^4	4×10^7	5×10^{-4}	Quad
Berkowitz and Chupka (1964)	1	400	2×10^3	2×10^7	10^{-4}	SF
Lincoln (1965, 1969)	1	500	2×10^3	2×10^6	10^{-3}	TOF
Isenor (1964, 1965)		0.02	3×10^7	8×10^{14}	3×10^{-4}	TOF
Knox (1968)	0.1	200		10^6		TOF
Knox and Vastola (1966)	5	1000		10^7		TOF
Bernal et al. (1966)	0.1	0.03	3×10^6	6×10^7	5×10^{-2}	TOF
Eloy and Dumas (1966); Eloy (1969)	$\leqslant 0.1$	5	2×10^4			SF
Fenner and Daly (1966, 1968)	0.01	0.03	3×10^5	2×10^{10}	2×10^{-5}	TOF
Langer et al. (1966, 1968)	1	0.03	3×10^7	10^{11}	3×10^{-4}	TOF
Namba et al. (1966)	0.3	300	10^3	10^6	10^{-3}	TOF
Bykovskii et al. (1969, 1970)	$\leqslant 0.05$	0.05	10^7		10^{-3}	TOF

[a] DF = double focusing; Quad = quadrupole; SF = single focusing; TOF = time-of-flight.

The second generation of laser microprobe mass analyzers were developed from the mid-1970s to the beginning of the 1980s. All LMMS instruments commercially available are based on essentially identical functional principles. They use short laser pulses of about 10 ns duration at wavelengths in the far-UV, typically the quadrupled wavelength of Q-switched Nd:YAG lasers at 265 nm. Optical microscopes are used for sample observation and focusing of the laser beam with a spatial resolution of approximately 0.5 μm in the analysis of thin specimens and about 3 μm for surface analysis of compact solids. Typical irradiances in the laser focus range from 10^7 to 10^{11} $W \cdot cm^{-2}$.

Substantial effort has been directed toward understanding high-temperature phenomena. A particularly challenging problem has been characterization of those chemical and physical processes that occur at elevated temperatures and are not predictable by extrapolation from lower-temperature data. In recent years the study of high-temperature processes has been facilitated by developments in instrumentation and measurement techniques (Hartford, 1984).

The first two commercial instruments were distinguished by two types of processes: (1) photon absorption by *transmission mode*, and (2) photon absorption by *reflection mode*. In the first laser interaction configuration, the laser focusing beam induces an absorption zone behind the solid thin layer. The laser-generated microplume emerges from the opposite side of the thin samples. In the second case, the focused laser beam induces an absorption zone and plasma creation before the solid surface as regards the laser incidence axis. Thus the expanding plasma spreads in the direction of the focusing lens. Among the main advantages of a laser transmission mode system is the fact that it is possible to reduce the distance from lens to sample to a minimum. In effect, the thin film of the specimen is located in an evacuated specimen chamber directly underneath a thin cover slide made of quartz that serves as both an optical window for the microscope and as a vacuum seal. An additional advantage lies in the fact the plasma expansion is possible only when the sample, facing the analyzer, is punctured by the laser beam. This gives a better definition of the zero point of the time of flight. In this configuration the smaller laser spot sizes are $\leqslant 0.8$ μm in diameter and 2–5 μm in depth vis-à-vis the sample.

Among the main advantages of laser reflection mode systems are the possibility of studying the bulky solid materials directly without preparation, the fact that the analytical useful laser power density is more easily defined when one is considering varied material, and the fact that the mass spectra are simplified because of less contamination by organic fragments, polyatomic ions, and doubly charged ions. The interaction time can be reduced to reach a volatilized material thickness limit of 0.005 μm.

Bibliographies of applications by Kaufmann (1986) and Moenke-Blankenburg (1989, pp. 236–257) are available, as are reports of three international workshops (Proceedings of the 1st, 2nd, and 3rd LAMMA Symposia, 1980, 1984, and 1986).

Quantitative analysis still has some limitations. Since the precision of every analytical device depends on the reproducibility of the processes involved, and since the interaction of laser light with a specimen is a highly nonlinear process, the system is rather sensitive to statistical fluctuations of the laser parameters and to inhomogeneity of the specimen. Quantitative analysis is possible only if the laser output is stabilized as far as possible and the specimen is relatively homogeneous. Relative standard deviation in such cases could be smaller than 5% at higher concentrations and smaller than 15% at low concentration levels.

4B.4. LASER ABLATION–INDUCTIVELY COUPLED PLASMA–MASS SPECTROMETRY (LA-ICP-MS)

Gray (1985) first reported the development and application of a laser sampling system for ICP-MS analysis. His paper clearly demonstrated the feasibility of ruby laser sampling for ICP-MS. His work is distinct from that extensively reported using a laser beam both to extract sample atoms from a solid and to ionize them, which has led to LMMS instruments described earlier in this volume (Chapters 2 and 3). The essential differences are that the sample remains at atmospheric pressure and ionization is performed separately from the laser ablation in a second step in the ICP, thus permitting separate optimization of the two successive processes.

In this way the ablated material is carried to the plasma in the argon stream joining the central jet of the ICP torch. Although the material cloud is introduced into the plasma for a period of 5–10 s, the rapid scan ability of a quadrupole mass spectrometer and multichannel scaler enables scans of 0.1 s duration to be made over the full mass range of 0–250 amu so that many scans are run during the transient sample presence in the plasma. Whereas LMMS uses a small focused beam for microanalysis, the work described by Gray (1985) has used a larger diameter of about 1 mm, which more approaches bulk analysis, although the reduction of the focus diameter would also provide spatial information.

Arrowsmith (1987) reported the first application of LA-ICP-MS using a high-repetition Nd:YAG laser, which opened the way to practical steady-state signal measurement (the laser may be used either in a single-pulse mode to give a transient signal or at a 10 Hz repetition rate, resulting in a continuous signal). The continuous signal may be maintained constant over long periods by transfer of the sample material, resulting in improved precision and lowered detection limits. The precision can be about 5%, and detection limits can be in the ppm range.

Table 4B.3. Technical Parameters of Two Commercially Available LA-ICP-MS Instruments

	PE ELAN 5000 Model 320 Laser Sampler	VG PlasmaQuad Laserlab
ICP spectrometer:		
Frequency	40 MHz	27.12 MHz
Frequency stability		0.05%
Power	1.2 kW	1.3–2.5 kW
Plasma gas	12 L/min	13 L/min
Auxiliary gas flow	1 L/min	0.5 L/min
Carrier gas flow	0.85 L/min	1 L/min
Sample uptake	0.7 mL/min	
Sampler/load coil separation	17 mm	
Sampling orifice	1.1 mm	
Skimmer orifice	0.9 mm	
Mass range		4–245 amu
Scanning speed		2500 amu/s
Dwell time per channel		160–320 μs
No. of channels		2048–4096
No. of sweeps per acquisition		320
Laser:		
Type	Nd:YAG	Nd:YAG
Wavelength	1064 nm	1064 nm
Output energy, free-running	0.5 J	0.5 J
Output energy, *Q*-switched	0.32 J	0.2 J
Pulse length, free-running	140 ms	140 ms
Pulse length, *Q*-switched	2.5 and 8 ns	8–10 ns
Pulse repetition rate	1–15 Hz	1–15 Hz
Aquisition time		65 s
Laser spot size		10–300 μm
Sampling cell:		
Construction	Glass	Glass/quartz
Maximum area sampled	11.5 × 11.5 cm	0.3–1 cm^2
Sampling stage	*xy*: 11.5 cm; *z*: 1.2 cm	
Sample aerosol transport	1–10 m	
Viewing optics:		
Sample viewing	Video camera, fiber-optic illumination of the sample, 25–250 × magnification	

Table 4B.4. Applications in Mineralogy and Petrography

Elements	Matrices	Object of the Analysis	Ranges of Concentration	Limits of Detection	Reproducibility (RSD)	Refs.
66 elements: Ag to Zr	USGS-GXR-1 to 6	Semiquantitative survey analysis	Major, minor, and trace elements		3–44%	Broadhead et al. (1990)
^{200}Hg	BGS granite GNI	Concentration and precision on heavy trace elements		10–29 μg/g	49%	Gray (1985)
^{202}Hg					51	
^{203}Tl					14	
^{205}Th					14	
^{208}Pb					10	
^{209}Bi					28	
^{232}Th					24	
^{235}U					55	
^{238}U					28	
Li	Basalt JBI	Quantitative analysis of original and powdered rocks and mineral inclusions			6.6–36%	Imai (1990)
V					4.7–29	
Cr					12.5–50	
Co					6.5–29	
Ni					16–29	
Cu					16–45	
Zn					3.7–31	
Rb					5.2–26	
Sr					1.9–27	
Sn					8.9–37	
Ba					3.5–28	
La					4.3–31	
Ce					5.7–29	
Pr					3.4–31	
Pb					2.4–22	
Th					1.7–28	
U					6.3–34	
Li to U	Basalt JB3	Powdered rock			5–13% 5–27	

^{139}La	Basalt JB3		9.1 ppm	—	—	Mochizuki et al. (1988)
^{140}Ce			20.5	0.75 ppm	3.2%	
^{141}Pr			3.2	0.09	2.4	
^{146}Nd			16.6	0.86	3.8	
^{152}Sm			4.3	0.33	3.7	
^{153}Eu			1.3	0.07	3.7	
^{158}Gd			4.6	0.21	4.2	
^{159}Tb			0.8	0.02	5.8	
^{163}Dy			4.4	0.57	5.9	
^{165}Ho			0.8	0.09	6.8	
^{166}Er			2.5	0.16	10.6	
^{169}Tm			0.5	0.02	8.3	
^{172}Yb			2.4	0.24	6.4	
^{175}Lu			0.4	0.06	6.4	
^{232}Th			1.3	0.11	4.2	
^{238}U			0.5	0.09	10.0	
Li	Chromite	Semiquantitative analysis	6–720 ppm	sub-ppm to ppb		Technical information (VG ELEMENTAL)
Mg	US1		330–65000			
Sc	Pyrite		1–184			
Ti	US2		20–22000			
Zn	Zircon		13–11000			
As	US3		3–2600			
Sr	Tourmaline		14–1100			
Y	US4		2–2700			
Nb	Chlorite		0.3–2000			
Ag	US5		0.1–115			
Sn	Scheelite		2.6–440			
Nd	US6		2.1–2200			
Yb	Silica		0.4–680			
W	US7		4.5–1100			
Au			0.1–100			
Pb			26–11000			
U			0.1–174			

[a] RSD = relative standard deviation.

Table 4B.5. Applications in Metallurgy and Related Fields

Elements	Matrices	Object of the Analysis	Ranges of Concentration	Limits of Detection	Reproducibility (RSD)	Refs.
^{75}As	Steel, NBS microprobe SRMs 661–665	Detection limits and response factors to ^{60}Ni	Trace	2 μg/g	~5%	Arrowsmith (1987)
^{65}Cu				2		
^{123}Sb				2		
^{96}Mo				0.9		
^{90}Zr				0.3		
^{52}Cr				0.3		
^{93}Nb				0.2		
^{75}As	Copper, Outokumpu Oy, Finland	Detection limits and response factors to ^{65}Cu	Trace	6 μg/g	~5%	Arrowsmith (1987)
^{121}Sb				4		
^{130}Te				0.3		
^{107}Ag				0.3		
^{55}Mn				0.2		
^{114}Cd				0.2		
^{206}Pb				0.05		
^{209}Bi				0.2		
^{11}B	Steel standard		0.003%	5.9 ppm	8.3%	Mochizuki et al. (1988)
^{27}Al			0.024	10	3.5	

^{48}Ti	JSS169 and JSS173	0.013	6.4	5.5	
^{51}V		0.035	17	2.1	
^{52}Cr		0.094	18	4.4	
^{59}Co		0.030	5.3	0.8	
^{60}Ni		0.038	15	4.8	
^{75}As		0.005	9.9	15.8	
^{90}Zr		0.019	29	15.6	
^{93}Nb		0.032	16	2.6	
^{98}Mo		0.067	11	5.4	
^{120}Sn		0.011	7.4	8.7	
^{121}Sb		0.005	1.7	2.3	
Fe	F1-A (high-purity Au and Au/Ag samples)		15 ppb	3.8%	Technical information (VG)
Ni			28	3.6	
Cu			32	3.02	
Zn			56	3.12	
Pd			42	3.6	
Mg			29	4.35	
Pt			26	2.56	
Pb			7	7.03	
Bi			4	7.34	

Table 4B.6. Applications in the Field of Refractory Materials

Elements	Matrices	Object of the Analysis	Ranges of Concentration	Limits of Detection	Reproducibility (RSD)	Refs.
Fe Ni Zr Ba Nd Hf W Th U	Al_2O_3–TiC–ZrO_2 ceramics	Semiquantitative analysis in comparison to XRF	0.1 ppm to 1%	0.1–1 ppm	~5%	Arrowsmith (1989)
B Co Sr Pb Th U	Glass matrix NIST 612		ppm range	0.3 ppm 0.02 0.006 0.02 0.006 0.006		Denoyer et al. (1991)
Pb Rb Ag Sr Th U	Glass matrix NBS 612	Application report	38.5 ppm 31.5 22.0 78.4 37.8 37.4	10s of ppb	9.03% 7.9 5.2 5.3 7.7 5	Technical information (VG)
^{11}B ^{59}Co ^{88}Sr	Soda-lime glass	Instrumentation		0.3 ppm 0.02 0.006	2–10%	Denoyer et al. (1991)

^{208}Pb			0.02		
^{232}Th			0.006		
^{238}U					
^{27}Al	Fluorophos-	Quantitative		1.24%	Moenke-
^{139}La	phate glass	analysis;		3.97	Blankenburg
^{40}Ca		comparison		0.95	et al. (1992)
^{86}Sr		of ICP-AES		0.36	
^{24}Mg		LM-ICP-AES;		2.48	
^{19}F		ICP-MS and		5.43	
^{31}P		LA-ICP-MS		0.63	
Rb	NBS 612	Semiquanti-	25 ppb	7.9%	Tye et al. (1989)
Sr	glass	tative	21	3.34	
Ag		analysis	30	6.14	
Pb			8	9.03	
Th			10	7.69	
U			5	5.93	
^{7}Li	U_3O_8		0.8 μg/g	4.4%	Tye (1987)
^{48}Ti	powder		0.13	7.9	
^{52}Cr			0.08	7.5	
^{56}Fe			0.15	6.8	
^{58}Ni			0.06	7.3	
^{59}Co			0.15	7.5	
^{88}Sr			0.05	8.8	
^{90}Zr			0.05	9.1	
^{93}Nb			0.15	8.6	
^{98}Mo			0.15	5.0	
^{133}Cs			0.02	8.1	
^{142}Nd			0.05	7.6	
^{151}Eu					

Table 4B.7. Applications in the Field of Environment, Biology, and Medicine

Elements	Matrices	Object of the Analysis	Ranges of Concentration	Limits of Detection	Reproducibility (RSD)	Refs.
Na, Mg, Si, Ca, Cr, Mn, Fe, Ni, Cu, Zn, As, Se, Rb, Cd, Pb, Th, U	NIST SRM 1633a coal fly ash	Sampled loose powder directly; pressed pellets with or without a binder	0.5–70 μg/g	0.001–0.1 μg/g		Denoyer and Fredeen (1991)
Cd, Cu, Pb, Mn, Ni, Tl, Th, U, Zn, Fe, Sb, As, Co, La, Sc	NIST SRM 1645 river sediment	Stabilized in an adhesive formulation				
Be, Cr, Mn, Co, Ni, Cu, Zn, As, Cd, Sb, Pb	NRC Mess 1 marine sediment	Prepared alkali fusion disk				
B	NBS SRM 1571 orchard leaves		53 ± 7 μg/g			Gray (1989)
Na			174 ± 29			
Mg			6284 ± 725			
Al			494 ± 17			
Si			472 ± 142			
P			2978 ± 127			

S	3675 ± 535
Cl	655 ± 84
K	16487 ± 1257
Ca	23951 ± 1445
Ti	37 ± 4
V	0.34 ± 0.05
Cr	1.46 ± 0.54
Mn	70 ± 10
Ni	0.86 ± 0.18
Co	0.20 ± 0.06
Cu	13 ± 1.9
Zn	31 ± 3
Br	11.1 ± 1.8
Se	0.16 ± 0.08
Rb	16.1 ± 1.7
Sr	36 ± 14
Mo	0.66 ± 0.26
Cd	0.64 ± 0.58
Sn	0.24 ± 0.10
J	0.28 ± 0.21
Cs	0.032 ± 0.028
Ba	49 ± 8.6
La	1.33 ± 0.14
Ce	0.86 ± 0.61
Pb	48 ± 11
U	0.013 ± 0.005

LA-ICP-MS may be used for quantitative analysis and for semiquantitative screening applications. The accuracy that can be achieved in quantitation depends to a large extent on the availability of internal and external standards for calibration and/or the ability to perform calibration by the following special means: Thompson et al. (1989/1990) and Moenke-Blankenburg et al. (1989/1990) suggested (for ICP-AES) the use of nebulized aqueous standards to overcome some of the disadvantages of solid standards (this may be useful too in ICP-MS). Hager (1989) recommended a method for determining relative elemental response factors for LA-ICP-MS. This model uses response factors determined from solution nebulization and modifies them based on element-dependent volatilization efficiencies. The approach for semiquantitative analysis is about 50%. Another capability of semiquantitative LA-ICP-MS is a method where the instrument response factor is used, i.e., dating of the instruments response functions with three response elements (at a minimum). The mass/response relationship has been found to be applicable to a wide range of matrices and can therefore be used in conjunction with an internal standard to generate standardless semiquantitative analyses (Date and Gray, 1989, p. 21).

Parameters of two commercially available instruments are shown in Table 4B.3. Examples of applications are presented in Table 4B.4. to 4B.7.

4B.5. CONCLUSIONS

The use of laser ablation for sample introduction in mass spectrometry will indubitably continue to be investigated. Its use will be complementary to traditional sample introduction techniques and will afford the analyst greater flexibility in choosing an appropriate method for a particular problem.

One of the most attractive methods using laser for ablation of nonconducting as well as conducting material in micro and macro regions is LA-ICP-MS.

The field of laser ablation in mass spectrometry is still growing: LA-ICP-MS may well soon become the method of choice for many special analytical problems.

REFERENCES

Abercrombie, F. N., Silvester, M. D., and Stoute, G. S. (1977a). *Proc. 28th Pittsburgh Conf., Cleveland, Ohio*, Paper 406.

Abercrombie, F. N., Silvester, M. D., and Stoute, G. S. (1977b). *ICP Inf. Newsl.* **2**, 309–312.

Adams, F., and Vertes, A. (1990). *Fresenius' Z. Anal. Chem.* **337**, 638–647.

Arrowsmith, P. (1987). *Anal. Chem.* **59**, 1437–1444.

Arrowsmith, P. (1989). *Ceram. Trans.*, pp. 87–101.

Berkowitz, J., and Chupka, W. A. (1964). *J. Chem. Phys.* **40**, 2735–2738.

Bernal, E., Levine, L. P., and Ready, J. F. (1966). *Rev. Sci. Instrum.* **37**, 938–941.

Broadhead, M., Broadhead, R., and Hager, J. W. (1990). *Atom. Spectrosc.* **11**, 205–208.

Bykovskii, Yu. A., Dorofeev, V. I., Dymovich, V. I., Nikolaev, B. I., Ryzhikh, S. V., and Silnov, S. M. (1969). *Sov. Phys.—Tech. Phys.* (*Engl. Transl.*) **13**, 986–989.

Bykovskii, Yu. A., Dorofeev, V. I., Dymovich, V. I., Nikolaev, B. I., Ryzhikh, S. V., and Silnov, S. M. (1970). *Sov. Phys.—Tech. Phys.* (*Engl. Transl.*) **14**, 955–959.

Capellen, J. M., and Conzemius, R. J. (1980). *Bibliography of Publications on Laser Interaction with Solids and Laser Mass Spectrometry of Solids*, IS-4715. Ames Laboratory, Iowa State University, Ames.

Carr, J. W., and Horlick, G. (1980). *Proc. 31st Pittsburgh Conf., Atlanta City, NJ*, Paper 56.

Date, A. R., and Gray, A. L. (1989). *Applications of Inductively Coupled Plasma Mass Spectrometry*. Blackie, Glasgow and London; Chapman & Hall, New York.

Denoyer, E. R., and Fredeen, K. J. (1991). *Eur. Winter Conf. Plasma Spectrochem., Dortmund, Germany*, Poster.

Denoyer, E. R., Fredeen, K. J., and Hager, J. W. (1991). *Anal. Chem.* **63**, 445A–457A.

Eloy, J. F., (1969). *Rev. Method. Phys. Anal.* **5**, 157–162.

Eloy, J. F., and Dumas, J. L. (1966). *Rev. Method. Phys. Anal.* **2**, 251–255.

Fabbro, R., Fabre, E., Amiranoff, F., Garbeau-Labaune, C., Virmont, J., Weinfield, M., and Marx, C. E. (1980). *Phys. Rev., Ser. A* **26**, 2289–2294.

Fenner, N. C., and Daly, N. R. (1966). *Rev. Sci. Instrum.* **37**, 1068–1072.

Fenner, N. C., and Daly, N. R. (1968). *J. Mater. Sci.* **3**, 259–263.

Giory, F. A., McKenzie, L. A., and McKinney, E. J. (1963). *Appl. Phys. Lett.* **3**, 25–30.

Gray, A. L. (1985). *Analyst* **110**, 551–556.

Gray, A. L. (1989). In *Advances in Mass Spectrometry*, pp. 1674–1693. Heyden, London.

Hager, J. W. (1989). *Anal. Chem.* **62**, 1243–1248.

Hartford, A., Jr. (1984). *Pure. Appl. Chem.* **56**, 1555–1558.

Honig, R. E., and Woolston, J. R. (1963). *Appl. Phys. Lett.* **2**, 138–142.

Houk, R. S. (1986). In *Analytical Applications of Lasers* (E. H. Piepmeier, ed.), pp. 587–625. Wiley, New York.

Imai, N. (1990). *Anal. Chim. Acta* **235**, 381–391.

Isenor, N. R. (1964). *Can. J. Chem.* **42**, 1413–1417.

Isenor, N. R. (1965). *J. Appl. Phys.* **36**, 316–319.

Kaufmann, R. (1986). *LIMS, Reference and Abstract Index*. University of Düsseldorf, Germany.

Knox, B. E. (1968). *Mater. Res. Bull.* **3**, 329–332.

Knox, B. E., and Vastola, F. J. (1966). *Chem. Eng. News* **44**, 48–52.

Langer, P., Tonon, G., Floux, F., and Ducauze, A. (1966). *IEEE J. Quantum Electron.* **2**, 499–503.

Langer, P., Pin, B., and Tonon, G. (1968). *Rev. Phys. Appl.* **3**, 405–409.

Lincoln, K. A. (1965). *Anal. Chem.* **37**, 541–545.

Lincoln, K. A. (1969). *Mass Spectrom. Ion Phys.* **2**, 75–81.

Mitchell, P. G., Sneddon, J., and Radziemski, L. J. (1987). *Appl. Spectrosc.* **40**, 274–279.

Mochizuki, T., Sakashita, A., Iwata, H., Kagaya, T., Shimamura, T., and Blair, P. (1988). *Anal. Sci.* **4**, 403–409.

Moenke-Blankenburg, L. (1989). Laser Microanalysis. In *Chemical Analysis: A Series of Monographs on Analytical Chemistry and Its Applications* (P. J. Elving and J. D. Winefordner, eds.), Vol. 9, Wiley, New York.

Moenke-Blankenburg, L., Gäckle, M., Günther, D., and Kammel, J. (1989/1990). In *Plasma Source Mass Spectrometry* (K. E. Jarvis, A. L. Gray, and J. G. Williams, eds.), Spec. Publ. No. 85, pp. 1–17. Royal Society of Chemistry, Cambridge, UK.

Moenke-Blankenburg, L., Schumann, T., Günther, D., Kuss, H.-M., and Paul, M. (1992). *J. Anal. At. Spectrom.* **7**, 251–254.

Namba, S., Kim, P. H., and Mitsuyama, A. (1966). *J. Appl. Phys.* **37**, 3330–3335.

Piepmeier, E. H., and Malmstadt, H. V. (1969). *Anal. Chem.* **41**, 700–705.

Ready, J. F. (1971). *Effects of High-Power Laser Radiation.* Academic Press, New York and London.

Salin, E. D., Carr, J. W., and Horlick, G. (1979). *Proc. 30th Pittsburgh Conf., Cleveland, Ohio*, Paper 563.

Thompson, M., Goulter, J. G., and Sieper, F. (1981a). *Jena Rev.* 202–210.

Thompson, M., Goulter, J. G., and Sieper, F. (1981b). *Analyst* **106**, 32–40.

Thompson, M., Chenery, S., and Brett, L. (1989/1990). *J. Anal. At. Spectrom.* **4**, 11–16; **5**, 49–55.

Tye, C. T., and Barett, P. (1988). *Steel Times* **216**, 240–253.

Tye, C. T., Henry, R., Abell, I. D., and Gregson, D. (1989). *Res./Dev.* April, pp. 1–4.

Vertes, A., Juhasz, P., DeWolf, M., and Gijbels, R. (1988). *Scanning Microsc.* **2**, 1853–1877.

Vertes, A., DeWolf, M., Juhasz, P., and Gijbels, R. (1989a). *Anal. Chem.* **61**, 1029–1035.

Vertes, A., Juhasz, P., Balazs, L., and Gijbels, R. (1989b). In *Microbeam Analysis—1989* (P. E. Russell, ed.), pp. 273–276. San Francisco Press, San Francisco.

Vertes, A., Juhasz, P., DeWolf, M., and Gijbels, R. (1989c). *Int. J. Mass Spectrom. Ion Processes* **94**, 63–85.

Vertes, A., Gijbels, R., and Adams, F. (1990). *Mass Spectrom. Rev.* **9**, 71–113.

CHAPTER

4

THE HIGH LASER IRRADIANCE REGIME

C. INORGANIC TRACE ANALYSIS BY LASER-INDUCED MASS SPECTROMETRY

HANS-JOACHIM DIETZE and JOHANNA SABINE BECKER

Central Department for Chemical Analysis
Research Centre, Jülich GmbH
Jülich, Germany

4C.1. INTRODUCTION

Inorganic mass spectrometry for trace analysis is widely used in all fields of modern science and technology: in materials research (e.g., high-purity materials), in metallurgy, in semiconductor production and microelectronics, in geology or mineralogy, in environmental and biological research, and in medical science. Being a universal multielement analysis method, it permits the simultaneous determination of all chemical elements and their isotopes in solid samples. Among the different spectrometric techniques for trace analysis (X-ray spectrometry, optical emission spectrometry, or atomic emission spectroscopy, spark source mass spectrometry, inductively coupled plasma or glow discharge mass spectrometry), laser-induced mass spectrometry is well established as a trace analytical method with a wide coverage. Figure 4C.1 shows a comparison between the sensitivity of laser mass spectrometry and the most usual mass spectrometric methods for inorganic trace analysis.

Its special features qualify laser-induced mass spectrometry for the quantitative determination of trace elements in inorganic compounds. The advantages of laser ionization for analyses of inorganic materials are as follows: (1) high efficiency of evaporation and ionization; (2) high absolute and relative sensitivity; (3) capacity for a direct analysis of any kind of solids (e.g., metals or alloys, semiconductors or nonconducting materials, ceramics, thin films or biological subjects, minerals, rocks, soils, ashes); (4) simplicity of the mass spectra obtained;

Laser Ionization Mass Analysis, Edited by Akos Vertes, Renaat Gijbels, and Fred Adams. Chemical Analysis Series, Vol. 124.
ISBN 0-471-53673-3

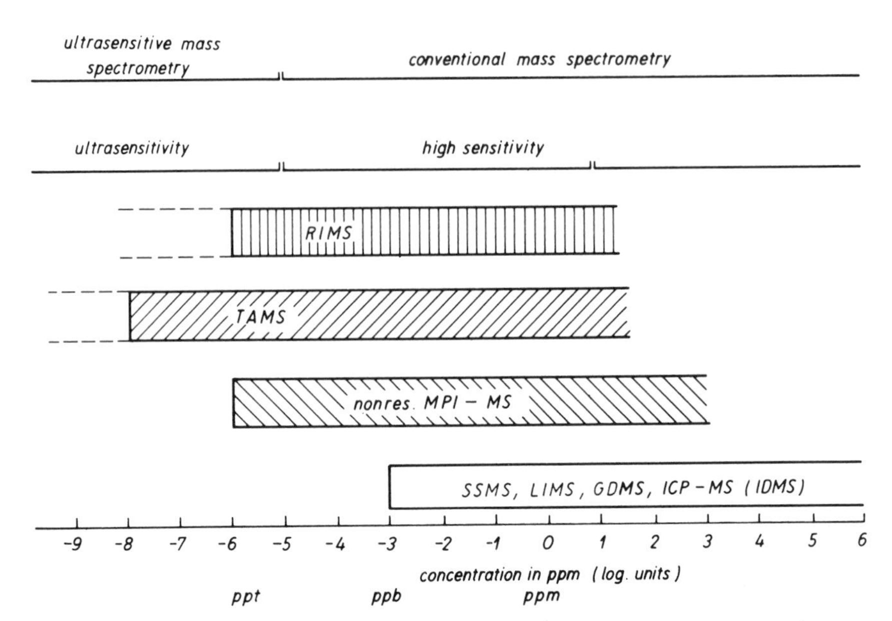

Figure 4C.1. Comparison of sensitivity of mass spectrometric methods for inorganic trace analysis of solids: RIMS = resonance ionization mass spectrometry; TAMS = tandem accelerator mass spectrometry; nonres. MPI = nonresonant multiphoton ionization mass spectrometry; SSMS = spark source mass spectrometry; LIMS = laser ionization mass spectrometry; GDMS = glow discharge mass spectrometry; ICP-MS = inductively coupled plasma mass spectrometry; IDMS = isotope dilution mass spectrometry in combination with other specified mass spectrometric methods.

(5) capacity for depth profile (layer-by-layer) analysis; and (6) determination of the distribution of inorganic trace elements in solids. Besides these most attractive features for trace analysis by laser-induced mass spectrometry, the technique can be used for equally accurate and precise multielement analyses of minor and even major components in inorganic samples. Further, the power density that is effective in the spot area can be varied by adjustments of the laser parameters (e.g., laser energy, or wavelength) and can be better controlled in comparison to the ionization parameters of other analytical mass spectrometric methods. It can be more easily applied as an absolute method without the use of standards under certain experimental conditions. These features and the general advantages of mass spectrometric methods favor laser-induced mass spectrometry for analysis of the composition and distribution of traces or impurities in solids.

The generation of high-temperature plasmas produced by focusing laser pulses onto a solid surface—basic principles of laser–solid interaction—has been the subject of numerous theoretical and experimental publications (Demtröder and Jantz, 1970; Zahn and Dietze, 1976; Kovalev et al., 1978; Shibanov, 1985; Vertes et al., 1990; see Chapter 4, Part A, in this volume). Special instrumental aspects of laser ionization mass spectrometry of inorganic trace analysis are discussed in Chapter 2.

The development of laser ionization mass spectrometry was started by Honig and Woolston (1963) with studies of ionization in the interaction of a focusing laser radiation (using a ruby laser) with the surface of a solid, where positive ions are produced from metals, semiconductors, and insulators. The authors describe the formation of ions, electrons, and neutrals in a laser cloud. Around the same time and in subsequent years Honig (1963), Dumas (1967), Ban and Knox (1969), Dietze and Zahn (1972), Dietze et al. (1976), and Beam (1973) published their papers on laser ion sources and on mass analysis by means of time-of-flight mass spectrometry or double-focusing static mass spectrometers. The application of laser-induced mass spectrometry to solids has been reviewed by Maksimov and Larin (1976), Kovalev et al. (1978), and Conzemius and Capellen (1980).

The following discussion is broken down into four main sections: 4C.2. laser mass spectrometric techniques; 4C.3, instrumentation; 4C.4, analytical features; and 4C.5, applications. It presents an overview of a special field in inorganic laser-induced mass spectrometry, i.e., trace analysis of chemical elements of inorganic solids. Section 4C.6 is the conclusion.

4C.2. LASER MASS SPECTROMETRIC TECHNIQUES FOR INORGANIC TRACE ANALYSIS

All laser mass spectrometric methods are suitable for the trace analysis of different materials, but only the three types of techniques—LIMS, RIMS, and LAMS—considered in the following subsections have been used in the analysis of impurities in inorganic solid materials. These techniques differ in the formation of ions in the laser ion source.

4C.2.1. Laser Ionization Mass Spectrometry (LIMS)

LIMS is based on the evaporation and atomization of sample material by means of a focused pulsed laser beam and on the ionization of the evaporated atoms, clusters, and molecules in a laser microplasma, formed in the spot area of the irradiated sample (see Figure 4C.2). This process depends on the intensity of the laser beam, the physical and chemical properties of the sample

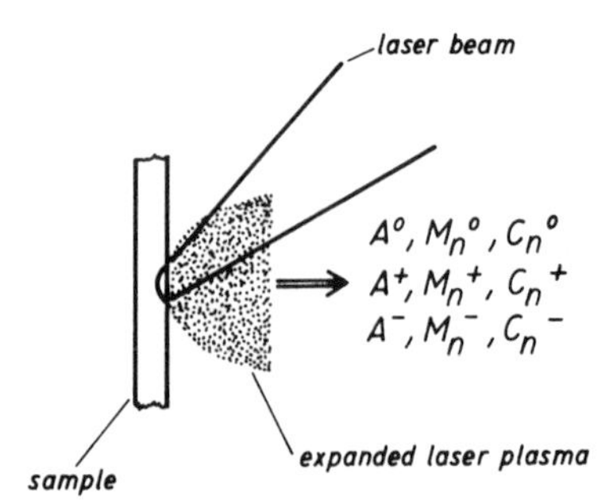

Figure 4C.2. Scheme of the laser ionization process. In a laser microplasma, atoms A, molecules M, clusters C, and the ionic species are present.

material, and the conditions of plasma formation. The fact that the ionization process can be influenced by the laser power density in the spot area is the great advantage of this ionization method. All types of solid materials are accessible to this ionization method, independent of their electrical conductivity, reflection properties, or types of chemical bonds. Details of the ion formation processes are described elsewhere in this volume (in Chapter 3, Parts A and B).

LIMS is a multielement method comparable to spark source mass spectrometry (SSMS). All chemical elements in the sample materials are evaporated and ionized in the laser plasma and can be detected by electrical or photographic ion detection methods. The mass separation of the laser-induced ion beam is carried out by dynamic systems [time-of-flight (TOF) mass spectrometers] or magnetic sector analyzers (both single- and double-focusing mass spectrometers).

4C.2.2. Resonance Ionization Mass Spectrometry (RIMS)

RIMS is a highly sensitive and element-specific laser mass spectrometric technique. It is an ultrasensitive method for the trace analysis of inorganic materials. Its analytic fundamentals and application are described in reviews by Hurst et al. (1979), Letokhov (1978), and Smith et al. (1989). A scheme of the resonance ionization of evaporated sample material is shown in Figure 4C.3. In the resonance ionization process the sample material is vaporized and atomized by various methods such as thermal evaporation, laser vaporization, or sputtering. Ions are then generated by interaction of atoms with a single laser beam or several laser beams.

In the resonance ionization process, a laser is tuned precisely to the wavelength required for the excited state, which in turn is unique to the element being measured. Five alternative basic schemes of the resonance ionization process are illustrated in Figure 4C.4. One of them has a pulsed laser beam produce photons of just the correct energy to excite an atom from its ground

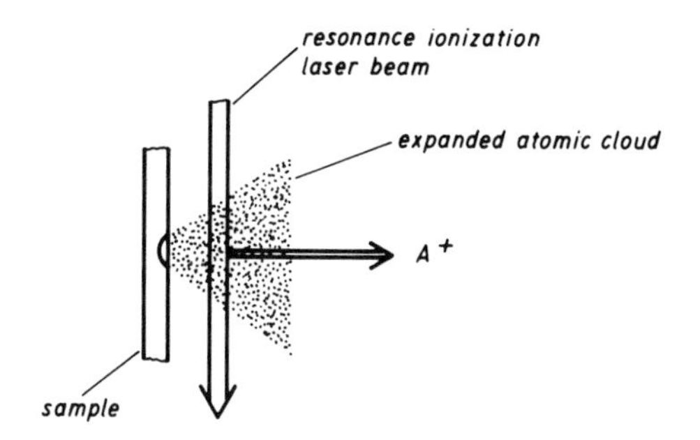

Figure 4C.3. Scheme of resonance ionization of evaporated sample material.

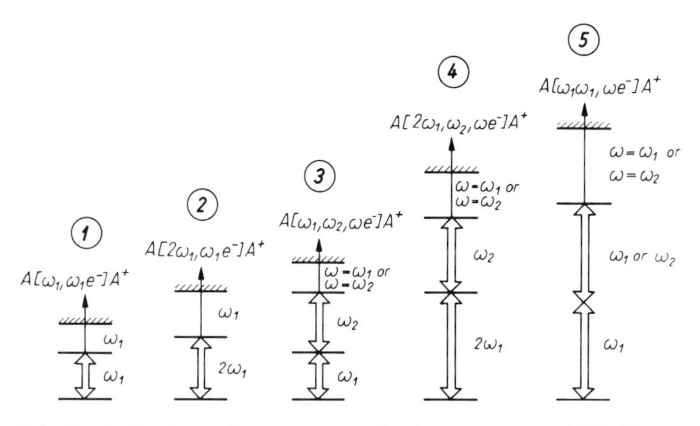

Figure 4C.4. Five basic schemes for resonance ionization spectroscopy (RIS) (Hurst et al., 1979).

state to an excited level. Additional photons of the same or another energy can further excite or photoionize this excited atom into the ionization state. The simplest resonance ionization process is a two-step procedure with photons of the same wavelength. This excitation process can be used for many but not all chemical elements. A second laser in combination with the first laser makes it possible to excite two states sequentially prior to ionization. Still another resonance ionization process involves a two-photon transition to an intermediate state prior to ionization. Together these five types of RIMS allow for the detection and analysis of every chemical element in the periodic table (Figure 4C.5).

The advantages of RIMS as follows:

- It is extremely selective, efficient, and sensitive because *only* the atoms of the one chemical element to be analyzed are ionized and *all* atoms of the selected element present in the laser beam are ionized.

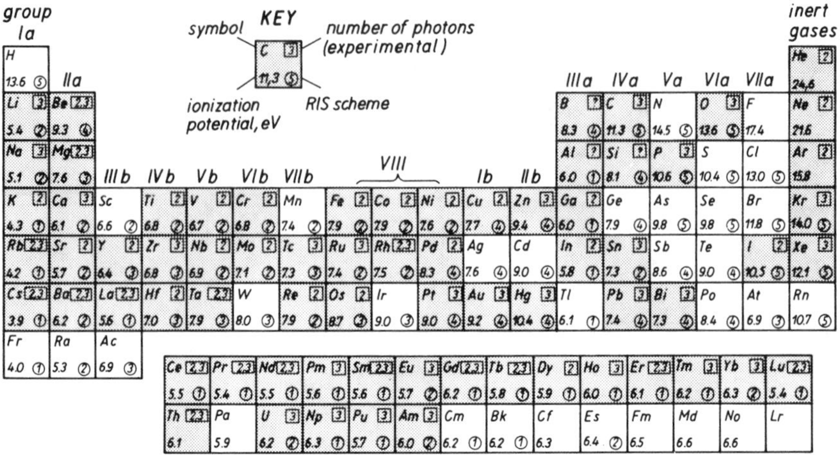

Figure 4C.5. Application of RIMS to elemental analysis (Young et al., 1989). The applied RIS schemes are depicted in Figure 4C.4 and described in detail by Hurst et al. (1979). The shaded blocks represent the elements for which experimental RIMS results have been obtained. The top right-hand corner of these blocks shows the number of photons absorbed to generate an ion.

- It is predictable because the physical process can be calculated (excited energy levels of all atoms in the periodic table are known).
- Usually, the chemical elements present in trace amounts in a matrix can be analyzed without need for a chemical separation.

4C.2.3. Laser Ablation Mass Spectrometry (LAMS)

LAMS is a combination of laser evaporation for the atomization of sample materials and various ionization methods and well-known mass spectrometric separation techniques, such as the connection of a plume generated by laser ablation and ionization of atoms in plasma torch inductively coupled with a gas flow, by electron impact ionization, by photo-ionization, or in a glow discharge cell.

The common trend in manufacturing of laser mass spectrometers is the separation of evaporation (atomization) and ionization steps, e.g., the LASERLAB LA-ICP-MS (laser ablation–inductively coupled plasma–mass spectrometer; produced by VG Instruments) allows LA of solid samples and ICP to be combined with direct analysis of solid samples without time-consuming pretreatment.

LAMS has been described in the preceding reviews by Phipps and Dreyfus (Part A of this chapter) and Moenke-Blankenburg (Part B).

4C.3. INSTRUMENTATION AND EXPERIMENTAL CONDITIONS

The instrumentation of laser-induced mass spectrometry for trace analysis differs from the other mass spectrometric trace analysis methods only in its use of an alternative type of ion source.

In general, two modes are used for the geometric configuration of laser beam, sample, and ion-optical axis in the laser ionization ion source (see Figure 4C.6): the reflection mode (a) and the transmission mode (b). The angle between the laser beam and the sample surface can be varied between 45° and 90°, and the same range of angles is used between the ion-optical axis and the sample surface. The disadvantage of the transmission configuration is that it does not allow for the maximum possible interaction between the laser beam and the sample area and that it requires a special sample form. Furthermore, owing to this ion source geometry, in situ investigation of many trace analysis problems, for example, of geological or natural samples, it not possible; however this ionization method is an excellent laser microprobe technique and a convenient complement to other microprobe techniques in the analysis of thin foils, small particles, and inclusions in technical, biological, and other transparent samples. The analytical properties and applications of the LAMMA technique are discussed in detail by Verbueken et al. (1988) and in Chapter 2 of this volume.

The most often used configuration for the trace analysis of inorganic sample material in commercial and laboratory instruments is the reflection mode (Dietze and Becker, 1985a; Vogt et al., 1981; Jochum et al., 1988; Eloy, 1986; Boriskin et al., 1983). The LAMMA® 1000 (manufactured by Leybold-Hereaus AG, Cologne, Germany) has been designed with a laser ion source (reflection geometry) in combination with a TOF drift tube. The application is described by Heinen et al. (1983) and Feigl et al. (1983, 1984). The features of LAMMA 1000 include both positive ion and negative ion detection capabilities, although there have been very few reports on its negative ion detection

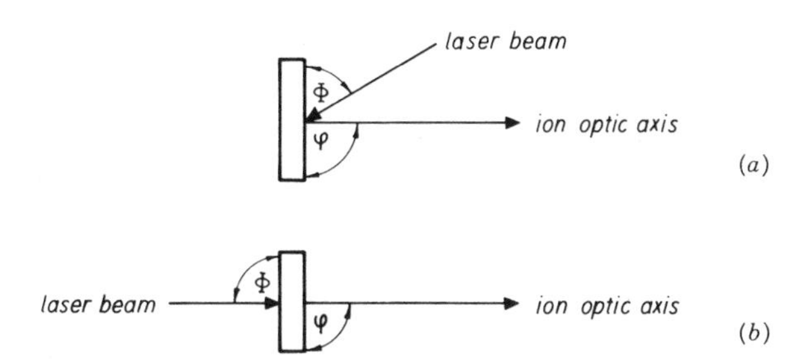

Figure 4C.6. Geometric configurations of laser ion sources for (a) the reflection mode and (b) the transmission mode.

capability. In general, all mass spectrometers can be used for positively and negatively charged ion detection, depending on the polarity of the potentials at the electrodes of the accelerating system and the field direction of the analyzers. Trace analysis uses the single positively charged atomic ions formed in laser plasmas.

The TOF mass analyzer is more popular than the static sector mass analyzer because it is most suitable for application to pulsed laser experiments and because it provides high ion transmission. A disadvantage of most dynamic TOF mass separation systems is their low mass resolution. However, the limited mass resolution of conventional TOF instruments has been essentially improved by insertion of an ion reflector by Mamyrin et al. (1973). Furthermore the application of soft multiphoton ionization produces ions with a low energy spread and low initial energies. Walter et al. (1986) in collaboration with the Bruker-Franzen Analytik firm have developed a new laser ionization TOF-MS and have been able to obtain a mass resolution of 10,000 (Frey et al., 1985).

Another commercially manufactured mass spectrometer is the LIMA® (Laser Ionization Mass Analyzer, Cambridge Mass Spectrometry Company, Cambridge, UK). With a short-pulse-length frequency-quadrupled Nd:YAG laser (*Q*-switched), a microscopic volume of the sample material is evaporated and ionized in the transmission mode for thin samples or the reflection mode for bulk materials (Dingle et al., 1981; Utley, 1990). In this arrangement the sample is viewed through a high-resolution optical microscope system (for local microanalysis), which also serves to focus the laser pulse onto the sample surface. The typical LIMA applications in microelectronics include identification of impurities in dielectrics, microlocal analysis, depth profiling, and analysis of thick films and hybrid circuits. Both TOF mass spectrometer types with laser ion source (LAMMA and LIMA) are valuable tools for the qualitative characterization of microsamples and allow for elemental localization. The disadvantage of LAMMA and LIMA instruments for the trace analysis of inorganics is low accuracy due to the inconstancy of ion currents during measurement. Therefore these mass spectrometers have been used mainly for analysis of organic compounds and in biochemistry. A new instrumental development of LAMMA, the LASERMAT (manufactured by Finnigan MAT), can be applied with limitation for inorganics.

A new development in laser-induced mass spectrometry by Muller et al. (1989) is the configuration of a pulsed laser ion source with a Fourier transform mass spectrometer (FTMS) featuring a remarkable mass resolution of up to 400,000. Application of a single KrF excimer laser shot at 249 nm on a ceramic has allowed for separation of doubly charged ^{88}Sr ions from singly charged ^{44}Ca ions. Laser energy is about 0.5 μJ, and mass resolution 160,000. With this laser FTMS, it is possible to separate all spectral interference.

An interesting combination of a laser ion source with a mass spectrometer was proposed by Eloy (1978, 1984) and Stefani (1981). The laser probe mass spectrograph (see Chapter 2; Section 2.4.1), is a combination of a reflection mode laser ion source with a single-focusing magnetic sector instrument. The photoplate has been replaced by a special device combining an ion–electron converter system with a scintillator and allowing the photons emitted from the whole image plane to be transmitted to a photomultiplier via a guide of conventional shape. This arrangement allows for the simultaneous determination of mass lines in a selected mass range.

Higher mass resolution in laser ionization mass spectrometry than is found in commercial LAMMA-TOF systems or single-focusing static mass spectrometers can be obtained using a double-focusing mass separation, e.g., a Mattauch–Herzog geometry. This is realized commercially in the Russian EMAL 2 laser ionization mass spectrometer (by Gladskoi and Belousov, 1980). The experimental details, features, and some applications of this static mass spectrometer are discussed by Basova et al. (1987). A further instrumental development on the basic setup of EMAL 2 is realized in the laser ionization mass spectrometer MC 3101. This instrument worked with a Nd:YAG laser at a higher laser power density ($\Phi \sim 10^{10}$ W/cm^2) in comparison to EMAL 2 ($\Phi \sim 10^9$ W/cm^2) and is the only commercial mass spectrometer with Mattauch–Herzog geometry in combination with a laser ion source and photoplate detection.

In some cases, commercial spark source mass spectrometers have been expanded into laser ionization mass spectrometers (e.g., Dietze and Zahn, 1972; Dietze et al., 1976, 1981; Dietze and Becker, 1985a; Jansen and Witmer, 1982; Sanderson et al., 1984; Matus et al., 1988). Such a combination of a laser ion source with a double-focusing mass spectrometer of Mattauch–Herzog geometry (MX 3301—originally with a spark ion source) is reported by Dietze and Becker (1985a). The corresponding reflection mode laser ion source for microlocal and bulk analysis is shown in Figure 4C.7. With the ruby, Nd:glass, or Nd:YAG pulsed lasers in single shot regimes, the local analysis or depth profile analysis of inorganic samples or thin films is possible, whereas the Nd:YAG laser in the high laser pulse repetition rate regime (in the range of 50 Hz to a few kilohertz) allows for efficient mass spectrometry of trace elements. The experimental arrangement of a laser ion source in connection with a double-focusing mass spectrometer (with a mass resolution of about 17,000) is suitable for interference-free trace analysis and also for the study of cluster ion formation processes.

One of the important features of the reflection mode is that ordinarily no special sample preparation is necessary. In general, the sample can be inserted in the ionization chamber of the ion source in all geometric forms—in the form of plates, sample fragments, or pellets, e.g., in a plane of 10 mm × 10 mm.

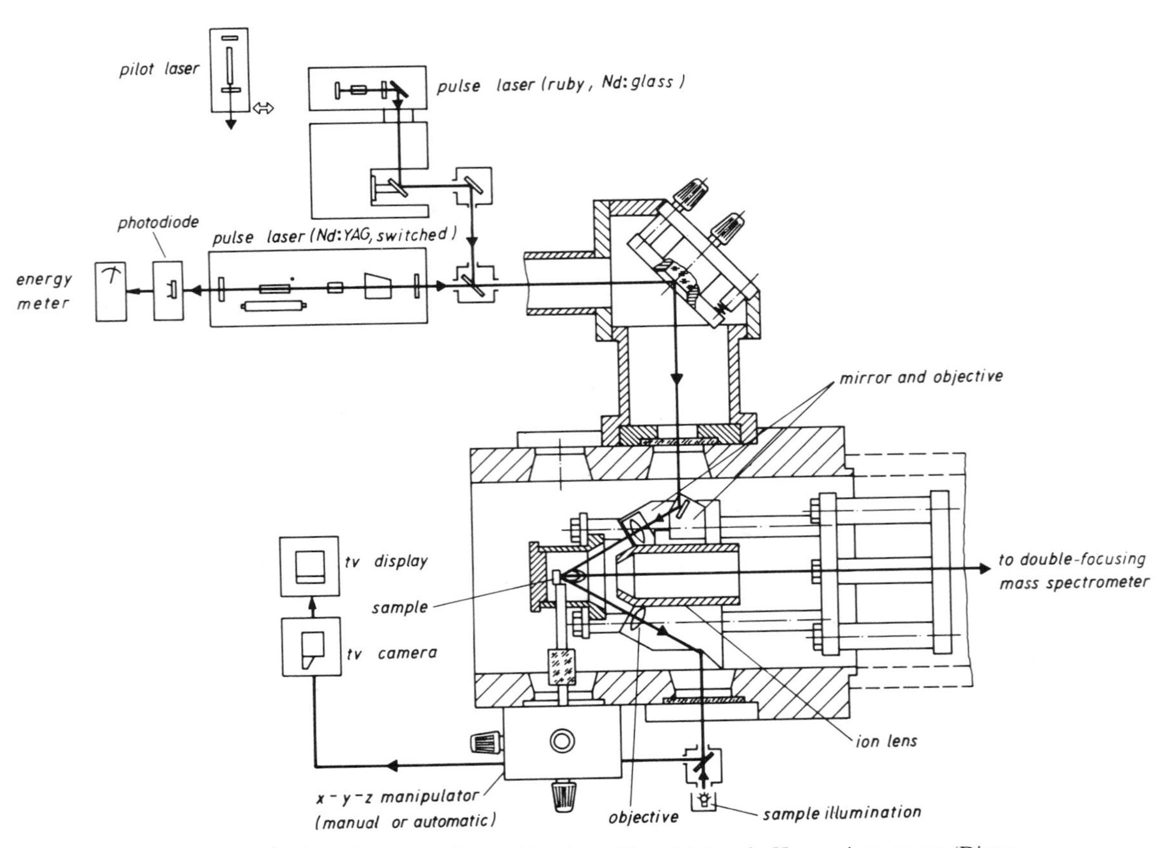

Figure 4C.7. Scheme of a laser ion source in combination with a Mattauch–Herzog instrument (Dietze and Becker, 1985a).

The size of the sample is in essence only limited by the size of the ionization chamber. If the sample is a powder, an internal standard can be added and pelletized. In the case of high reflectivity powdered material, a small amount of graphite can be added to increase the absorption of the sample material. This method can be used only if high mass resolution of the mass spectrometric system is available because there is a likelihood of interference of analyte ions with molecular and cluster ions of sample atoms in the applied laser power density range. In the case of carbon addition, carbide cluster ions and carbon cluster ions are formed.

For the sensitive trace analysis of inorganics the aforementioned mass spectrometric systems developed from commercial spark source instruments with Mattauch–Herzog geometry and photoplate detection were applied successfully (see Section 4C.5, below). However, the complicated mass spectrometers with static sector fields are very expensive compared to instruments using dynamic separation systems [TOF-MS and QMS (quadrupole mass spectrometry)]. A further disadvantage is the time-consuming analysis: whereas a trace analysis in the ppb range for about 60 chemical elements needs several hours (from sample pretreatment to measurements of blackening curves on the photoplate and calculation of trace concentrations), other mass spectrometric methods for trace analysis (e.g., GDMS or ICP-MS) are more effective. In spite of the remarkable advantages of ion detection by means of a photoplate in comparison with electrical detection, the commercial development of more user-friendly mass spectrometers is still necessary.

A comparison of the efficiency of laser ionization mass spectrometry with spark source, glow discharge, and secondary ion mass spectrometry for multi-element trace analysis of solids is given by Beske (1988) and by Adams and Vertes (1990).

An important part of laser-induced analysis is the coupling of the laser beam energy into the solid surface to vaporize and ionize the sample material. Effective coupling is influenced by a number of parameters of the laser beam as well as by the properties of the sample material: wavelength of the laser radiation; duration and form of the laser pulse; laser power density in the spot area; reflectivity of the sample surface and thermal properties of the sample material at the laser wavelength; and absorption properties of the sample surface.

The energy distribution of positively charged n^+ ions can be varied as a function of laser power density in the range of < 1 eV to some kilo-electron volts. Figure 4C.8 shows the energy spectra of atomic Mn ions in the laser power density range of 2.5×10^8 to 1×10^{10} W/cm^2 (Bykovskii et al., 1971). Abundance distributions of atomic ions with different charge states are determined via the applied laser power density in the spot area. At $\Phi < 10^8$ W/cm^2 in laser plasmas, only singly charged atomic ions are detected. The mean

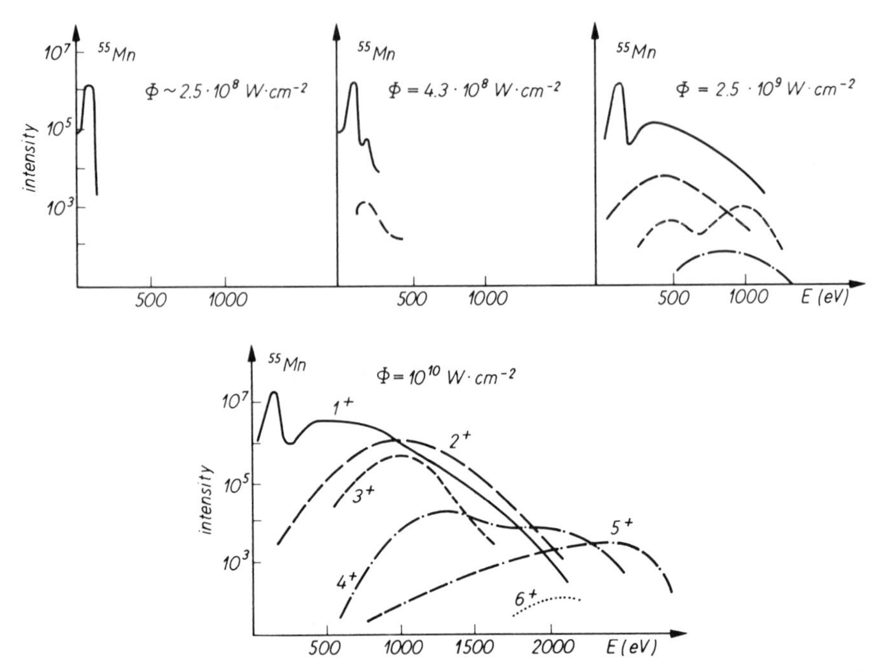

Figure 4C.8. Energy spectra of Mn atomic ions as a function of laser power density (Bykovskii et al., 1978a).

charge state and the mean of the energy distribution of ions formed in the laser plasma increase with increasing laser power density. The preferable laser power density in trace analysis of inorganic samples is located in the interval between 10^9 and 10^{10} W/cm^2. In this laser power density range, there is no fractional evaporation of chemical elements or compounds in the ion source. The ion beam composition corresponds to the composition of the analyzed target material. In this laser power density range Bykovskii et al. (1978a), Arefjev et al. (1986), Ramendik (1986), and Belousov (1984) observed uniform element sensitivities [relative sensitivity coefficient (RSC) near 1; see also Section 4C.4.2 and Figure 4C.12, below].

An interesting result of laser-induced mass spectrometry is shown in Figure 4C.9. Here, the ion current ratios are constant in an interval of about 5×10^9 to 5×10^{10} W/cm^2 laser power density. For chemical compounds with very different physical and chemical properties [(I) CsI; (II) GaAs; (III) PbSn], when an Nd:YAG laser is used a different ion formation behavior is observed for low laser power densities (Bykovskii et al., 1978a). At low laser power densities the intensity ratio (A_1^+/A_2^+) of Cs^+ to I^+ ions is fairly high owing to the low ionization potential of Cs; however, the semiconductor GaAs

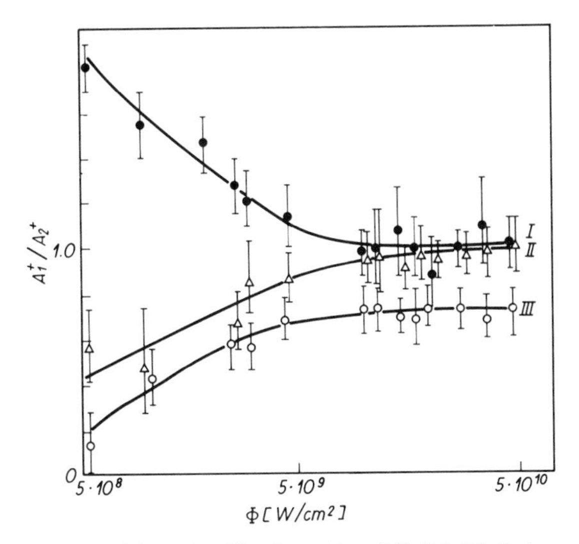

Figure 4C.9. Dependence of the ratio of ion intensities of (I) CsI, (II) GaAs, and (III) PbSn on laser power density.

shows the opposite behavior. On the other hand, for the stoichiometric alloy PbSn, where both metals have similar ionization potentials, the ion intensities are influenced by the different heats of evaporation.

The experimental results of investigations on ion abundance distributions by Zahn and Dietze (1976) and Bykovskii et al. (1976) in laser plasmas in the 10^8–10^{13} W/cm^2 laser power density range yielded about 10^{14}–10^{15} n^+ ions per laser shot. Ionization efficiencies for the metallic elements were found to be of the same order of magnitude.

The second laser-induced mass spectrometric method, RIMS, is becoming an established tool for the trace analysis of most chemical elements of the periodic system [see Figure 4C.5 (Young et al., 1989)]. It also promises to overcome traditional limitations of sensitivity and selectivity in trace analysis. The RIMS system consists of a pump laser (for example, an N_2 or Nd:YAG laser), a tunable dye laser, a mass analyzer, and an ion detector. The Nd:YAG laser pumps the tunable dye laser with pulses of a few nanoseconds pulse length at a 10–100 Hz repetition rate. The dye laser produces a pulsed laser beam at a different wavelength, which can be directed into the source chamber of the mass spectrometer directly or via a frequency doubler. The laser beam is focused onto a neutral atom cloud generated by a conventional single- or multiple-filament assembly, which is normally used in thermal ionization mass spectrometry. In the focusing area of the laser beam atoms of the sample material are ionized, and subsequently these ions are accelerated into the

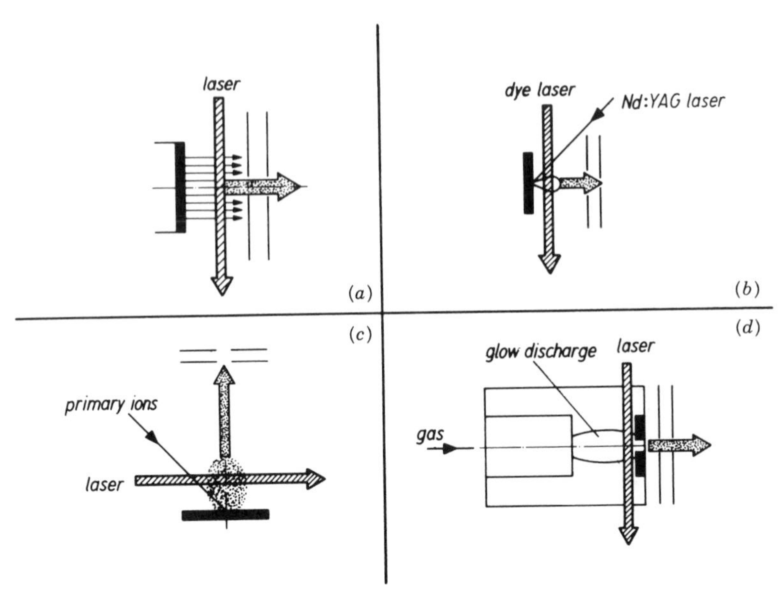

Figure 4C.10. Four possible basic layouts for the atomization and ionization by RIS of inorganic samples: (a) electrothermal heating; (b) laser ablation or laser-induced sputtering; (c) ion sputtering; and (d) glow discharge.

mass analyzer (Fassett et al., 1983; for the instrumental arrangement see Chapter 2, Section 2.4.1, in this volume).

Resonance ionization is a postionization technique for neutral atoms. The atomization of the elements is of fundamental significance. Figure 4C.10 shows four possible basic layouts for the atomization and resonance ionization of inorganic solids. The sample material can be atomized (a) by electrothermal heating, (b) by laser ablation or laser-induced sputtering, (c) by ion sputtering, or (d) in a glow discharge. The neutral atoms are photoionized by laser beams of one to three tunable pulsed dye lasers, depending on the photoionization process used (see Figure 4C.4). Most analytical applications of RIMS have used thermal atomization and ionization by simple resonance ionization (see Figure 4C.4—RIS processes 1–3).

The combination of laser ablation and resonance ionization (LARIS) was successful in detecting extremely low impurity levels in solid materials (Mayo et al., 1982). The combined application of the well-known secondary ion mass spectrometry (SIMS) and multiphoton resonance ionization of sputtered neutrals has been termed sputter-initiated resonance ionization spectrometry (SIRIS) by Parks et al. (1983a) or surface analysis by resonance ionization of sputtered atoms (SARISA) by Pellin et al. (1988). The SIRIS and SARISA concept combines the multielement atomization of an ion-sputtering process with the optimal selectivity and sensitivity of the RIS process. This SIRIS technique allows for an ultrasensitive analysis with sensitivities up to 1 part in 10^{12}, that is, about 3–6 orders of magnitude higher than in a normal SIMS apparatus. The apparatus consists of a primary ion source, a mass separation, and a beam-focusing system to form a focused ion beam for the bombardment of the sample surface. The resulting cloud of sputtered material from the sample contains secondary ions and neutral atoms. The secondary ions disturb the analysis because they can interfere with resonant ions and therefore must be repelled to the sample surface. Only the resonantly ionized atoms leave the ionization region and are accelerated into a double-focusing mass spectrometer. A resonant ionization mass spectrometer that can ablate the sample by either an argon ion gun or an Nd:YAG laser is described by Towric et al. (1990). The instrument is fitted with a TOF mass spectrometer of reflectron type.

A method of surface analysis by nonresonant multiphoton ionization in combination with TOF-MS (namely, surface analysis by laser ionization, or SALI) is described by Becker and Gillen (1984, 1985). Here the sample surface is bombarded with a pulsed 3 keV Ar^+ beam. Then the sputtered material is photoionized with a KrF laser beam at about 10^9 W/cm^2. In nonresonant multiphoton ionization, an untuned laser ionizes the neutrals through virtual electronic states. The ionizing laser is an excimer laser operating at 193 or 248 nm, with a repetition rate of 5–10 Hz and with pulses of 13 nsec pulse length. The nonresonant SALI method offers greater versatility than the resonant technique because the ionization is nonspecific and all masses are detected simultaneously. SALI should achieve sensitivities of 1 part in 10^9. An example of a study that has used such sensitivity is the 10^{-15} g sampling of a uranyl salt by Becker and Gillen (1985).

Laser ablation and post-laser ionization have been combined by Schueler et al. (1986). They add a second Nd:YAG pulsed laser, which is focused above and parallel to the sample surface and which ionizes the ablated neutrals via a nonresonant multiphoton ionization process. Separating the ablation and ionization steps and operating the ionizing laser at short wavelength (265 nm) and high-power density (up to 10^{12} W/cm^2) yielded high uniform ionization efficiency.

Today, LIMS essentially uses *Q*-switched Nd:YAG lasers at doubled, tripled, or quadrupled frequencies. Only in a few cases, CO_2 lasers, Nd:glass lasers, or excimer lasers are applied to evaporate and ionize the sample materials.

The selected laser parameters depend on the nature of the measurement. In the case of laser ionization mass spectrometry, typical parameters for the energy of the laser pulse are from 10 mJ to a few 100 mJ; for the pulse width, 10–100 ns; for the laser power density in the spot, from 10^9 to 10^{11} W/cm^2; and for the repetition rate of the laser pulses, from the single-shot regime to a few hundred kilohertz. By means of a large working distance objective the high-power laser beam is focused onto the sample surface to a diameter of only a few micrometers. The diameter and depth of the resulting crater for a single shot are in the same order of magnitude. In the case of a laser system with a higher repetition rate, for instance, 100 Hz or more, the sample position must be changed automatically after each shot by the movement of the sample holder with the aid of stepping motors. This is necessary because usually several shots fired at the same spot result in different focusing conditions, each with a different laser power density.

Most RIMS systems use Nd:YAG or excimer pumped dye lasers in the wavelengths range of 100–500 nm. Typical parameters of the dye lasers are energies from 0.1 to 200 mJ per pulse, pulse widths in the range of 10–20 ns, and pulse repetition rates from 1 to 100 Hz. In principle, any mass spectrometer type can be combined with a laser ion source. Because of the high spread of the initial energy of laser induced ions, the mass spectrometric separation system must be used in conjunction with an electrostatic energy compression device or filter, e.g., a TOF instrument with a reflectron or double-focusing sector field instrument. The high mass resolution necessary for an interference-free trace analysis can only be obtained with such instruments.

The ion current pulses produced in the ion source are on the order of picoamperes or less, and the pulse duration of a few microseconds requires high-speed electrical detection systems with ultralinear specifications and nanosecond recovery times.

The best ion recording method for the combination of laser ionization and double-focusing mass spectrometer is a photographic recording technique using an ion-sensitive photo emulsion, e.g., Q photoplates. The detection of ions by means of such photoplates allows for a simultaneous determination of the concentrations of all chemical elements in one step.

4C.4. ANALYTICAL FEATURES

4C.4.1. Fundamentals

In recent years, laser-induced mass spectrometry has become the approach of choice for very sensitive multielement and isotopic analyses. Owing to its marked evaporation and ionization capability, laser-induced mass spectrometry offers high, uniform sensitivity. In addition the low background facilitates low detection limits of the chemical elements. The mass separation of laser-produced ions and the easy identification of trace elements even at very low concentration levels—in the ppm or sub-ppm range—are characteristic features of laser-induced mass spectrometry.

This analytical technique can be applied in trace determination in solids, liquids, and gases. Gas analytical mass spectrometry (Megrue, 1970), the analysis of liquids (Dennemont et al., 1985), or trace analysis in organic samples (Yamamoto et al., 1984) are special fields distinct from trace analysis of inorganic solids and do not fall within the scope of this review.

An important advantage of laser-induced mass spectrometry is that this analytical method does not require extensive sample preparation, since laser ionization is possible for most of the materials with few limitations.

The identification of trace elements is very simple and is accomplished by qualitative analysis of their isotopes at given masses. Orientation in the mass spectra is easy, since the lines of different charge states (z) of atomic ions of major elements of the sample material appear at the corresponding m/z ratio. The mass spectrum of the sample to be analyzed can be predicted if the atomic, molecular, and cluster ions of the major, minor, and trace elements formed in the laser plasma are known. Thus, all line interferences appearing between the analysis ions of the analyte—usually these are single-charged atomic ions of the chemical element to be determined—and the disturbing ions, e.g., molecular and cluster ions formed in the laser plasma, charge exchange lines, and others, can also be determined. In quantitative trace analysis of inorganic samples, all important line interferences must be detected unambiguously. A mathematical correction of the line interferences is possible only in a few cases.

The quantitative mass spectrometric analysis is carried out by measuring the ion current extracted from the laser plasma. The ion current is proportional to the element concentration c_x of trace element x. A quantitative determination of element concentration c_x is carried out in comparison with an internal standard element with well-known concentration c_v. A main component (e.g., in the analysis of high-purity metals or semiconductors) can be used as an internal standard or known amount of a rare chemical element in the sample (e.g., Re or In) can be admixed to the sample material. The following analysis equation is applied for the evaluation of the concentration of chemical elements:

$$c_x = c_v \left(\frac{I_x}{I_v} \frac{E_v}{E_x} \frac{H_v}{H_x} \right)$$

Here I_v and I_x are the ion currents (or the numbers of ions) measured for the internal standard v and for trace element x; H_v and H_x are the respective isotopic abundances; and E_v and E_x are the respective relative element sensitivities. The relative sensitivity coefficient (RSC) E_x/E_v is a function of the physical and chemical properties of the sample and the analyte (ionization potential, melting point of the trace element, dissociation energy of analyzed compounds, sample composition, etc.) and of the laser ion source parameters (laser power density, ion focusing conditions, initial ion energy, etc.). By applying the RSC, all these effects can be considered indirectly and the accuracy (see Section 4C.4.2, next) of the analytical results can be improved.

4C.4.2. Detection Limits, Sensitivity, Precision, and Accuracy

The element-specific *detection limits* of laser-induced mass spectrometry are very similar for different elements and vary between 10 ppb and 1 ppm.

The absolute detection limits of metals deposited on thin epon foils measured with a TOF mass spectrometer are given by Heinen et al. (1979) (see Table 4C.1). The lowest detection limits (down to 10^{-20} g) were found for elements with a low ionization potential, i.e., for the alkali metals. Remarkably high absolute sensitivities—observed in the analysis of an ideal thin-film sample—are associated with a low relative sensitivity. An increase in the mass resolution of any mass spectrometric system (e.g., a TOF system with an electrostatic reflector or a double-focusing Mattauch–Herzog instrument) in order to resolve line interferences of atomic and molecular ions always leads to a loss of sensitivity. Both bulk and surface analyses of inorganic compounds can be performed with a spatial resolution of 0.5–1 μm.

The *absolute sensitivity* of laser–induced mass spectrometry using a classical static ion separation system varies between 10^{-8} and 10^{-12} g. This results in

Table 4C.1. Detection Limits of Thin-Film Analysis by LAMMA

	Detection Limit	
Element	Absolute (g)	Relative (ppm)
Li	1×10^{-20}	0.07
Na	2×10^{-20}	0.2
K	2×10^{-19}	0.1
Ca	2×10^{-19}	1.0
Cu	4×10^{-18}	20.0
Rb	5×10^{-20}	0.5
Cs	3×10^{-20}	0.3
Sr	4×10^{-19}	20.0
Ag	4×10^{-18}	1.0
Pb	1×10^{-19}	0.6
U	2×10^{-18}	20.0

Source: From Heinen et al. (1979).

an ion transmission reduced 10^3 times. Jansen and Witmer (1982) coupled an Nd:YAG laser with an AEI MS 702 double-focusing mass spectrometer. At a crater depth of 1 μm, about 10^{-9} g of sample material is vaporized. The detection limit per laser shot is 10^{-12} g. The relative standard deviation per shot is about 10%.

The *relative sensitivity* of a laser-induced mass spectrometer using a Mattauch–Herzog configuration is generally somewhat better than that of a TOF mass spectrometer and is about 10 ppb. Laser-induced mass spectrometry features a sensitivity variation between different elements and also a matrix-dependent variation in sensitivity.

To estimate the relative sensitivity of their double-focusing laser ionization mass spectrometer (EMAL 2), Briukhanov et al. (1983) analyzed a copper standard. All concentrations were calculated without taking into account mass correction. The intensities of the trace elements were compared to the intensity of Cu-63.

As is seen from Table 4C.2, the error of trace analysis for most elements is not more than 35% when no standards are used. Errors between 20% and 50% in the ppm and ppb range are normal in standardless mass spectrometry.

The *accuracy* of laser-induced mass spectrometry depends on the calibration procedure used. Measuring standard samples, it is possible to obtain relative sensitivity coefficients (RSCs). The RSC of a chemical element RSC(x) is determined by dividing the measured concentration c_x by the certified concentration c_s of standard samples for a given element. Usually, relative sensitivity coeffi-

Table 4C.2. Trace Analysis of Elements in a Copper Sample with EMAL-2

Element	$c_{certified}$ (ppm)	$c_{experimental}$ (ppm)	Error (%)
Fe	116	120	3.4
Mg	81	57	30
Cr	151	110	27
Mn	50	85	66
Ni	47	62	31
Zn	57	31	45
Ag	34	38	12
As	52	34	35
Sn	30	37	21
Sb	32	21	34
Pb	16	13	18
Bi	14	9	35
Si	81	97	19

Source: From Briukhanov et al. (1983).

cients are defined on the basis of a certain matrix and are applicable only to a particular analysis technique with specified analytical parameters. Correcting the concentration with a measured RSC provides analysis results with improved accuracy. The RSC can also be applied in trace analyses using ultrasensitive mass spectrometry, provided that the RSCs have been determined at higher concentrations of the same element in the calibration sample. Any influence of the matrix should be taken into account, but homogeneous standard reference materials are not available for many different matrices. In such a case the preparation of a mixed synthetic standard using ultrapure chemicals with given elemental concentrations is necessary in order to determine chemical elements in a specific matrix. The maximum differences of experimental RSCs using different matrices do not exceed a factor of 10.

Table 4C.3. Determination of the Concentration of Trace Elements in Zr-NBS SRM 1235[a]

Element	c_{NBS} (ppm)	$c_{experimental}$ (ppm)	RSC
Na	270	—	—
Mg	8	—	—
Al	(810)[b]	105	(7.71)
Si	96	95	1.01
P	42	44	0.95
S	46	—	—
K	(1000)[b]	—	—
Ca	70	—	—
Ti	230	90	2.55
Cr	340	60	5.66
Mn	26	25	1.04
Cu	44	80	0.55
Fe	930	850	1.09
Co	20	20	1.00
Ni	70	65	1.08
Nb	210	200	1.05
Mo	38	40	0.95
Sn	25	25	1.00
Hf	52	95	0.55
Ta	70	280	0.25
W	14	50	0.28

Source: From Dietze and Becker (1985a).

[a]Only the concentration of Hf in the Zr-NBS standard was certified (NBS = U.S. National Bureau of Standards).

[b]Contamination possible.

Ramendik et al. (1987) and Ramendik (1986, 1990) showed that RSCs can be derived by considering the atomization energy and ionization potential of the elements. A theoretical calculation of RSCs on the basis of physical and chemical properties such as melting point, boiling point, heat of sublimation, and ionization potential of the elements has been proposed. These empirical relations are often not satisfactory because indeed differences between the theoretical and the respective experimental RSCs have been observed. RSCs were also calculated by Vertes et al. (1989).

Table 4C.3 summarizes the experimental results on the laser ionization mass spectrometry of a zirconium standard (Zr-NBS SRM 1235) and the resulting RSC values at laser power density 5×10^8 W/cm^2. In an exposure of the ion-sensitive photoplate to 100 nC, a detection limit of 10 ppb was obtained.

The *precision* of the analysis results was, on average, 25–30% RSD (relative standard deviation). For many elements, the RSCs were in the vicinity of unity. In agreement with laser-induced mass spectrometric investigations by Bingham and Salter (1976), who used a CO_2 laser, and in accordance with the SSMS of elements with high boiling points (e.g., Hf, Ta, or W), a poor sensitivity (low RSC) was observed. Using a ruby or Nd:YAG laser, these investigators compared the relative sensitivity factors for some elements as a function of the laser wavelength and also found that the RSCs are in the vicinity of unity.

When a CO_2 laser was used, the sensitivity varied considerably in some cases. In laser-induced mass spectrometry for the trace analysis of inorganic sample materials, usually Nd:YAG laser systems (sometimes with frequency

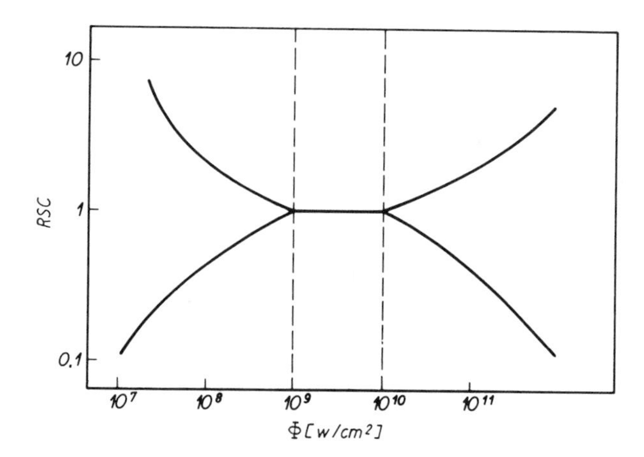

Figure 4C.11. General dependence of relative sensitivity of chemical elements (RSC) on laser power density (Belousov, 1984).

multiplication) have been applied because these are easy to handle and their cost is relatively low.

The relative sensitivities of elements can also vary as a function of laser power density. This dependence is shown in a general diagram in Figure 4C.11. In the laser power density interval of 10^9–10^{10} W/cm^2, the RSCs are nearly constant around 1 (Jansen and Witmer, 1982; Bykovskii et al., 1978b; Belousov, 1984). This range of laser power density is favorable for mass spectrometric analysis.

Several authors have found a correlation of the RSCs with the ionization potentials of elements (Ramendik et al., 1987; Adams, 1983; Surkyn and Adams, 1982; Beusen et al., 1983). Furthermore, the literature attests to the use of semiempirical correction procedures for the quantitative analysis of laser mass spectra such as the local thermal equilibrium model proposed by Anderson and Hinthorne (1973) for SIMS analysis.

Matus et al. (1988) compared the RSCs of LIMS and SSMS for a steel standard NBS 1161 and an oceanic basalt ML-3. As can be seen from Figure

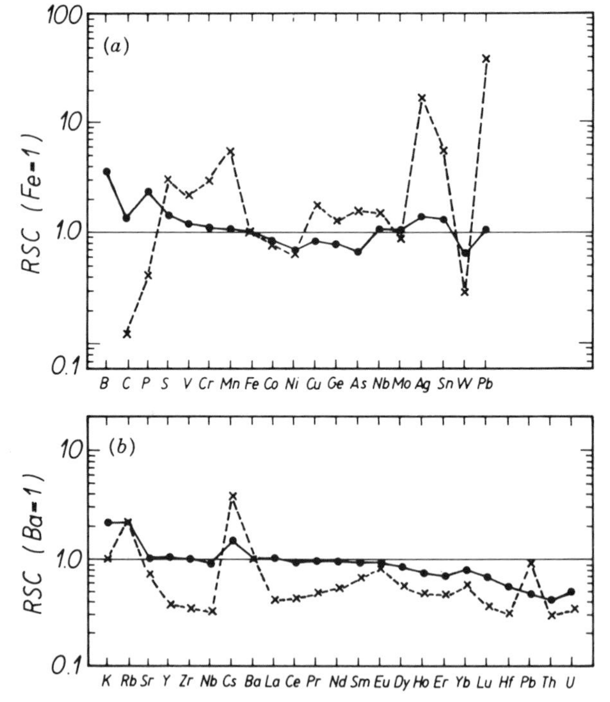

Figure 4C.12. Comparison of relative sensitivity coefficients obtained from the analysis of (a) steel standard NBS 1161 and (b) ocean island basalt ML-3 (Hawaii) by means of LIMS (solid circles) and SSMS (crosses) (Matus et al., 1988).

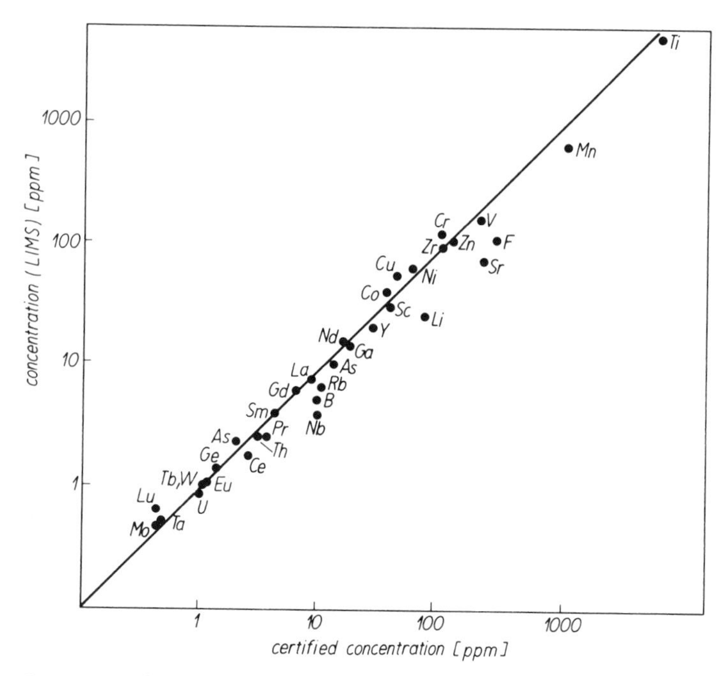

Figure 4C.13. Comparison of concentration measured by LIMS and certified values of the geological ZGI-BM standard (Bykovskii et al., 1978b).

4C.12, the difference in the sensitivity of elements in LIMS is much smaller than in SSMS. Figure 4C.13 shows the correlation of measured values (of Bykovskii et al., 1978b) and certified values of a geological standard sample (ZGI-BM). Similar results were obtained in the analysis of geological standards by Jansen and Witmer (1982).

The critical factor responsible for the sensitivity, accuracy, and precision of the analytical method is a reproducible ion formation rate in the laser ion source. This can be influenced by the variation of laser parameters, e.g., of the laser power density during the interaction of the laser beam with the sample. The main reason for the fluctuation of ion currents in laser-induced mass spectrometry is the inhomogeneity of the sample material. The problems of inhomogeneities are greater in LIMS than in SSMS because, with the laser, a smaller sample volume is evaporated and ionized. To reduce analytical errors in mass spectrometric measurements, laser excitation is carried out by scanning the target surface with the laser beam or by movement of the target (e.g., rotation). On the other hand, inhomogeneities can be reduced by thorough powdering and mixing of the sample material, e.g., in the trace analysis of geological samples. Inhomogeneities can also be avoided by dissolving the sample chemically. The sample must be handled very carefully to avoid possible contamination of the sample material. The sensitivity can be also increased significantly by additional chemical separation of the matrix elements, but this approach is only rarely followed.

Further improvement of trace analyses can be attained by application of the powerful isotope dilution technique (ID-MS). However, ID-MS is costly because of the extensive sample preparation processes required. ID-MS has only long been used as a single or multielement analytical method in combination with thermal ionization (Heumann, 1988), laser ionization (Jochum et al., 1988), or spark ionization (Dietze, 1979; Dietze and Opauszky, 1979). The sample preparation is similar in all cases. The sample material is dissolved completely in acids, and an isotopically spiked solution is added. In laser-induced mass spectrometry a small amount of the solution is dried on a target surface (e.g., high-purity Ag). To avoid matrix effects, the matrix elements can also be separated before adding the spiked solution. An accuracy of 2–5% can be expected in this case. The isotopic spikes can also be used as internal standards in trace analysis. Jochum and co-workers (1988) have described some geochemical applications of isotope dilution techniques in combination with SSMS and LIMS. The lengthy dissolution step can be avoided: it is sufficient to mix the rock powder with "spiked" graphite. A different ion yield of the analyte and the spike (a mixture of the sample powder with spiked graphite) as a result of the existence of different chemical compounds of analyte and spike has recently been observed (K. P. Jochum and H. M. Seufert, private communications, 1991). By dissolving the mixture of sample powder with spiked graphite, the analyte and spike can be compounded in the same chemical form and so uniform atomization and ionization of analyte and spike can be expected.

4C.4.3. Appearance of Molecular and Cluster Ions in Laser-Induced Mass Spectra

Molecular and cluster ions are well known in laser-induced mass spectrometry but play a subordinate role in common elemental analysis because of their minor intensities. In trace analysis of inorganic sample materials, however, the molecular and cluster ions M_n^+ and C_n^+ (and also the multiply charged ions A^{z+}) formed in the laser plasma disturb the atomic ions of analyte by interferences. A mass resolution of up to 10,000 is required for the separation of all of these interfering ions from the atomic ions A^+.

The problem of cluster formation is similar in the various kinds of ionization. The disturbing interferences between the ions of analyte and clusters or molecules lead to a decrease of sensitivity of the mass spectrometric analysis method. The cluster formation rate in plasma sources [e.g., inductively coupled plasma (ICP) and glow discharge] is much higher than in laser plasma. Under certain experimental conditions (see below) the cluster formation in laser plasma can be suppressed. A systematic study of the types of clusters formed in a laser plasma is useful for estimating mass spectral interferences of cluster and atomic ions of the same nominal mass. Therefore, the knowledge of cluster formation and abundance distributions is of considerable importance for mass spectrometric analysis as well as for the understanding of chemical and physical processes in laser plasmas.

Characteristically, for example, during the mass spectrometric analysis of boron compounds, one detects high cluster formation rate in laser plasma (Becker and Dietze, 1986a). During trace analysis of boron compounds, one must investigate and consider all line interferences between the analysis and cluster ions in order to obtain high accuracy of the mass spectrometric method. This problem may be especially complicated for ultratrace analysis of high-purity boron compounds. In this case the intensities of disturbing cluster ions are a few orders of magnitude higher than the intensities of atomic ions of traces in the investigated sample material. In the laser mass spectra of boron nitride at the maximum mass resolution, mass line multiplets are observed. At a maximal experimental mass resolution of 17,000 of the laser ion mass spectrometric system (Dietze and Becker, 1985b) in some mass spectra up to 5 lines at one mass can be detected.

Possible line interferences in the laser mass spectra of boron nitride for trace analysis of magnesium and sulfur are summarized in Table 4C.4. The abundance of cluster ions (respectively to B^+) given in Table 4C.4 are measured experimentally for the clusters with the isotopes ^{11}B and ^{14}N. The table shows that the mass resolution for the mass separation of atomic ions and neighboring cluster ions at the same mass in every case is lower than 1000.

The necessary mass resolution for the trace elements and boron nitride clusters as a function of mass number is shown in Figure 4C.14. In contrast,

Table 4C.4. Possible Line Interferences in the Laser Mass Spectrum of a Boron Nitride Sample for Trace Analysis of Magnesium and Zinc

Analysis Ion	m (amu)	Cluster Ion	m (amu)	A^a	Abundance of Cluster
$^{24}Mg^+$	23.9850	$^{10}B^{14}N^+$	24.0160	770	4.7×10^{-4}
$^{25}Mg^+$	24.9858	$^{11}B^{14}N^+$	25.0124	940	1.9×10^{-3}
		$^{10}B^{15}N^+$	25.0130	14670	1.4×10^{-6}
$^{26}Mg^+$	25.9826	$^{11}B^{15}N^+$	26.0094	970	5.7×10^{-6}
$^{32}S^+$	31.9721	$^{10}B^{11}B_2^+$	32.0316	540	2.3×10^{-2}
$^{33}S^+$	32.9715	$^{11}B_3^+$	33.0279	585	9.2×10^{-2}
$^{34}S^+$	33.9679	$^{10}B_2{}^{14}N^+$	34.0289	560	1.4×10^{-1}

[a]Mass resolution: $A = m/\Delta m$.

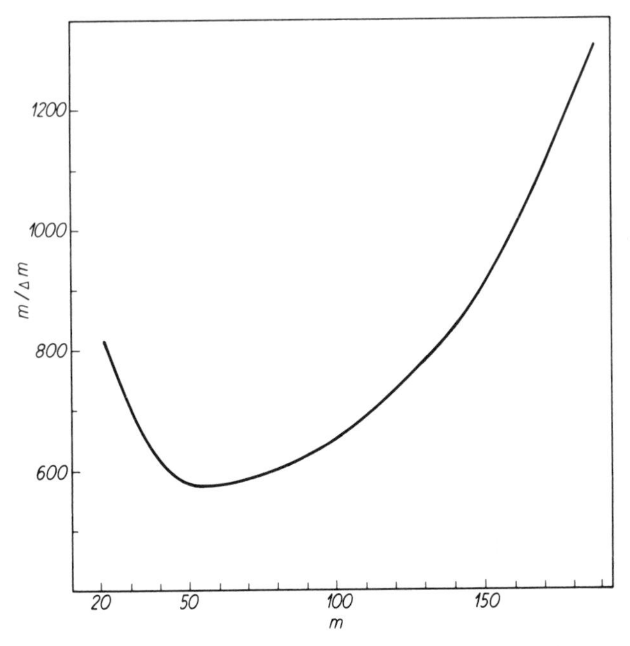

Figure 4C.14. Necessary mass resolution for trace analysis in boron nitride as a function of mass number.

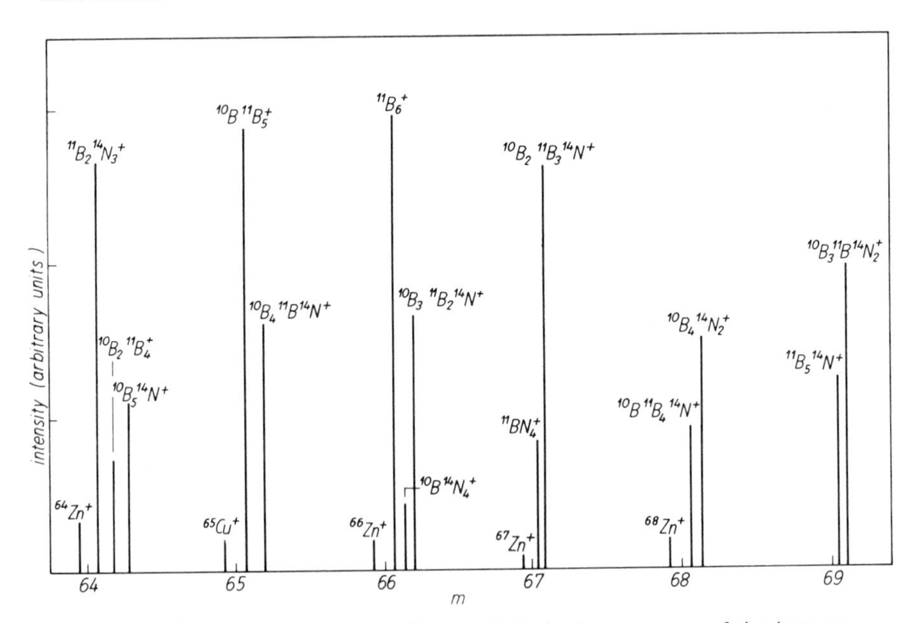

Figure 4C.15. Laser mass spectrum of boron nitride in the mass range of zinc isotopes.

the line interferences between the different cluster ions of boron nitride require a significantly higher mass resolution. The necessary mass resolution is especially high for separation of clusters with the isotope ^{15}N of the neighboring cluster ions. But these cluster species are expected to have very low intensities since the abundance of the natural isotope ^{15}N is 0.365%. The abundance of $^{10}B^{15}N^+$ in Table 4C.4 has been estimated. Figure 4C.15 shows a part of a laser mass spectrum of boron nitride in the mass range of zinc isotopes. The necessary mass resolution to separate atomic ions of zinc and the neighboring boron nitride cluster ion is not more than 650.

Cluster formation is also of interest for several practical purposes. For example, new clusters can be synthesized via laser-induced reactions of atomic and molecular species, which eventually, by their deposition on substrates, can result in thin films with interesting properties (Martin, 1986; Duncan and Rouvray, 1989; Dietze and Becker, 1987; Becker and Dietze, 1990; Becker et al., 1990).

For inorganic trace analysis by laser-induced mass spectrometry, those experimental conditions are of interest under which the cluster formation rates are minimal. The influence of laser power density on cluster intensities is shown in Figure 4C.16 (Becker and Dietze, 1987; for a theoretical interpretation of cluster formation processes of carbon and boron nitride, see Seifert

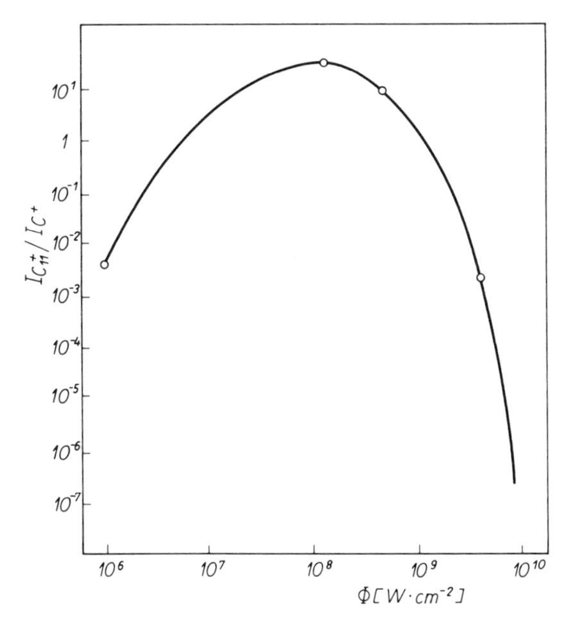

Figure 4C.16. Relative intensity of C_{11}^+ cluster ions as a function of laser power density.

Table 4C.5. Investigation of Cluster Formation of Inorganic Species in Laser Plasmas

Reference	Laser	MS	Target	Detected Clusters
Fürstenau et al. (1979); Fürstenau (1981)	Ruby	LAMMA	Graphite	C_n^+
Fürstenau and Hillenkamp (1981)	Ruby ($\lambda = 347$ nm)	LAMMA	Ge Foil Si foil Ag foil Au foil Al foil	Ge_n^+ Si_n^+ Ag_n^+ Au_n^+ Al_n^+
Michiels et al. (1982); Michiels and Gijbels (1983); Michiels et al. (1983)	Nd:YAG ($\lambda = 265$ nm)	LAMMA	Si SiO_2	Si_m^+ $Si_mO_n^+$
Michiels and Gijbels (1983)	Nd:YAG ($\lambda = 265$ nm)	LAMMA	Ti_nO_m	$Ti_nO_m^+$
Michiels and Gijbels (1984)	Nd:YAG ($\lambda = 265$ nm)	LAMMA	Sc_2O_3, Ho_2O_3, BeO, ZnO, CuO, Ag_2O, As_2O_3, Nb_2O_5, Ta_2O_5, V_2O_3, V_2O_5, TeO_2, MnO_2, Sb_2O_4, Bi_2O_3	$M_nO_m^+$

Dietze et al. (1981, 1983); Dietze and Becker (1985b, 1987, 1988a,b); Becker and Dietze (1985, 1986a,b, 1987, 1988); Becker et al. (1986)	Nd:YAG ($\lambda = 1064$ nm)	MH	Graphite, metal oxide	C_n^+, UC_n^+, ThC_n^+, WC_n^+, WO_n^+
			High-T_c superconductors ($YBa_2Cu_3O_x$)	$Y_nO_m^+$, $Ba_nO_m^+$, $YBa_nO_m^+$, $Ba_nCuO_m^+$, $YCuO_n^+$
			Boron nitride	$B_nN_m^+$
			Boron carbide	$B_nC_m^+$
			REE[a] oxides/graphite mixture	Oxide and carbide clusters of REE
Surkyn and Adams (1982)	(Nd:YAG ($\lambda = 265$ nm)	LAMMA	Glass	BaO^+, $BaAlO_n^+$, $BaCaO^+$, $BaFeO^+$, $FeAlO^+$, $CaAlO^+$, $Si_mO_n^+$, $Si_mAlO_n^+$
Dennemont et al. (1982, 1989)	Nd:YAG	LAMMA	$NaNO_3$	$Na_nN_mO_r^{+/-}$
			Aluminosilicate	$Al_nSi_mO_r^{+/-}$
			Alkali halide	$M_nX_m^{+/-}$
Kroto et al. (1985)	Nd:YAG	TOF	Graphite	C_n^+, C_{60}^+
Nadahara et al. (1985)	Nd:YAG ($\lambda = 265$ nm)	LAMMA	Airborne particulates	C_n^-, $Si_2O_n^+$, $NaAl_2Si_2O_n^+$, $N_nO_m^-$, SO_n^+, HSO_4^+, TiO^+
Vanderborgh and Jones (1983)	Nd:YAG ($\lambda = 265$ nm)	LAMMA	Coals	$FeAl^+$, FeS_n^+, MgO_n^+, FeO^+, TiO^+, SiO^+

[a]REE = rare earth elements.

et al., 1988a,b). A maximum of the cluster formation rate is observed at 10^8 W/cm^2, similar to other systems [e.g., in carbide formation using rare earth oxide and a graphite mixture target (Becker and Dietze, 1987)]. At this laser power density, the intensities of the cluster ions are higher than those of the atomic ions to be analyzed. The abundance distribution of cluster ions in laser plasma under the influence of weak electrical fields has been investigated by Dietze and Becker (1987). A significant variation of ion intensities of tungsten carbides and carbon clusters as a function of the retarding potential was observed.

At high laser power density, $\Phi > 10^{10}$ W/cm^2, cluster formation is negligible because of the high dissociation rate of possible clusters formed in the laser plasma. In laser mass spectra, using a graphite target at a laser power density of about 10^{10} W/cm^2, Bykovskii et al. (1978a) observed only C_2^+ and C_3^+ clusters with intensities of 10^{-5} and 10^{-7}, respectively, relative to atomic ions. From the point of view of the cluster formation rate and the constancy of relative sensitivity coefficients of chemical elements for inorganic trace analysis, this laser power density is most suitable.

Essentially, contributions to our understanding of the cluster formation in laser plasmas have been established by the Hillenkamp group in Frankfurt and later in Münster (see Fürstenau and Hillenkamp, 1981), by Gijbels and Michiels's group in Antwerp (see Michiels et al., 1982), and by Dietze and Becker in Leipzig (1987).

Michiels et al. (1983) found a strong correlation between the positive and negative cluster ion distribution and the valence electron configuration of the metal in the oxide. Also, the intensities of positively charged monocarbide and dicarbide ions of rare earth elements correlate with the dissociation energies of the molecules, depending directly on the transition energy from the $4f^{n-1}5d6s^2$ electronic state to the $4f^n6s^2$ electronic state (Becker and Dietze, 1985).

Inorganic cluster ions were also studied by Linton et al. (1987) in attempts to explain the relationships between the sample composition and the characteristic cluster ion distributions. Table 4C.5 presents a systematic list of studies of exemplary cluster formation in laser plasmas.

4C.5. APPLICATIONS OF LASER-INDUCED MASS SPECTROMETRIC METHODS

Laser-induced mass spectrometric methods can be applied in versatile ways. Laser source mass spectrometers have been used in mineralogy, geology, chemistry, biology, materials research, environmental studies, and other research areas (Yokozawa et al., 1987; Mauney and Adams, 1984; Bruynseels et al., 1990; Valerio, 1984). LAMMA instruments were initially developed for biological and medical research (Hirche et al., 1981; Kupka et al., 1981) and are most frequently applied in the structural analysis of organic compounds (Hillenkamp et al., 1975; Verbueken et al., 1985). TOF-MS is well suited for the laser desorption of large organic molecules, since its mass range is unlimited and it records the complete spectrum simultaneously and instantaneously. LAMMA provides good facilities for the analysis of particulate materials, e.g., of aerosol particles deposited on a foil. The particles are visible under the microscope and can be analyzed individually. The application of laser-induced mass spectrometry (especially of LAMMA) in aerosol research has been discussed by Wieser et al. (1981), Nadahara et al. (1985), Kaufmann and Wieser (1980), and Kaufmann et al. (1980). These papers report on the analysis of individual aerosol particles with LAMMA for their major, minor, and trace elemental components.

Laser source mass spectrometry of solids is carried out mostly without prior chemical treatment and covers the whole periodical table. Some applications in inorganic analytical mass spectrometry are given by Conzemius and Capellen (1980) and Conzemius et al. (1983). An extensive reference and abstract index of laser-induced mass spectrometry research can be found in Kaufmann (1991). The material is also available as a data base on diskette.

4C.5.1. Application to Metals

A typical field of the application of laser-induced mass spectrometry in trace analysis is the purity control of metal samples. Metallic specimens appear to be ideal for analysis by laser ionization mass spectrometry. Furthermore, metals with a high boiling point can be analyzed.

Commercialization of laser-induced TOF mass spectrometers (especially LAMMA and LIMA) has made them fit for use in a wide range of areas. For example, in materials research and in metallurgy, they are utilized in the analysis of light elements in alloys and segregates formed during brazing and in corrosion studies (Kohler et al., 1989).

Table 4C.6 lists examples of analytical investigations in which laser ionization mass spectrometers were used to characterize metals and alloys.

4C.5.2. Application to Semiconductors and Insulators

Among the rapid developments of microelectronics technology, laser-induced mass spectrometry has provided a new possibility for analytical investigations of semiconducting materials for their trace impurities and for the concentrations of doping elements.

Table 4C.6. Applications of LIMS to Metals

	Laser						
Ref.	System (MS)	Wave-length (nm)	Power density ($W \cdot cm^{-2}$)	Energy (J)	Pulse Length (μs)	Target	Element
Heinen et al. (1979)	Nd:YAG LAMMA	265 353	10^8–10^{11}	10^{-4}	—	Steel alloys	Si, Mo, Co, Cu, W, Sn, Nb, Pb, Zr
Bykovskii et al. (1978a)	Nd:YAG EMAL 2	1064	10^9–10^{10}	—	0.015	Cu Sn	Fe, Mg, Cr, Mn Ni, Zn, Ag, As, Sb, Bi, Si
Bingham and Salter (1976)	Nd:YAG	1064	10^9–10^{11}	10^{-2}	0.015	Steel	Pb, W, Ta, Sn, Mo, Nb, Zr, Ag
	CO_2, MH[a]	10600		3×10^{-1}	0.1		As, Ge, Cu, Ni, Co, Mn, Cr, V, Ti, S, B
	Nd:YAG	1064	10^9–10^{11}	10^{-2}	0.015	Cu	Pb, Bi, Sb, Sn, Ag, Ga, Cr
Kovalev et al. (1978, 1985)	Nd:YAG LAMMA	1064	5×10^9	—	—	Er, Ti Zn, Y Mg, Cd	C, Na, Al, Si Cu, Ca, Fe, K Mn, Mo, Y, Mg
Conzemius and Svec (1978)	Nd:YAG MH	1064	2×10^9	10^{-3}	0.2	Brass Fe	Be, Al, Si, P Mn, Ni, As, W, Ag, Cd, Sn, Pb, Ta, Co

Huang et al. (1987)	Nd:YAG MH	1064	2×10^{9}	10^{-3}	0.1	Steel standard	Mn, Ni, Cu, Zn, As, Co, Cr
Dietze and Becker (1985a, 1991)	Nd:YAG MH	1064	10^{9}	10^{-3}	0.1	Zr NBS, Ta	Na, Mg, Al, P, Si, S, Cl, Ca, K, Ti, Cr, Mn, V, Cu, Fe, Co
						Steel standard	Ni, Nb, Mo, W, Sn, Hf, Ta
Hamer et al. (1981)	Nd:YAG LAMMA	1064				Fe–Ti alloys	Li, B, Mg, Na, Al, Si, K, Ca, Mn, Fe, Ba
Jansen and Witmer (1982)	Nd:YAG MH	1064	10^{10}–10^{11}	10^{-1}	0.015	Al standard	Pb, Sb, Sn, Ga, Zn, Cu, Co, Fe, Mn, Ni, Cd, Mg, Si
Matus et al. (1988)	Nd:YAG MH	1064	2×10^{10}	1.5×10^{-2}	0.015	Steel NBS standard	B, C, P, S, Ni, V, Cr, Mn, Co, Cu, Ge, As, W, Nb, Mo, Ag, Sn, Pb
Singh (1987)	Nd:YAG LIMA	532 265	$<10^{13}$	1.5×10^{-1}	0.006	Al–Li alloy	Na, Mg, K, Ca, Fe, Cu, Ga
Svec (1984)	Nd:YAG MH	1064	10^{9}		0.1	Steel standard	C, N, O

[a]MH = Mattauch-Herzog-type instrument.

Nonconducting materials that are not completely transparent to laser radiation can be analyzed directly. For visually transparent samples, an Nd:YAG or ruby laser with multiplied frequencies is used.

LIMS is undoubtedly destined to analyze insulators, especially ceramics. For most mass spectrometric methods the analysis of such samples is problematic, e.g., in SSMS the insulating sample material must be mixed with a extremely pure conducting material (powdered graphite, gold, or silver).

The concentrations of trace elements in a boron nitride sample used for laser-induced plasma deposition in preparation of thin boron nitride films are listed in Table 4C.7 (Becker and Dietze, 1991). The alkali and earth alkali elements with the highest concentrations in the boron nitride sample are included. The laser ionization was carried out with a Nd:YAG laser at 1064 nm. For a sensitive laser ionization of boron nitride, a laser system with short wavelength and high laser energy should be utilized.

Semiconducting or insulating powders can be pressed in a Teflon or polyethylene die, and the pellet is mounted in the ion source chamber of the laser ionization mass spectrometer. Thus, the microanalysis of high-purity GaAs and $Hg_{0.78}Cd_{0.22}Te$ samples (Schueler and Odom, 1987) was carried out by means of laser ablation in combination with laser ionization using two Nd:YAG laser systems in a LIMA 2A instrument. Daniel et al. (1988) performed semiconductor analyses with the aid of a LAMMA instrument to characterize ionic contaminations of integrated circuits. In Table 4C.8 we

Table C.7. Results of Trace Analysis of a Boron Nitride Sample by LIMS

Element	Concentration (ppm)
F	35
Na	590
Mg	90
Al	280
P	240
S	160
Cl	240
K	300
Ca	210
Ti	490
V	14
Fe	920
Co	30
Zn	110
Cu	195

Table 4C.8. Applications of LIMS to Insulators

Ref.	Laser: System (MS)	Laser: Wave-length (nm)	Laser: Power density ($W \cdot cm^{-2}$)	Laser: Energy (J)	Laser: Pulse Length (ms)	Target	Element
Eloy (1978)	Nd:YAG single-foc. MS	353	10^8–10^{10}	10^{-7}–10^{-5}	3×10^{-2}	Glass	O, Si, Na, Al, Ca, Cr, Mn, Zn, Sr, Zr, Mo
Bingham and Salter (1976)	Nd:YAG MH	1064	10^9–10^{11}	10^{-2}	1.5×10^{-2}	Glass	Pb, B, U, Sr, Th, Ag, Zn, Co, Fe, Cu, Mn, Rb, Ti, K
Surkyn and Adams (1982)	Nd:YAG LAMMA	265	10^8–10^{10}	5×10^{-8}–2.5×10^{-5}	1.5×10^{-2}	Glass	Na, Mg, Al, P, Si, K, Ca, Ti, Mn, Fe, Ba
Spurny et al. (1981)	LAMMA	265	10^8–10^{10}	—	1.5×10^{-2}	Glass fibers	Mg, K, Ca, Fe
Leybold-Hereaus (1982)	Nd:YAG LAMMA	265	10^8–10^{10}	—	1.5×10^{-2}	Glass NBS standard	K, Ca, Sc, Ti, Mn, Fe, Ni, Co, Cu, Zn, Ga, Ge, As, Sb, Rb, Si, Cr, Y, Zr, Nb, Mo, Ag, Cd, In, Sn, Se, Te, Cs, Ba
Sanderson (1985)	Nd:YAG MH	1064	2×10^{11}	—	—	Glass NBS standard	K, Ti, Fe, Co, Ni, Cu, Rb, Ag, Ba, La, Ce, Nd, Eu, Pb, Th, U
Michiels et al. (1984)	Nd:YAG LAMMA	265	10^8–10^{11}	—	1.5×10^{-2}	Glass NBS standard	Li, Be, B, Mg, Si, K, Ti, Y, Nb, In, Cs, Ba, La, Ce, Pr, Ho, Tm, Lu, Ta, Pb, Th, U
Becker and Dietze (1991)	Nd:YAG MH	1064	10^9	10^{-3}	0.1	Boron compounds: carbide, oxide, & nitride	F, Na, Mg, Al, Si, S, Cl, Ca, K, Ti, Cr, Mn, V, Cu, Fe, Co, Ni, Nb, Mo, W, P, Sn, Hf, Ta, REE

have summarized some applications of laser-induced mass spectrometry in insulator research.

Some additional recent applications of laser microprobe mass spectrometry to semiconductors (Si, GaAs, InAs, and GaP) and electronic materials are described in Nishikawa et al. (1986a,b), Cerezo et al. (1986), and Adachi et al. (1986).

4C.5.3. Geological Applications

Laser-induced mass spectrometry has been applied to the analysis of a series of geological samples (Jochum et al., 1988; Eloy et al., 1983; Eloy, 1985; Weinke et al., 1983; Vanderborgh and Jones, 1983; Bykovskii et al., 1978b). Mostly, the samples were analyzed directly by scanning the target using constant laser parameters. By local analysis in single-shot regimes, it was possible to determine inclusions and special mineral phases in these samples. For the bulk and trace analysis of geological materials, inhomogeneous samples must be homogenized by pulverization or chemical dissolution. The direct analysis of geological samples is always preferred because the possibility of contamination of the sample is low. When Jochum et al. (1988) applied the multielement isotope dilution technique to the mass spectrometric analysis of geological samples, inhomogeneity of the samples was avoided by dissolving the rock samples in acids and drying them onto a quartz disk. Figure 4C.17 compares

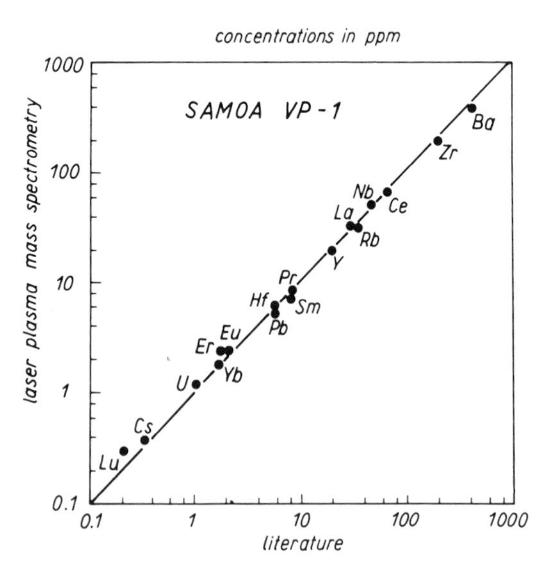

Figure 4C.17. Results of laser ionization mass spectrometry of an oceanic basalt in comparison with data in the literature (Jochum et al., 1988).

the experimental results of an oceanic basalt from Somoa with values reported in the literature. About 90% of the data deviate less than ±15%.

The concentration of trace elements uranium, thorium, and lead in zirconium deposits were determined by Steel et al. (1984). Weinke et al. (1983) investigated meteoritic troilite and terrestrial pyrite. The concentrations of the trace elements were calculated using a method based on the local thermodynamic equilibrium (LTE) model. Englert and Herpes (1980) analyzed osmium-containing specimens from different geological deposits and found isotopic anomalies for ^{187}Os as a result of the β-decay of ^{187}Re, using the Re–Os method for age determination of the minerals. The advantage of laser-induced mass spectrometry in geological research is the combination of trace element analysis with accurate isotopic abundance measurements. It is used for determining ages of minerals, and the local analysis of inclusions and mineral phases in geological samples based on high spatial resolution of the method.

Eloy et al. (1983) carried out laser-induced mass spectrometric analysis of microfluid inclusions ($< 10\,\mu m$) inside geological crystals. Some applications of LIMS in geological research are summarized in Table 4C.9.

4C.5.4. Applications of RIMS

Although RIMS has only been used for a few years, numerous applications have already been reported, which is indicative of the fact that this method is a valuable supplement to the well-known trace analytical methods. Figure 4C.5 gives a survey of the analytical applicability of RIS and demonstrates the wide potential of RIMS for the trace analysis of inorganic sample materials.

Table 4C.10 lists the results of applications of RIMS to trace analysis, along with the pertinent references. Further references to the application of elemental RIMS are given by Smith et al. (1989) for all elements of the periodic table.

RIMS is a method nearly free from matrix effects, with good linearity and limited mass interferences, e.g., it is possible to separate isobaric interferences (see Figure 4C.18), and the detection limits are in the ppb or sub-ppb range.

For example, Bekov et al. (1989) investigated boron impurities in the ultra-low concentration range (less than 1 ppb) in highly pure germanium and measured noble metal traces in environmental samples. The selective resonance photoionization was combined with a TOF mass spectrometer. The concentrations determined for four elements and the detection limits are summarized in Table 4C.11.

Detection limits in the ppt range were obtained, e.g., for Au, Pt, and Rh in seawater (Bekov et al., 1988); for Ba in aqueous samples containing 290 fg of this element (Bushaw and Gerke, 1988); and for Pu (Ruster et al., 1989;

Table 4C.9. Geological Applications of LIMS

Ref.	Laser					Target	Element
	System (MS)	Wave-length (nm)	Power density (W·cm^{-2})	Energy (J)	Pulse Length (ms)		
Eloy et al. (1983)	Nd:YAG magnetic sector	353	10^8–10^{10}	10^{-7}–10^{-5}	3×10^{-2}	Zircons meteorite	Y, Ce, La, U
Weinke et al. (1983)	Nd:YAG LAMMA	265	10^8–10^{11}	—	1.5×10^{-2}	Fe-meteorite pyrites	Ti, V, Ni, Mn, Co, Cr, Cu
Vanderborgh and Jones (1983)	Nd:YAG LAMMA	265	10^7–10^{11}	—	1.5×10^{-2}	Coal shale	Na, Mg, Al, Si, P, Ca, K, Ti, Fe, Mn, Ni, Co, Cu, Zn, Sr, Mo
Bykovskii et al. (1987b)	Nd:YAG MH	1064	10^9–10^{10}	—	1.5×10^{-2}	Geological standards: meteroties Allende	Li, B, F, P, K, Cl, Sc, Ti, V, Mn, Co, Cu, Zn, Ga, Ge, As, Se, Br, Rb, Sr, Y, Nb, Mo, Ru, Cd, Te, Cs, Ba, La, Ce, W, Os, Pt, Ta

Jochum et al. (1988)	Nd:YAG MH	1064	10^9–10^{10}	1.5×10^{-3}	1.5×10^{-2}	Basalt (ocean island)	Ba, Zr, Ce, Nb, La, Rb, Y, Pr, Sm, Hf, Pb, Eu, Er, Yb, U, Cs, Lu
Jansen and Witmer (1982)	Nd:YAG MH	1064	10^{10}–10^{11}	1×10^{-1}	1.5×10^{-2}	Standard rock AGV-1	Zr, Ba
Spurny et al. (1981)	LAMMA	265	10^8–10^{11}	—	—	Zeolites	Mg, K, Ca, Fe
Beusen et al. (1983)	Nd:YAG LAMMA	265	—	—	1.5×10^{-2}	Horn-blende, augite, lepidolite	Al, Fe, K, Ca, Li, F, Rb, Cs, Ti, Cr, Zr, Mn, Mg, Na
Dietze and Becker (1985b)	Nd:YAG MH	1064	10^9	10^{-3}	1×10^{-1}	Geological standard BM-ZGI	Na, Mg, Al, Si, P, S, Cl, K, Ca, V, Ti, Cr, Mn, Fe, Co, Ni, Cu, Zn, Ga, Ge, Sr, Ba, Zr, Y, La, Ca
Morelli et al. (1988)	Nd:YAG LAMMA	265	—	—	2×10^{-2}	Coal vitrinite	Ba, Cr, Ga, Sr, Ti, V
Kosztolanyi et al. (1986)	Nd:YAG magnetic sector	353	10^8–10^{10}	10^{-7}–10^{-5}	3×10^{-2}	Zircons	Y, Ce, La, U, Fe, Mn, Ca, Ti, Al, Th, Pb, P

Table 4C.10. Applications of RIMS

RIS Process	Analyzed Element	Matrix	Ref.
$A(2\omega_1, 2\omega_1 e^-)A^+$	V, Mo, Re	Pure metal oxides	Fassett et al. (1983)
$A(\omega_1, 2\omega_1 e^-)A^+$	K	—	Beekmann et al. (1980)
$A(2\omega_1, 2\omega_1 e^-)A^+$	Ga, In, Al, V B	Silicon Silicon	Parks et al. (1983b) Parks et al. (1985)
$A(2\omega_1, 2\omega_1 e^-)A^+$	Lu, Yb, Tc	—	Miller et al. (1983)
$A(2\omega_1, 2\omega_1 e^-)A^+$	Nd, Sm	geological material	Young and Donohue (1983); Donohue and Young (1983); Donohue et al. (1982, 1984)
$A(2\omega_1, \omega_1 e^-)A^+$	Pu	—	Krönert et al. (1985)
Two-photon sub-Doppler resonance	Sr	—	Lucatorto et al. (1984)
$A(\omega_1, \omega_2, \omega_3 e^-)A^+$	Tc	—	Ames et al. (1990)
$A(\omega_1, \omega_2 e^-)A^+$	V	Silicon-based material	Mayo et al. (1990)
$A(\omega_1, \omega_2, \omega_3 e^-)A^+$	Os, Re	Pure metals, meteorites	Blum et al. (1990)
$A(\omega_1, \omega_2 e^-)A^+$	Mg, Fe Ti, Co, Ni, Cu P	Al Fe Si	Gelin et al. (1989)
$A(2\omega_1, 2\omega_2 e^-)A^+$ $A(\omega_1, \omega_1 e^-)A^+$	Ca Rb	NBS standards	Towric et al. (1990)
—	Ga Al	Silicon steel NBS standards	Beekmann and Thonnard (1988)
$A(2\omega_1, \omega_2 e^-)A^+$	I	Biological material	Fassett et al. (1984)
$A(\omega_1, \omega_2 e^-)A^+$	Na, Al, B, Ga	Pure Ge	Bekov et al. (1989)
$A(\omega_1, \omega_2, \omega_3 e^-)A^+$	Pu, Am, Cm	Nuclear fallout	Trautmann et al. (1986); Ruster et al. (1988)
$A(\omega_1, \omega_2 e^-)A^+$	Fe	Fluoride glasses	Gilbert et al. (1990)

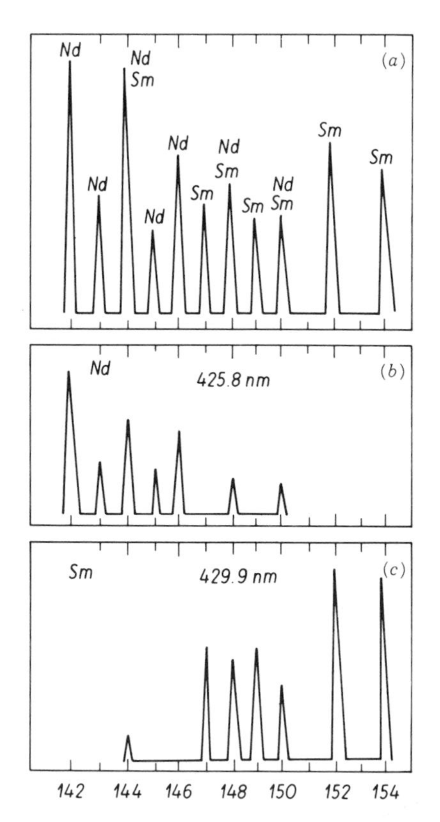

Figure 4C.18. Mass spectra of the following mixtures: (a) natural Sm and Nd obtained by LIMS or SSMS; (b) Sm and Nd by RIMS at wavelength 425.8 nm for Nd detection; (c) Sm and Nd by RIMS at wavelength 429.98 nm for Sm detection.

Table 4C.11. Results of RIMS Measurements for High-Purity Germanium

Element	Concentration (ppb)	Detection Limit (ppb)	Ref.
Na	0.2	0.05	Bekov and Letokhov (1988)
Al	1	0.01	Bekov and Letokhov (1988)
B	0.2	0.05	Bekov et al. (1984); Bekov and Letokhov (1986)
Ga	0.2	0.005	Bekov et al. (1988)

Peuser et al., 1985). The precision and accuracy of RIMS results are typically in the 1–5% range for isotopic analysis and higher than 5% for the elemental concentrations, depending on sample size and analytical conditions. Higher precision results were obtained by combining the RIS technique with the isotope dilution method (ID-RIMS).

Fassett et al. (1990) determined iodine in biological sample and standard material by ID-RIMS. With a precision of 2–3%, they analyzed concentration levels of about 0.1 ppm in water. With the ID-RIMS technique, Mayo et al. (1990) analyzed vanadium impurities in silicon-based materials, in films, and in water from various sources. The sensitivity of ID-RIMS to vanadium is in the ppt range (pg/g) and is limited by the blank level of a few picograms per gram in the acid.

Resonance ionization of ion-sputtered or laser-ablated neutral atoms is an effective way to separate the atomization step and the ionization step in the analysis. The essential advantage of this technique in contrast to ID-RIMS is that it does not require any chemical sample preparation. Beekmann and Thonnard (1988) combined the atomization of sample material by LA using a frequency-doubled Nd:YAG laser and RIS-TOF-MS analysis. They determined the Ga concentration in silicon samples and the Al content of steel (NBS Standard Reference Materials). Detection limits were in the neighborhood of 15 ppb.

Gelin et al. (1989, 1990) studied the matrix effects in RIMS analyses of metallic alloys utilizing ion-sputtered neutral atoms. In comparison to SIMS analyses they have found that the matrix effects are small in RIMS, much smaller than with the SIMS technique. Table 4C.12 shows the results of detection limit measurements for some elements using Ar^+ beam sputtered neutral atoms and resonance ionization.

The main advantage of combining the ion-sputtering process with RIMS (see SIRIS and SARISA described earlier in Section 4C.3 and in Chapter 2) is the high efficiency of the method. RIMS uses a neutral atom plume for the analysis, and neutral species dominate the sputtered plume with more than 99% of all sputtered species. The same is true in the case of LA in combination with RIMS. Many authors have demonstrated the application of sputtered-initiated (Parks et al., 1983a, 1988; Blum et al., 1990; Gelin et al., 1988, 1989; Towrie et al., 1990) and laser ablation (Beekmann and Thonnard, 1988; Bengtsson et al., 1988) RIMS for a great number of elements. Nogar

Table 4C.12. Experimental Detection Limits[a] for RIMS of Sputtered Neutral Atoms

Number of Laser Shots	Fe Matrix					Al Matrix			Si Matrix
	P	Ti	Cr	Co	Ni	Cu	Mg	Fe	P
1	—	3	2	10	10	5	0.3	3	—
1.8×10^4	0.1	0.01	0.007	0.03	0.03	0.015	0.001	0.01	0.5

Source: From Gelin et al. (1989).

[a]Detection limits in atom ppm.

et al. (1988) applied laser desorption and laser ablation RIMS to the analysis of small iron-containing microinclusions in multilayered glass substrates.

4C.5.5. Application to Depth Profiling and Local Analysis

In thin-film and depth profile analysis, laser-induced mass spectrometric methods play a subordinate role in comparison with other surface analytical methods [e.g., secondary neutral mass spectrometry (SNMS), SIMS, atomic emission spectroscopy (AES), Rutherford back scattering (RBS)]. In most cases, the concentration change of the bulk elements in the depth profile is measured. With laser-induced mass spectrometry, the depth resolution is about 0.1 μm with a detection limit of 10^{-6}% and providing that the detection surface impurities are at 10^{10} atoms/cm^2. The behavior of trace impurities in thin film is of interest for the analysis of electronic materials [e.g., GaAs (Mahavadi et al., 1985)] or microelectronic devices (Grasserbauer et al., 1986). Laser-induced mass spectrometry in a single-shot regime can determine contaminations on microelectronic devices with a lateral resolution of about 2–10 μm and with a depth resolution of 0.1–1 μm. The detection limits of trace elements in thin films are in the ppb range. A description of laser microprobe mass spectrometry and selected applications is given by Simons (1988). Measuring surface layers of stainless steel with different colors using a LAMMA instrument, Heinen et al. (1984) found different trace compositions.

The LAMMA analysis of particles requires a special calibration with particle standards. Thus, the preparation of particles with well-defined size and chemical composition is necessary. This method was used by Kaufmann et al. (1980) to calibrate the dependence of ion intensity on the particle size for NaCl particles.

Van Doveren (1984) describes the application of LIMS for microanalysis in relatively small volumes near the surface of gold layers on nickel and on glass substrates. The experimental setup of Philips Research Laboratories in Eindhoven, Netherlands, consists of a laser ion source (Nd:YAG, wavelength 1064 nm, laser power density 10^{11} W/cm^2, pulse duration 15 ns) and a double-focusing mass spectrograph with Mattauch–Herzog geometry and a photoplate detector.

Laser-probe microanalysis of aluminum–lithium alloys was carried out with a LIMA 2A instrument by Southon et al. (1985). Chemical inhomogeneities in the samples and impurities of Ga, Mg, Cu, and Fe in the interface regions were examined by Smith et al. (1985) in thin films, amorphous semiconductors, and insulators using LIMA.

Conzemius et al. (1981) studied the migration of doped traces in solids by means of scanning laser mass spectrometry. The advantage of this method over other scanning techniques (Hall et al., 1976) lies in its higher sensitivity, minimal matrix effects, and insensitivity to surface effects. Scanning laser mass spectrometry has excellent analytical capabilities for measuring concentration profiles of trace level solutes in metal systems. The authors quantified the chemical analysis of different traces (Re, Mo, W, Zr, Fe, Co, and Ni) in Th and Y matrices with standards, and the relative deviation of the mean calibration factor was 1.6%.

Trace element analysis of inclusions and mineral phases in geological samples and meteorites was carried out by Jochum et al. (1991) using a double-focusing laser ionization mass spectrometer. To determine trace element concentrations down to about 1 ppm, a sample surface of about 1 mm^2 must be scanned by the laser beam. The concentrations of major, minor, and trace elements (Al, Mg, Fe, Cr, Ga, V, Ge, K, Na, P, Co, Cl, Ti, Ni, S, Mn, Ca, Si, Sc, and Cu) in spinel and periodotite rock were determined.

Surface analysis by laser ionization (SALI) in combination with reflection TOF-MS was introduced by Pallix et al. (1988). In their studies, an excimer laser operating with KrF (248 nm, 5 eV per photon) was used. The depth profiles of Au films on GaAs, the diffusion of Ga and As in the Au film, and the compositional variation at the interface after annealing at 450 °C were investigated.

Another interesting application of SALI is the analysis of high-T_c superconducting films. Laser mass spectra taken from different depths within a superconducting film of the $YBa_2Cu_3O_x$ type on $SrTiO_3$ substrate show a variety of impurity compounds such as BaCl, BaF, CO, CO_2, and several hydroxides (indicating reactions with atmospheric water) present on the film surface. The application of RIMS to depth profiling in III–V semiconductor devices is also reported by McLean et al. (1988).

Further applications of laser microprobe mass spectrometry in inorganic microanalysis with LAMMA and LIMA instruments are reviewed by Van Vaeck and Gijbels (1990).

4C.6. CONCLUSION

The great analytical potential of laser-induced mass spectrometry lies in its capability to simultaneously determine all chemical elements and their isotopes in solid samples, in its potential for direct analysis of any kind of solid, in its high efficiency of evaporation and ionization, in its high absolute and relative sensitivity, and in its applicability to microlocal and depth profile analyses. Furthermore laser-induced mass spectrometric methods can be used in materials research, the semiconductor industry, microelectronics, atomic physics, geology, chemical analysis, biomedicine, environmental research, etc. In comparison with other spectrometric techniques for trace analysis, laser-induced mass spectrometry can be applied equally well to the multielement analysis of both minor and major components in inorganic materials with great accuracy and precision. Besides commercial laser mass spectrometers, which use ion dynamic mass spectrometers for mass separation, some double-focusing mass spectrometers with a Mattauch–Herzog geometry have also been coupled with laser ion sources for ubiquitous use in the trace analysis of inorganic sample materials. In general, the disadvantage of laser-induced mass spectrometry is the high cost of the instrument, especially for systems utilizing more than one laser, as in RIMS.

Nevertheless, the development of laser-induced mass spectrometry will surely continue to move forward.

REFERENCES

Adachi, T., Kuroda, T., and Nakamura, S. (1986). *J. Phys., Colloq.* **C47**, 293–296.

Adams, F. (1983). *Spectrochim. Acta* **38B**, 1379–1393.

Adams, F., and Vertes, A. (1990). *Fresenius' Z. Anal. Chem.* **337**, 638–647.

Ames, F., Brumm, T., Jäger, K., Kluge, H.-J., Suri, B. M., Rimke, H., Trautmann, N., and Kirchner, R. (1990). *Appl. Phys.* **51**, 200–206.

Anderson, C. A., and Hinthorne, J. K. (1973). *Anal. Chem.* **45**, 1421–1437.

Arefjev, I. M., Benjajev, N. E., Komleva, A. A., Ramendik, G. I., and Tjurin, D. A. (1986). *Zh. Anal. Khim.* **41**, 50–56.

Ban, V. S., and Knox, B. E. (1969). *Int. J. Mass Spectrom. Ion Phys.* **3**, 131–141.

Basova, T. A., Boriskin, A. I., Brjuchanov, A. S., Bykovskii, Yu. A., Jeremenko, V. M., and Nevolin, V. H. (1987). *High Purity Mater.* (*USSR*) **3**, 49–55.

Beam, E. C. (1973). Ph.D. Thesis, Pennsylvania State University, University Park, University Microfilms Order No. 74–4215.

Becker, C. H., and Gillen, K. T. (1984). *Anal. Chem.* **56**, 1671–1674.

Becker, C. H., and Gillen, K. T. (1985). *J. Opt. Soc. Am.* **B2**, 1438–1443.

Becker, S., and Dietze, H.-J. (1985). *Int. J. Mass Spectrom. Ion Processes* **67**, 57–65.

Becker, S., and Dieze, H.-J. (1986a). *Int. J. Mass Spectrom. Ion Processes* **73**, 157–166.

Becker, S., and Dietze, H.-J. (1986b). *ZfI-Mitt.* **115**, 123–130.

Becker, S., and Dietze, H.-J. (1987). *ZfI-Mitt.* **134**, 25–48.

Becker, S., and Dietze, H.-J. (1988). *Int. J. Mass Spectrom. Ion Processes* **82**, 287–298.

Becker, S., and Dietze, H.-J. (1990). *Physica C* (*Amsterdam*), **167**, 509–514.

Becker, S., and Dietze, H.-J. (1991). Unpublished results.

Becker, S., and Dietze, H.-J., and Pompe, W. (1986). *Isotopenpraxis* **26**, 453–454.

Becker, S., Dietze, H.-J., Kessler, G., Bauer, H.-D., and Pompe, W. (1990). *Z. Phys. B: Condensed Matter* **81**, 47–51.

Beekmann, D. W., and Thonnard, N. (1988). *Conf. Ser.—Inst. Phys.* **94**, 163–166.

Beekmann, D. W., Callcott, T. A., Kramer, S. D., Arakawa, E. T., and Hurst, G. S. (1980). *Inst. J. Mass Spectrom. Ion Phys.* **34**, 89–97.

Bekov, G. I., and Letokhov, V. S. (1986). *Laser Analytical Spectrochemistry*, Vol. 99. Adam Hilger, Bristol.

Bekov, G. I., and Letokhov, V. S. (1988). *Conf. Ser.—Inst. Phys.* **94**, 331–336.

Bekov, G. I., Radaev, V. N., Maksimov, G. A., and Nikogosyan, D. N. (1984). *Sov. J. Quantum Electron.* (*Engl. Transl.*) **11**, 825–827.

Bekov, G. I., Letokhov, V. S., and Radaev, V. N. (1988). *Spectrochim. Acta* **43B**, 491–493.

Bekov, G. I., Letokhov, V. S., and Radaev, V. N. (1989). *Fresenius Z. Anal. Chem.* **335**, 19–24.

Belousov, V. I. (1984). *Anal. Pure Mater., Gorki*, p. 28.

Bengtsson, L., Lucatorto, T. B., and Kreider, K. (1988). *Conf. Ser. Inst. Phys.* **94**, 167–170.

Beske, H. E. (1988). *Fresenius' Z. Anal. Chem.* **331**, 150–153.

Beusen, J. M., Surkyn, P., Gijbels, R., and Adams, F. (1983). *Spectrochim. Acta* **38B**, 843–851.

Bingham, R. A., and Salter, P. Z. (1976). *Anal. Chem.* **48**, 1735–1740.

Blum, J. D., Pellin, M. J., Calaway, W. F., Young, C. E., Gruen, D. M., Hutcheon, I. D., and Wasserburg, G. J. (1990). *Anal. Chem.* **62**, 209–214.

Boriskin, A. I., Eremenko, V. M., Ljalko, I. S., Brjuchaov, A. S., Cmijan, O. D., and Bykovskii, Yu. A. (1983). *Prib. Sist. Upr.* **1**, 26–29.

Briukhanov, A. S., Boriskin, A. I., Bykovskii, Yu. A., Briomenko, V. M., and Yarimenko, V. M. (1983). *Int. J. Mass Spectrom. Ion Phys.* **47**, 35–38.

Bruynseels, F. J., and Van Grieken, R. E. (1983). *Spectrochim. Acta* **38B**, 853–858.

Bruynseels, F. J., Storms, H., Tavares, T., and Van Grieken, R. E. (1990). *J. Environ. Anal. Chem.* **337**, 755–762.

Bushaw, B. A., and Gerke, G. K. (1988). *Conf. Ser.—Inst. Phys.* **94**, 277–280.

Bykovskii, Yu. A., Degtjapenko, N. N., Elesin, W. F., Kosyrev, Yu. P., and Silnov, S. M. (1971). *Zh. Eksp. Teor. Fiz.* **60**, 1306–1319.

Bykovskii, Yu. A., Basova, T. A., Belousov, V. I., Gladskoi, V. M., Gorshkov, V. V., Degtjarev, V. G., Laptev, J. D., and Nevolin, V. N. (1976). *Zh. Anal. Khim.* **31**, 2092–2096.

Bykovskii, Yu. A., Schuralev, G. I., Belousov, V. I., Gladskoi, V. M., Degtjarev, V. G., Kolosov, Yu. N., and Nevolin, V. N. (1978a). *Fiz. Plazmy* (*Moscow*) **4**, 323–331.

Bykovskii, Yu. A., Schuravlev, G. I., Gladskoi, V. M., Degtjarev, V. G., and Nevolin, V. N. (1978b). *Zh. Tekh. Fiz.* **48**, 382–385.

Cerezo, A., Grovenor, C. R. M., and Smith, G. D. W. (1986). *J. Phys., Colloq.* **C2**, 309–314.

Conzemius, R. J., and Capellen, J. M. (1980). *Int. J. Mass Spectrom. Ion Phys.* **34**, 197–271.

Conzemius, R. J., and Svec, H. J. (1978). *Anal. Chem.* **50**, 1854–1860.

Conzemius, R. J., Schmidt, F. A., and Svec, H.-J. (1981). *Anal. Chem.* **53**, 1899–1902.

Conzemius, R. J., Simon, D. S., Shankai, Z., and Byrd, G. D. (1983). *Microbeam Analysis—1983* (R. Gooley, ed.), pp. 301–328. San Francisco Press, San Francisco.

Daniel, W. M., DeLorenzo, D. J., and Wilson, H. R. (1988). In *Microbeam Analysis—1988* (D. E. Newbury, ed.), pp. 301–328. San Francisco Press, San Francisco.

Demtröder, W., and Jantz, W. (1970). *Plasma Phys.* **12**, 691–703.

Dennemont, J., Landry, J.-C., and Jaccard, J. (1982). *Chimica* **42**, 405–412.

Dennemont, J., Jaccard, J., and Landry, J.-C. (1985). *Int. J. Environ. Anal. Chem.* **21**, 115–127.

Dennemont, J., Landry, J.-C., Chevalley, J.-Y., and Jaccard, J. (1989). *Analysis* **17**, 139–142.

Dietze, H.-J. (1979). *Isotopenpraxis* **15**, 41–51.

Dietze, H.-J., and Becker, S. (1985a) *Fresenius' Z. Anal. Chem.* **302**, 490–492.

Dietze, H.-J., and Becker, S. (1985b). *ZfI-Mitt.* **101**, 5–60.

Dietze, H.-J., and Becker, S. (1987). *Beitr. Klusterforsch., ZfI-Mitt.* **134**, 5–174.

Dietze, H.-J., and Becker, S. (1988a). *Int. J. Mass Spectrom. Ion Processes* **82**, 47–53.

Dietze, H.-J., and Becker, S. (1988b). *Int. J. Mass Spectrom. Ion Processes* **82**, R1–R5.

Dietze, H.-J., and Becker, S. (1991). Unpublished results.

Dietze, H.-J., and Opauszky, I. (1979). *Isotopenpraxis* **15**, 309–312.

Dietze, H.-J., and Zahn, H. (1972). *Exp. Tech. Phys.* **20**, 389–400.

Dietze, H.-J., Zahn, H., and Schmidt, W. (1976). *Int. J. Mass Spectrom. Ion Phys.* **21**, 231–240.

Dietze, H.-J., Becker, S., Opauszky, I., Matus, L., Nyary, I., and Frescka, J. (1981). *ZfI-Mitt.* **48**, 3–48.

Dietze, H.-J., Becker, S., Opauszky, I., Matus, L., Nyary, I., and Frescka, F. (1983). *Mikrochim. Acta* **3**, 263–270.

Dingle, T., Griffiths, B. W., and Ruckman, J. C. (1981). *Vacuum* **31**, 571–573.

Donohue, D. L., and Young, J. P. (1983). *Anal. Chem.* **55**, 378–379.

Donohue, D. L., Young, J. P., and Smith, D. H. (1982). *Int. J. Mass Spectrom. Ion Phys.* **43**, 293–307.

Donohue, D. L., Smith, D. H., Young, J. P., McKown, H. S., and Pritchard, C. A. (1984). *Anal. Chem.* **56**, 379–381.

Dumas, J. L. (1967). *Methods Phys. Anal.* **64**, 47–49.

Duncan, M. A., and Rouvray, D. H. (1989). *Sci. Am.* **261** (6), 60–65.

Eloy, J. F. (1978). *Microsc. Acta, Suppl.* **2**, 307–317.

Eloy, J. F. (1984). *J. Phys. Colloq.* **45**, (C2), 265–269.

Eloy, J. F. (1985). *Scanning Electron Microsc.* **2**, 573–576.

Eloy, J. F. (1986). *Scanning Electron Microsc.* **4**, 1243–1253.

Eloy, J. F., Leley, M., and Unsöld, E. (1983). *Int. J. Mass Spectrom. Ion Phys.* **47**, 39–42.

Englert, P., and Herpes, U. (1980). *Inorg. Nucl. Chem. Lett.* **16**, 37–43.

Fassett, J. D., Travis, J. C., Moore, L. J., and Lytle, F. E. (1983). *Anal. Chem.* **55**, 765–770.

Fassett, J. D., Powell, L. J., and Moore, L. J. (1984). *Anal. Chem.* **56**, 2228–2233.

Fassett, J. D., Powell, L. J., and Moore, L. J. (1990). *Anal. Chem.* **62**, 386–389.

Feigl, P., Schueler, B., and Hillenkamp, F. (1983). *Int. J. Mass Spectrom. Ion Phys.* **47**, 15–18.

Feigl, P., Krueger, F. R., and Schueler, B. (1984). *Mikrochim. Acta* **2**, 85–96.

Frey, R., Weiss, G., Kaminski, H., and Schlag, E. W. (1985). *Z. Naturforsch.* **A40**, 1349–1350.

Fürstenau, N. (1981). *Fresenius' Z. Anal. Chem.* **308**, 201–205.

Fürstenau, N., and Hillenkamp, F. (1981). *Int. J. Mass Spectron. Ion Phys.* **37**, 135–151.

Fürstenau, N., Hillenkamp, F., and Nitsche, R. (1979). *Int. J. Mass Spectrom. Ion Phys.* **31**, 85–91.

Gelin, P. Gobert, O., Dubrenil, B., Debrun, J. L., and Inglebert, R. L. (1988). *Conf. Ser. Inst. Phys.* **94**, 201–204.

Gelin, P., Debrun, J. L., Gobert, O., Inglebert, R. L., and Dubrenil, D. (1989). *Nucl. Instrum. Methods* **B40/41**, 290–292.

Gelin, P., Barthe, M. F., Debrun, J. L., Gobert, O., Gilbert, T., Inglebert, R. L., and Dubrenil, B. (1990). *Nucl. Instrum. Methods* **B45**, 580–581.

Gilbert, T., Barthe, M. F., Dubrenil, B., and Debrun, J. L. (1990). *Conf. Ser.—Inst. Phys.* **114** (Sect. 11, Pap. RIS 90).

Gladskoi, V. M., and Belousov, V. I. (1980). *Elektron. Ind.* **11**, 95–98.

Grasserbauer, M., Stingeder, G., Pötzl, H., and Guerrero, E. (1986). *Fresenius' Z. Anal. Chem.* **323**, 421–449.

Hall, P. M., Morabito, J. M., and Poate, J. M. (1976). *Thin Solid Films* **23**, 107–109.

Hamer, E., Gerhard, W., Plog, C., and Kaufmann, R. (1981). *Fresenius' Z. Anal. Chem.* **308**, 287–289.

Heinen, H.-J., Wechsung, R., Vogt, H., Hillenkamp, F., and Kaufmann, R. (1979). *Acta Phys. Austriaca, Suppl.* **20**, 257–272.

Heinan, H.-J., Meier, S., Voigt, H., and Wechsung, R. (1983). *Int. J. Mass Spectrom. Ion Phys.* **47**, 19–22.

Heinen, H.-J., Holm, R., and Storp, S. (1984). *Fresenius' Z. Anal. Chem.* **319**, 606–610.

Heumann, K. G. (1988). In *Inorganic Mass Spectrometry* (F. Adams, R. Gijbels, and R. Van Grieken, eds.), pp. 301–348. Wiley, New York.

Hillenkamp, F., Unsöld, E., Kaufmann, R., and Nitsche, R. (1975). *Appl. Phys.* **8**, 341–348.

Hirche, H., Heinrichs, J., Scharfer, H. E., and Schramm, M. (1981). *Fresenius' Z. Anal. Chem.* **308**, 224–228.

Honig, R. E. (1963). *Appl. Phys. Lett.* **3**, 8–11.

Honig, R. E., and Woolston, J. R. (1963). *Appl. Phys. Lett.* **2**, 138–139.

Huang, L. Q., Conzemius, R. J., and Houk, R. S. (1987). *Appl. Spectrosc.* **41**, 667–670.

Hurst, G. S., Payne, M. G., Kramer, S. D., and Young, J. P. (1979). *Rev. Mod. Phys.* **51**, 767–819.

Jansen, J. A. J., and Witmer, A. W. (1982). *Spectrochim. Acta* **3B**, 483–491.

Jochum, K. P., Matus, L., and Seufert, H. M. (1988). *Fresenius' Z. Anal. Chem.* **331**, 136–139.

Jochum, K. P., Seufert, H. M., and Matus, L. (1991). *Proc. 12th Int. Mass Spectrom. Conf., Amsterdam, 1991*, p. 331.

Kaufmann, R. (1991). *LIMS, Reference and Citation Index '91*. University of Düsseldorf, Germany.

Kaufmann, R., and Wieser, P. (1980). *NBS Spec. Publ. (U.S.)* **533**, 199–223.

Kaufmann, R., Wieser, P., and Wurster, R. (1980). *Scanning Electron Microsc.* **2**, 607–622.

Kohler, V. L., Harris, A., and Wallach, E. R. (1989). In *Microbeam Analysis—1989* (P. E. Russell, ed.), pp. 359–363. San Francisco Press, San Francisco.

Kosztolanyi, C., Eloy, J. F., and Bertrand, J. M. (1986). *Bull. Mineral.* **109**, 265–274.

Kovalev, I. D., Maksimov, G. A., Suchkov, A. I., and Larin, N. V. (1978). *Int. J. Mass Spectrom. Ion Phys.* **27**, 101–137.

Kovalev, I. D., Larin, N. V., Potapov, A. M., and Sutschkov, A. I. (1985). *Zh. Anal. Khim.* **40**, 1971–1977.

Krönert, U., Bonn, J., Kluge, H.-J., Ruster, W., Wallmeroth, K., Paiser, P., and Trautmann, N. (1985). *Appl. Phys.* **B38**, 65–70.

Kroto, H. W., Health, J. R., O'Brien, S. C., Curl, C. F., and Smalley, R. E. (1985). *Nature (London)* **318**, 162–163.

Kupka, K. D., Schropp, W. W., Schiller, C., and Hillenkamp, F. (1981). *Fresenius' Z. Anal. Chem.* **308**, 229–233.

Letokhov, V. S. (1978). *Usp. Fiz. Nauk* **125**, 57–60.

Leybold-Hereaus GmbH, Cologne, Germany (1982). "*LAMMA 1000-Laser Microprobe*" *Application*, pp. 12–18.

Linton, R. W., Musselman, I. H., Bruynseels, F., and Simons, D. S. (1987). *Microbeam Anal.* **22**, 161–166.

Lucatorto, T. B., Clark, C. W., and Moore, L. J. (1984). *Opt. Commun.* **48**, 406–410.

Mahavadi, K. K., Smith, G., and Milne, W. I. (1985). *Thin Solid Films* **124**, 237–240.

Maksimov, G. A., and Larin, N. V. (1976). *Usp. Khim.* **45**, 2121–2125.

Mamyrin, B. A., Karataev, V. I., Shmikk, D. V., and Zagulin, V. A. (1973). *Sov. Phys.—JETP (Engl. Transl.)* **37**, 45–48.

Martin, T. P. (1986). *Angew. Chem.* **98**, 197–212.

Matus, L., Seufert, M., and Jochum, K. P. (1988). *Int. J. Mass Spectrom. Ion Processes* **84**, 101–111.

Mauney, T., and Adams, F. (1984). *Sci. Total Environ.* **36**, 215–224.

Mayo, S., Lucartorto, T. B., and Luther, G. G. (1982). *Anal. Chem.* **54**, 553–556.

Mayo, S., Fassett, J. D., Kingston, H. M., and Walker, R. J. (1990). *Anal. Chem.* **62**, 240–244.

McLean, C. J., Marsh, J. H., Cahill, J. W., Drysdale, S. L. T., Jennings, R., McCombes, P. T., Land, A. P., Ledingham, K. W. D., Singhal, R. P., Smyth, M. H. C., Stewart, D. T., and Towrie, M. (1988). *Conf. Ser.—Inst. Phys.* **94**, 193–196.

Megrue, G. H. (1970). *Recent Dev. Mass Spectrom., Proc. Int. Conf. Mass Spectrosc. Kyoto, Japan, 1969*, pp. 654–655.

Michiels, E., and Gijbels, R. (1983). *Spectrochim. Acta* **38B**, 1347–1354.

Michiels, E., and Gijbels, R. (1984). *Anal. Chem.* **56**, 1115–1121.

Michiels, E., Celis, A., and Gijbels, R. (1982). In *Microbeam Analysis—1982* (K. F. J. Heinrich, ed.), pp. 383–388. San Francisco Press, San Francisco.

Michiels, E., Celis, A., and Gijbels, R. (1983). *Int. J. Mass Spectrom. Ion Phys.* **47**, 23–26.

Michiels, E., Van Vaeck, L., and Gijbels, R. (1984). *Scanning Electron. Microsc.* **3**, 1111–1128.

Miller, C. M., Nogar, N. S., and Dowey, S. W. (1983). *Proc. SPIE—Int. Soc. Opt. Eng.* **426**, 8–12.

Morelli, J. J., Hercules, D. M., Lyons, P. C., Palmer, C. A., and Fletcher, J. D. (1988). *Mikrochim. Acta* **3**, 105–118.

Muller, J. F., Pelletier, M., Krier, G., Weil, D., and Campana, J. (1989). In *Microbeam Analysis—1989* (R. P. Russell, ed.), pp. 311–316. San Francisco Press, San Francisco.

Nadahara, S., Kikuchi, T., Furuya, K., Furuya, S., and Hoshino, S. (1985). *Mikrochim. Acta* **1**, 157–166.

Nishikawa, O., Nomura, E., Kawada, E., and Oida, K. (1986a). *J. Phys., Colloq.* **C2**, 297–302.

Nishikawa, O., Nomura, E., Yanagisawa, M., and Nagai, M. (1986b). *J. Phys. Colloq.* **C2**, 303–308.

Nogar, N. S., Estler, R. C., Rowo, M. W., Fearey, B. L., and Miller, C. M. (1988). *Conf. Ser.—Inst. Phys.* **94**, 147–150.

Pallix, J. B., Becker, C. H., and Newman, N. (1988). *J. Vac. Sci. Technol.* **A6**, 1049–1054.

Parks, J. E., Schmitt, H. W., Hurst, G. S., and Fairbank, M. (1983a). *Thin Solid Films* **108**, 69–78.

Parks, J. E., Schmitt, H. W., Hurst, G. S., and Fairbank, W. M. (1983b). *Proc. SPIE—Int. Soc. Opt. Eng.* **426**, 32.

Parks, J. E., Beekmann, D. W., Schmitt, H. W., and Taylor, E. H. (1985). *Nucl. Instrum. Methods, Phys. Res.* **B10/11**, 280.

Parks, J. E., Spaar, M. T., Beekmann, D. W., and Moore, L. J. (1988). *Conf. Ser.—Inst. Phys.* **94**, 197–200.

Pellin, M. J., Young, C. E., and Gruen, D. M. (1988). *Microscopy* **2**, 1353.

Peuser, P., Herrmann, G., Rimke, H., Sattelberger, P., Trautmann, N., Ruster, W., Ames, F., Bonn, J., Kroenert, U., and Otten, E.-W. (1985). *J. Appl. Phys.* **B38**, 249.

Ramendik, G. I. (1986). *ZfI-Mitt.* **115**, 39–48.

Ramendik, G. I. (1990). *Fresenius' Z. Anal. Chem.* **337**, 772–776.

Ramendik, G. I., Manzon, B. M., Tjurin, D. A., Benyaev, N. E., and Komleva, A. A. (1987). *Talanta* **34**, 61–62.

Ruster, W., Ames, F., Mang, M., Mühleck, C., Rehklau, D., Rimke, H., Sattelberg, P., Herrmann, G., Kluge, H.-J., Otten, E.-W., and Trautmann, N. (1988). *Fresenius' Z. Anal. Chem.* **331**, 182–185.

Ruster, W., Ames, F., Kluge, H.-J., Otten, E.-W., Rehklau, D., Scheerer, F., Herrmann, G., Mühleck, C., Riegel, J., Rimke, H., Sattelberger, P., and Trautmann, N. (1989). *Nucl. Instrum. Methods Phys. Res.* **A281**, 547–558.

Sanderson, T. K. (1985). *Anal. Proc.* **22**, 118–119.

Sanderson, T. K., Mapper, D., and Farren, J. (1984). "Laser Source Solid State Mass Spectroscopy." *AERE Harwell Rep.* **AERE-R 11113**.

Schueler, B. W., and Odom, R. W. (1987). *J. Appl. Phys.* **61**, 4652–4661.

Schueler, B. W., Odom, R. W., and Evans, C. A., Jr. (1986). *Proc. 3rd Int. Laser Microprobe Mass Spectrom. Workshop, Antwerp, Belgium, 1986.*

Seifert, G., Becker, S., and Dietze, H.-J. (1988a). *Int. J. Mass Spectrom. Ion Processes* **84**, 121–133.

Seifert, G., Schwab, B., Becker, S., and Dietze, H.-J. (1988b). *Int. J. Mass Spectrom. Ion. Processes* **85**, 327–338.

Shibanov, A. N. (1985). In *Laser Analytical Spectrochemistry* (V. S. Letokhov, ed.), pp. 353–376. Adam Hilger, Bristol.

Simons, D. S. (1988). *Appl. Surf. Sci.* **31**, 103–117.

Singh, S. (1987). *Nature (London)* **329**, 183–184.

Smith, D. H., Young, J. P., and Shaw, R. W. (1989). *Mass Spectrom. Rev.* **8**, 345.

Smith, G. J., Eagle, D. J., and Milne, W. I. (1985). *Appl. Surf. Sci.* 22/23, 930–934.

Southon, M. J., Harris, A., Kohler, V. Mullock, S. J., Wallach, E. R., Dingle, T., and Griffiths, B. W. (1985). *Springer Ser. Chem. Phys.* **44**, 198.

Spurny, K. R., Schörmann, J., and Kaufmann, R. (1981). *Fresenius Z. Anal. Chem.* **308**, 274–279.

Steel, E. B., Simons, D. S., Small, J. A., and Newbury, D. E. (1984). In *Microbeam Analysis—1984* (A. D. Romig and J. I. Goldstein, eds.), p. 27. San Francisco Press, San Francisco.

Stefani, R. (1981). *Trends Anal. Chem.* **1**, 84.

Surkyn, P., and Adams, F. J. (1982). *Trace Microprobe Tech.* **1**, 79–114.

Svec, H.-J. (1984). *Anal. Chem. Symp. Ser.* **19**, 89–101.

Towrie, M., Drysdale, S. L. T., Jennings, R., Land, A. P., Ledingham, K. W. D., McCombes, P. T., Singhal, R. P., Smyth, M. H. C., and McLean, C. L. (1990). *Int. J. Mass Spectrom. Ion Processes* **96**, 309–320.

Trautmann, N., Peuser, P., Rimke, H., Sattelberger, P., Herrmann, G., Ames, F., Krönert, U., Ruster, W., Bonn, J., Kluge, H.-J., and Otten, E.-W. (1986). *J. Less-Common Met.* **122**, 533–538.

Utley, A. (1990). *Microelectron. Manuf. Test.*, February, pp. 27–28.

Valerio, F. (1984). *Spectrosc. Int. J.* **3**, 427–430.

Vanderborgh, N. E., and Jones, C. E. R. (1983). *Anal. Chem.* **55**, 527–532.

Van Doveren, H. (1984). *Spectrochim. Acta* **39B**, 1513.

Van Vaeck, L., and Gijbels, R. (1990). *Fresenius' Z. Anal. Chem.* **337**, 743–754.

Verbueken, A. H., Bruynseels, F. J., and Van Grieken, R. E. (1985). *Biomed. Mass Spectrom.* **12**, 438–463.

Verbueken, A. H., Bruynseels, F. J., Van Grieken, R., and Adams, F. (1988). In *Inorganic Mass Spectrometry* (F. Adams, R. Gijbels, and R. Van Grieken, eds.), pp. 173–194. Wiley, New York.

Vertes, A., Juhasz, M., De Wolf, M., and Gijbels, R. (1989). *Int. J. Mass Spectrom. Ion Processes* **94**, 63–85.

Vertes, A., Gijbels, R., and Adams, F. (1990). *Mass. Spectrom. Rev.* **9**, 71–113.

Vogt, H., Heinen, H.-J., Meier, S., and Wechsung, R. (1981). *Fresenius' Z. Anal. Chem.* **308**, 195–200.

Walter, K., Boesl, U., and Schlag, E. W. (1986). *Int. J. Mass Spectrom. Ion. Processes* **71**, 309–313.

Weinke, H. H., Michiels, F., and Gijbels, R. (1983). *Int. J. Mass Spectrom. Ion Processes* **47**, 43–46.

Wieser, P., Wurster, R., and Haas, U. (1981). *Fresenius' Z. Anal. Chem.* **308**, 260–269.

Yamamoto, T., Munakata, T., Nomiya, Y., Tsukakoshi, M., and Kasuya, T. (1984). *Jpn. J. Appl. Phys.* **23**, 1336.

Yokozawa, H., Kikuchi, T., Furuya, K., Ando, S., and Hoshino, K. (1987). *Anal. Chim. Acta* **195**, 73–80.

Young, J. P., and Donohue, D. L. (1983). *Anal. Chem.* **55**, 88–91.

Young, J. P., Shaw, R. W., and Smith, D. H. (1989). *Anal. Chem.* **61**, 1271A–1279A.

Zahn, H., and Dietze, H.-J. (1976). *Int. J. Mass Spectrom. Ion Phys.* **22**, 111–120.

CHAPTER

5

EXOTIC INSTRUMENTS AND APPLICATIONS OF LASER IONIZATION MASS SPECTROMERY IN SPACE RESEARCH

G. G. MANAGADZE and I. Yu. SHUTYAEV

Space Research Institute
Russian Academy of Sciences
Moscow, Russia

5.1. INTRODUCTION

The title for this chapter was proposed by R. Gijbels. Two laser mass spectrometers are described in it: LIMA-D, an on-board instrument of the *Phobos* space mission; and LASMA, the prototype of an instrument for future exploration of Mars' surface. The ion flight path-length of LIMA-D reaches 80 m and is the largest of the modern time-of-flight (TOF) instruments, whereas it is only 25 cm in the LASMA unit and can be considered the smallest TOF instrument. Therefore, use of the word "exotic" in the title to describe these instruments would seem to be fully justified.

LIMA-D was included in the on-board scientific payload of the interplanetary space mission *Phobos* to study the chemical and isotopic composition of Mars' satellite Phobos during a fly-by over its surface at a 30–80 m altitude. Two spacecraft were launched toward Phobos in June–July 1988. Both were lost—the first in September 1988; the second, 7 days before reaching its goal. These dramatic mishaps do not obviate the need for a detailed account of how these unusual units came to be designed.

Let us briefly touch on the scientific goals of the *Phobos* mission. The two Mars satellites Phobos and Deimos have long been at the center of planetologists' attention. A hypothesis exists that these two satellites are asteroids trapped by the Martian gravitational field. For that reason Phobos and Deimos have attracted our particular interest as possible representatives of

Laser Ionization Mass Analysis, Edited by Akos Vertes, Renaat Gijbels, and Fred Adams. Chemical Analysis Series, Vol. 124.
ISBN 0-471-53673-3

the primordial material from which the solar system was formed. The possible implementation of such a relatively cheap "asteroid" mission is considered a very important and worthy project. In this respect the American press has called the *Phobos* mission a "poor man's asteroid mission." The *Viking* optical data showed that the surface of the Martian satellites is similar in appearance to that of carbonaceous chondritic meteorites, giving indirect proof in support of the captured asteroid hypothesis. One of the main *Phobos* mission goals was to have been direct analysis of the regolith's chemical composition. The basic requirement for such measurements is maintaining the experiment's purity since samples should not be polluted while being taken, which might occur during the vehicle landing. That is why development of remote methods was of top priority.

The birth of the LIMA-D remote laser mass analyzer was closely connected with the *Phobos* project and its evolution. First options of the *Phobos* project began to be worked out in 1980. It was originally planned to be implemented by only our country's technical resources—without any need for international cooperation. (That later changed, as we shall relate below.)

The project's culmination was to be soil sampling from the Phobos surface by means of special penetrator (harpoon) at the moment when the spacecraft was to hover at a 30–50 m altitude. Then, soil analysis was to be carried out on board the spacecraft. Absence of other countries in the experiment is most probably explained by the complexity of the fly-by maneuver and the soil-sampling process: it was to be done for the first time; the spacecraft was of a new type; troubles might well occur.

The start of the *Phobos* project coincided with the end of development of the on-board scientific payload for the most successful international project, *Vega*. The *Vega* spacecraft was designed for study of Halley's comet and Venus (Sagdeev, 1988). A PUMA dust mass analyzer was included in the *Vega* scientific payload (Sagdeev et al., 1987), a development which has a direct bearing on this historical account. PUMA was the unit designed to determine the elemental and isotopic composition of microparticles in the comet's tail. These particles, with an approximate velocity of 70 km/s when bombarding a silver target, were evaporated and ionized. Then, the analysis was carried out by the TOF method with a reflectron. It was due to the PUMA unit that we became acquainted with the TOF method.

At about that time the *Phobos* project took some new turns. It was suggested that the scientific payload also include a laser mass spectrometer for analyzing soil composition after collection of soil samples. We soon developed an engineering model of a laser mass spectrometer, LIMA, based on the PUMA analytical unit. [This LIMA is not to be confused with LIMA-D (mentioned earlier) and with the LIMA mass spectrometer series of Kratos Analytical (Manchester, UK).] Unlike the PUMA unit, the new instrument had a laser ion source and, what was very important, no additional ion acceleration in the target vicinity. Because this approach essentially changed the physics of the ion formation process, it requires a detailed explanation. Practically all laser mass spectrometers have been designed with an additional ion acceleration unit; the target is at a potential on the order of 1 kV with respect to the ground. Even in the presence of primary ion energy up to 100 eV, this makes it possible to increase the resolution. LAMMA-1000 (see Hillenkamp et al., 1975), PUMA, and many other TOF mass reflectrons operate on this principle.

However, if the density of the plasma formed in the acceleration area is higher than the critical density, $\sim 10^7\,\mathrm{cm}^{-3}$, the plasma screens the electric field, peaks are destroyed, and the resolution is decreased (Sagdeev et al., 1987). In the Chemistry Institute of our Academy of Sciences (in Gorki, now Nizhni Novgorod) an original laser ion source was suggested in which ions are not accelerated but move at the primary velocities through a field-free flight space (Kovalev et al., 1985). In this operational mode the plasma density does not affect the mass resolution, and so the simplicity and reliability of the instrument become higher. We therefore naturally preferred this operational mode. We realized that this crucial plasma density effect might decrease the resolution in the PUMA instrument while increasing the mass of dust microparticles, and some changes were introduced in the PUMA operational mode before *Vega* was launched, thus permitting the unit to avoid this problem.

Meanwhile, work on the *Phobos* project was proceeding apace. The LIMA engineering model was functioning normally, and the first satisfactory spectra were obtained. Then something unexpected occurred. Since the *Phobos* project had begun to achieve considerable international stature, the Program Committee decided not to take any unnecessary risks: the penetrator was excluded from the spacecraft payload, as were all units for on-board soil analysis including the LIMA analyzer.

5.2. LIMA-D: A REMOTE LASER MASS ANALYZER FOR SPACE APPLICATIONS

5.2.1. General Description of LIMA-D and the History of Its Design

Nonetheless, the fly-by over the Phobos surface at a 30–50 m altitude remained in the program. This presented us with an opportunity to suggest a new methodology for determining soil composition with the LIMA-D remote laser mass analyzer (see Sagdeev et al., 1985); various other remote techniques were considered as well. The physical concept of LIMA-D was based upon the following line of reasoning: In a free-flight regime, location of the device just near the surface was not required; indeed, the distance could be considerable in this case. It was suggested that we include a laser with a tunable focusing lens, an altimeter, a reflector with a detector, and data-processing and control units. At a distance of 30–80 m the altimeter controlled the lens; it could correct the focusing of the laser radiation to a spot 1 mm in diameter. The ions thus produced would be scattered in the hemisphere and, with the effective area for particle collection equal to 300 cm^2, 10^{-6} of the ions could be trapped by a detector. It was known that if such laser radiation is focused in a 1 mm spot, the maximum ion energy distribution is 500–600 eV (Devyatikh et al., 1976). The total quantity of ions formed owing to laser radiation at a power density of 10^9 W/cm^2 and an ion yield of 0.1% will be 10^{12} particles. This estimate was made using the data obtained in laboratory tests with the small LIMA (Managadze et al., 1984).

This only gave us the basic physical idea of how to design the new device, however, and we had to start practically from scratch with an acute lack of time to design and manufacture one of the most complicated on-board instruments, the main units of which were not known at all, or known very little. After official discussions of our proposal, the Scientific Technical Council allotted a volume of ~ 1 m^3 and a weight of ~ 80 kg for the LIMA-D experiment. According to our preliminary estimates this allotment seemed to be enough, though many specialists here and abroad viewed them with some skepticism. This was only natural, because these estimates had not yet been tested experimentally.

Some experts, for example, had doubts that it was possible to focus laser radiation to a 1 mm diameter spot from a distance of 50–80 m and with a lens diameter of only 20 cm. They stated that the lens diameter should be not less than 1 m.

To reach high mass resolution an ion reflector was required. According to our estimate, the 30 cm characteristic size (depth) of a multigap reflector is enough to reach a resolution of ~ 200 if the free flight of the ions is equal to 50–80 m. The ratio of the reflector's depth and the free-flight path for laboratory instruments is 0.1–0.25, whereas in the LIMA-D case this ratio is less by an order of magnitude. It was hard to believe that the TOF focusing of ions into narrow peaks at such a low ratio was possible. On the initiative of our Academy of Sciences a committee on verification of the LIMA-D scientific concept was established. The committee was headed by B. N. Mamyrin, inventor of the mass reflectron (see Mamyrin, 1967a,b). At the initial stage of the committee's work, Mamyrin could not believe in the wide capabilities of the technique. It was only after a detailed check of the calculations that he expressed his confidence in the new possibilities of reflectrons not assumed until then; by that time the first experimental results were also obtained. Mamyrin's review turned out to be most favorable, and he suggested cooperation on the experiment and became our best friend and adherent.

To provide the necessary sensitivity, the reflector input area should be $\sim 300\,cm^2$ and it should have high mass resolution. The required measurement time of the altimeter was 20 ms with an accuracy of ~ 30 cm. There were no major difficulties, except that the allowed weight and overall dimensions were too small. However, with the increase of research activity in this direction specialists appeared who joined the work since they believed in the idea and were convinced that it could be implemented.

The overall responsibility for the LIMA-D instrument and its technical implementation lay with our laboratory. We were responsible, too, for general guidance of the experiment during the space mission.

A significant amount of work was carried out by our colleagues in Germany who developed the unique ion detector system and the on-board computer for spectral processing: J. Kissel of the Max Planck Institute for Nuclear Physics, Heidelberg, together with H. von Hoerner (Von Hoerner & Sulger Electronics GmbH, Schwetzingen) proved to be reliable partners despite the project's significant financial constraints.

Design and manufacture of the reflector turned out to be extremely laborious. It was made in Bulgaria, at the National Observatory of Kirdjali, by S. Zlatev and P. Belyakov, from special composite materials. The Space Research Institute of the Bulgarian Academy of Sciences was also involved in the project.

Specialists from the Institute of General Physics of our Academy of Sciences took part in the development and manufacture of the on-board laser. The on-board laser suggested by G. B. Altshuler [Leningrad (now Saint Petersburg) Institute for Fine Mechanics and Optics (LITMO)], with a tunable focusing lens and altimeter unit, was operating excellently till the last days of the space station's existence.

Several important functional units and the control system were developed by the Special Design Bureau of the Space Research Institute (in Frunze, the Kirghiz Republic). Responsibility for the instrument assembly lay with S. R. Tabaldyev, the head of that organization, and V. I. Terentiev, the leading designer of the instrument.

Invaluable assistance was rendered by specialists from Finland: R. Pellinen's team (Finnish Meteorological Institute, Helsinki) undertook the design of the ground support equipment and its software and provided financial support for part of that activity. At different stages of the project the following specialists contributed to the work: W. Riedler (Space Research Institute of the Austrian Academy of Sciences, Graz); J. Silen (Finnish Meteorological Institute, Helsinki); G. Dulnev (LITMO), W. Boynton (University of Arizona); J.-L. Bertaux (Service d'Aeronomie, Verrieres-le-Buisson, France); J. Geiss (University of Bern, Switzerland), D. Rumyantsev (LITMO), specialists from the VPZ company, Prague, Czechoslovakia, and from the Space Research Institute of the former German Democratic Republic.

Young scientists of our laboratory responsible for various units—A. Bondarenko, P. Timofeev, and V. Ter-Mikaelyan—spent many sleepless nights during development and testing of the full-scale models as well as separate units together with the present authors and G. Zubenko, A. Vladykin, and many other specialists of the Space Research Institute of our Academy of Sciences.

During the fly-by over the Phobos surface one more remote mass spectrometric experiment, DION, was to be conducted in parallel with LIMA-D (Managadze and Sagdeev, 1987a,b; Hamelin et al., 1988). The experiment was a space modification of the secondary ion mass spectrometry (SIMS) technique broadly used in laboratory practice. The on-board accelerator was to act upon the surface with krypton ions of 3 keV energy. Secondary ions, knocked out from the surface, were to be registered by a quadrupole mass analyzer. The DION experiment was implemented by four countries: our own, France, Finland, and Austria. Responsibility for its guidance was also laid upon our laboratory. The LIMA-D and DION experiments advantageously complemented one another, and much attention was given to them by the press at home and abroad from the first days of the project announcement.

Covault (1985), in one of the first technical articles devoted to the *Phobos* mission, in the American weekly *Aviation Week and Space Technology*, highly evaluated the LIMA-D and DION experiments. That served as an essential support, especially in the initial stages of the project. The constant attention of the press did not abate later on.

After this somewhat lengthy historical prologue, we come to the description of LIMA-D to which the remainder of Section 5.2.1 is dedicated.

Principles of LIMA-D Operation. As mentioned earlier, the idea of the remote laser mass analyzer LIMA-D is based on the mass reflectron proposed by Mamyrin (1967a,b; see also Karataev et al., 1972). What was new in LIMA-D was the proposal to control the parameters of the reflector and the focusing lens in accordance with the distance to the surface (Managadze et al., 1987a,b). The TOF throughout some specified distance is used for identifying ion masses. The time dispersion due to some specified energy spread is corrected by the reflector with a specially designed electric field. The device measures the ions formed by the laser microburst.

Figure 5.1 shows the units of the LIMA-D system. It consists of a laser with a power supply unit, a tunable focusing lens, a laser altimeter, a reflector with a detector, and data-processing and control blocks.

The instrument operates as follows: During the fly-by, the altimeter measures the distance to Phobos' surface with a frequency of 20 Hz. The data are used

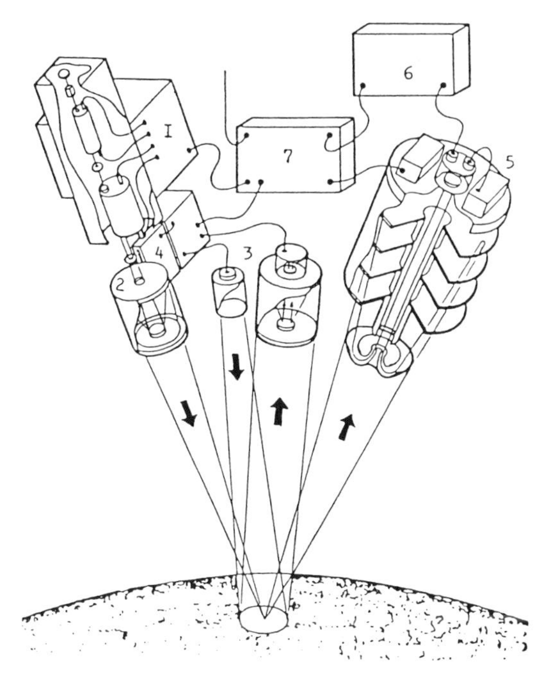

Figure 5.1. LIMA-D experiment schematic: (1) laser with power supply unit; (2) focusing lens; (3) altimeter; (4) servodrive; (5) reflector with ion detector; (6) data-processing unit; (7) control unit.

by the servodrive control unit, which continuously corrects the lens focusing. The distance data are also used to correct the electric field in the reflector. The laser fires every 5 s. The ions formed owing to the interaction of the laser radiation with the surface fly away in all directions, and they partly enter the input window of the reflector. Ions are focused into narrow packages according to their masses while flying throughout the reflector, and then they come into the detector. The information is processed by a microprocessor in the data unit and is sent via a telemetry system to the Earth.

The construction and parameters of the constituent units permit recording about 10^5 ions per laser shot. Given in the accompanying tabulation (on next page) are the most important physical characteristics of the LIMA-D system.

Such characteristics as accuracy and reproducibility, which depend on many features of the device and on the substance to be measured, are not specified. (For further discussion of laboratory tests of the instrument, see Section 5.2.5, below.)

Mass range	1–200 amu
Mass resolution, M/dM	150–200
Relative sensitivity	0.1%
Dynamic range of measurements	$>10^4$
Laser radiation wavelength	1.06 μm
Laser pulse energy	0.5 J
Pulse duration	10 ns
Laser pulse repetition rate	0.2 Hz
Focusing spot diameter	1–2 mm
Crater depth after the shot	0.5–1 μm
Power density in the spot	$>2 \times 10^9$ W/cm^2
Working range	30–80 m
Range measurement rate	20 Hz
Range measurement accuracy	± 30 cm
Measurement cycle	5 s

5.2.2. The Reflector: Calculations and Construction

The main problem in designing the LIMA-D–TOF analyzer was the necessity to use a reflector of relatively short length. The TOF analyzer of the mass reflectron type usually has a reflector of a size comparable with the length of the free-flight space. In this case, the TOF of ions through the free space, which is proportional to the inverse square root of energy, is comparable with the TOF through the reflector, which is proportional to the square root of energy. The sum of two inverse dependences allows us to make the total TOF independent of ion energy in some specified energy range. The time focusing is of the first order for the simplest one-gap reflectron (that is, the reflector has one region with a constant electric field). It is of the second order for the widely used two-gap reflectron, which has two regions with constant but different electric fields (Karataev et al., 1972).

It is easy to calculate that to satisfy the aforementioned focusing condition the reflector length must be equal to 1/4 of the free-flight space length for the simplest reflectron. For the two-gap reflectron the lengths ratio must be equal to approximately 1/10 (Shmikk and Dubensky, 1984). LIMA-D uses the space between the vehicle and Phobos' surface as a free-flight space. The planned flight height was 30–80 m, and it was impossible to employ a reflector more than 3 m long. So the problem arose as to whether it was possible to design a TOF mass analyzer with a reflector having a length of no more than 50 cm. The mass analyzer had to be of high resolution (not less than

200) and sensitivity (at least 0.1%), which meant that we could not use a narrow energy range for the ions recorded: though it would increase the resolution, it would also decrease the sensitivity.

The simplest estimation of laser plasma ion scattering shows that the device has to register ions with $\pm 5\%$ energy dispersion to obtain 0.1% sensitivity. It is assumed for this calculation that the laser has a focal spot 1 mm in diameter, a distance of 50 m to the surface, and the reflector has a 400 cm^2 entrance window and 5% total efficiency (including the "transparency" of the reflector and the efficiency of a secondary electron multiplier). A large entrance window is possible because of the special cylindrical shape of the reflector; details will be discussed later.

As a result of theoretical considerations, the following problem is posed:

The ion of mass M, charge q, and energy $q\varepsilon$ flies through the field-free region of length L, is repelled inside a reflector with potential $U(x)$, and is registered. Calculate the position dependence of the potential in the reflector, $U(x)$, which makes the total time of flight, T, independent of the initial energy of the ion (for energy from $q\varepsilon_1$ to $q\varepsilon_2$).

The equation for $U(x)$ is

$$\left(\frac{M}{2q}\right)^{1/2}\left[\frac{L}{\varepsilon^{1/2}}+2\int_0^{\varepsilon}\frac{dU}{(\varepsilon-U)^{1/2}\cdot\dfrac{dU}{dx}}\right]=T=\text{constant} \tag{1}$$

The equation can be solved by multiplying both sides by $d\varepsilon/[(V-\varepsilon)^{1/2}]$ and integrating for ε going from ε_1 to ε_2. The solution can be found in elementary functions for $x(U)$—the inverse function with respect to $U(x)$ (this solution was initially found by A. M. Natanzon in a somewhat incomplete form):

$$\begin{aligned}\pi\cdot x(U) &= k\cdot L\cdot\left(\frac{U-\varepsilon_1}{\varepsilon_1}\right)^{1/2}-L\cdot\arctan\left(\frac{U-\varepsilon_1}{\varepsilon_1}\right)^{1/2}\\ &\quad -2\int_0^{\varepsilon_1}\frac{dx}{dV}\cdot\arctan\left(\frac{U-\varepsilon_1}{\varepsilon_1-V}\right)^{1/2}dV\end{aligned} \tag{2}$$

The reference point for x is the point at which $U=\varepsilon_1$; here $k\geqslant 1$ is an arbitrary parameter. The integral on the right-hand side means that the potential distribution is arbitrary while $U<\varepsilon_1$. One should substitute the real $x'(V)$ and perform the integration to get the final solution. Some restriction on the function's smoothness might exist in reality, but this question has not yet been clarified. A small gap with a constant field might be proposed, which seems to be the simplest construction. In that case $x' = d/\varepsilon_1 = \text{constant}$, where d is the gap thickness, and the dependence $x(U)$ is

$$\pi \cdot x(U) = (k \cdot L - 2d)\left(\frac{U-\varepsilon_1}{\varepsilon_1}\right)^{1/2} + \pi \cdot d \cdot \frac{U-\varepsilon_1}{\varepsilon_1} - \left(L + \frac{2d \cdot U}{\varepsilon_1}\right)\arctan\left(\frac{U-\varepsilon_1}{\varepsilon_1}\right)^{1/2} \tag{3}$$

The ideal $U(x)$ is shown in Figure 5.2 (solid line). The following parameters were chosen: $L = 50.3$ m; $k = 1.0174$; $\varepsilon_1 = 450$ eV. It can be seen that the potential distribution for a two-gap reflectron is a very good approximation of the ideal distribution. Three and more gaps can give a better approximation, but the gain will not be as significant.

We used this formula to estimate the minimal size of the reflector needed to focus ions with $\pm 5\%$ energy spread. It was found to be equal to 17 cm (when $k = 1$). We chose a three-gap reflector 30 cm long. The grids were used to separate gaps so as to get a uniform field inside each gap. The potential distribution inside the LIMA-D reflector is shown in Figure 5.2 by the dashed line for $L = 50$ m. Grid positions are shown by vertical dotted lines (grids at $x = 0$ and $x = 30$ cm are not shown). It is impossible to use an ideal field in

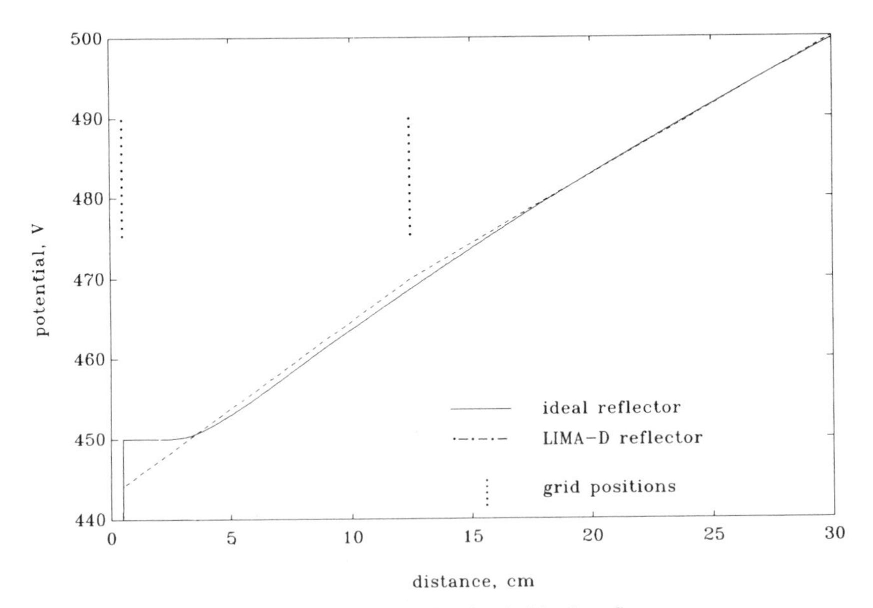

Figure 5.2. Potential distribution inside the reflector.

practice because of the nonzero divergence of the field (div $E \neq 0$). Free charges in the reflector volume are needed to produce this field (which means an infinite number of grids). We did not manage to use this ideal field in designing LIMA-D, but the concept was useful for estimating the size of the reflector.

Optimization of the positions of the grids and their voltages was made by computer. The grids' positions were optimized for a 50 m free-flight distance, and the voltages were adjusted for an actual distance with a step of 5 m. The reflector voltage source was controlled by means of the altimeter data. Calculations showed that this could give a mass resolution of 200 or higher within the distance range of 30–80 m. Theoretically the resolution must not depend on the mass, though practically it does so.

The reflector has a cylindrical shape. Ions come in through the ring-shaped entrance window. Its diameter is 36 cm, and its width is 4 cm. The ions are gathered toward the center of the reflector owing to a small inclination of the grids (0.6°). Next the reflected ions come to an additional toroidal deflector. It is necessary to decrease the ultraviolet (UV) radiation that could reach the detector. This deflector has two sections. It consists of three half-toroids cut along the equator and put one into another. In each section a transverse electric field is used to deflect ions and the total deflection angle is 180°. The toroidal deflector lets ions through with an energy spread of $\pm 10\%$ relative to the central energy for which it is adjusted. The ions fly throughout the 30 cm long field-free space after the toroidal deflector and come to the secondary electron multiplier. A Cu–Be multiplier of the type MM-1 produced by Johnston Laboratories (Townson, Maryland, USA) was used in the device. An additional three-grid assembly is placed just before the detector to reflect the ions of energies less than some lower limit. The voltage supplied to the medium grid defines this limit. The voltage on the last reflector's grid defines the upper energy limit of ions to be detected.

The reflector as a whole was made of metallized plastics. It was designed, manufactured, and mechanically tested in Bulgaria (at the National Observatory, Kirdjali, and the Space Research Institute, Sofia). The reflector passed all the tough qualification tests though it had large dimensions (42 cm diameter and 50 cm length) and the grids occupied a large area in it. Its weight was 10.5 kg. The detector system and signal-processing system were made in Germany (at the Max Planck Institute of Nuclear Physics, Heidelberg, and at Von Hoerner & Sulger Electronics GmbH, Schwetzingen).

The UV radiation attenuation was one of the problems to be solved. The solar UV flux near Mars is approximately 10^{11} photons/cm^2·s. We had to decrease the flux by a factor of 10^8 so as to have no more than 1000 photons registered per second—that is, 1 per spectrum. This was achieved by using a special carbon coating for all parts inside the reflector. The reflection coefficient of the coating is no more than 1% for wavelengths less than 2000 nm. The toroidal deflector transmits the ions but attenuates the UV radiation by a factor of 10^6.

The collection efficiency of the reflectron was estimated. The reflectron transmits 35% of the flux coming to the entrance window. Only those ions with an energy spread of ±5% are taken into account in this calculation. Multiplying this coefficient by the transparency of the grids (0.95 for each grid), we estimated the total collection efficiency to be equal to 25%. A scattering on the field roughness near the grid wires decreases the efficiency, but we did not calculate this effect. We used a collection efficiency of 5% for estimating the sensitivity of the device. The ion energy distribution extends up to 1 keV, with the maximum at approximately 500 eV when the plasma is produced by the laser beam focused onto a 1 mm spot (Devyatich et al., 1976). Ions with an energy spread of ±5% and with the average energy near the maximum (500 eV) constitute nearly 5% of the total number of ions generated after the laser burst. Data on the crater depths published by Devyatich et al. (1976) and an assumption of spherical symmetrical scattering allow the LIMA-D sensitivity to be evaluated. It turns out to be equal to 0.01% (10 ions of the impurity might be detected in a single spectrum).

5.2.3. The Laser and Focusing System

Preliminary estimates of the main energetic characteristics of the laser showed that to provide a power density of about 10^9 W/cm^2 in a spot 1 mm in diameter with a pulse duration of 10 ns, the pulse energy should be ~0.5 J. It was also clear that the laser should work in a *Q*-switched mode. These calculations were followed by estimates of the feasibility of concentrating laser radiation onto small areas at a distance of 50–100 m from the emitter. Estimates obtained in collaboration with the Institute of General Physics of our Academy of Sciences showed that to minimize the focusing lens parameters a laser should operate in the TEM_{00} mode.

Two variants of laser and focusing lens were suggested. The laser emitter suggested by the Institute of General Physics was a solid-state laser of neodymium-phosphate glass with a passive shutter of lithium fluoride in the generator and a complicated amplifier with seven passes of a beam throughout the crystal. In the flight model liquid cooling of the crystals was envisaged. The laser provided a pulse energy up to 0.5 J when operated in the TEM_{00} mode regime. This laser became the first laboratory model that permitted us to start experiments with laboratory mock-ups of the reflector in a large vacuum chamber. Yet the laser still required fine tuning and was not ready for use in space. Moreover, the model's utilization of liquid cooling with forced pumping raised serious concerns for the whole emitter.

In this connection it was decided to design as an insurance backup a new version of a simpler and more reliable laser of a well-known and tested configuration without liquid cooling. LITMO specialists took responsibility for designing such an instrument. The concept they suggested was simple, reliable, and not without originality (Altshuler et al., 1988). The laser emitter with focusing lens was developed in a relatively short period of time thanks to intensive work of the LITMO team. It was convenient that production of the focusing lens and design of the laser were carried out by the same laboratory, which avoided the complications occurring when one must assemble components produced in different places.

The laser emitter was a Gaussian mode laser with a passive Q-modulator and a two-stage one-way amplifier, on yttrium aluminum garnet (YAG) activated with neodymium ions. The minimum angular divergence in solid-state lasers with a short base resonator is achieved by decreasing the aperture; hence the radiation energy decreases. A resonator with a large effective length was used in laser instruments for the first time. That was accomplished by placing a two-lens telescope with minimized spherical aberration and magnifying factor ~ 2.5 inside an ordinary flat resonator. The laser resonator Q-factor was modulated by the LiF shutter with F^2 color centers. The size of the generator crystal was $\varnothing$ 6.3 × 65 mm. With a pumping energy of 25 J the Gaussian mode pulse energy was up to 85 mJ, and pulse duration was 7–10 ns. The energy was restricted by the stability of optical elements placed in the narrow part of the beam. The energy was enhanced in the two-stage amplifier with crystals of $\varnothing$ 8 × 80 mm size. Mathematical modeling of the amplifier made it possible to determine its dynamic characteristics and to optimize the parameters of active cells, pumping, and transverse beam distribution.

Specially designed high-quality crystals were used in the laser that were produced in the laboratory headed by X. S. Bagdasarov of the Crystallography Institute of our Academy of Sciences. They permitted us to have an amplifier output pulse energy up to 0.6 J in the TEM_{00}-mode regime.

The laser generator and amplifiers employed a two-lamp pumping system with natural cooling. Their optimal design and parameters were determined by mathematical modeling of optical and thermal processes. The Monte Carlo method was used to determine the bulk distribution of the lamps' radiation in the active cells and the dynamics of thermo-optical processes. Based on this calculation, a monoblock reflector configuration was proposed to provide maximum energy efficiency with minimum distortions. The electric energy–radiation energy conversion coefficient reached 1.1% when crystals with 0.8% neodymium ion concentration were employed.

One of the most complicated problems in designing the focusing system was related to the presence of noncontrolled amplitude-phase inhomogeneities in the radiation of solid-state lasers. The whole task was made more complex by the fact that the focusing system should be following the distance from the laser to the surface in the range 30–80 m, in accordance with altimeter indications. It should also provide a constant transverse beam distribution at the focusing length. Several modifications of focusing lens were suggested. A four-component pancratic lens proved to be best, in which the first and fourth components are stationary. Transfer of the second and third component with the velocity ratio 2:1 provided the refocusing of radiation. The fourth (largest) component was 160 mm in diameter. The constant transverse beam distribution in the constriction region was ensured by preserving the convergence angle after the fourth component.

To provide the transfer of mobile components a special servodrive was designed in Czechoslovakia (by VPZ, Prague). This servodrive control unit transferred the second and third components according to a program based on altimeter data with an accuracy better than 10 μm. It guaranteed the accuracy of focusing onto the surface to be not worse than 50 cm, which was enough taking into account the constriction region length ($\pm$1.5 m).

Special equipment and techniques were developed at our Space Research Institute for calibration of the laser with the focusing system (Altshuler et al., 1989a). Such calibrations were made on a horizontal base. A surface simulator was installed at a given distance, e.g., 50 m. On altimeter command the focusing lens was retuned in accordance with a program in the memory of the servodrive control unit. The beam diameter was determined automatically by computer from the radiation distribution data. Several similar measurements in the focusing zone enabled us to determine the length of the constriction region within which the beam diameter is increased by 5%, as well as to determine power density and its distribution. The minimum diameter of the beam and the exact focus position were determined, too. Similar measurement cycles were repeated at distances of 30, 60 and 80 m.

The optical calibration results showed that all physical characteristics were very close to those calculated, and required parameters were provided for successful implementation of the LIMA-D experiment. All three flight models (two main and one spare) have a pulse energy (460, 500, and 480 mJ) and a focused beam diameter at 50 m (1.7, 1.5, and 1.8 mm) sufficient to produce a laser plasma. The length of the 5% constriction zone within which the operation was permitted was, on the average, about 1.5 m. This overlaps the total error that could appear owing to the instability of separate units—with an ample reserve. The stability of the beam diameter in focus within 30–80 m was shown.

A few words should be said about the altimeter. Altimeters of various types were considered, and the simplest one was chosen. The altitude was determined by measuring a time interval between the emitted signal and the echo.

The new instrument was based on an altimeter of this type previously designed in LITMO. Besides the altimeter itself, a focusing control unit was included in the instrument.

A semiconductor pulse laser was used as an emitter operated at 0.89 μm wavelength. The reflected signal was recorded by two *p–i–n* photodiodes. The receiving optical system was a Cassegrain telescope with a 1.06 μm optical filter for suppression of the main laser radiation. A frequency of about 3.2 kHz was selected for the light pulse emission. This allowed 64 pulses to be collected during 20 ms and the statistical processing to be carried out. This procedure increased the accuracy of the measurement by a factor of 8. The accuracy of the altitude measurements was ± 30 cm in this mode. The analysis of various possible profiles of Phobos' surface during the fly-by and the data on expected vehicle velocities showed that the accuracy and operation rate of the altimeter were a little bit higher than the allowable limit.

5.2.4. Control and On-board Data Processing

The fly-by conditions did not allow real-time control of the device from the earth. As the time of radio wave propagation is approximately 40 min (two ways) and the fly-by duration had to be only 15–20 min, it was very important to design an autocontrolled system of high reliability. Some parameters were not known in advance, and we needed a flexible system with the capability of being tested and readjusted during the flight.

We had two principal ways to control the device and readjust its parameters. The first is via *relay commands*, i.e., high current lines used to switch various units on and off. These commands were used for switching the whole device and the laser heating system on and off, for opening the covers of the reflector and the laser-objective unit, and for starting the main or test programs execution. The other means of control was to send *digital command words* (CW), i.e., 1- or 2-byte commands that could be processed and executed by the LIMA-D microprocessors. We will describe the control procedures used for each unit in the following subsections.

Control of the Laser Power Supply. Apparently, as of this writing, our experience with LIMA-D is the only one with a laser after 300 days of space flight. We had to foresee the need to readjust a maximal number of its parameters. The commands to the laser were as follows:

- Switch on the heating system.
- Adjust the laser generator voltage.
- Adjust the laser amplifier voltage.

- Adjust the delay between triggering the generator lamp and triggering the amplifier lamp.
- Set the voltage boundaries for the laser generator self-adjustment.

The autocontrol system should look after the number of pulses generated in one laser shot during the fly-by. The system should increase the voltage on the laser generator capacitor if there are no pulses, and decrease the voltage if two or more pulses are generated. The output laser energy must change according to a fixed program, by increasing or decreasing over eight energy levels. The output energy is controlled by changing the voltage on the laser amplifier capacitor.

The Focusing System with an Altimeter. Some elements in this unit were duplicated (the altimeter laser diode and control subunit). We could choose either the main or spare element by command from Earth according to the results of test sessions. One command was available to switch on the hardware that adjusted the focusing system at a distance of 50 m—the average height of the fly-by. This command was to be used in case of any trouble with the altimeter or control subunit.

During the fly-by the high-speed servodrive would readjust the focusing lens in accordance with the distance measured by altimeter.

The Reflector. Two commands were available to check voltage sources. The arbitrary voltages could be set on the reflector grids and checked via the telemetry system.

The only control of the reflector during fly-by should be changing of the reflector voltages depending on the distance to the surface to maintain high resolution.

The Data Unit and Detector Module: Control. The following commands could be sent to the data unit:

- SHV—Set the new high-voltage value on the detector.
- EIT—Execute the in-flight test. This command switched on the generator that simulated the spectrum signal to check the entire signal-processing line.
- HVR—Execute the high-voltage routine, i.e., the special routine that could check the detector gain by using a built-in alpha-source.

The data unit should execute several control functions during the fly-by. It should look through the obtained spectra and, if the peak amplitudes are too low (less than one-fourth of the full scale), increase the high voltage on the detector.

Another function was to control the output laser energy. We had various proposals for the criteria of this control during the preparation of the LIMA-D experiment. The problem is that the optimal laser power density in the focal spot—that power density at which plasma occurs stably and with not too many multiply ionized ions—strongly depends on the sample composition and the surface condition. This could not be predicted for the Phobos material. Two different approaches were elaborated. One, an adaptive approach, assumed working out a program that could estimate the "quality" of the spectrum. The program should be able to determine the amount of multiply ionized ions. For example, a program was developed that counts the number of peaks in various mass ranges (1–12, 12–20, 20–60 amu), and the criterion of spectrum evaluation has been worked out based on these values. From the spectrum quality the program decides to increase or decrease the laser energy or to leave it unchanged for the next shot. Another "spectrum quality" approach was developed by scientists from the Finnish Meteorological Institute (Helsinki). The criterion was based on the Fourier transformation of the spectrum, that is, also on the number of peaks in various time (mass) ranges. This procedure is more vague in a physical sense, but it is much easier to realize in a program.

Two difficulties occurred in the development of these methods. The first, a technical one, was associated with possible interference from the laser. Strong interference occurred during the high-power circuit discharge. It sometimes caused triggering of the detection system in spite of strenuous efforts undertaken to avoid the laser effect on other units and devices on board (optocouples, etc.). This led to uncertainty in determination of a reference point for the mass scale calculation and to failures in the applications. The other difficulty was due to methodological problems. Both criteria are essentially based on assumptions about the composition of Phobos material. We assumed, in accordance with the *Viking* data (Pang et al., 1978; Pollack et al., 1978), that Phobos' composition is analogous to that of carbonaceous chondrites. The elements of masses 2–23 amu were expected to have abundance below the device's sensitivity. Only oxygen and carbon could occur in the spectra. On the other hand, many peaks of multiply charged ions appear in this mass range, and this fact is the physical basis for the proposed algorithms. The program would control laser power incorrectly if our assumption of Phobos' composition proved not to be true.

The second approach proposes a simple changing in the laser energy according to a fixed program. Eight energy levels were chosen in advance. So, 10% of spectra would be of good quality, and 20–40% of average quality. The rest of the spectra could also contain some useful information, but additional efforts would be necessary to get it out. According to our estimate, the adaptive algorithm would provide 80% of good-quality spectra in the event of success. Nevertheless, the fixed program was chosen for the reasons mentioned above.

The Data Unit: Data Processing. After we had rejected the adaptive algorithm for laser power control the only purpose for the data-processing unit was to compress the information to be transmitted to Earth. The restriction of telemetry system capacity is one of the most important, as well as the restrictions of power consumption and weight. The total amount of raw information was 16 kbyte per spectrum. These data contained three spectra obtained from different dynodes of the MM-1 multiplier and data from the so-called single event unit (SEU). The spectra from different dynodes corresponded to different gain coefficients. The SEU allowed single ion peaks to be recorded with the simplest amplitude analysis. The dynodes gain ratio was equal to approximately 1:5:60 and depended on the multiplier voltage. The combination of all these data made it possible to increase the dynamic range of the registering section to 10^5. The length of the experiment data frame (EDF) had to be no more than 1187 bytes. It started with some housekeeping information (the header), followed by the spectrum data.

The Compression Procedure. The offset levels for each channel are calculated initially. The samples with amplitudes higher than the offset and lower than some specific high level (FA hex, FF hex is a saturation level) are considered transferable. The data to be stored in the EDF are selected from the transferable data with priority decreasing from the low- to the high-sensitivity channel. Only the sums of each two subsequent samples, starting with odd sample numbers, are at first stored in EDF. The summing up decreases time resolution of the registering section but also decreases the body of information: 9-bit values are converted to 6-bit values according to a special table; the error produced by this conversion is not more than 3% (the logarithmic law is used). The left two bits in a byte are used for channel identification. The intermediate nontransferable samples (too low or too high) are omitted, and only their number is written down along with a special mark. This main information block allows us to restore the initial spectrum. Additional information is written down to the EDF if there is some free space left. It depends on the number of peaks present in the spectrum, and it was expected that the main block would take no more than half of EDF. The data from the medium-sensitivity channel are added to EDF after the main block. Also sums of two subsequent samples but started with the even sample numbers are stored in the same compressed form. Then, the procedure is repeated with data from the low- and high-sensitivity channels. This information permits us to restore the spectrum with the initial time resolution and to determine the relative sensitivity of the channels. Finally, SEU data are added to EDF.

The compression procedure described above was tested with Phobos-like substances. Normally 5–15 peaks were present in a spectrum and the EDF size was enough to pack in all the information.

The Control Block. The control block looked after the synchronization of the units. It also was responsible for the interaction with spacecraft systems. It received and distributed the commands, sent prepare signals to units before a laser shot, and executed all necessary housekeeping routines. If the EDF received from the data unit was not too large, the digital information from the altimeter was added to it. This information would show the profile of the Phobos surface along the spacecraft trajectory with an accuracy of 30 cm, which is an important scientific by-product of the LIMA-D device. The altimeter information would also be transmitted via the analog telemetry channel, but with lower accuracy and with some omissions.

LIMA-D Operation During the Fly-by. To complete the description of LIMA-D control, we shall now describe the planned operation of the device during the fly-by near Phobos. The so-called 9th session would have started for the LIMA-D 2 h before the Phobos encounter with switching on of the laser heating system. One and a half hours later, the relay commands for opening the reflector and lens covers would have come. At time T_0, the beginning of the encounter, the device would receive the relay command "PROG3" and all necessary digital commands. After 30 s, used to initialize routines, the device comes to a mode when it makes one laser shot every 5.5 s. Shots are synchronized with the spacecraft engine operations. The engines have a break of 2.5 s during each period when their operation is allowed only if some danger for the spacecraft should occur. The laser fires during these breaks to have good vaccum conditions for the measurements. Another mass spectrometer, the secondary ion quadrupole analyzer DION, works during the "safety" interval, too. The spacecraft was to be situated over the same region of the surface with the transverse velocity of lower than 2 m/s for the first 5 min of the fly-by. Then it was to fly with a velocity of 10 m/s during the next 5 min. The displacement was to have been between 1 and 3 km. The spacecraft was to have "stayed" in one place over the Phobos surface for the last 5 min.

The digital information would have been read every 11 s—two EDFs at a time. Some parameters were to be transmitted via analog telemetry: the voltages of the reflector and laser circuits; the high voltage on the detector; the distance measured by the altimeter and the position of the lens-adjusting elements; and the temperatures. The total amount of information was estimated as 2 Mbit of digital information and 1.5 Mbit of analog information. The information was to be transcribed by the on-board tape recorder. The information would have been transmitted to Earth after the completion of the fly-by session.

5.2.5. Laboratory Experiments with LIMA-D and In-flight Tests

Laboratory Tests. The big problem was to test physical parameters of the reflector (the TOF mass spectrometer). The main difficulty was that we had failed to locate a vacuum chamber with the necessary parameters: 50 m long and pressure of no more than 10^{-5} torr, though we had looked all over the world. We used the 10 m vacuum chamber available at the Space Research Institute, Moscow.

As a first step we tested a 6 cm long scale model of the reflector (scale factor 1:5). The voltages calculated for 50 m distance were used. A resolution of 400 (for Ag) and better was obtained (50% peak height, Figure 5.3). Later on, the full-scale mock-up and the flight models were tested. We calculated optimal voltages for 10 m free-flight distance and the given grids' disposition. The calculated resolution in this case was 150 (for all masses). An additional decrease of resolution was expected owing to the toroidal deflector. The ions that go through the toroidal deflector near the walls have different velocities due to the local potential variations, and also the ions that go through the different sections of the toroidal deflector have different paths to pass. These effects lead to the decrease of resolution. The decrease was calculated. It is of no importance for the 50 m free-flight distance, as the time dispersion is on the order of a few microseconds and the resolution is limited to 350. The resolution limit was 60–80 in our tests with the 10 m vacuum chamber. The value was achieved easily in experiments by using the calculated voltages. So, we had good conformity of the theoretical and experimental results for resolution.

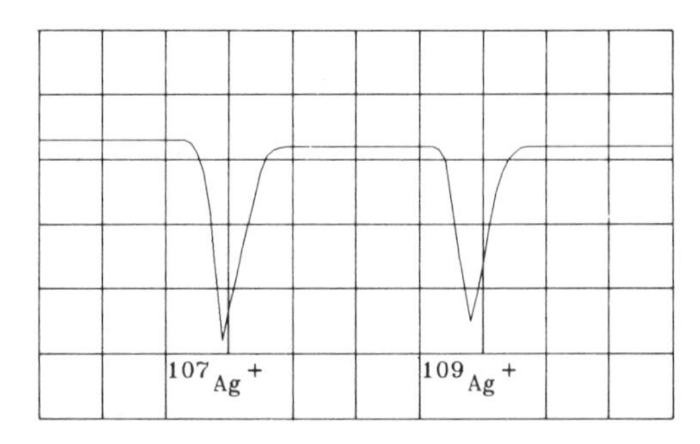

Figure 5.3. Ag spectrum obtained by the 6 cm reflector and for the 10 m free-flight distance.

We employed three different lasers during the device tests. The radiation of all of them was focused on a spot of 1–2 mm diameter, and the laser plasma produced similar numbers of ions as we expected in the real experiment near Mars. We had to reduce the ion flux to imitate the current that was expected during the flight experiment. This was accomplished by using a special mask at the entrance of the reflector with small holes in it. The mask attenuated the ion flux by a factor of 125—to imitate the spherical scattering over 50 m (instead of 10 m) and to get the same value of the current as for 50 m while all the times of flight are 5 times shorter.

Spectra of various materials were obtained during the tests. We used a silver target to confirm the resolution of the models. Phobos-like materials were used to check the possibility of peak identification in a complicated real spectrum. We often used a piece of the Allende meteorite as a target (the sample of this meteorite was presented to us by A. A. Barsukov and Yu. A. Shukolyukov, the Vernadskii Institute of Geochemistry and Analytical Chemistry of our Academy of Sciences). The Allende meteorite is classified as a carbonaceous chondrite of type CV and is supposed to be of a composition rather similar to that of the Phobos surface. The carbonaceous chondrites of type CI or CM are thought to be even more similar, but these types of meteorites are much more rare and we did not manage to get a sample. The Phobos surface is composed of *regolith*, a rocklike material powdered and mixed by micrometeorite impacts. The powdering process was expected to increase the value of data to be obtained because it makes the outer layer more representative of the bulk composition. Otherwise the LIMA-D, which measures only to a depth of a few micrometers, would yield data distorted by interaction with the solar wind. The regolith depth is on the order of 5 m to hundreds of meters (Veverka, 1978). The conditions in the plasma processes for this state of surface might differ from that for a hard rock surface. We used a solid target of Allende material as well as the powder target to look for the difference in the spectra. The specially prepared homogenized powder sample was kindly presented by V. Boynton of the University of Arizona.

The artificial samples were prepared by the Finnish scientists engaged in the LIMA-D project for the detailed investigation of the quantitative analysis possibilities. Olivines $(Mg,Fe)_2SiO_4$ were used as a matrix with various Mg/Fe ratios. The olivine with Mg/Fe equal to 0.54:0.46 has an elemental composition very close to that of the carbonaceous chondrite. Some other elements of known amount were added to the matrix for the quantitative measurements. The samples were prepared with carbon, sulfur, and metallic iron. The samples were homogenized up to the scale of a few tens of micrometers. Unfortunately, only some preliminary stages of the work were completed when the second *Phobos* spacecraft was lost, and the work has since been stopped.

We used the various working modes of the device to investigate their influence on the spectra recorded. The effect of the laser power density was of special interest. We found that by changing the laser power density one can obtain spectra of various "quality." No spectra could be obtained at low energy, and the threshold of the spectra appearance depends strongly on the target material and to some degree on the surface conditions. One can see the very poor spectra with, say, only the Fe peak in it at higher energy. Good spectra with the peaks of all elements expected and not many peaks of multiply charged ions could be obtained in some rather narrow energy range. The range seems to be some $\pm 20\%$ from the optimal energy, which was also specific for each sample used. Further increase of the laser energy leads to overly complicated spectra with multiply charged ion peaks. These peaks can coincide with the main (singly charged) peaks, making a correct quantitative analysis impossible. Such analysis is problematic even for "good" spectra.

The spectrum of the Allende target registered by laboratory equipment is shown in Figure 5.4. Here, the peaks of titanium and phosphorus can be seen. The contents of these elements are 0.047 and 0.083 atom%, respectively. The spectrum shows that the planned accuracy was achieved. The spectrum of the Allende target registered by the LIMA-D in-flight system is shown in Figure 5.5. The chromium peak (mass = 52) is seen in the spectrum; its content is 0.2 atom%.

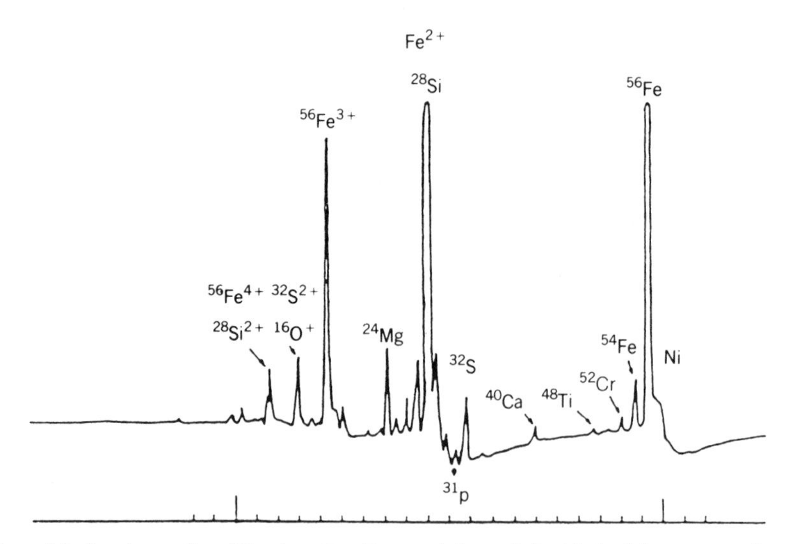

Figure 5.4. Spectrum of an Allende meteorite sample recorded with the laboratory registering system.

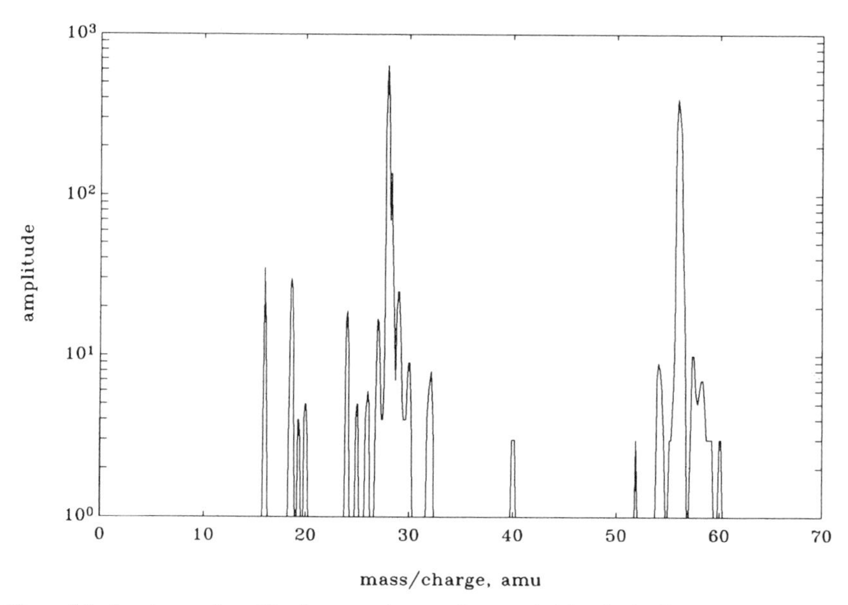

Figure 5.5. Spectrum of an Allende meteorite sample recorded by the in-flight data-processing system (x-axis—mass, amu; y-axis—amplitude, logarithmic scale).

All necessary mechanical and climatic tests were made, and the LIMA-D worked perfectly.

In-flight Tests. There was no possibility of switching on the LIMA-D device before the *Phobos* spacecraft came to the Mars orbit. The movable elements of the lens were caged, and they were to be uncaged only after the maneuvers near Mars had been finished and the cruise engines had been separated. Two test runs of the LIMA-D were made on the Mars orbit in March 1989.

Two test modes were provided in the LIMA-D. Their functions are as follows:

"PROG 1" mode—the electric tests of the units without switching on the high voltage power supplies (of laser and detector). The control system, the data unit, the focusing system, and the altimeter are tested.

"PROG 2" mode—the tests with the high voltages on. The same tests as in "PROG 1" are carried out as well as a check of the laser and servodrive operation, a check of the MM-1 multiplier gain with the help of the alpha-source, and a check of the reflector power supply.

The tests showed that the device operated normally with no glitches. The test results are described in detail in the preprint available in Russian (Altshuler et al., 1989b). A short description is presented below.

The Laser. The first switch-on of the device was carried out almost 300 days after the launch. Degradation of the optical coatings and the passive Q-modulator could occur owing to the influence of cosmic radiation, vacuum, and temperature fluctuations. Besides, disadjustment of the laser generator could take place because of the spacecraft maneuvers.

During the tests the following parameters were checked: (a) the lower limit of the generator monopulse mode (generation threshold voltage); (b) the upper limit of the generator monopulse mode; (c) the delay between the flashlamp initiation and the laser pulse; and (d) the laser energy. The generation threshold decreased by 5%. This was explained by a decrease of the temperature (10 °C, instead of 20 °C in the laboratory tests). The dependence of the time delay between the flashlamp initiation and the generation of laser radiation on the laser generator voltage was approximately the same as during laboratory tests. The laser energy stability measured in flight tests was 10%—the same as in laboratory. So, the conclusion was reached that the laser operated satisfactorily. No influence of the space flight conditions was detected.

The Altimeter and Focusing System. The main problem was to check the dynamic characteristics and the accuracy. All parameters corresponded to the technical specifications. This confirmed the correctness of choosing molybdenum dioxide grease (used for the movable elements of the lens) for space conditions. The diode lasers used in the altimeter had no glitches either.

The Data Unit. Two types of tests were made. The simple test using the in-flight test generator showed that the whole section dealing with spectrum processing worked all right. The more difficult question was connected with the multiplier gain factor. We had provided a special alpha-particle source for its measurement. The idea was to measure the pulse height distribution depending on multiplier voltage. The single event unit (SEU) has three amplitude discriminators and allows us to count the number of pulses in each channel during some time interval. The high voltage routine (HVR) makes it possible to measure the pulse height distribution for four voltages differing by 200 V. Thirty-four measurements were made after the HVR of various initial voltages. The diagram showing the count number for each channel as a function of the voltage was compared with that obtained during laboratory tests. It was found that the data coincided if we shifted the flight data horizontally by 400 V (see Figure 5.6). This meant that if we increased the voltage

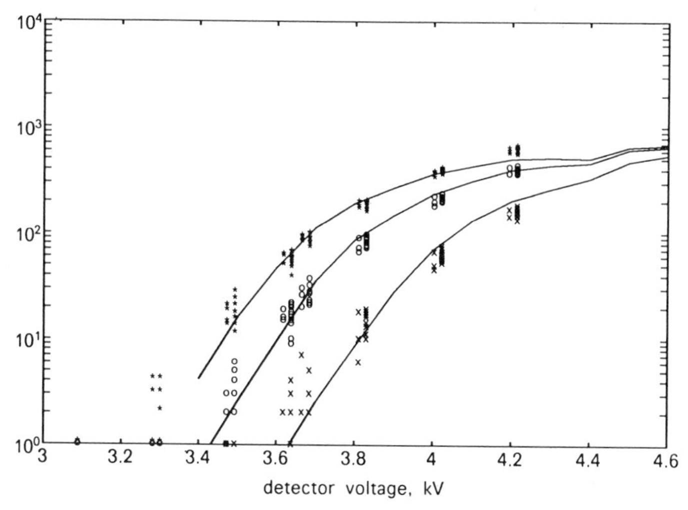

Figure 5.6. The alpha-test data. Count rates after three discriminator settings. Points: in-flight test data. Lines: laboratory calibration data shifted by 400 V along the x-axis. *Key:* $*$, $\bigcirc$, and $\times$ correspond to three discriminators, with discrimination level ratios 1:e:e^2, respectively.

by 400 V, the pulse height distribution and, we assumed, the gain factor would be the same. During the laboratory tests that simulated the real, in-flight ion current (flux), it had been found that the optimal multiplier gain corresponds to 3000 V. So, we chose the multiplier voltage of 3400 V for the fly-by. The gain degradation during the flight was by a factor of 5–15 (at the same voltage). The MM-1 multiplier had been subjected to some special activation during manufacture, and such gain changes are possible owing to negative factors like water vapor or high current operation, according to the manufacturer's data. The gain change can be easily adjusted by the high voltage change.

Other Units. Only simple electric checking was required for the control system and the reflector. The control system had an autotesting routine that was executed at the beginning of every switch-on. No error was detected during the tests.

5.2.6. A Promising Instrument for the Future Asteroid Mission

A proposal more impressive than LIMA-D, not to say fantastic, was suggested by one of coexecutors of LIMA-D—the German firm Von Hoerner & Sulger Electronics GmbH (Schwetzingen). They carried out a study, "Facility for

Remote Analysis of Small Bodies" (FRAS), under contract to the Max Planck Institute for Nuclear Physics, Heidelberg (see D'Orazio et al., 1986). It was proposed to integrate various remote analysis techniques in one complicated device. Only a brief description is given below.

To further study the evolution of the solar system it is necessary to determine the chemical composition of many objects in the system, ranging from the major planets to small bodies, especially comets and asteroids. It looks as if no sample-returning missions will be organized in the near future owing to financial constraints both in our country and abroad. So, there will be no laboratory measurements available to answer the related questions. Two possibilities exist for such measurements in situ. One, using landing vehicles and various rovers, is feasible for major planets. It was used for investigations of the Moon, and projected Mars rover missions are under study in some countries. The other possibility is the remote analysis technique similar to that used in the *Phobos* project. This seems to be the most suitable approach for asteroids and comets. Only a close fly-by at distances up to hundreds of meters would be required, not a landing. During a fly-by mission some scientific objectives can be achieved that are not possible by laboratory analysis of the returned sample—say, a multitarget mission with an inspection of variations in composition over all the asteroid surface. The fly-by measurements are also more clean with respect to the landing. The contamination problem is real, and some research in this field has been conducted [see the paper of Brahič (1985) and others in the same issue of *Advances in Space Research*].

FRAS, multipurpose scientific instrument able to solve the aforementioned problems, is proposed in the Von Hoerner & Sulger report (D'Orazio et al., 1986). The following techniques are combined in the instrument:

- TOF laser ionization mass spectrometry (LIMS)
- Secondary ion mass spectrometry (SIMS); this uses a TOF registering system, too
- Laser-induced fluorescent and UV spectrometry
- Raman spectrometry
- Surface profile measurements.

Detailed discussion of the analytical capabilities of and problems with these techniques is given in the FRAS report. An overview of the FRAS package concept is presented in Figure 5.7. Three large components can be distinguished—the ion gun unit, the optical unit, and the TOF-reflector unit.

The ion propulsion system RIT-35 is proposed to be employed as the ion gun. It has been built and approved (as an engineering model) by MBB-

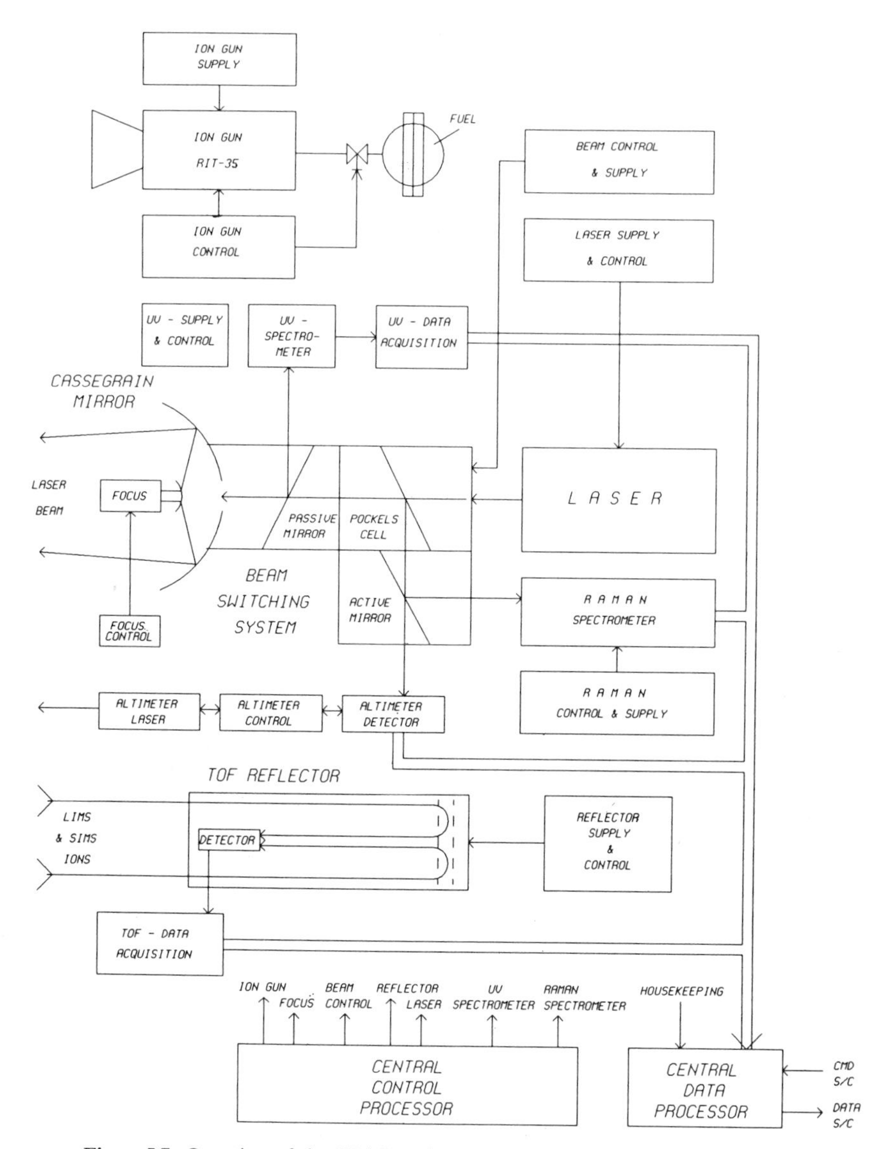

Figure 5.7. Overview of the FRAS package concept. From D'Orazio et al. (1986).

ERNO in Ottobrunn, Germany. It has two roles in the instrument: it is used for producing short ion pulses for the secondary ion mass spectrometer and also as a thruster.

The optical unit produces the laser ionization of the surface and includes optical spectrometers (UV and Raman) and an altimeter. All these subunits use the same Cassegrain telescope with controlled focal length. The focus control uses a small mirror built around a piezoelement. The telescope has a diameter of 60 cm. It focuses the radiation of the main laser on to the asteroid surface from a distance of 100 m. The focal spot diameter is 1 mm. The laser parameters are energy, 2 J; pulse duration, 20 ns. The backscattered radiation is distributed by a set of active and passive mirrors to the Raman spectrometer, fluorescent spectrometer, and altimeter detector.

The proposed reflector is 2 m long and has an axial symmetry. It is used for registering TOF spectra of ions from the laser-induced plasma, and secondary ions. Inside is a changing electric field to provide a good time focusing for the ions with wide energy spread (see Section 5.2.2, above). Unlike the LIMA-D reflector, the FRAS reflector contains no grids and uses the occurring transverse electric field for the transverse focusing of the ion flux.

The control and data-processing system is common to the whole instrument. According to the aforementioned report, the FRAS instrument can operate at distances from tens up to a few hundreds of meters from an asteroid's (or comet's) surface.

If the ion gun and main laser work at 0.1 Hz, the power consumption will be 112 W. The total mass for the experimental package is estimated to be 80 kg. The required telemetry rate is 7 kbit/s. The outer dimensions are mainly determined by the Cassegrain telescope, with 60 cm diameter and 55 cm length; the TOF reflector, with the same diameter and a length of about 2.5 m; and the ion gun, with about 40 cm diameter and a length of approximately 22 cm. All these technical parameters are acceptable for modern spacecraft (only the multitarget asteroid mission is doubtful—or at least very expensive).

The estimated cost reaches 83 million DM (prices of 1986).

5.3. LASMA: A MINIATURE LASER MASS ANALYZER

5.3.1. Description of the Instrument

After the main LIMA-D work was completed, we decided to go on with the design of mass spectrometers for space research. We already had some experience with the LIMA designed at the initial stage of the *Phobos* project, as well as good experience with LIMA-D. We started development of a large complex instrument for chemical, isotopic, and mineralogical analysis of a planet's soil and atmosphere using the ideas we had evolved. It would combine the various techniques of in situ measurements, such as LIMS, SIMS, and SNMS (secondary neutral mass spectrometry). The first stage was to design a miniature laser mass spectrometer.

The device could find application to terrestrial problems, also. There already existed large laboratory mass spectrometers such as the LAMMA-1000 produced by Leybold-Heraeus (Cologne, Germany) and the LIMA series produced by Kratos Analytical (Manchester, UK). Our idea was to produce a small instrument that could be used not only in the laboratory but also under field conditions. The device must be capable of making an analysis of various substances with on-line data processing. According to our estimate, the accuracy (which is in any case not very high in laser TOF-MS) and sensitivity could be made approximately the same as those of the existing devices. Such an instrument could be utilized in geology, in environmental pollution control, and in various industries (metallurgy, semiconductors, etc.). The LASMA device was our first attempt to make such a device.

LASMA consists of the following functional units. The sample ionization section includes the laser with a power supply unit and focusing lens. The device is provided with an observation system that uses the same optics as the laser and with a lamp for the sample illumination. A special system is used for the introduction of samples. Eight or ten samples can be installed at a time, and only a few minutes are required for sample changing. The TOF analyzer is 24 cm long and contains a 4 cm reflector. Its diameter is only 10 cm. The analyzer also contains four metal rings used as the electrostatic lenses, a three-grid section for reflecting the ions with energies lower than some specified limit. The detector consists of two microchannel plates of 56 mm diameter. The large input area of the detector and the electrostatic focusing system provide the high sensitivity of the device.

The TOF mass analyzer is placed into a small vacuum chamber of 30 cm height and 15 cm maximum diameter. A high-vacuum turbomolecular pump and a forevacuum pump provide a vacuum of less than 5×10^{-6} torr. The signal is processed with a high-speed A/D (analog-to-digital) converter and an IBM AT-compatible computer. The A/D converter has 10 ns sampling time and 16 kB RAM (random access memory). The power supply unit provides all necessary voltages, and it has approximately the same dimensions as a standard personal computer. The whole instrument can be easily placed on a desk. The appearance of the LASMA instrument is shown in Figure 5.8.

Complete automatic control of electric circuits is planned. Only the vacuum system will be manually controlled in order to have it as small as possible. The computer-aided control for the laser energy, laser start-up, voltages, and automatic data processing will provide high serviceability.

There is no need to describe in detail the operation of the device. The principles are the same as for LIMA-D. The obvious simplification is that the distance is constant. Nevertheless, the capability to manually refocus the lens is provided. That is necessary because the processes in the laser-produced

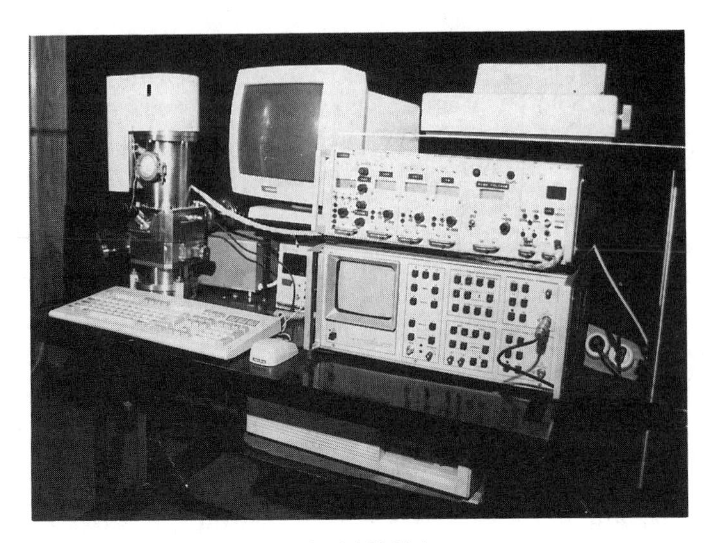

Figure 5.8. The LASMA instrument.

plasma depend on the position of the focal plane: exactly on the surface, or below or above it.

The capability of laser refocusing and the ease of manual and automatic control make the device a laboratory instrument convenient for investigation of laser–solid interaction.

The analytic characteristics of the instrument are given in the accompanying tabulation (see facing page).

We should explain just why we like the TOF system so much. Its advantages as compared with other systems, e.g., a radio-frequency analyzer or a quadrupole analyzer, are obvious when one is working with a pulsed ion source. The quadrupole analyzer is considered to have excellent analytical characteristics in the continuous mode of operation. However, when state-of-the-art electronics are taken into account, a careful comparative analysis of mass spectrometers used in space research (Niemann, 1977; Carignan, 1981; Arnold and Henschen, 1981; Hunton and Trzcinski, 1986; Grechnev et al., 1979; Kochnev, 1980) reveals that almost all physical characteristics of the TOF analyzer are better by 1–2 orders of magnitude than those of radio-frequency or quadrupole analyzers if the TOF analyzer works in the semi-continuous mode, registering 10^4 spectra per second. These properties of the TOF analyzer in combination with its relatively high aperture, construction simplicity, and low required accuracy of constituent parts are decisive in opting for this analyzer.

Mass range	1–250 amu
Mass resolution	⩾200
Maximum resolution obtained (for Pb)	600
Relative sensitivity of a single analysis	1–10 ppm
Relative sensitivity of the spectrum accumulation mode	0.1–1 ppm
Absolute detection limit	5×10^{-14} g
Diameter of the focus spot	10–50 μm
Depth resolution during layer-by-layer analysis	0.1–3 μm
Maximum depth of layer-by-layer analysis	⩾1 mm
Reproducibility of Ag isotope ratio	10%
Laser wavelength	1.06 μm
Laser pulse energy	15 mJ
Laser pulse duration	5 ns
Power density at the focal spot	10^{10} W/cm^2
Range of attenuation of laser radiation	1–100
Overall dimensions of the units:	
Analytical system	50 × 40 × 40 cm
Pumping system	50 × 50 × 40 cm
Power supply unit	60 × 40 × 30 cm
Personal computer	Standard
Weight	⩽75 kg
Power consumption	⩽600 W

The most important physical characteristics of MS analyzers are listed in Tables 5.1 and 5.2. All these analyzers are supposed to have comparable weights and overall dimensions suitable for use in space research. The same analysis could be provided for the larger laboratory systems. The result would surely be the same—the TOF system is the most suitable.

5.3.2. Test Results: Resolution, Sensitivity, and Accuracy of Analysis

The design of LASMA is not yet completed. We have two prototypes: one designed using Western components (electronics and vacuum pumps), and the other assembled using only components made in our country. The laboratory investigations of the analytical characteristics were obtained on these two devices. We were interested mainly in such characteristics as resolution and sensitivity.

Table 5.1. Relative Sensitivity of Various Types of Mass Spectrometers[a,b]

MS Type	Radio Frequency (RF)	Quadrupole	TOF
Collection area (cm^2), S	2.5	0.25	20
Relative energy transmission, E	0.2	0.2 (0.04[c])	1
Grids' transparency, G	0.48 (18 grids)	1 —	0.72 (4 grids)
Accumulation time per spectrum (s), T^d	3×10^{-2}	3×10^{-2}	10^{-5}
Number of spectra per second, F	1	1	10^4
Sensitivity decrease factor, $\Sigma = S \cdot E \cdot G \cdot T \cdot F$	7.2×10^{-3}	1.5×10^{-3}	1.44
Relative sensitivity, Σ_{TOF}/Σ	200	1000	1

[a]For the electron ionization ion source.
[b]Each instrument is assumed to work for 1 s, the TOF in the semicontinuous (spectra accumulation) mode.
[c]For the high-resolution (1000) mode.
[d]Per one peak. The spectrum is assumed to contain 30 peaks, and RF and quadrupole analyzers measure each peak during equal time.

Table 5.2. The Most Important Physical Characteristics of Mass Spectrometers

MS Type	Resolution	Mass Range	Relative Sensitivity[a]	Dynamic Range	Measurement Time per Mass	Time per Spectrum	Registration of elements
RF	40	1–100	200	10^6	1–30 ms	0.1–1 s	Sequential
QUAD	200–1000	1–200	1000–5000	10^5	1–30 ms	0.1–1 s	Sequential
TOF	200–1000	1–3500	1	10^7	0.5 μs	50 μs	Simultaneous

[a]See Table 5.1.

Resolution. The resolution, not lower than 200, is enough for all main applications of the laser TOF mass analysis of inorganic substances [we calculate the resolution from the full width at half maximum (fwhm)]. It allows us to distinguish all available elements and isotopes. To resolve isobaric interferences a resolution of more than 2000–5000 is required, and such problems (especially in geochronology) seem beyond the range of LIMS. We used a lead sample for the resolution determination. The voltages were adjusted to get maximal resolution. They did not coincide with the calculated ones. This could be explained by the inaccuracy of the manufacturing and assembly of constituent parts due to technological limitations. Such inaccuracy plays a noticeable role for the miniature spectrometer, whereas it may be less critical for a normal-size device. Nevertheless, in optimal conditions a resolution of up to 600 was obtained in some spectra. An example is shown in Figure 5.9. The resolution is, however, not constant from spectrum to spectrum, but in 80% of the spectra it is better than 200.

Sensitivity. For many applications the sensitivity is an even more important parameter than resolution. The problem of whether an impurity is present above some lower limit or not commonly arises in pollution control, the semiconductor industry, etc. The desired sensitivity for a general purpose

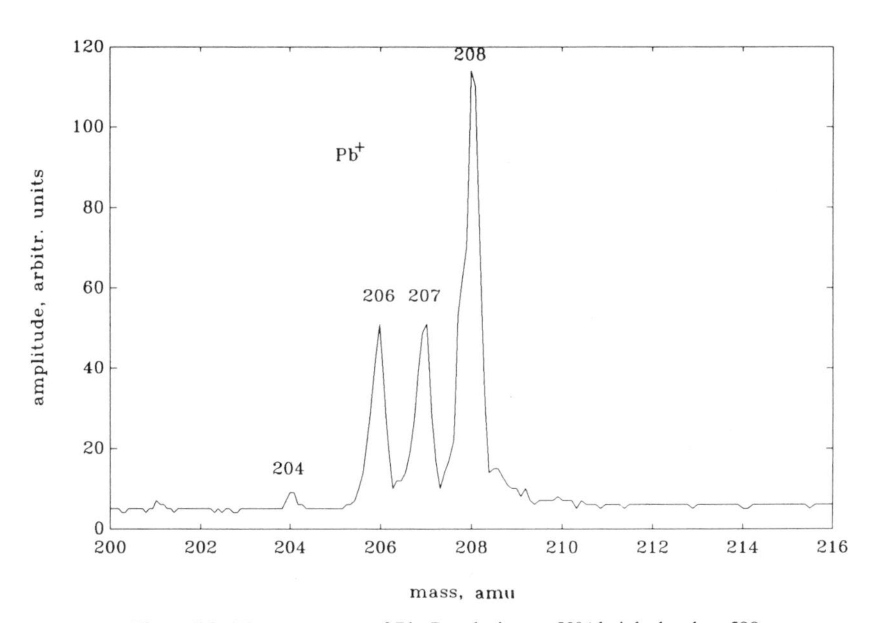

Figure 5.9. Mass spectrum of Pb. Resolution at 50% height level: ~500.

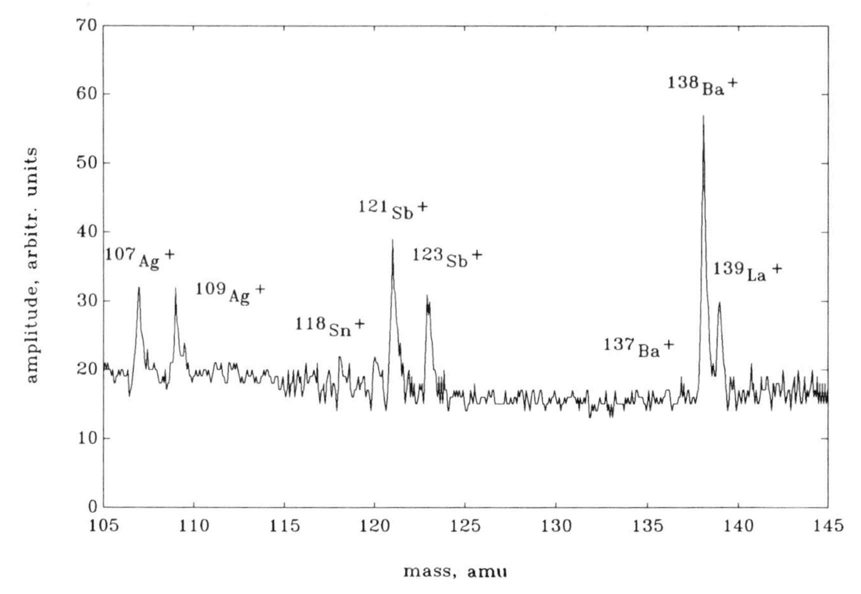

Figure 5.10. Spectrum of impurities in aluminium. According to the certificate, the sample has 25 ppm Ag, 9 ppm Sb, 12 ppm La, and 12 ppm Sn; Ba was not mentioned in the certificate.

device is on the order of 1–10 ppm. This was the value we wanted to achieve. Adjustment of the electrostatic focusing system allows the sensitivity to be increased to this level. An example is shown in Figure 5.10. The problem was that the electrostatic focusing system decreases the resolution while increasing the sensitivity. A compromise can usually be found, depending on the problem under study. The resolution in the spectrum shown in Figure 5.10 is 180 (at mass range 100–140 amu). The spectrum was recorded by a standard laboratory transient recorder with no preliminary amplification of the signal. To obtain high sensitivity a preamplifier of wide bandwidth and low noise is required. The ability to withstand the saturation is of importance, too. A new preamplifier based on a hybrid operational amplifier of 2 GHz unity gain bandwidth is now under development. It should permit us to increase the sensitivity 10-fold.

Precision and Accuracy of Analysis. Though threshold measurements are widely needed and seem to best suit LIMS, we are also very interested in checking the possibility of direct quantitative analysis. We have measured the composition of iron–cobalt–nickel alloy (Kovar). The composition was calculated using peak areas, and the relative sensitivity coefficients were assumed

Table 5.3. Element Concentration[a] for the Kovar Target

Element	Concentration from Metal Reference Book	Measured Concentration	
		Sample 1	Sample 2
Fe	52.5–54.5	53	53
Ni	28.5–29.5	30.7	30.5
Co	17–18	16.4	16.5

[a]Concentration: atom %.

to be equal to unity. The result was better than expected (see Table 5.3; also the spectrum in Figure 5.11).

The laser energy really affects the result strongly. It was chosen at a level that provides the singly ionized ions with a small number of doubly ionized ones. The latter constitute less than 10%. Methodological investigation of a particular quantitative analysis problem concerns mainly selection of the right laser energy. Also some selection of the spectra is possible based on such spectral features as the number and amplitudes of multiply charged ion peaks. We have started investigating the possibility of controlling the steel hardening (carbonizing) process, where all these problems arise.

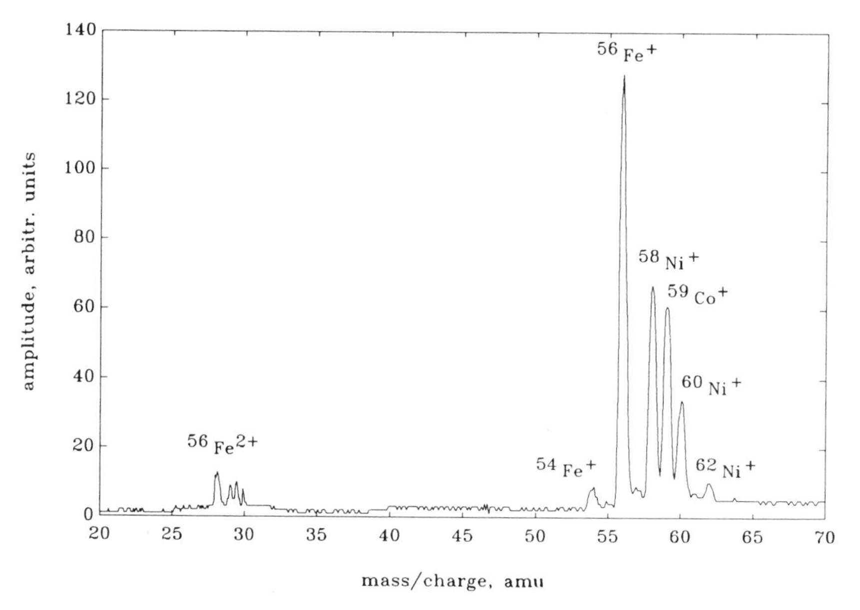

Figure 5.11. Spectrum of Kovar.

Also the spectrum reproducibility is of great importance for laser analysis. The process of laser–solid interaction is rather complicated, as discussed earlier in this volume (see Chapter 4, Parts A and B). The ions to be registered with a mass spectrometer are the remnants that are left after the process of recombination is complete. They are approximately a thousandth part of the initial number of ions. Even very high stability of the recombination process (say, 99.90–99.95% recombination) leads to a large spread in peak amplitudes. The peak area ratios, nevertheless, are obviously more stable. The measured reproducibility of the absolute values of peak areas for Fe, Ni, Co, Zn, and Sn is 20–30%. This estimate of σ was made by processing 10–20 single shot spectra without selection. As to minor components (with amplitudes of about 0.1 of the full scale of A/D conversion), the reproducibility is up to 50%. The spread of the peak area ratio is ± 10–20% (or 30–50% for low peaks).

The problems as to reproducibility could be partially explained by the sample's heterogeneity. For example, our best domestic standards do not guarantee the homogeneity at contents of 10^{-2}% and lower. This value is significantly larger than the amount analyzed by LASMA. In this connection Kovar spectra are of great interest because of the high homogeneity and the correctness of analysis reached. The elements that compose the alloy (Co, Fe, Ni) have similar mass, chemical, and other properties, and this can provide the high homogeneity. This alloy can be recommended when one is looking for standards that are homogeneous at the level of 30 μm and 10^{-11} g.

5.3.3. Applications of Mini-MS: Advantages and Disadvantages

The main disadvantage and also advantage of the LASMA device is that it is not as sophisticated an instrument as are the LAMMA-1000 and LIMA. We shall first examine some of the positive aspects.

Advantages. The device needs no specially trained personnel to work with it. We are going to produce a fully automated version that can be used in pollution control via local or global monitoring networks. Fully automated control seems to be possible for such operational modes as the voltage adjustment to get high resolution, the sensitivity control, and automatic registering of spectra and processing. The vacuum system control is not yet designed, but that is only a question of size and weight.

The small size represents another advantage. The device can be used in an industrial laboratory, a small pollution control station, a clinic, or anywhere where compactness is important. It can be mounted on a jeep for use by a geological team. Hard conditions such as transport over rough terrain do not pose a problem for the instrument.

The LASMA device's accuracy of analysis is enough for many applications. Let us consider some of them.

Ecology. Mostly threshold analyses are required. The detection limit for main pollution substances must be 1–100 ppm. An accuracy of $\pm 50\%$ is usually sufficient. Mini-MS seems to best suit the ecological applications. This seems to be a common point of view; mini-MSs for air pollution minitoring produced by various firms (Pernica, USA; VG Instruments and Kratos Analytical, UK; Bruker, Germany; Sciex, Canada) are available on the market. As an example of a LASMA application, Figure 5.12 shows the spectra taken from a powdered sample of industrial wastes. The contribution from Cr and Pb was found to be more than 0.1%.

Space Research. Only miniature automatic devices can be used nowadays in space. One of the purposes of the LASMA design was to propose it for the Mars mission. Of course, much more accurate data are desirable, but in a competition between greater accuracy, on the one hand, and compactness, on the other, only the latter can be preferred.

Geology. Elemental and isotopic analyses in the field and in the laboratory are needed in geological studies. The elements that could not be found with optical spectrometry (e.g., Li, B, S, Cl, Sc, Rb, In, Eu, or Er) could be easily detected with LIMS. We have shown the possibility of finding gold in sand at a concentration of some 40 g/t (metric ton).

Metallurgy. We have tried to design a procedure for control of the steel-hardening process. LASMA allows us to see that the hardened surface contains 5–10 times more carbon than the normal one. Performing layer-by-layer and spatial analyses we have found that the carbon distribution is inhomogeneous; it gathers in conglomerates. Investigation by an electron microscope shows that granule sizes are on the order of tens of micrometers. The size and spatial distribution of conglomerates may influence steel properties, and it can be measured by mini-MS.

Microelectronics. The analysis of semiconductors is widely needed nowadays. A quick and easy-to-handle analyzer such as LASMA can become a useful instrument for an engineer. The amount, depth, and spatial distribution of implanted atoms can be studied.

Medicine and Biology. For diagnosis of various diseases, especially ones caused by environmental pollution, elemental analysis is required. The mass range of LASMA could be extended to thousands of atomic mass units, and

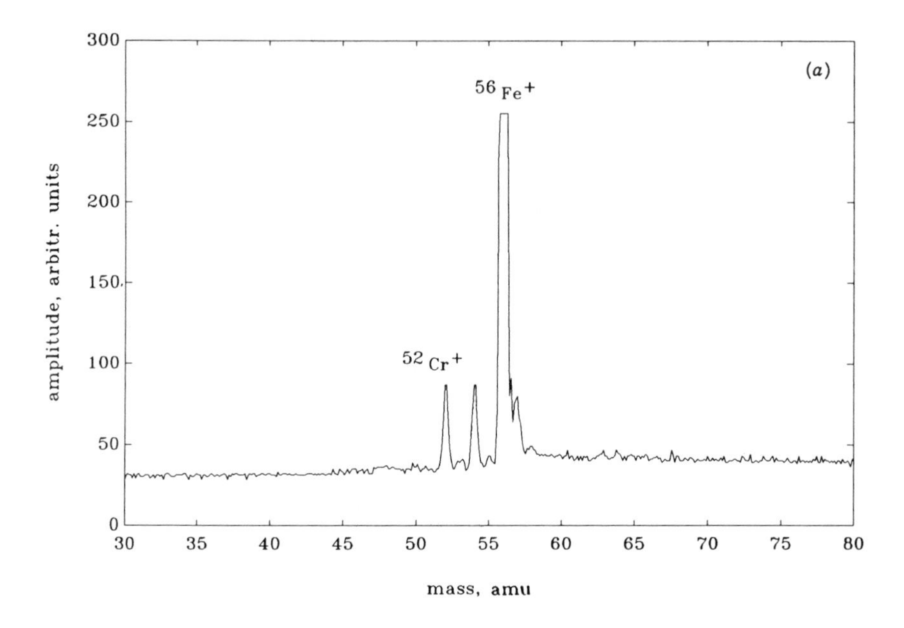

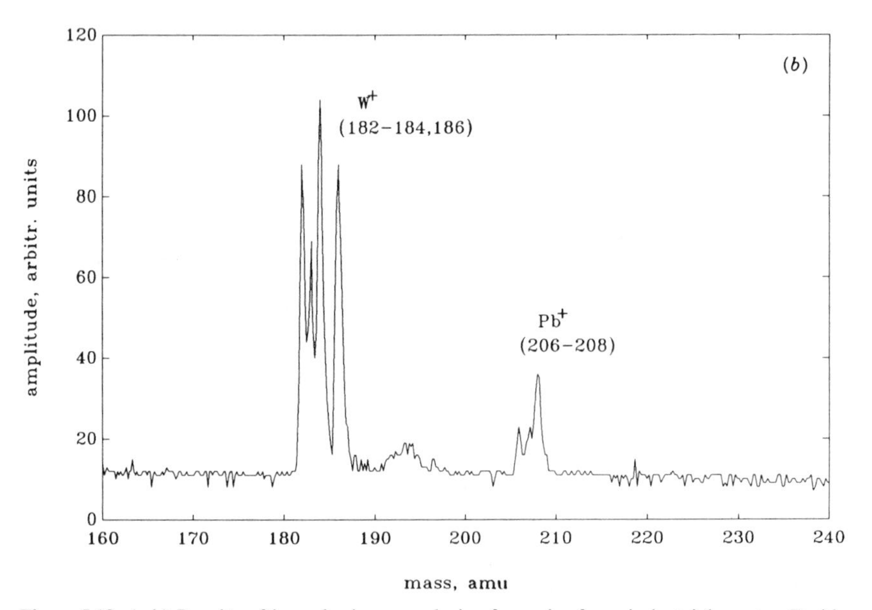

Figure 5.12. (a,b) Results of layer-by-layer analysis of powder from industrial wastes. Besides Fe and W, a contribution (>0.1%) from Cr and Pb was found.

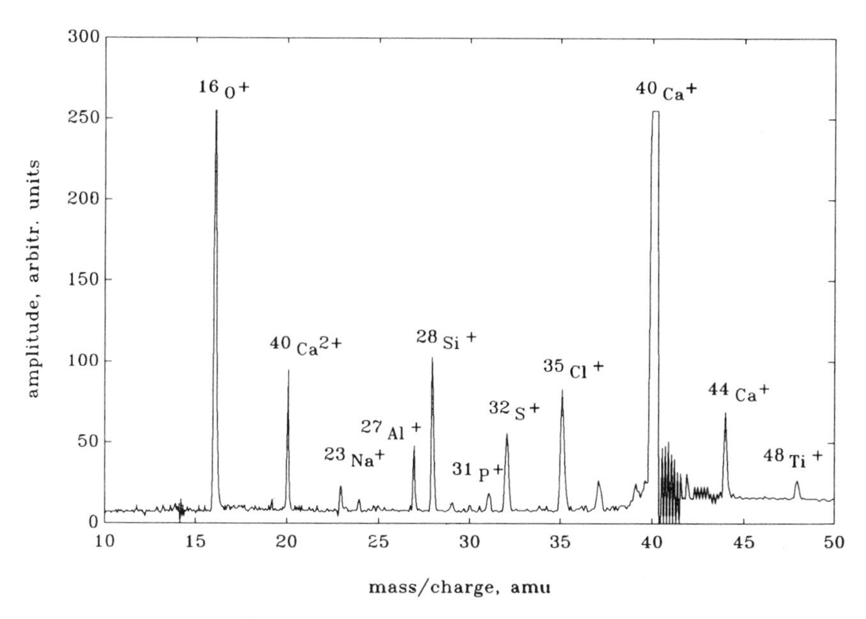

Figure 5.13. Mass spectrum of a renal stone.

some analysis of organic components is possible if the proper laser mode and power density can be found that do not lead to the complete destruction of organic molecules. As an example of elemental analysis, the spectrum of a renal stone is shown in Figure 5.13. Elements like O, Na, Cl, Ca, and P are normally involved in metabolism, but we also find peaks of Al and Ti. A weak indication of heavy metals was found during layer-by-layer analysis.

Just a few preliminary attempts to use the mini-MS in this way have been made as yet. The list of possible applications could easily be lengthened. Whenever a quick analysis is needed and simplicity is preferred without extreme sensitivity and accuracy, LASMA could be of use. The device will no doubt become more complex and sophisticated with time.

Disadvantages. The medium-size mass spectrometer LIMA of Kratos Analytical, say, the 401LS LIMA-SIMS model, combines laser-induced and secondary ion mass spectrometric techniques and an electron beam post-ionization approach. This combination results in a very broad-purpose device for complex investigations. Such great complexity seems beyond the capacity of mini-MS, though some attempts might be made to combine different techniques in the manner of, say, the TOF analytical unit. We might well try to unite LIMS with SNMS and/or SIMS; such a combination is necessary for ecological monitoring stations and space research. The dilemma is obvious: any increase in complexity means an increase in size. Only new ideas on the combination of these techniques might help resolve this dilemma.

The device does not have very high accuracy; it cannot be compared with the accuracy of "wet" chemical analysis. This restricts the field of applications, but in the common field of chemical and mini-MS analysis the latter nevertheless has many advantages.

Our choice is mini-MS, and we appeal for collaboration in its development.

5.3.4. Perspectives of Mini-MS Design in Our Laboratory

Three LASMA devices have been produced up to now. Two of them were manufactured with Western components, and a 100 MHz sampling rate of the registering system was provided. The third device was manufactured with only domestically made components except for a personal computer. The sampling rate for this unit is 20 MHz now, but an A/D converter with 10 ns sample time is under development. It is designed in PC/AT-board size standard. We should note that this device is in considerable demand in our country, and it seems to be finding wide application abroad. Five additional units are now being manufactured after some modification and improvement.

The Western version was designed with the financial support of Nukem GmbH (Hanau, Germany). The work was carried out with the help of Isotop (Kaiserslautern, Germany). One of the units is now being demonstrated in the laboratory of the latter company.

The physical idea and construction plan of LASMA are protected by two patents in seven major countries (Managadze, 1984). Without being able to completely reveal technical know-how, we can say that the analytical part of the device has an axial symmetry, which allows some problems to be solved associated with the plasma expansion. These processes need more detailed investigations and are now under study.

The main idea—the elaboration of a multipurpose universal complex for planetary research—requires an accumulation of experimental experience involving several diverse techniques. The current promising activities of our laboratory are directly related to this: the design of miniature TOF mass analyzers for organic molecule, gas, and volatile substances. One of these instruments, integration of which is finished now, is a laser analyzer with multiphoton UV ionization made in LASMA format (by the use of the same TOF analyzer).

A neutral gas TOF analyzer with resonant ionization is under development, too. It is expected to be of high selectivity and sensitivity.

A new type of TOF analyzer—a foil mass reflectron—seems to be very promising for the supersensitive analysis of trace elements with the registering of single ions (Managadze, 1986). The development of the idea and its realization are of some interest. In 1985 work began on the international project SOHO (the Solar and Heliospheric Observatory). The aim of this project is to investigate the Sun and solar wind. We realized that such an instrument is the optimal (and may be the only possible) device for analyzing the isotope composition of the high-energy component of the solar plasma. There are no other devices able to analyze ions of 50 keV energy with the required resolution. We also understood that the technological level in our country does not guarantee the possibility to develop the detecting unit of the device. It requires a time resolution better than 1 ns. It was impossible to buy the necessary components abroad because of the U.S. embargo policy.

The optimal solution was found. It was decided to produce the device by joint efforts of the laboratories of D. Hovestadt, the Max Planck Institute of Extraterrestrial Physics, Garching, Germany, and of G. Gloeckler, the University of Maryland, with the use of our invention. Hovestadt is the principal investigator of the CELIAS (charge, element, and isotope analysis system) experiment of the SOHO project. The laboratory mock-up of MTOF (solar wind mass time-of-flight sensor)—the foil mass reflectron—was produced and tested very quickly at the University of Maryland. The mock-up has a resolution of more than 200 (Hovestadt et al., 1989). The instrument passed a hard and exacting competition and was included in the scientific payload of the SOHO project. The flight models are now being manufactured, and the launch is planned for 1994.

We have considered two LIMS instruments designed for space research. We think that LASMA will find wide applications on the Earth. As for LIMA-D, we are sure that this or a similar instrument will be successfully used in a mission like *Phobos* and will provide valuable scientific results. We hope to participate in such a project.

NOTE ADDED IN PROOF

A few months have gone by since we sent this chapter to the editors. During this period we have carried out more qualitative tests with the LASMA instrument, which showed that it has better characteristics than we assumed. These tests became possible owing to the standard glass SRM No. 610 of the National Institute of Standards and Technology (Gaithersburg, Maryland), which was kindly placed at our disposal by D. Simons. The homogeneous distribution of more than 10 elements with an average concentration of about 500 ppm (by weight) provided the success. We failed to determine correctly

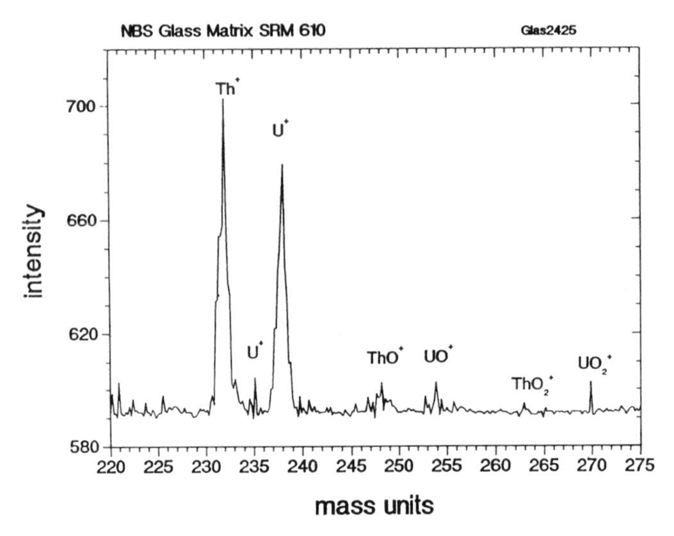

Figure 5.14. Segment of the spectrum of the standard glass SRM No. 610.

such important parameters as sensitivity and reproducibility because of the absence of good quality standards in our country.

The region with U and Th of the glass spectrum is presented in Figure 5.14. According to the certificate, the concentrations of Th and U are 457 and 461 ppm (by weight), respectively. We have estimated the sensitivity to be equal to 1–3 atom ppm. This coincides with the previously estimated value, but using the standard glass sample we get this result in every measurement. The peak of the U isotope with mass = 235 can be seen in the spectrum shown. It accounts for 0.72% in natural U, or around 0.3 atom ppm in the sample. So, the sensitivity is equal to 0.2–0.5 atom ppm, but the peak is not present in every spectrum. These results are obtained using one-shot spectra. The sensitivity could probably be increased up to 0.1 ppm by using signal averaging.

We calculated the ratios Th/U and Th/Pb for 20 spectra. The values proved to be equal to 1.24 ($\pm 13\%$) and 0.53 ($\pm 27\%$), respectively (the standard deviation for one shot is presented in brackets; the deviation for the mean values is 4.4 times less). One can see that the precision of measurements is rather high for Th and U, which have similar chemical properties. Previously we got 20–30% deviations for various elements. It may be assumed that the inhomogeneous distribution of elements in our previous samples was responsible for half of this value. The relative sensitivity coefficients are equal to 0.82 for U and 1.8 for Pb (with respect to Th).

It must be mentioned that the vacuum in our instrument is rather poor: the pressure is around 8×10^{-6} torr. The miniature instrument for field conditions could not be equipped with a powerful pumping system. The residual gases might be partially responsible for the noise occurring in the detector, i.e., because of scattering of ions in the drift region and reflector. This cannot be helped until we improve the vacuum system. We do not want to increase the dimensions of the instrument; thus we are looking for other possibilities to reduce the noise. The signal-to-noise ratio could be improved by using a large peak cutoff. Now we are trying to put this device into operation. It is not difficult to cut off preselected peaks, but we prefer a system that will automatically cut off the large peaks when they occur in the spectrum. The signal-to-noise ratio could also be increased when signal averaging is used.

These tests have strengthened our confidence in the possibility of using the instrument for quantitative analysis. The measurements were carried out in cooperation with Isotop, Kaiserslautern, Germany.

ACKNOWLEDGMENTS

We would like to acknowledge our colleagues from the Laboratory of Active Experiments in Space for their help during preparation of this paper, with special thanks to Anatoly Bondarenko, the responsible scientist for the LASMA project.

REFERENCES

Altshuler, G. B., Balebanov, V. M., Bondarenko, A. L., Grimm, V. A., Dulnev, G. N., Karasev, V. B., Korkin, A. I., Managadze, G. G., Parfenov, V. G., Rumyantzev, D. M., Sagdeev, R. Z., Khloponin, L. V., Khramov, V. Y., and Shchedrov, M. V. (1988). In *PHOBOS: Scientific and Methodological Aspects of the Phobos Study—Proceedings of the International Workshop*, pp. 231–236. Space Research Institute, USSR Academy of Sciences, Moscow.

Altshuler, G. B., Arumov, G. P., Bondarenko, A. L., Valah, P., Zubenko, G. I., Lyash, A. N., Managadze, M. M., Novotny, A., Pershin, S. M., Rumyantsev, D. M., Khloponin, L. V., and Yuzgin, A. V. (1989a). *Physical and Meteorological Calibration of Laser System of LIMA-D Remote Mass Analyzer* (in Russian), Pr-1595, pp. 1–22. Space Research Institute, USSR Academy of Science, Moscow.

Altshuler, G. B., Bondarenko, A. L., Bubnov, A. E., Gritc, Z., Zubenko, G. I., Managadze, M. M., Ruzichka, I., Terentiev, V. I., Timofeev, P. P., Tuktarov, V. A., Khloponin, L. V., Khramov, V. Yu., and Shutyaev, I. Yu. (1989b). *Results of LIMA-D Remote Mass Analyzer Flight Tests* (in Russian), pr-1613, pp. 1–28. Space Research Institute, USSR Academy of Science, Moscow.

Arnold, A., and Henschen, G. (1981). *Geophys. Res. Lett.* **8**, 83.

Brahič, A. (1985). *Adv. Space Res.* **5**(2), 97–106.

Carignan, G. R. (1981). *Space Sci. Instrum.* **5**, 429.

Covault, C. (1985). *Aviat. Week Space Technol.* **122**(13), 18–20.

Devyatikh, G. G., Gaponov, S. V., Kovalev, I. D., Larin, N. V., Luchin, V. I., Maksimov, G. A., Pontus, L. I., and Suchkov, A. I. (1976). *Sov. Tech. Phys. Lett.* (*Engl. Transl.*) **1**, 356; *Pis'ma Zh. Tekh. Fiz.* **2**, 906–910 (1975).

D'Orazio, M., Feigl, P., Grix, R., von Hoerner, H., Krueger, F. R., Li, G., Mildner, G., Rohr, W., Tinschmann, A., and Wollnik, H. (1986). *Facility for Remote Analysis of Small Bodies F.R.A.S.*, Parts I and II, pp. 26–273. Von Hoerner & Sulger Electronics GMBH, Schwetzingen, Germany.

Grechnev, K. V., Istomin, V. G., Kochnev, V. A., and Ozerov, L. V. (1979). *Kosm. Issled.* **17**(5), 70.

Hamelin, M., Balebanov, V. M., Beghin, C., Belousov, K. G., Bujor, M., Evlanov, E. N., Inal-Ipa, A., Khromov, V. I., Kochnev, V. A., Langevin, Y., Lespagnol, J., Managadze, G. G., Martinson, A. A., Michau, J. L., Minkala, J. L., Pelinen, R., Podkolzin, S. N., Pomathiod, L., Riedler, W., Sagdeev, R. Z., Schwingenschuh, K., Steller, M., Thomas, R., and Zubkov, B. V. (1988). In *PHOBOS: Scientific and Methodological Aspects of the Phobos Study—Proceedings of the International Workshop*, pp. 245–260. Space Research Institute, USSR Academy of Sciences, Moscow.

Hillenkamp, F., Kaufmann, R., Nitsche, R., and Unsold, E. (1975). *Appl. Phys.* **8**, 341–348.

Hovestadt, D., Geiss, J., Gloeckler, G., Mobius, E., Bochsler, P., Gliem, F., Ipavich, F. M., Wilken, B., Axford, W. I., Balsiger, H., Burgi, A., Coplan, M., Dinse, H., Galvin, A. B., Gringauz, K. I., Grunwaldt, H., Hsieh, K. C., Klecker, B., Lee, M. A., Managadze, G. G., Marsch, E., Neugebauer, M., Rieck, W., Scholer, M., and Studemann, W. (1989). In *The SOHO Mission: Scientific and Technical Aspects of the Instruments*, ESA SP-1104, pp. 69–74. ESA Publication Division, Noordwijk, The Netherlands.

Hunton, D. E., and Trzcinski, E. (1986). *Proc. Int. Sample Tech. Conf., Covina, Seattle*, p. 1027.

Karataev, V. I., Mamyrin, B. A., and Shmikk, D. V. (1972). *Sov. Phys.—Tech. Phys.* (*Engl. Transl.*) **16**, 1177; *Zh. Tekh. Fiz.* **41**, 1498 (1971).

Kochnev, V. A. (1980). Ph.D. Thesis, Space Research Institute, Moscow (in Russian).

Kovalev, I. D., Larin, N. V., Suchkov, A. I., Voronov, A. M., and Shmonin, P. A. (1985). *Prib. Tek. Eksp.* **6**, 139–142.

Mamyrin, B. A. (1967a). Soviet Patent No. 198,034.

Mamyrin, B. A. (1967b). *Sov. Invent. Bull.* **13**, 148.

Managadze, G. G. (1984). Soviet Patent No. 1,118,289.

Managadze, G. G. (1986). U.S. Patent No. 4,611,118.

Managadze, G. G., and Sagdeev, R. Z. (1987a). Soviet Patent No. 1,190,849.

Managadze, G. G., and Sagdeev, R. Z. (1987b). *Sov. Invent. Bull.* **5**, 279.

Managadze, G. G., Shutyaev, I. Yu., and Bondarenko, A. L. (1984). In *Acquisition and Analysis of Clean Substances* (in Russian) (A. D. Zorin, ed.), pp. 50–51. Gorki University, Gorki (now renamed Nizhni Novgorod, Russia).

Managadze, G. G., Sagdeev, R. Z., and Shutyaev, I. Y. (1987a). Soviet Patent No. 1,218,852.

Managadze, G. G., Sagdeev, R. Z., and Shutyaev, I. Y. (1987b). *Sov. Invent. Bull.* **17**, 273.

Niemann, H. B. (1977). *Space Sci. Rev.* **20**, 489.

Pang, K. D., Pollack, J. B., Veverka, J., Lane, A. L., and Ajello, J. M. (1978). *Science* **199**, 64–66.

Pollack, J. B., Veverka, J., Pang, K., Colburn, D., Lane, A. L., and Ajello, J. M. (1978). *Science* **199**, 66–69.

Sagdeev, R. Z. (1988). In *Exploration of Hally's Comet* (M. Grewing, F. Praderie, and R. Reinhard, eds.), pp. 959–964. Springer-Verlag, Berlin and New York.

Sagdeev, R. Z., Managadze, G. G., Shutyaev, I. Yu., Szego, K., and Timofeev, P. P. (1985). *Adv. Space Res.* **5**, 111–120.

Sagdeev, R. Z., Kissel, J., Evlanov, E. N., Fomenkova, M. N., Inogamov, N. A., Khromov, V. N., Managadze, G. G., Prolutski, O. F., Shapiro, V. D., Shutyaev, I. Y., and Zubkov, B. V. (1987). *Astron. Astrophys.* **187**, 179–182.

Shmikk, D. V., and Dubensky, B. M. (1984). *Zh. Tekh. Fiz.* **54**(5), 912–916.

Veverka, J. (1978). *Vistas Astron.* **22**, 2.

INDEX